Büttner

Lecture Notes in Applied and Computational Mechanics

Volume 10

Series Editors

Prof. Dr.-Ing. Friedrich Pfeiffer
Prof. Dr.-Ing. Peter Wriggers

Springer

Berlin
Heidelberg
New York
Hong Kong
London
Milan
Paris
Tokyo

Deformation and Failure in Metallic Materials

Kolumban Hutter
Herbert Baaser (Eds.)

Springer

Professor Dr. Kolumban Hutter
Dr. Herbert Baaser
Technical University of Darmstadt
Department of Mechanics
Hochschulstr. 1
D-64289 Darmstadt
GERMANY

e-mail: hutter@mechanik.tu-darmstadt.de
e-mail: baaser@mechanik.tu-darmstadt.de

With 130 Figures

Cataloging-in-Publication Data applied for
Bibliographic information published by Die Deutsche Bibliothek
Die Deutsche Bibliothek lists this publication in the Deutsche Nationalbibliografie; detailed bibliographic data is available in the Internet at <http://dnb.ddb.de>

ISBN 3-540-00848-9 Springer Verlag Berlin Heidelberg New York

Springer-Verlag Berlin Heidelberg New York
a member of BertelsmannSpringer Science+Business Media GmbH

http://www.springer.de

Printed in Germany

Cover design: design & production GmbH, Heidelberg
Typesetting: Digital data supplied by author

Printed on acid-free paper 62/3020Rw-5 4 3 2 1 0

Preface

A "**S**onder**f**orschungs**b**ereich" (SFB) is a programme of the "Deutsche Forschungsgemeinschaft" to financially support a concentrated research effort of a number of scientists located principally at one university, research laboratory or a number of these situated in close proximity to one another so that active interaction among individual scientists is easily possible.

Such SFBs are devoted to a topic, in our case "*Deformation and Failure in Metallic and Granular Materials*", and financing is based on a peer reviewed proposal for three (now four) years with the intention of several prolongations after evaluation of intermediate progress and continuation reports. An SFB is terminated in general by a formal workshop, in which the state of the art of the achieved results is presented in oral or/and poster communications to which also guests are invited with whom the individual project investigators may have collaborated. Moreover, a research report in book form is produced in which a number of articles from these lectures are selected and collected, which present those research results that withstood a rigorous reviewing process (with generally two or three referees).

The theme *deformation and failure of materials* is presented here in two volumes of the *Lecture Notes in Applied and Computational Mechanics* by Springer Verlag, and the present volume is devoted to metallic continua. The complementary volume (*Lecture Notes in Applied and Computational Mechanics*, vol. 11, Eds. K. HUTTER & N. KIRCHNER) is dedicated to the *Dynamic Response of Granular and Porous Materials under Large and Catastrophic Deformations.*

The SFB "Deformation and Failure in Metallic and Granular Materials" lasted from October 1994 until December 2002, thus a total of slightly more than eight years, and had an interdisciplinary focus: Teachers, researchers, including Ph.D. students from various University Departments were involved, namely Mathematics, Mechanics, Material Sciences, Civil and Mechanical Engineering. Many projects were headed by researchers from two different departments. Each project had one – sometimes two – principal researchers who either as Ph.D. students or postdoctoral assistants would perform the actual research under the supervision of the proposers of the individual proposals.

This volume tries to summarise the obtained results not so much in a form as it would appear in specialised peer reviewed periodicals, but such that a broader community is able to follow the red lines of the arguments in each article. This required that the authors were asked to include also items that may be obvious to specialists. We hope that this endeavour has been achieved.

This book on the *Deformation and Failure in Metallic Materials* is divided into four major parts:

(I) Constitutive behaviour - modelling and numerics,
(II) Constitutive behaviour - experiments and verification,
(III) Constitutive behaviour - mathematical foundation and
(IV) Damage and fracture.

These four topics define the main classes into which the 16 different articles may fit. They reflect a fairly objective cross section of the research that was conducted during the eight years this SFB existed.

Part I. Constitutive Behaviour - Modelling and Numerics

This largest part of the book consists of five articles which all deal in one way or an other with numerical integration of systems of differential equations, mixed with partial and ordinary differential equations and in two specific cases also containing algebraic equations. The equations derive from descriptions of material equations involving inelastic material behaviour in the spirit of plasticity, viscoplasticity and elasto-viscoplasticity, generally involving a yield surface. Numerical integration techniques of material equations of the mentioned class has been a center of activity of the SFB.

- BÜTTNER & SIMEON study the numerical properties of materials having constitutive relations of the elastoplastic and viscoplastic type with yield surfaces and classify these equations in the context of differential-algebraic equations and thereby motivate the temporal discretization by implicit RUNGE–KUTTA schemes. They apply the methods to deduce existence and uniqueness results for coersive methods and demonstrate that algebraically stable methods preserve the contractivity of the elastoplastic flow. By combining the fully implicit RUNGE-KUTTA methods with the two-level NEWTON method of computational plasticity they deduce a return mapping technique that delivers improved accuracy and efficiency of the numerical solution
- In their article on elasto–visco–plasticity SCHERF & SIMEON provide a specific example that an initial boundary value problem of elasto–visco-plasticity leads to a system of differential–algebraic equations. They show that the above mentioned integration technique can be improved by using the second order time integration method BDF-2. They further propose an alternative to the standard two-stage NEWTON process and use it in simulation experiments to demonstrate the possible gain in accuracy and efficiency.
- GRUTTMANN & EIDEL present a constitutive model for orthotropic elasto-plasticity at finite plastic strain and propose a scheme of its numerical implementation. The essential ingredients of the model are the multiplicative decomposition of the deformation tensor into elastic and plastic

parts and an invariant setting of the elastic free energy and the yield function by means of structure tensors reflecting the privileged directions of the material. The exponential map approach is used to integrate the associated flow rule, thus preserving plastic incompressibility. Implementation of the constitutive relations into a brick–type shell element paired with special interpolation techniques allows prediction of locking free deformations, even for very thin structures. Typical examples illustrate the suitability of the method.

- Of very similar intent is the article by TSAKMAKIS & WILLUWEIT devoted to the time integration algorithms for finite deformation plasticity. Because of the ordinary differential–algebraic–equation structure of the classical plasticity theories, special elastic predictor–plastic corrector integration schemes are proposed that preserve the plastic incompressibility condition and employ the exponential map approach. The model equations also incorporate isotropic and kinematic hardening.
- SANSOUR, KOLLMANN & BOCKO propose a non-linear model of finite strain multiplicative inelasticity with an anisotropic elastic constitutive law. Since complete proposals for anisotropic strain energy functions are not known to date the authors motivate their own proposal and only afterwards direct their attention to numerical integration which is considerably complicated by the anisotropy. The numerical details are presented in full and then illustrated with applications to shells.

Part II. Constitutive Behaviour - Experiments and Verification

Experiments can be used in connection with constitutive models of continuous systems in basically two different forms. First, one may verify a theoretical model by comparing computational results for a typical boundary value problem with experimental results for a realistic arrangement of the same problem. Second, experiments may be used to validate a continuum mechanical model by identifying its phenomenological parameters.

- The article by HARTH, LEHN & KOLLMANN is of this second type. The material parameters of their inelastic constitutive model are identified by minimization of the distance between the model response and the experimental data using an optimization algorithm. Because the amount of test data is not sufficient for a statistical analysis, a method of stochastic simulation is used to generate artificial data with the same behaviour as the experimental data. This procedure requires some care in the application of the loading history of the experiments when the parameters are identified, but the method is a suitable tool to enlarge data sets. Parameters of a number of inelastic constitutive models are thus identified.
- DAFALIAS, SCHICK & TSAKMAKIS address the problem of induced anisotropy in large deformation plasticity. In metallic materials, deformation induced anisotropy is reflected by the translation, rotation and distortion of the yield surface. The authors present a thermodynamically consistent

model describing the evolving anisotropy and show its imprint in the deformation of the yield surface. Various load histories, applied in an experimental arrangement by ISHIKAWA, are analysed and it is shown how the yields surfaces obtained from the experiments and the computations compare with one another.

- EMMEL, STIEFEL, GROSS & RÖDEL compare experimental and numerical results in a problem of damage and failure of ceramic metal composites due to temperature induced stresses. Two different structures, a metal layer between two ceramic supports and an inter–penetrating network, correlated to two different failure modes are studied. Cavitation is found to occur as is interface debonding in conjunction with crack branching at the interface. Experimental observations together with numerical calculations lead to a theoretical description for the prediction of the failure mode and the critical stresses in the composite.

Part III. Constitutive Behaviour - Mathematical Foundation

It is well known that continuum mechanics paired with inelastic constitutive models offers a wealth of nonlinear initial boundary value problems for existence and uniqueness tests. The mathematicians in our SFB devoted their activities to such questions, ideally by taking up models dealt with by the engineers in a more applied context. This fruitful interplay has made the models designed by engineers and material scientists more reliable and more trustworthy.

- One central activity among the mathematicians within our SFB has been with the existence theory of initial boundary value problems for the class of monotone operators. The article by CHELMINSKI deals with such an existence theory of global in time solutions to systems describing the inelastic behaviour of metals. Geometrically linear deformations are studied and the inelastic constitutive models are of the monotone type, leaving nonmonotone constitutive relations to a few closing remarks.
- The initial boundary value problems that are presented by NEFF derive from models of finite elastoplasticity and apply the multiplicative decomposition of the deformation gradient. For such a model, based on the ESHELBY tensor, the behaviour of the systems at frozen plastic flow is studied. Extending the analysis to viscoplastic flows and introducing stringent elastic stability assumptions and a nonlocal extension in space the author proves local existence in time of the corresponding initial value problems. A new model is also introduced that is suitable for small elastic strains. Its key feature is an independent elastic rotation for which a closure involving the elastic deformation tensor is proposed. For this case, the equilibrium equations at frozen plastic flow are now linear elliptic leading straight forwardly to existence and uniqueness results without additional stability assumptions.
- EBENFELD attacks in his analysis of initial boundary value problems in continuum mechanics those material formulations which automatically

lead to problems that admit smooth solutions. The models, constrained this way yield equations of hyperbolic or parabolic type and the conditions constitute the mathematical counterpart to those conditions obtained by the second law of thermodynamics. The mathematical formulation of the problem is presented but no proofs are given.

- A challenging mathematical question in the context of homogenization is the simultaneous proof of existence of solutions of the homogenized problem at the macrolevel and a corresponding problem on the microlevel. This problem is attacked by ALBER in his article. Within the context of viscoplasticity (with monotone constitutive operators) it is shown that the derived homogenized initial boundary value problem possesses a solution. From this, an asymptotic solution of the microproblem is constructed with the property that the difference between the exact and asymptotic solutions tends to zero if the length scale of the microscale becomes vanishingly small.

Part IV. Damage and Fracture

Both are important notions in a balanced treatment of metallic and composite materials. Failure occurs because of damage of the material and requires the introduction of a model for its evolution. The four articles in this part address such questions.

- The article by THOMAS et al. gives a review of the effect of the stress state on deformation, roughness and damage in metal forming. The overview presents elasto–plastic constitutive relations including anisotropy induced during metal forming and by crack propagation. The arising material parameters are identified by experiments. This leads to a modified GURSON model. The article also discusses the feasibility of different experimental techniques by which the local strain is determined.
- BAASER & GROSS realize 3D simulations of ductile damage and crack propagation processes by the use of continuum damage mechanics such as the models proposed by GURSON or ROUSSELIER. The numerical implementation in the finite element method (FEM) is based on locking–free 20–noded brick elements in the finite strain regime. The main attention is directed to the well–known disadvantage of modeling strain–softening behaviour, the mesh–dependence occuring due to the loss of ellipticity of the leading differential equations. The initially given well–posedness of the system is checked permanently during the iteration process by evolving the determinant of the acoustic tensor in a very efficient way. They claim, that softening behaviour can be treated very well until the type of diffential equations changes and the results become questionable due to mesh–dependence. However, the point of loosing ellipticity can not be computed a priori, the testing has to be exhibited in a steady manner.
- SCHMIDT, RICHTER & GROSS discuss the modelling and simulation of mode I crack growth in metal foames. Its initiation and subsequent re-

sistance to propagation are numerically explored with two and three-dimensional beam networks as models for the microstructure around the crack tip. The cell wall material is elasto-plastic with linear hardening, a critical fracture stress and fracture energy per unit area. Using a finite element implementation parameter studies are performed by which the influence of the cell–wall parameters on the crack resistance curves was estimated under small scale yielding for both 2D and 3D models.

- In the final article of this part on numerical techniques RECKWERTH & TSAKMAKIS present a model for continuum damage mechanics involving an elastic plastic response and isotropic damage formulation. The approach is based on the concept of effective stress combined with a principle of generalized energy equivalence. A family of yield functions is proposed one of which yields the damage model based on the strain equivalence hypothesis, valid for damage in metals. Numerical finite element results obtained with it as well as with a model established in the framework of the strain equivalence principle indicate deviating results whenever inhomogeneous deformations arise.

Most of the research reported in this book has wholly or partially been supported by the Deutsche Forschungsgemeinschaft through our SFB. Salaries for Ph.D. students, post-doctoral fellows, experimental equipments, computer soft– and hardware were financed as were subsidies for our travel and guest programmes. Complementary support came also from the Darmstadt University of Technology. All this has helped us in advancing in our scientific work which has in many respects also supported us in improving our teaching. Supporting in the achievement of this improvement were the referees who accompanied us through the eight years. To the Deutsche Forschungsgemeinschaft, to the University and to them we express our gratitude.

Darmstadt,
November 2002

Kolumban Hutter
Herbert Baaser

Contents

Part I. Constitutive Behaviour — Modelling and Numerics

Numerical Treatment of Material Equations with Yield Surfaces 3
Jörg Büttner, Bernd Simeon

Differential-Algebraic Equations in Elasto-Viscoplasticity 31
Oliver Scherf, Bernd Simeon

Finite Element Analysis of Anisotropic Structures at Large Inelastic Deformations 51
Bernhard Eidel, Friedrich Gruttmann

Use of the Elastic Predictor–Plastic Corrector Method for Integrating Finite Deformation Plasticity Laws 79
Charalampos Tsakmakis, Adrian Willuweit

A Model of Finite Strain Viscoplasticity with an Anisotropic Elastic Constitutive Law 107
Carlo Sansour, Franz Gustav Kollmann, Jozef Bocko

Part II. Constitutive Behaviour — Experiments and Verification

Identification of Material Parameters for Inelastic Constitutive Models: Stochastic Simulation 139
Tobias Harth, Jürgen Lehn, Franz Gustav Kollmann

A Simple Model for Describing Yield Surface Evolution During Plastic Flow 169
Yannis F. Dafalias, David Schick, Charalampos Tsakmakis

Damage and Failure of Ceramic Metal Composites 203
Thomas Emmel, Ulrich Stiefel, Dietmar Gross, Jürgen Rödel

Part III. Constitutive Behaviour — Mathematical Foundation

Monotone Constitutive Equations in the Theory of Inelastic Behaviour of Metals - A Summary of Results 225
Krzysztof Chełmiński

Some Results Concerning the Mathematical Treatment of Finite Plasticity. ... 251
Patrizio Neff

Initial Boundary Value Problems in Continuum Mechanics ... 275
Stefan Ebenfeld

Justification of Homogenized Models for Viscoplastic Bodies with Microstructure ... 295
Hans-Dieter Alber

Part IV. Damage and Fracture

Models and Experiments Describing Deformation, Roughness and Damage in Metal Forming ... 323
Sven Thomas, Stefan Jung, Michael Nimz, Clemens Müller, Peter Groche, Eckart Exner

Remarks on the Use of Continuum Damage Models and on the Limitations of their Applicability in Ductile Fracture Mechanics ... 345
Herbert Baaser, Dietmar Gross

Ductile Crack Growth in Metallic Foams ... 363
Ingo Schmidt, Christoph Richter, Dietmar Gross

The Principle of Generalized Energy Equivalence in Continuum Damage Mechanics ... 381
Dirk Reckwerth, Charalampos Tsakmakis

Index

Index ... 407

Acknowledgements

The articles published in this book have all been peer reviewed by internationally known specialists. The editors wish to acknowledge the help of the following individuals in judging the scientific contents of the articles:

- Otto T. Bruhns
- Wolfgang Ehlers
- Dietmar Gross
- Friedrich Gruttmann
- Stefan Hartmann
- Peter Haupt
- Gerhard Holzapfel
- Marc Kamlah
- Pavel Krejci
- Meinhard Kuna
- Matthias Lambrecht
- Jean Lemaitre
- Rolf Mahnken
- Anton Matzenmiller
- Robert M. McMeeking
- Christian Miehe
- Pia Redanz
- Daya Reddy
- Stefanie Reese
- Jörg Schröder
- Karl Schweizerhof
- Bernd Simeon
- Bob Svendsen
- Peter Wriggers
- Tarek Zohdi

Part I

Constitutive Behaviour — Modelling and Numerics

Numerical Treatment of Material Equations with Yield Surfaces

Jörg Büttner[1] and Bernd Simeon[2]

[1] IWRMM, University of Karlsruhe
Engesser Str. 6, D–76128 Karlsruhe, Germany
buettner@iwrmm.math.uni-karlsruhe.de
[2] Centre for Mathematical Sciences, Munich University of Technology
Boltzmannstr. 3, D–85748 Garching, Germany
simeon@mathematik.tu-muenchen.de

received 30 Jul 2002 — accepted 23 Sep 2002

Abstract. This paper discusses material equations with yield surfaces and their numerical simulation from several viewpoints. First, the constitutive equations of elastoplasticity and viscoplasticity are classified in the sense of differential-algebraic equations, which motivates the time discretization by implicit Runge-Kutta methods. Next, these methods are applied to the weak dual problem in time, with the space variables kept continuous. We obtain an existence and uniqueness result for the class of so-called coercive methods and show in addition that algebraically stable methods guarantee the preservation of the contractivity of the elastoplastic flow. Finally, we combine fully implicit Runge-Kutta methods with the two-level Newton method of computational plasticity. In particular, we demonstrate how the return mapping strategy for local material integration can be extended to this class of discretization schemes, leading to both improved accuracy and efficiency of the numerical solution.

1 Introduction

Various material equations in solid mechanics feature yield surfaces that model the transition from elastic to plastic behavior. Concentrating on the example of classical elastoplasticity, we analyse the problem class with emphasis on time integration methods. Inspired by an interpretation as unilaterally constrained differential-algebraic equation (DAE), we consider both the local formulation of the material law as well as the weak global boundary value problem and show how implicit Runge-Kutta methods can be applied.

The material or so-called constitutive equations of elastoplasticity treated in this paper arise as first order necessary conditions from a convex optimization principle, the principle of maximum plastic work [18,27]. Already Papadopoulos and Taylor [20] pointed out that these equations can be viewed as an evolution equation under an algebraic constraint or, more precisely, as a DAE of so-called index two, see [3,15] for a definition of the index. We take this idea and discuss advantages and problems arising from the DAE approach. In this context, the Perzyna formulation of viscoplasticity appears in a new light since it represents not only a penalty regularization of the principle of maximum plastic work but equally a regularization technique for the present class of DAEs.

Implicit Runge-Kutta methods have proved to be among the methods of choice in numerous application fields. Even the return mapping strategy of computational plasticity uses implicit Euler and generalized midpoint rules [25,27,29], which belong to this method class, and only recently Ellsiepen and Wieners [10,32] extended this approach to so-called DIRK methods (Diagonally Implicit Runge-Kutta).

Fully implicit methods feature better stability properties and also seem promising for an implementation with stepsize control. To analyse their potential when applied to material equations with yield surfaces, we present here a grid-independent existence and uniqueness result for the time discretization of the weak form where the space variables are left continuous. In this way, we lay the foundation for the reverse method of lines, also known as Rothe's method. The result is based on a certain formulation of the so-called dual problem that is closely related to a mixed or saddle point formulation [2,12]. It turns out that the class of coercive Runge-Kutta methods preserves the coercivity within the saddle point problem. Thus the existence of the numerical solution can be shown.

In addition to uniqueness, numerical solutions should be stable. For the class of problems at hand, stability is defined in terms of the contractivity of the elastoplastic flow, i.e. the dissipation of energy during deformation [26,27]. Algebraically stable Runge-Kutta methods preserve this property [26,32], i.e., they do not add numerical energy to the system. The classes of coercive and of algebraicly stable methods almost coincide, which implies that various implicit Runge-Kutta methods are applicable in elastoplasticity, among them the Radau IIa, Gauss, Lobatto, and several DIRK schemes [15].

For the implementation, we combine FEM in space and Runge-Kutta discretization in time, with particular emphasis on the structure of the equations. Though the return mapping scheme is straightforward to generalize to this time discretization, several questions deserve particular attention, among them the solution of the nonlinear equations by means of the two-level Newton method with consistent tangent operator. Our numerical results show both clear improvements for the overall accuracy of the numerical solution as well as a convergence order of two for several selected schemes.

The outline of this paper is as follows. In Sect. 2, we give an interpretation of the local constitutive equations as DAE and relate them via a singular perturbation approach to the viscoplastic model. Sect. 3 introduces the global weak form in terms of generalized stress and conjugated forces. We focus in particular on the relation to saddle point problems and prepare in this way Sect. 4, which introduces the time discretization of the weak form and states the main existence result. Sect. 5 is devoted to a stabilitiy analysis in terms of the contractivity of the elastoplastic flow while Sect. 6 dicusses the computational solution of the elastoplastic boundary value problem by FEM in space and implicit Runge-Kutta discretization in time. Finally, Sect. 7 presents some numerical examples.

2 Classical Elastoplasticity and Viscoplasticity

In this section, we focus on the constitutive equations in their classical formulation. We consider small strain rate-independent elastoplasticity and rate-dependent viscoplasticity. For the interested reader, we mention that references [18,27] contain introductions to continuum mechanics and constitutive equations whereas references [3,14,15] give overviews on ODEs, DAEs and singularly perturbed problems.

2.1 Kinematic Relations and Balance Law

Let $\Omega \subset \boldsymbol{R}^3$ be the unconstrained reference configuration of the body. The displacement of a particle $x \in \Omega$ at time t is described by the vector field $u(x,t) \in \boldsymbol{R}^3$ while the total strain tensor $\varepsilon(x,t)$ measures changes of local length elements. For small deformations, the strain tensor is given by

$$\varepsilon = \tfrac{1}{2}(\nabla u + (\nabla u)^{\mathrm{T}}) \,. \tag{1}$$

Here and in the following, we mostly drop the arguments x and t for convenience. The strain ε belongs to the symmetric rank two tensors. Using a fixed Cartesian coordinate system, the symmetric rank two tensors are identified with symmetric 3×3-matrices [2].

Stress, on the other hand, is defined as the density of surface force per area. Like the strain tensor ε, the stress tensor σ is a symmetric rank two tensor. The constitutive equations relate stress and strain, see the next subsection. For the stress, the balance law of momentum holds, i.e., in the quasi static case we have

$$\operatorname{div} \sigma + f = 0, \tag{2}$$

where f denotes some time dependent body force.

Note that the kinematic relation (1) and the momentum balance (2) appear again in the weak formulation of the elastoplastic boundary value problem in Sect. 3. There, the so far missing boundary conditions will also be specified. For the moment, however, we turn to the local constitutive equations.

2.2 Constitutive Equations of Elastoplasticity

For elastic materials, the stress-strain relation is linear and expressed by Hooke's law

$$\sigma = C : \varepsilon \,.$$

Here C denotes the symmetric forth order elasticity tensor and the ":" sign stands for interior tensor multiplication. Hooke's law is characterized by reversibility, which means that as soon as ε becomes zero, the stress vanishes, too. Moreover, the stress does not depend on the strain history but merely on the current strain state.

In contrast to elastic materials, the stress response of elastoplastic and viscoplastic materials depends on the strain history. We first consider elastoplasticity. As long as the stress σ remains smaller than some yield stress σ_0, the material reacts elastically. Thereafter, irreversible deformation occurs. To model this behaviour, the total strain is split additively into an elastic part, ε^{e} and a plastic part, ε^{p},

$$\varepsilon = \varepsilon^{\mathrm{e}} + \varepsilon^{\mathrm{p}} \,. \tag{3}$$

Then, for a prescribed strain history $\varepsilon(t)$, the constitutive equations read

$$\sigma = C : [\varepsilon - \varepsilon^{\mathrm{p}}], \tag{4}$$

$$\dot{\varepsilon}^{\mathrm{p}} = \gamma \mathrm{r}(\sigma, q), \tag{5}$$

$$\dot{q} = -\gamma \mathrm{g}(\sigma, q), \tag{6}$$

augmented by the Kuhn-Tucker conditions

$$0 = \gamma\phi(\sigma, q), \quad \gamma \geq 0, \quad \phi(\sigma, q) \leq 0 \,. \tag{7}$$

The internal state variables $q \in \boldsymbol{R}^m$ describe additional effects such as hardening. They are either strain like or stress like, see Sect. 3, and satisfy, as the plastic strain ε^{p}, a differential equation.

The convex yield function ϕ seperates elastic from plastic deformations. As long as ϕ remains smaller than zero, the Lagrange multiplier γ vanishes and there is no evolution of plastic strain or internal variables. Thus, the stress is determined from Hooke's law (4) as an algebraic relation. As soon as ϕ becomes zero, the constraint becomes active and the evolution of plastic strain and internal variables starts. We denote by t_s the switching time where the deformation process switches from the elastic to the plastic phase.

Before we state additional assumptions on the functions r, g, and ϕ, we give an example [27]. For linear isotropic hardening, the yield function reads

$$\phi(\sigma, q) = ||\mathrm{dev}\ \sigma|| - \sqrt{\frac{2}{3}}(\sigma_0 + Kq)$$

with positive hardening modulus K, strain like q and initial yield stress σ_0. Further, dev σ denotes the deviator of the stress and $||\tau||^2 := \tau^T : \tau$. The evolution equations read in associated form

$$\dot{\varepsilon}^{\mathrm{p}} = \gamma \frac{\mathrm{dev}\ \sigma}{||\mathrm{dev}\ \sigma||}, \quad \dot{q} = \sqrt{\frac{2}{3}}\gamma \,.$$

This means

$$\mathrm{r} = \phi_\sigma =: \partial\phi/\partial\sigma, \quad \mathrm{g} = K^{-1}\phi_q =: K^{-1}\partial\phi/\partial q\,.$$

Equations (4) to (7) are called the strain space formulation if q is strain like. For the following treatment, however, there is no difference, whether the internal variables are strain or stress like. This difference becomes important in the weak formulation, see Sect. 3.

As already stated, the yield function ϕ is convex. This implies that the elastic domain

$$\mathcal{E} = \{(\sigma, q) \mid \phi(\sigma, q) \le 0\} \tag{8}$$

is a closed and convex set. Additionally, we postulate that

$$\phi_\sigma : C : \mathrm{r} + \phi_q \cdot \mathrm{g} > 0 \tag{9}$$

must hold for (σ, q) close to $\partial\mathcal{E}$. This condition is fullfilled for various materials in important application fields. We will give a physical interpretation of Ineq. (9) in the sense of dissipation when we treat the Perzyna model of viscoplasticity.

In this section, we explained constitutive equations by means of elastic and plastic phases. In Sect. 3, we will derive the model from a constraint optimization principle, the principle of maximum plastic work. In this context, Eqs.(4)–(7) are necessary conditions of optimization under constraints. Thus the term Kuhn-Tucker conditions becomes obvious.

2.3 The Index of the Constitutive Equations

The constitutive equations show the following properties: In the elastic deformation phase, the stress strain relation is simply an algebraic relation, since plastic strain remains constant. During plastic deformation, the evolution equations for plastic strain and internal variables are constrained by the algebraic equation $\phi = 0$. Considering σ as a function of ε^{p}, the constitutive equations form a differential-algebraic equation (DAE)

$$\dot{\varepsilon}^{\mathrm{p}} = \gamma \mathrm{r}(\varepsilon^{\mathrm{p}}, q), \tag{10}$$
$$\dot{q} = -\gamma \mathrm{g}(\varepsilon^{\mathrm{p}}, q), \tag{11}$$
$$0 = \phi(\varepsilon^{\mathrm{p}}, q)\,. \tag{12}$$

Plastic strain and internal variables are called differential variables and the Lagrange multiplier is referred to as algebraic variable.

How can we characterize the DAE (10)–(12)? One concept is the differentiation index. If we differentiate the constraint (12) with respect to time t, we get

$$0 = \phi_\sigma : C : (\dot{\varepsilon} - \dot{\varepsilon}^{\mathrm{p}}) + \phi_q \cdot \dot{q} \quad \Rightarrow \quad \gamma = \frac{\phi_\sigma : C : \dot{\varepsilon}}{\phi_\sigma : C : \mathrm{r} + \phi_q \cdot \mathrm{g}}\,. \tag{13}$$

Due to (9), the last equation is well defined. Another differentiation of the Lagrange multiplier leads to a differential equation for γ. We therefore need two differentiations to extract an explicit ODE from the DAE, and this number of differentiations is called the index [3,15]. Summing up, the constitutive equations of elastoplastiticty are a DAE of index two.

What are the consequences of the index being two? If we search for a solution of an ODE, we can prescribe arbitrary initial values, in contrast to DAEs. Any solution of (10)–(12) has to satisfy the constraint $\phi = 0$ and in addition the hidden constraint

$$0 = \phi_\sigma : C : (\dot{\varepsilon} - \dot{\varepsilon}^{\mathrm{p}}) + \phi_q \cdot \dot{q} \; .$$

This holds especially for the initial values at time $t = t_s$.

2.4 Viscoplasticity and Singularly Perturbed ODEs

Closely related to elastoplasticity is viscoplasticity where the material shows an additional viscous behavior like, e.g., relaxation or rate-dependence. The viscosity is usually denoted by a parameter η and as $\eta \to 0$, the viscoplastic effects vanish and the material reacts purely elastoplastically.

Viscoplasticity can be derived as a regularization of the elastoplastic equations. We remark that the regularization can be done in two ways. The classical procedure [9,17,18,27] is a penalty regularization of the principle of maximum plastic work and thus a technique of convex optimization theory. The penalty term contains a viscosity parameter η. As η tends to zero, the material behaviour changes to elastoplasticity. We treat this approach in Sect. 3. Depending on the penalty term, we obtain either the Perzyna model or the generalized Duvaut-Lions model [9,27].

For now, we stay in the world of DAEs and ODEs and consider the Perzyna model. This model can be seen as singularly perturbed ODEs, derived from a standard regularization technique of the DAE theory [19]. First, we recall the constitutive equations of elastoplasticity during plastic deformation

$$\begin{aligned}
\dot{\sigma}^{\mathrm{p}} &= C : (\dot{\varepsilon} - \gamma \mathrm{r}), \\
\dot{q} &= -\gamma \mathrm{g}, \\
0 &= \phi_\sigma : C : \dot{\varepsilon} - \gamma \left(\phi_\sigma : C : \mathrm{r} + \phi_q \cdot \mathrm{g} \right) \; .
\end{aligned}$$

This is the index one formulation since we substituted (12) by (13). Now we replace this system by the singularly perturbed system

$$\dot{\sigma}_\eta = C : (\dot{\varepsilon} - \gamma_\eta \mathrm{r}), \tag{14}$$

$$\dot{q}_\eta = -\gamma_\eta \mathrm{g}, \tag{15}$$

$$\eta \dot{\gamma}_\eta = \phi_\sigma : C : \dot{\varepsilon} - \gamma_\eta \left(\phi_\sigma : C : \mathrm{r} + \phi_q \cdot \mathrm{g} \right) \; . \tag{16}$$

If we set η equal to zero, we obtain the elastoplastic equations.

Of course, the question arises at this point whether in the limit as $\eta \to 0$ the perturbed solutions σ_η, q_η and γ_η converge to the solution of the elastoplastic equations or not. The theory of singular perturbation problems gives an answer. Setting

$$\mathrm{p}(\sigma, q, \gamma) := \phi_\sigma : C : \dot{\varepsilon} - \gamma_\eta \left(\phi_\sigma : C : \mathrm{r} + \phi_q \cdot \mathrm{g}\right)$$

the solution of the viscoplastic equations has to fullfill the dissipation inequality

$$\frac{\partial}{\partial \gamma}\mathrm{p} \leq -c^2 < 0$$

close to the elastoplastic solution. This means

$$-(\phi_\sigma : C : \mathrm{r} + \phi_q \cdot \mathrm{g}) \leq -c^2$$

holds in the neighborhood of the yield surface. The latter equation is condition (9), which represents thus a dissipation condition on the material laws. As the singular perturbation theory ensures, the viscoplastic solutions converge to the elastoplastic solution for $\eta \to 0$ as long as the initial values in the switching time are the same.

Note that usually the Perzyna model is formulated in a different way as

$$\dot{\sigma}_\eta = C : (\dot{\varepsilon} - \gamma_\eta \mathrm{r}) \tag{17}$$

$$\dot{q}_\eta = -\gamma_\eta \mathrm{g} \tag{18}$$

where

$$\gamma_\eta = \langle \phi(\sigma, q) \rangle / \eta \quad \text{with} \quad \langle x \rangle := (x + |x|)/2\,. \tag{19}$$

This is an index one formulation since the differentiation of the last equation yields Eq. (16). Hence we refer to (17)–(19) as index one formulation of the Perzyna model whereas (14)–(16) is the ODE formulation. Note, that (σ_η, q_η) are not required to lie inside the elastic domain. To sum up, we have shown the following:

The ODE formulation of the Perzyna model is a singular perturbation of the index one formulation of the elastoplastic constitutive equations.

To illustrate the convergence of the viscoplastic solutions and some other properties, we consider an example.

2.5 Example: Linear Isotropic Hardening

If we reduce the elastoplastic model to one dimension, the stress and strain tensors become scalars. Now the evolution equations for linear isotropic hardening read

$$\dot{\sigma} = C(\dot{\varepsilon} - \gamma \operatorname{sign}(\sigma)),$$

$$\dot{q} = \gamma,$$

augmented by the yield function

$$\phi(\sigma, q) = |\sigma| - (\sigma_0 + Kq) \; .$$

We prescribe vanishing initial values for plastic strain and internal variable as well as a linear total strain

$$\varepsilon(t) = kt, \quad k > 0 \; .$$

Due to the linearity of the elastoplastic and viscoplastic constitutive equations, these two systems can be solved by the standard approaches for linear ODEs. First, we look at the stress. The elastoplastic solution $\sigma(t)$ is given by

$$\sigma(t) = \begin{cases} Ckt, & t < t_s, \\ \sigma_0 + \frac{CK}{C+K} k(t - t_s), & t \geq t_s \end{cases}$$

with switching time $t_s = \sigma_0/(Ck)$. The viscoplastic solution is

$$\sigma_\eta(t) = \begin{cases} Ckt, & t < t_s, \\ \sigma_0 + \frac{CK}{C+K} k(t - t_s) + \eta \frac{C^2}{(C+K)^2} k(1 - e^{-\frac{K+C}{\eta}(t - t_s)}), & t \geq t_s. \end{cases}$$

Here we can see the typical η-expansion form of the solution of singularly perturbed problems. The expansion splits into an asymptotic and a transient part:

$$\sigma_\eta = \underbrace{\sigma(t) + \eta \frac{C^2}{(C+K)^2} k}_{\text{asymptotic}} \underbrace{- \eta \frac{C^2}{(C+K)^2} k e^{-\frac{K+C}{\eta}(t - t_s)}}_{\text{transient}} \; . \tag{20}$$

As t tends to infinity, the transient part vanishes. Therefore σ_η converges to the asymptotic part. The asymptotic part itself splits additively into the solution of the elastoplastic equations, which is the index one DAE, and a correction term for the perturbation. After all, as $\eta \to 0$, the viscoplastic solution converges to the elastoplastic solution.

For the internal variable and the Lagrange multiplier, similar results hold. Equation (20) also explains the physical background. The transient part represents relaxation effects that vanish for $t \to \infty$ while the η-expansion of the asymptotic part describes the rate-dependent behaviour. Moreover, the asymptotic solution depends on k, i.e., the velocity or rate of deformation.

Another typical phenomenon is the stiffness of the singularly perturbed problem. We rewrite the viscoplastic equations as

$$\begin{pmatrix} \dot{\sigma}_\eta \\ \dot{q}_\eta \\ \dot{\gamma}_\eta \end{pmatrix} = \begin{pmatrix} 0 & 0 & -C \\ 0 & 0 & 1 \\ 0 & 0 & -\frac{C+K}{\eta} \end{pmatrix} \begin{pmatrix} \sigma_\eta \\ q_\eta \\ \gamma_\eta \end{pmatrix} + \begin{pmatrix} C\dot{\varepsilon} \\ 0 \\ \frac{\operatorname{sign}(\sigma_\eta)}{\eta} C\dot{\varepsilon} \end{pmatrix} .$$

The matrix on the right hand side possesses eigenvalues zero and $-\frac{C+K}{\eta}$. The latter converges to $-\infty$ as $\eta \to 0$, which is typical for stiff ODEs.

3 Weak Formulation as Dual Problem

In this section, we introduce a weak formulation of the elastoplastic boundary value problem, the so-called dual problem. For this global point of view, the local constitutive equations in form of the DAE (4)–(7) are no longer appropriate. Instead, we make use of a thermodynamical approach where the stress is derived from the principle of maximum plastic work [12,18]. We refer to [18] for the physical background and to [9,31,12] for a mathematical analysis.

3.1 Modelling Thermomechanical Processes

As in Sect. 2, the total strain of (1) splits into an elastic part ε^{e} and a plastic part ε^{p}. The internal state variables are strain like and we denote them by $\xi = (\xi_1, \ldots, \xi_r)$. Thus the state of a material point is modelled by the generalized strain $P := (\varepsilon^{\mathrm{p}}, \xi)$.

A crucial quantity is the internal free energy function $\psi := \psi(\varepsilon^{\mathrm{e}}, \xi)$, which comprises the usual stored energy function of elasticity and additional contributions from processes like hardening. Furthermore, the free energy defines the generalized stress $\Sigma = (\sigma, \chi)$ with stress tensor σ and conjugate force χ by

$$\sigma = \tfrac{\partial}{\partial \varepsilon^{\mathrm{e}}} \psi \tag{21}$$

$$\chi = -\tfrac{\partial}{\partial \xi} \psi. \tag{22}$$

One also says that the generalized stress is conjugate to the generalized strain.

In this paper, we restrict the discussion to linear hardening materials where the quadratic internal free energy reads

$$\psi(\varepsilon^{\mathrm{e}}, \xi) = \underbrace{\tfrac{1}{2}\varepsilon^{\mathrm{e}} : C\varepsilon^{\mathrm{e}}}_{=:\psi^{\mathrm{e}}(\varepsilon^{\mathrm{e}})} + \underbrace{\tfrac{1}{2}\xi \cdot H\xi}_{=:\psi^{\mathrm{p}}(\xi)}$$

with a regular $r \times r$ symmetric positive-definite matrix H of hardening moduli. The derivative on the right hand side of (21) yields Hooke's law

$$\sigma = C : (\varepsilon - \varepsilon^{\mathrm{p}})$$

and (22) becomes

$$\chi = -H\xi.$$

At this point, all relevant variables of the constitutive equations have been introduced.

However, as in Sect. 2, the generalized stress Σ is required to lie in the admissible set $\mathcal{E}$ of elastic behavior. As example, consider so-called linear J_2-isotropic hardening [27] with only one scalar conjugate force χ and a scalar hardening modulus H. There, the yield function reads

$$\phi(\sigma, \chi) = \|\mathrm{dev}\ \sigma\| - \sqrt{\frac{2}{3}}(\sigma_0 - \chi) \tag{23}$$

with given initial yield stress $\sigma_0 > 0$.

The constitutive equations follow now from the local *Principle of Maximum Plastic Work.* Every deformation stores the plastic work

$$W(\dot{P}) = \Sigma : \dot{P} := \sigma : \dot{\varepsilon}^p + \chi \cdot \dot{\xi}\,.$$

Given a rate of change $\dot{P}$ in the generalized strain, the actual stress $\Sigma \in \mathcal{E}$ maximizes the plastic work. In other words, every plastic deformation process stores as much energy as possible:

$$W(\dot{P}) = \max_{T\in\mathcal{E}} \{T : \dot{P}\}. \tag{24}$$

For the current generalized stress Σ, therefore, the following inequality must hold:

$$\Sigma : \dot{P} \geq T : \dot{P} \qquad \forall T \in \mathcal{E}, \tag{25}$$

which is a convex optimization problem. Below, we use inequality (25) to formulate a variational inequality that leads directly to the dual problem.

Remark: Using the method of Lagrange multipliers from convex optimization theory, it is easily concluded that

$$\begin{aligned}
\dot{\varepsilon}^{\mathrm{p}} &= \gamma \frac{\partial}{\partial \sigma}\phi(\sigma,\chi)\\
\dot{\xi} &= \gamma \frac{\partial}{\partial \chi}\phi(\sigma,\chi)\\
0 &= \gamma\phi(\sigma,\chi), \quad 0 \geq \phi(\sigma,\chi), \quad 0 \leq \gamma.
\end{aligned}$$

Here the last equation corresponds to the usual Kuhn-Tucker conditions of unilateral constraint optimization theory. In case of linear isotropic hardening we thus have

$$\begin{aligned}
\dot{\sigma} &= C : \dot{\varepsilon} - C : \frac{\partial}{\partial \sigma}\phi(\sigma,\chi)\gamma\\
\dot{\chi} &= -H \frac{\partial}{\partial \chi}\phi(\sigma,\chi)\gamma\\
0 &= \gamma\phi(\sigma,\chi), \quad 0 \geq \phi(\sigma,\chi), \quad 0 \leq \gamma.
\end{aligned}$$

This is just the classical formulation of the constitutive equations.

3.2 Viscoplastic Regularization of the Principle of Maximum Plastic Work

In Sect. 2 we derived the Perzyna formulation of viscoplasticity by a singular perturbation technique. In the context of thermodynamical modelling, we consider a different approach. The stress response in viscoplasticity is now defined by adding a penalty term J_η to the principle (24),

$$W_\eta(\dot{P}) = \max_{\Sigma}(\{\Sigma : \dot{P}\} - J_\eta(\Sigma)). \tag{26}$$

We end up with an unconstrained maximization principle since the generalized stresses are not required to be admissible.

A common choice for the convex function J_η is the Yosida regularization

$$J_\eta(\Sigma) = \frac{1}{2\eta}\|\Sigma - \Pi\Sigma\|^2,$$

where Π denotes the orthogonal projection onto the elastic domain $\mathcal{E}$ and η is the regularization parameter [9,12,18,27]. Obviously, for η tending to zero, the penalty term $J_\eta(\Sigma)$ tends to infinity if $\Sigma \notin \mathcal{E}$.

For further use below, we mention an important property of the Yosida regularization. Since J_η is convex, its Gâteaux derivative J'_η is monotone and given under the Riesz isomorphism by

$$J'_\eta(\Sigma) = \frac{1}{\eta}(\Sigma - \Pi\Sigma).$$

Note that the Yosida penalty does not yield the Perzyna formulation. Using the quadratic regularization

$$\begin{cases} \frac{1}{2\eta}\phi(\Sigma)^2, & \Sigma \notin \mathcal{E}, \\ 0 & \Sigma \in \mathcal{E} \end{cases}$$

instead, the Perzyna formulation is obtained from the necessary conditions of unconstrained optimization theory [27].

After this outline of the mathematical model, we pass now to the weak form and specify the appropriate function spaces.

3.3 The Dual Problem

Before we derive a variational formulation, let us point out that the generalized stress Σ and the displacement u are the unknowns we are interested in. We require stress, strain, internal variables and conjugate forces to be square-integrable on the domain Ω at any time. Consequently, we introduce the space

$$\mathcal{S} := \{\tau = (\tau_{ij})_{3\times 3} \,:\, \tau_{ij} = \tau_{ji},\, \tau_{ij} \in L^2(\Omega)\}$$

for the symmetric tensors and the space

$$\mathcal{M} := \{\mu = (\mu_j) \,:\, \mu_j \in L^2(\Omega),\, j = 1, \ldots, r\}.$$

for the conjugate forces. The product space

$$\mathcal{T} := \mathcal{S} \times \mathcal{M},$$

which, as the other spaces, is endowed with the natural L^2 inner product $(\cdot,\cdot)$, contains thus the square-integrable generalized stresses. Furthermore, we define the convex subset

$$\mathcal{P} := \{T = (\tau,\mu) \in \mathcal{T} \,:\, (\tau,\mu) \in \mathcal{E} \text{ almost everywhere in } \Omega\}$$

for admissible states T.

In order to derive the variational formulation, we multiply the balance of momentum (2) by test functions belonging to

$$V = \{v \in H^1(\Omega)^3 : v = 0 \quad \text{on } \partial\Omega\}.$$

Using the Gauss divergence theorem, we arrive at

$$-\int_\Omega \varepsilon(v) : \sigma\, dx = -\int_\Omega f(t)\cdot v\, dx \qquad \forall v \in V. \tag{27}$$

Here, without loss of generality, homogeneous boundary conditions are considered. The Principle of Maximum Plastic Work (25) implies the variational inequality

$$\begin{aligned} 0 &\leq \int_\Omega (\sigma-\tau) : \dot{\varepsilon}^p + (\chi-\mu)\cdot\dot{\xi}\, dx \\ &= -\int_\Omega (\sigma-\tau) : C^{-1}\dot{\sigma} + (\chi-\mu)\cdot H^{-1}\dot{\chi}\, dx + \int_\Omega (\sigma-\tau) : \dot{\varepsilon}\, dx \end{aligned}$$

for all $T = (\tau,\mu) \in \mathcal{P}$. To obtain a more compact notation, the integrals in the variational inequality are abbreviated by the bilinear forms

$$\begin{aligned} A : \mathcal{T}\times\mathcal{T} \to \boldsymbol{R}, \quad & A(\Sigma,T) := \int_\Omega \sigma : C^{-1}\tau\, dx + \int_\Omega \chi\cdot H^{-1}\mu\, dx \\ b : V\times\mathcal{S} \to \boldsymbol{R}, \quad & b(v,\tau) \;:= -\int_\Omega \varepsilon(v) : \tau\, dx \end{aligned}$$

for $\Sigma = (\sigma,\chi)$ and $T = (\tau,\mu)$. Finally, the integral on the right-hand side of the balance equation (27) is identified with a time-dependent linear operator on the dual space, that is

$$l(t) \,:\, V \to \boldsymbol{R}, \quad \langle l(t), v\rangle = -\int_\Omega f(t)\cdot v\, dx.$$

Now we can state the

Dual Problem: *Given $l \in H^1(0,T;V')$ with $l(0) = 0$, find $(u,\Sigma) = (u,\sigma,\chi) : [0,T] \to V\times\mathcal{P}$ with $(u(0),\Sigma(0)) = (0,0)$ such that for almost all $t \in (0,T)$ the following conditions hold true:*

$$b(v,\sigma(t)) \;= \langle l(t), v\rangle \qquad \forall v \in V, \tag{28}$$

$$A(\dot{\Sigma}(t), T-\Sigma(t)) + b(\dot{u}(t), \tau-\sigma(t)) \geq 0 \qquad \forall T = (\tau,\mu) \in \mathcal{P}. \tag{29}$$

The name "dual" is related to the use of the conjugate quantities, i.e., the generalized stress, instead of the primal unknowns, plastic strain ε^p and internal variables ξ [12]. Note that the dual problem can be viewed as an abstract constrained initial value problem. The generalized stress Σ is the main quantity of interest, whereas the displacement u is not unique since only the velocity $\dot{u}$ appears in the formulation.

3.4 Existence of Solutions

For the analysis of the dual problem, we require an assumption on the shape of the admissible domain $\mathcal{P}$, see [12].

SAFE LOAD CONDITION SLC: *There is a constant $c > 0$ such that for any $T_1 = (\tau_1, \mu_1) \in \mathcal{P}$ and any stress tensor $\tau_2 \in \mathcal{S}$, there exists a conjugate force $\mu_2 \in \mathcal{M}$ such that*

$$||\mu_2|| \leq c||\tau_2|| \quad \text{and} \quad (\tau_1, \mu_1) + (\tau_2, \mu_2) \in \mathcal{P}.$$

The safe load condition is fullfilled for various yield conditions, in particular for linear isotropic hardening (23). In some sense, such a condition is necessary to complete the dual problem, cf. [12]. The same holds true for the next assumption, which refers to the set $\mathcal{E}$ from (8).

ASSUMPTION A1: *For any $T \in \mathcal{E}$ and any $\kappa \in [0, 1)$, we have $\kappa T \in \mathcal{E}$ and*

$$\inf_{x \in \Omega} \operatorname{dist}(\kappa T(x), \partial \mathcal{E}) > 0.$$

Under these assumptions, existence of a solution of the dual problem can be demonstrated [12]:

Theorem 1. *If the safe load condition (SLC) and assumption (A1) are fullfilled, then the dual problem has a solution (u, Σ) and Σ is unique.*

We briefly outline the idea of the proof. The details can be found in [6,9,12]. The key idea is the viscoplastic penalty regularization (26), which is an unconstrained optimization problem. Therefore taking the derivative of (26) and setting it to zero yields the variational equality

$$A(\dot{\Sigma}(t), T) + \big(J'_\eta(\Sigma(t)), T\big) + b(\dot{u}(t), \tau) = 0 \qquad \forall T = (\tau, \mu) \in \mathcal{T} \tag{30}$$

$$b(v, \sigma(t)) \ = \langle l(t), v \rangle \qquad \forall v \in V. \tag{31}$$

The balance equation (28) has been added to show the connection to a saddle point problem. Next, for the bilinear form b there exists a constant $\beta_b > 0$ such that

$$\sup_{0 \neq \tau \in \mathcal{S}} \frac{|b(v, \tau)|}{||\tau||_{\mathcal{S}}} \geq \beta_b ||v||_V \quad \forall v \in V. \tag{32}$$

This is the Babuška-Brezzi condition, but note that b satisfies (32) only on $\mathcal{S}$ and not on $\mathcal{T}$. For this reason, we need in addition the safe load condition

SLC. This assumption guarantees a solution σ of (31) in the orthogonal complement of the kernel of b such that there exists χ with $(\sigma, \chi) \in \mathcal{P}$. Due to the coercivity of A, which means there exists a constant β_A such that

$$A(T,T) \geq \beta_A \|T\|_{\mathcal{T}}^2 \qquad \text{for all } T \in \mathcal{T},$$

and due to the monotonicity of J'_η, there exists a solution $(\Sigma, \dot{u})$ of the regularized problem (30)–(31).

In the second step of the proof, one lets a suitable subsequence of η tend to zero and shows for this sequence that the limit with respect to Σ and $\dot{u}$ exists and solves the dual problem. Uniqueness follows from the coercivity of the bilinear form A.

4 Runge-Kutta Methods

The dual problem stated in the last section can be viewed as an infinite-dimensional system of DAEs or as PDAE (Partial Differential-Algebraic Equation). With the numerical analysis of DAEs in mind, we focus now on the versatile class of implicit Runge-Kutta methods for the time integration. We have two objectives in mind. First, we want to discretize the local constitutive equations. This will become the basis of the implementation in combination with the FEM, see Sect. 6. Our second objective is the direct application of Runge-Kutta methods to the global infinite dimensional system, i.e., we want to leave the space variables continuous in order to obtain results independent from space discretization.

4.1 Runge-Kutta Methods for ODEs and DAEs

Runge-Kutta methods are one-step methods. Implicit Euler or the midpoint rule are well-known representatives of this class. Consider the autonomous ODE

$$\dot{z} = f(z)$$

with initial value $z(t_0) = z_0$. We discretize time by timesteps of stepsize h, set $t_n = t_0 + nh$, and denote the approximation of the solution z at time t_n by z_n. A Runge-Kutta method computes z_{n+1} from the weighted sum

$$z_{n+1} = z_n + h \sum_{i=1}^{s} b_i \dot{Z}_{ni} \tag{33}$$

with stage derivatives $\dot{Z}_{ni}$. They satisfy

$$\dot{Z}_{ni} = f(Z_{ni}), \tag{34}$$

$$Z_{ni} = z_n + h \sum_{j=1}^{s} a_{ij} \dot{Z}_{nj}. \tag{35}$$

Thus, the stage Z_{ni} is an approximation of the solution z at time $t_n + c_i h$ and the stage derivative $\dot{Z}_{ni}$ is an approximation of the derivative of the solution. For notational convenience, the coefficients (a_{ij}) are collected in the Runge-Kutta matrix $\mathcal{A}$. Note that the quadrature nodes c_i satisfy $c_i := \sum_{j=1}^{s} a_{ij}$.

For $a_{ij} = 0, j \geq i$, the method is explicit, since (35) defines Z_{nj} in terms of $Z_{n1}, \ldots, Z_{nj-1}$. Otherwise, it is called implicit and (34)–(35) defines a nonlinear system, which needs to be solved by Newton's method or variants thereof. The implicit Euler method is given by $s = 1$, $\mathcal{A} = (1)$ and $b_1 = 1$. In viscoplasticity, implicit methods are preferable because of the stiffness of the singularly perturbed ODE.

Runge-Kutta methods can be generalized and applied to DAEs, including the constitutive equations of elastoplasticity. We set $y := (\varepsilon^{\mathrm{p}}, q, t)$ and write equations (10)- (12) as an autonomous, semi-explicit system

$$\dot{y} = F(y, \gamma) \tag{36}$$

$$0 = \phi(y). \tag{37}$$

To apply a Runge-Kutta method to this DAE, we simply replace (34) by the corresponding relations (36) and (37) [14]. Thus we obtain the approximation

$$y_{n+1} = y_n + h \sum_{i=1}^{s} b_i \dot{Y}_{ni}, \qquad \gamma_{n+1} = \gamma_n + h \sum_{i=1}^{s} b_i \tilde{\Gamma}_{ni}, \tag{38}$$

with internal stages

$$Y_{ni} = y_n + h \sum_{j=1}^{s} a_{ij} \dot{Y}_{nj}, \qquad \Gamma_{ni} = \gamma_n + h \sum_{j=1}^{s} a_{ij} \tilde{\Gamma}_{nj}, \quad i = 1, \ldots, s, \tag{39}$$

and derivatives

$$\dot{Y}_{ni} = F(Y_{ni}, \Gamma_{ni}), \qquad 0 = \phi(Y_{ni}), \quad i = 1, \ldots, s. \tag{40}$$

Since the smoothness of the Lagrange multiplier γ is not clear, we write $\tilde{\Gamma}_{ni}$ instead of $\dot{\Gamma}_{ni}$. Each time step requires the solution of the nonlinear system (39)–(40).

As already pointed out, this method definition includes various schemes like implicit Euler, DIRK (Diagonally implicit RK) methods and further implicit discretizations. Our focus will be on the Gauss and the Radau IIA class, since they are extensions of the implicit midpoint and the implicit Euler method. The latter are already used in computational plasticity [25,27].

4.2 Convergence Order

What is the benefit of such an – at least at first sight – computationally expensive method? The answer is that we can improve the convergence order of the numerical solution. A method is said to possess global convergence order k if the error between the true solution $y(t_n)$ and its approximation satisfies

$$\|y(t_n) - y_n\| \leq K \cdot h^k.$$

In Tab. 1, we listed the convergence orders of implicit Runge-Kutta methods with s stages when applied to ODEs and DAEs of index two [15]. Observe that an order reduction occurs [15].

		ODE	DAE	
Class	s	z	y	γ
Gauss	odd	$2s$	$s+1$	$s-1$
	even	$2s$	s	$s-2$
Radau IIA		$2s-1$	$2s-1$	s

Table 1. Global convergence orders of the s-stage Gauss/Radau IIA methods for ODEs and DAEs of index two.

Numerical experiments show that the theoretical orders are obtained if during plastic deformation the constitutive equations of plasticity are integrated with these methods, see [5]. However, the inequality constraint of the Kuhn-Tucker condition involves additional discontinuities, which may effect the global convergence order.

4.3 Runge-Kutta Methods for the Dual Problem

The method definition of Subsect. 4.2 can be directly carried over to the dual problem. For this purpose, we replace the derivatives in (29) by the stage derivatives. One step of the Runge-Kutta method from t_n to $t_{n+1} = t_n + h$ is then defined by

$$\Sigma_{n+1} = \Sigma_n + h \sum_{i=1}^{s} b_i \Psi_i, \qquad u_{n+1} = u_n + h \sum_{i=1}^{s} b_i w_i$$

where Ψ_i and w_i are the stage derivatives at time $t_i = t_n + c_i h$. The stages $\Sigma_i = (\sigma_i, \chi_i) \in \mathcal{P}$ and the stage derivatives $w_i \in V$ satisfy the nonlinear system

$$b(v, \sigma_i) = \langle l_i, v \rangle \qquad \forall v \in V \tag{41}$$

$$A(\Psi_i, T - \Sigma_i) + b(w_i, \tau - \sigma_i) \geq 0 \qquad \forall T = (\tau, \nu) \in \mathcal{P} \tag{42}$$

for $i = 1, \ldots, s$, with right hand sides $l_i = l(t_n + c_i h)$. Stages Σ_i and stage derivatives Ψ_i are coupled via

$$\Sigma_i = \Sigma_n + h \sum_{j=1}^{s} a_{ij} \Psi_j. \tag{43}$$

Existence of Numerical Solution. In the ODE or DAE case, the numerical solution of the nonlinear system (39)–(40) exists for small stepsizes [15]. For the infinite dimensional dual problem (41)–(42), however, existence is not at all obvious. If one looks back to the proof of Theorem 1, one notices that the coercivity of the bilinear form A was a basic prerequisite. At first look, the coupling (43) destroys this property: Consider (42) and replace the stage deriviatives Ψ_i by the stages Σ_i using the relation (43). Thereby, the Σ_i are multiplied in a certain sense by the inverse of the Runge-Kutta matrix before the bilinear form A is evaluated, and we obtain a new bilinear form that depends on the method coefficients. The crucial point is now whether this new bilinear form is still coercive.

In the theory of stiff ordinary differential equations, similar problems occur [15]. In general, the Runge-Kutta matrix $\mathcal{A}$ is not positive definite, but in many cases it possesses itself a coercivity property.

Definition 1. We consider the inner product $(u, v)_D := u \cdot Dv$ on $\boldsymbol{R}^s$ where $D = \text{diag}\,(d_1, \ldots, d_s)$ with positive entries $d_i > 0$. The Runge-Kutta matrix $\mathcal{A}$ is coercive iff there exists a positive diagonal matrix D and a constant α_D such that

$$(u, \mathcal{A}^{-1}v)_D \geq \alpha_D(u, v)_D$$

for all $u, v \in \boldsymbol{R}^s$. In this case, we set

$$\alpha_0(\mathcal{A}^{-1}) := \sup_{D>0} \alpha_D.$$

From [15, Th. 14.5], we have the classification

Theorem 2. *If the Runge-Kutta method belongs to one of the classes Gauss, Radau IA or Radau IIA, then the Runge-Kutta matrix is coercive.*

We remark that DIRK (Diagonally Implicit Runge-Kutta) methods with positive coefficients a_{ii} on the diagonal are coercive, too [15, Th. 14.6]. With the notion of coercivity for a Runge-Kutta method at hand, we are able to state

Theorem 3. *Let the Runge-Kutta matrix be coercive and let the safe load condition (SLC) and Assumption (A1) be satisfied. Then the discretization scheme (41)–(43) possesses a solution (w_i, Σ_i) for $i = 1, \ldots, s$, and the stages Σ_i of the generalized stresses are unique.*

The proof follows the same lines as the proof of Theorem 1. Since we have to consider s stages, however, many technical details need to be addressed. Starting point is the Runge-Kutta discretization of the viscoplastic regularization (30) in combination with the balance equation (31). In order to distinguish between original and regularized problem, we use now the superscript η for the variables of the regularized problem and discretize it in the following way:

Find $(w_i^\eta, \Sigma_i^\eta) = (w_i^\eta, \sigma_i^\eta, \chi_i^\eta) \in V \times \mathcal{T}$ such that

$$\begin{aligned} b(v, \sigma_i^\eta) &= \langle l_i, v \rangle, \qquad \forall v \in V, \\ A(\Psi_i^\eta, T) + \left(J_\eta'(\Sigma_i^\eta), T\right)_{\mathcal{T}} + b(w_i^\eta, \tau) &= 0, \qquad \forall T = (\tau, \nu) \in \mathcal{T} \end{aligned}$$

for $i = 1, \ldots, s$. Stages Σ_i^η and stage derivatives Ψ_i^η are related to each other via

$$\Sigma_i^\eta = \Sigma_n + h \sum_{j=1}^{s} a_{ij} \Psi_j^\eta .$$

Note that the solution (u_n, Σ_n) of the last step is taken from the original but not from the regularized problem.

In a certain product space, the discretized regularization represents a saddle point problem and therefore the reasoning of the proof relies very much on the corresponding theory. The details of the proof can be found in [6,12].

5 Contractivity

A-stability and B-stability are the standard concepts for ODE integration schemes. For DAEs, however, it is so far not clear how to define a conclusive notion of stability. In the special case of linear hardening elastoplasticity, it is possible to find an analogue property to B-stability.

First we recall the ODE theory. The differential equation

$$\dot{z} = f(t, z)$$

is called contractive if it satisfies the one-sided Lipschitz condition

$$(f(t, y) - f(t, z), y - z) \leq 0.$$

The classical formulation of elastoplasticity also features a contractivity property. We consider the case of linear isotropic hardening.

$$\begin{aligned} \dot{\sigma} &= C : (\dot{\varepsilon} - \gamma \phi_\sigma), \\ \dot{\chi} &= -\gamma H \phi_\chi, \\ 0 &= \gamma \phi \end{aligned}$$

As in Section 3, we set $\Sigma = (\sigma, \chi)$ and choose the so-called complementary energy norm

$$||(\sigma, \chi)|| := \sqrt{\sigma : C^{-1} : \sigma + \chi \cdot H^{-1} \chi}. \tag{44}$$

Then we have, see [26],

Theorem 4. *For different initial values $\Sigma(t_0)$ and $\tilde{\Sigma}(t_0)$, the inequality*

$$||\Sigma(t) - \tilde{\Sigma}(t)|| \leq ||\Sigma(t_0) - \tilde{\Sigma}(t_0)||$$

holds true for all $t \geq t_s$.

Note, that the contractivity holds true regardless whether t_0 is in the elastic deformation phase or in the plastic phase.

The dual problem corresponds to linear isotropic hardening and as can be expected, the contractivity carries over to this weak formulation. The pointwise energy norm (44) is replaced by the norm

$$\|\cdot\|_A := \sqrt{A(\cdot,\cdot)}$$

on the function space $\mathcal{T}$. Then:

Theorem 5. *Assume (u, Σ) and $(\widehat{u}, \widehat{\Sigma})$ are solutions of the dual problem for different initial values. Then*

$$\|\Sigma(t) - \widehat{\Sigma}(t)\|_A \leq \|\Sigma(0) - \widehat{\Sigma}(0)\|_A$$

holds true for all times $t > 0$.

Proofs of this property can be found in [12,27].

Any numerical method should preserve this contractivity property, i.e., its approximations should not add numerical energy to the system. In the ODE case, algebraically stable Runge-Kutta methods preserve contractivity.

Definition 2. A Runge-Kutta method is called algebraically stable, iff

1. $b_i \geq 0$ for all $i = 1, \ldots, s$,
2. $M = (m_{ij}) = (b_i a_{ij} + b_j a_{ji} - b_i b_j)_{i,j=1}^{s}$ is non-negative definite.

Due to the special nature of the elastoplastic DAE, this is also true here, i.e., in the classical form it holds, see [5]

Theorem 6. *Let the Runge-Kutta method be algebraically stable. For given consistent inital values Σ_0 and $\tilde{\Sigma}_0$ at time t_s and arbitrary $n \geq 1$,*

$$\|\Sigma_n - \tilde{\Sigma}_n\| \leq \|\Sigma_0 - \tilde{\Sigma}_0\|.$$

holds true.

Furthermore in the weak form, see [6],

Theorem 7. *Algebraically stable Runge-Kutta methods preserve the contractivity of the elastoplastic flow, that is*

$$\|\Sigma_n - \widehat{\Sigma}_n\|_A \leq \|\Sigma_0 - \widehat{\Sigma}_0\|_A$$

for different initial values Σ_0 and $\widehat{\Sigma}_0$.

The notions of algebraic stability and coercivity of a Runge-Kutta method are not equivalent. Remarkably, however, there are several classes, in particular Gauss, Radau IA, and Radau IIA, that feature both properties.

6 Computational Treatment

We discuss now the computational solution of elastoplastic boundary value problems by combining the FEM in space and implicit Runge-Kutta integration in time. On the one hand, we have to consider the balance equation and the kinematic relation, which are partial differential equations in space and which we discretize by standard finite elements. On the other hand, in the quadrature nodes of each element the material law needs to be integrated in time. For the handling of the yield condition as unilateral constraint, we use the standard return-mapping strategy that applies a trial elastic predictor step followed by a corrector step whenever plastic deformations occur. The corrector step is based on the discretization by an implicit Runge-Kutta method as described in Sect. 4.

6.1 The Finite Element Method

Starting point is the weak form of the balance of momentum, see Sect. 3. We augment the equations here by some Neumann boundary conditions $\bar{\sigma}$ on Γ_σ and switch to a matrix-vector notation [2,25]. All symmetric rank two tensors are identified with 6-dimensional vectors and the elasticity tensor transforms to a 6×6 matrix. Thus the tensor product becomes a matrix-vector product.

In this notation, the balance of momentum in weak form reads

$$- \int_\Omega \varepsilon(v)^{\mathrm{T}} C(\varepsilon(u) - \varepsilon^{\mathrm{p}})\, dx = - \int_\Omega v^{\mathrm{T}} f(t)\, dx - \int_{\Gamma_\sigma} v^{\mathrm{T}} \bar{t}\, d\sigma.$$

The plastic strain ε^{p} is determined from the constitutive equations. Taking into account that the total strain is a function of the displacement and

$$\sigma = C(\varepsilon(u) - \varepsilon^{\mathrm{p}}),$$

we obtain

$$\begin{aligned} \dot{\varepsilon}^{\mathrm{p}} &= \gamma \mathrm{r}(u, \varepsilon^{\mathrm{p}}, q), \\ \dot{q} &= -\gamma \mathrm{g}(u, \varepsilon^{\mathrm{p}}, q), \\ 0 &= \gamma \phi(u, \varepsilon^{\mathrm{p}}, q), \quad \gamma \geq 0, \quad \phi(u, \varepsilon^{\mathrm{p}}, q) \leq 0. \end{aligned}$$

In short, the Ritz-Galerkin approach replaces the Sobolev space V by a finite dimensional subspace V_D and solves the weak form on V_D instead of V (for convergence results see [2,25]). We take the usual finite elements to construct V_D. Therefore, the reference configuration Ω is partitioned into elements and in each element we define the form functions. Assemblage of these form functions yields a basis of V_D, which we arrange columnwise in the matrix $N(x)$. Consequently, we seek a solution u_D of the form

$$u_D(x,t) = N(x)u(t) + u_u(x,t)$$

where $u(t)$ is the vector of the time-dependent so-called nodal variables an u_u a part that fullfills the Dirichlet boundary conditions. With

$$\varepsilon(N(x)u(t)) = \tfrac{1}{2}(\nabla N(x)u(t) + \nabla(N(x)u(t))^{\mathrm{T}}) =: B(x)u(t)$$

the Galerkin approach yields the nonlinear system

$$\begin{aligned} 0 = & \int_{\Omega} B^{\mathrm{T}}(x)CB(x)u(t)\,dx - \int_{\Omega} B^{\mathrm{T}}(x)C\varepsilon^{p}(x,t)\,dx \\ & - \int_{\Omega} N^{\mathrm{T}}(x)f(x,t)\,dx - \int_{\Gamma_\sigma} N^{\mathrm{T}}(x)\bar{\sigma}(x,t)\,d\sigma \\ & + \int_{\Omega} B^{\mathrm{T}}(x)C\varepsilon(u_u(x,t))\,dx. \end{aligned} \tag{45}$$

Recall that the plastic strain ε^p is a function of the displacement and therefore a function of $u(t)$. In this expression, the matrix

$$K := \int_{\Omega} B^{\mathrm{T}}(x)CB(x)\,dx$$

is called *stiffness matrix* and can be evaluated directly, since the form functions are known. The vector

$$\begin{aligned} b(t) := & \int_{\Omega} N^{\mathrm{T}}(x)f(x,t)\,dx + \int_{\Gamma_\sigma} N^{\mathrm{T}}(x)\bar{\sigma}(x,t)\,d\sigma \\ & - \int_{\Omega} B^{\mathrm{T}}(x)C\varepsilon(u_u(x,t))\,dx \end{aligned}$$

is called *load vector*. Some quadrature rule is needed to evaluate $b(t)$.

The remainig integral of (45) is of special interest since its integration in space needs to be performed by a quadrature rule. In each element Δ, the integral is replaced by a sum

$$\int_{\Delta} B^{\mathrm{T}}(x)C\varepsilon^{\mathrm{p}}(x,t)\,dx \approx \sum \beta_k B^{\mathrm{T}}(\nu_k)C\varepsilon^{\mathrm{p}}(\nu_k,t), \tag{46}$$

where the ν_k stand for the quadrature or Gauss points of the element Δ and β_k are the corresponding weights. Since the quadrature rule is a linear operation, we can assemble the weights of all elements in a matrix Q and obtain

$$\int_{\Omega} B^{\mathrm{T}}(x)C\varepsilon^{\mathrm{p}}(x,t)\,dx \approx Q\left(\varepsilon_1^{\mathrm{p}}(t),\ldots,\varepsilon_m^{\mathrm{p}}(t)\right)^{\mathrm{T}},$$

where m denotes the total number of Gausspoints and $\varepsilon_i^{\mathrm{p}}(t)$ the plastic strain in the i-th Gausspoint. Thus, the space discretization via the Galerkin approach is summarized in the time-dependent system

$$0 = Ku - b(t) - Q\left(\varepsilon_1^{\mathrm{p}},\ldots,\varepsilon_m^{\mathrm{p}}\right)^{\mathrm{T}}, \tag{47}$$

$$\dot{\varepsilon}_i^{\mathrm{p}} = \gamma_i \mathrm{r}(u,\varepsilon_i^{\mathrm{p}},q_i), \tag{48}$$

$$\dot{q}_i = -\gamma_i \mathrm{g}(u,\varepsilon_i^{\mathrm{p}},q_i), \tag{49}$$

$$0 = \gamma_i\phi(u,\varepsilon_i^{\mathrm{p}},q_i), \quad \gamma_i \geq 0, \quad \phi(u,\varepsilon_i^{\mathrm{p}},q_i) \leq 0. \tag{50}$$

Here, a permanent constraint (47), which represents the balance law, appears. Since the stiffness matrix is regular, it is an index one constraint for the nodal variables u. Thus, the space discretization yields a system with index one and during plastic deformation with additional index two constraints [11,24].

Remark: The introduction of the matrix Q is somehow confusing since it never appears in the codes explicitly. It is however hidden in the assembling and quadrature procedure. Thus, Q is merely a formal notation.

6.2 Two-Level Newton Method and Return Mapping

In most FE codes, the system (47)–(50) is solved by the implicit Euler method. The discretization in time then reads

$$0 = K u_{n+1} - b(t_{n+1}) - Q\left(\varepsilon^{\mathrm{p}}_{1,n+1}, \ldots, \varepsilon^{\mathrm{p}}_{m,n+1}\right)^{\mathrm{T}}, \tag{51}$$

$$\varepsilon^{\mathrm{p}}_{i,n+1} = \varepsilon^{\mathrm{p}}_{i,n} + h\gamma_{i,n+1}\mathrm{r}(u_{n+1}, \varepsilon^{\mathrm{p}}_{i,n+1}, q_{i,n+1}), \tag{52}$$

$$q_{i,n+1} = q_{i,n} - h\gamma_{i,n+1}\mathrm{g}(u_{n+1}, \varepsilon^{\mathrm{p}}_{i,n+1}, q_{i,n+1}), \tag{53}$$

$$\begin{aligned} 0 &= \gamma_{i,n+1}\phi(u_{n+1}, \varepsilon^{\mathrm{p}}_{i,n+1}, q_{i,n+1}), \\ &\gamma_{i,n+1} \geq 0, \quad \phi(u, \varepsilon^{\mathrm{p}}_{i,n+1}, q_{i,n+1}) \leq 0. \end{aligned} \tag{54}$$

The large dimension of this system can be greatly reduced. Consider the subsystem (52)–(54). The plastic strains in the Gauss points $\varepsilon^{\mathrm{p}}_{i,n+1}$ are determined from u_{n+1}. Hence, (51) can be viewed as a nonlinear equation in u_{n+1}, which is solved iteratively by Newton's method [11,22]. In each step of this so-called global iteration, the system (52)–(54) is solved for given iterates of u_{n+1}.

The return mapping strategy, a predictor corrector scheme, handles the unilateral constraint. First, the trial elastic step is defined by

$$\varepsilon^{\mathrm{p},tr}_{n+1} := \varepsilon^{\mathrm{p}}_{n}, \qquad q^{tr}_{n+1} := q_n.$$

Then the yield function is evaluated at the trial elastic step, i.e.

$$\phi^{tr}_{n+1} := \phi(u_{n+1}, \varepsilon^{\mathrm{p},tr}_{n+1}, q^{tr}_{n+1}).$$

If

- $\phi^{tr}_{n+1} < 0$, elastic deformation is assumed, i.e. the elastic trial step is taken as new iterate for plastic strain and internal variables.
- $\phi^{tr}_{n+1} \geq 0$, plastic deformation is assumed. Now the nonlinear system

$$\begin{aligned} \varepsilon^{\mathrm{p}}_{i,n+1} &= \varepsilon^{\mathrm{p}}_{i,n} + h\gamma_{i,n+1}\mathrm{r}(u_{n+1}, \varepsilon^{\mathrm{p}}_{i,n+1}, q_{i,n+1}) \\ q_{i,n+1} &= q_{i,n} - h\gamma_{i,n+1}\mathrm{g}(u_{n+1}, \varepsilon^{\mathrm{p}}_{i,n+1}, q_{i,n+1}) \\ 0 &= \gamma_{i,n+1}\phi(u_{n+1}, \varepsilon^{\mathrm{p}}_{i,n+1}, q_{i,n+1}) \end{aligned}$$

  is solved by a so-called local iteration.

For quadratic convergence of the global iteration, the computation of the Jacobian of (51) with respect to u_{n+1}, the so-called consistent tangent [27,28], is essential. We may call this scheme a two-level Newton's method, since it consists of a global iteration with respect to the balance equation and local iterations in each Gauss point for the material laws.

6.3 Extension to Runge-Kutta Methods

In the last subsections, we solved the spatially discretized boundary value problem (47)–(50) by the implicit Euler method. A direct application of arbitrary Runge-Kutta methods to this system runs into difficulties. Fully implicit methods do not allow the two-level Newton scheme and hence the dimension of the time discretized system becomes

$$\text{stages} \times (\#\ \text{Gausspoints} \times \#\ \text{Internal states} + \#\ \text{nodal variables}),$$

which is too large for practical computations. To overcome this problem, [11,32] considered the classical approach for diagonally implicit Runge-Kutta (DIRK) methods. It proceeds from stage to stage by weighted implicit Euler steps [15]. An example is given by the method class of Alexander [11].

We take a diffferent approach to preserve the two-level Newton scheme. Again, the nodal variables are discretized by the implicit Euler method. The time discretization of (48)–(50), however, is done by means of an implicit Runge-Kutta method where the internal stages of the nodal variables are interpolated linearily from u_n to u_{n+1}. In this way, the return mapping scheme is kept to separate elastic from plastic behaviour. For details, see 3.2. in [5].

The main difference to the classical DIRK approach is the partitioning of the variables. The nodal variables and the internal quantities are discretized with different Runge-Kutta methods. Note that implicit midpoint or generalized midpoint rules are already treated in the same fashion in [25–27]. The results of the following section show that higher accuracy and global convergence order two is obtained. However, due to the order one discretization of the nodal variables, no higher convergence order than two can be expected. Order two follows from the linear structure of the balance equation (51).

In the following, we will denote this approach NRK (Newton Runge-Kutta), whereas CDIRK stands for the classical DIRK approach. For implicit Euler, NRK and CDIRK are equivalent.

7 Numerical Results

We consider two kinds of numerical simulations. First, the behaviour of the return mapping strategy when applied to the constitutive equations in one material point [5] is studied. Second, we solve a typical two-dimensional example from the literature, a rectangular strip with hole.

7.1 Local Integration of Material Laws

Summarizing the results of [5], we have the following conclusions: For materials with nonlinear evolution such as the associated flow rules of combined kinematic and exponential isotropic hardening, the convergence order for higher order Runge-Kutta methods like the Radau IIA with $s = 2$ or $s = 3$ stages or the Gauss methods breaks down to two [5]. This can be explained by the discontinuity in the switching point and a missing detection of the switching point [14,15]. For DAEs of index two, the method has to start with consistent initial values to keep its convergence order. However, for materials with linear evolution like linear isotropic hardening or the polynomial hardening of the next example, the performed integration is almost perfect.

7.2 Two-Dimensional Example

A typical example from the literature is a rectangular strip with hole. Due to the symmetry, the geometry of the problem can be reduced to one quarter, see [27]. For our computations, we took the mesh and constraints as depicted in Fig. 1 with 6-noded quadratic elements. The plate has height 18 and width 10. The yield function is

$$\phi(\sigma, q) = ||\text{dev}(\sigma)|| - \sqrt{\frac{2}{3}}\sigma_0 \left(\frac{q}{\sigma_0}E + 1\right)^{1/N}$$

with $E = 2.1 \cdot 10^{11}\, Pa, \nu = 0.3, \sigma_0 = 4.6 \cdot 10^8\, Pa$ and $N = 7$, see [1]. On the upper edge, we prescribe a displacement of $0.002t$. We assume plane strain and remark that the simulation has been implemented in MATLAB with the aid of the SME/FEMLAB toolbox.

First, we took the common implicit Euler rule and compared the results with the two stage Radau IIA and Gauss method. The resulting convergence orders are depicted in Fig. 2. We computed here the relative error in each FE node or Gauss point, respectively, and printed the 1–norm of the error over all nodes or Gauss points. Radau IIA turns out to improve the accuracy compared to implicit Euler and Gauss. The latter lacks most in accuracy. Implicit Euler and Gauss show convergence order 1, Radau IIA shows order two.

Next, we investigated the performance of the two-level Newton method. As local time discretization schemes, we took implicit Euler, Radau IIa s=2 and Alexander (NRK implicit Euler, NRK Radau IIa s=2, NRK Alexander). These methods are compared to the classical DIRK approach with the method of Alexander (CDIRK Alexander). The latter is preferred by [11]. The results are depicted in Fig. 3. In all four cases, we used the consistent tangent and stopped the Newton iterations, whenever the relative increment of the residual remained less than 10^{-9} for the global iteration or less than 10^{-12} for the local iteration. The Jacobians are set up in each step.

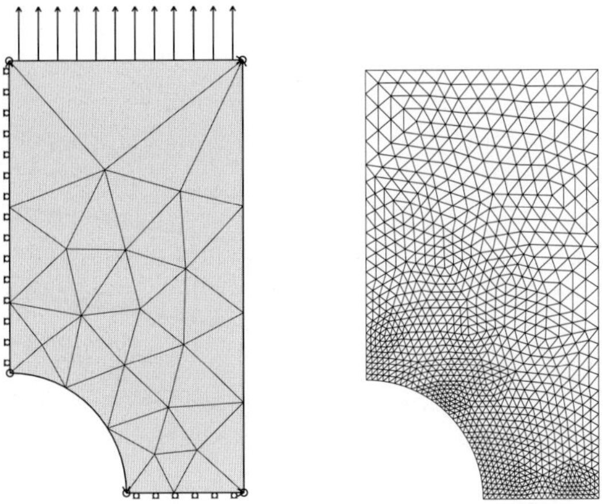

(a) 37 elements (b) 1968 elements

Fig. 1. Mesh and constraints of the rectangular strip with hole

It turns out, that NRK implicit Euler performs worst. All other three methods show about the same qualitiy. The classical CDIRK Alexander approach performs best, especially for large step sizes. Between NRK Radau IIa s=2 and the method NRK Alexander there is hardly any difference. Thus we can summarize that the two-level Newton approach in combination with the implicit Runge-Kutta methods Radau IIa s=2 or Alexander is easy to implement and yields a competitive extension of the standard implicit Euler method. Our simulations show that the accuracy increases significantly and that convergence order two is obtained.

8 Outlook

Inspired by an interpretation as differential-algebraic equation (DAE), we considered in this paper material equations with yield surfaces both in a local formulation as well as in a weak global boundary value problem and showed how implicit Runge-Kutta methods can be applied. In particular, the notions of coercivity and algebraical stability were sufficient to ensure well-posedness and stability of the time discretization.

The challenge ahead is an adaptive approach where the solution of the constitutive equations yields an additional error estimate that can be used for stepsize control. Due to the weak coupling of the evolution in different Gausspoints, it seems worthwhile to investigate multi-rate techniques to cap-

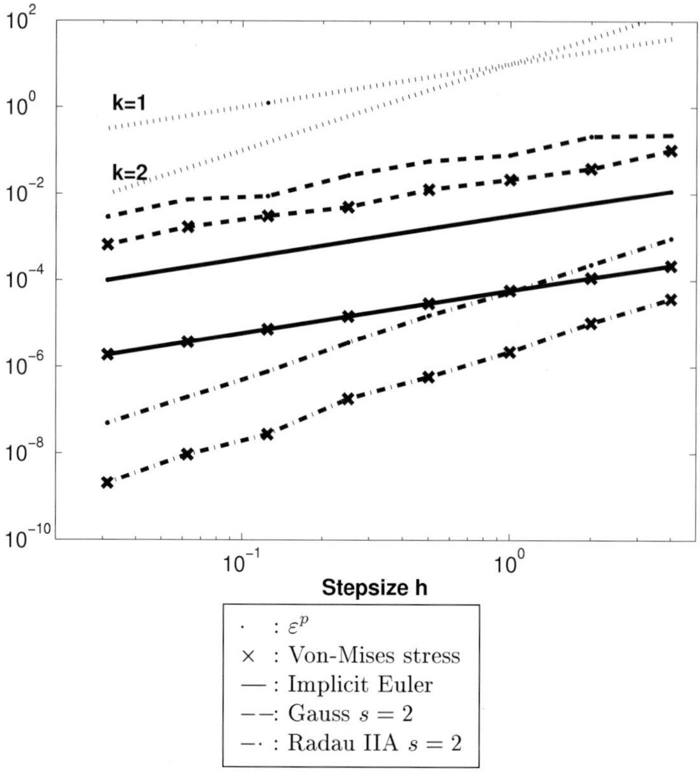

Fig. 2. Average relative error of differential variables in plastic points at time $t = 32$ for 37 elements

ture the elastoplastic flow. Furthermore, the results of Section 4 suggest an alternative implementation by the reverse method of lines.

Acknowledgement: The authors would like to thank SFB 298 for the fruitful cooperation and the support during the past years.

References

1. Baaser H., Tvergaard V. (2000) A new algorithmic approach treating nonlocal effects at finite rate-independent deformation using the Rousselier damage model. Report, No. 647, Kgs. Lyngby
2. Braess D. (2001) Finite Elements: Theory, Fast Solvers and Applications in Solid Mechanics. Cambridge University Press, Cambridge
3. Brenan K., Campbell S., Petzold L. (1996) The Numerical Solution of Initial Value Problems in Ordinary Differential-Algebraic Equations. 2. Ed., SIAM, Philadelphia

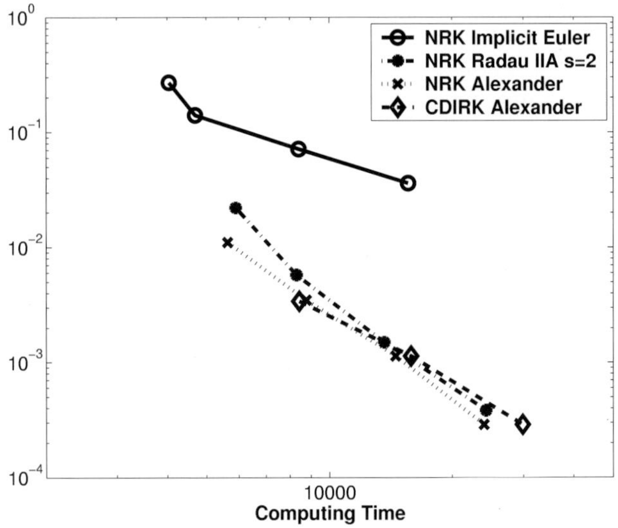

Fig. 3. Average relative error of nodal variables at time $t = 32$ compared to computing time in seconds on an Intel PIII 650 MHz under Matlab 6 for 1968 elements

4. Büttner J. (2002) Computational Inelasticity – How to implement an elastoplastic problem with SME/FEMLAB. Preprint 01/19, IWRMM, Universität Karlsruhe
5. Büttner J., Simeon B. (2002) Runge-Kutta Methods in Elastoplasticity. App. Numer. Math. 41:443–458
6. Büttner J., Simeon B. (submitted) Time Integration of the Dual Problem of Elastoplasticity by Runge-Kutta Methods. submitted to SINUM
7. Carstensen C. (1996) Coupling of FEM and BEM for interface problems in viscoplasticity and plasticity with hardening. SIAM J. Numer.Anal. 33:171–208
8. Cameron F. (1999) A class of low order DIRK methods for a class of DAEs. Appl. Numer. Math 31(1):1–16
9. Duvaut G., Lions J. L. (1976) Inequalities in Mechanics and Physics. Springer, Berlin New York
10. Ellsiepen P. (1999) Zeit- und ortsadaptive Verfahren angewandt auf Mehrphasenprobleme poröser Medien. Bericht Nr. II-3, Institut für Mechanik, Universität Stuttgart
11. Ellsiepen P., Hartmann S. (2001) Remarks on the interpretation of current non-linear finite element analyses as differential-algebraic equations. Int. J. Num. Meth. Engng., 51:679–707
12. Han W., Reddy B. D. (1999) Plasticity, Matehematical theory and numerical analysis. Springer, Berlin New York
13. Higueras I., García-Celayeta B. (1999) Logarithmic Norms for Matrix Pencils. SIAM J. Matrix Anal. Appl., 20(3):646–666

14. Hairer E., Lubich Ch., Roche M. (1989) The numerical solution of differential-algebraic systems by Runge-Kutta methods. Lecture Notes in Math. 1409, Springer, Berlin
15. Hairer E., Wanner G. (1996) Solving ordinary differential equations II. Springer, Berlin
16. Hairer E., Zennaro M. (1996) On error growth functions of Runge-Kutta methods. Appl. Numer. Math. 22(1-3):205–216
17. Haupt P., Kamlah M., Tsakmakis C. (1992) On the Thermodynamics of Rate-Independent Plasticity as an Asymptotic Limit of Viscoplasticity for Slow Processes. In: Besdo D., Stein E.: Finite Inelastic Deformations - Theory and Applications, IUTAM Symposium Hannover/Germany 1991:107–116, Springer, Berlin
18. Lemaitre J., Chaboche J. L. (1990) Mechanics of Solid Materials. Cambridge University Press, Cambridge
19. O'Malley R. M. (1974) Introduction to Singular Perturbations. Academic Press, New York
20. Papadopoulos P., Taylor R. (1994) On the application of multistep integration methods to inifinitesimal elastoplasticity. Int. J. Num. Meth. in Engng., 37:3169–3184
21. Rabier P. J., Rheinboldt W. C. (to appear) Theoretical and Numerical Analysis of Differential-Algebraic Equations. to appear in Handbook of Numerical Analysis, Elsevier Science
22. Rabbat N. B. G., Sangiovanni-Vincentelli A. L., Hsieh H. Y. (1979) A Multilevel Newton Algorithm with Macromodelling and Latency for the Analysis of Large-Scale Nonlinear Circuits in the Time Domain. IEEE Transactions on Circuits and Systems, CAS-26:733–741
23. Rentrop P., Scherf O., Simeon B. (1999) Numerical Simulation of Mechanical Multibody Systems with Deformable Components. : In Bungartz H.-J., Durst F., Zenger Chr. (Ed.), *High Performance Scientific and Engineering Computing.* LNCSE 8:143–156, Springer, Berlin
24. Scherf O. (2000) Numerische Simulation inelastischer Körper. VDI Fortschrittberichte, Reihe 20, Nr.321, VDI-Verlag, Düsseldorf
25. Simo J. C. (1998) Numerical Analysis and Simulation of Plasticity. Handbook of Numerical Analysis, vol. VI, Elsevier Science
26. Simo J. C., Govindjee S. (1991) Nonlinear B-stability and symmetry preserving return mapping algorithms for plasticity and viscoplasticity. Int. J. Num. Meth. in Engng., 31:151–176
27. Simo J. C., Hughes T. J. R. (1998) Computational Inelasticity. Springer, New York
28. Simo J. C., Taylor R. L. (1985) Consistent Tangent Operators for Rate Independent Elasto-Plasticity. Comp. Meth. in Appl. Mech., 48:101–118
29. Simo J. C., Taylor R. L. (1986) Return Mapping Algorithm for Plane Stress Elastoplasticity. Int. J. Num. Meth. in Engng., 22:649–670
30. COMSOL (2000) Structural Mechanics Engineering Module, For use with FEMLAB. COMSOL AB, Stockholm
31. Temam R. (1986) A generalized Norton-Hoff Model and the Prandtl-Reuss Law of Plasticity. Arch. Rational Mech. Anal., 95:137–183
32. Wieners C. (2000) Theorie und Numerik der Prandtl-Reuß-Plastizität. Habilitation Thesis, Universität Heidelberg

Differential-Algebraic Equations in Elasto-Viscoplasticity

Oliver Scherf[1] and Bernd Simeon[2]

[1] Opel Powertrain GmbH, Powertrain Engineering
D–65423 Rüsselsheim, Germany
oliver.dr.scherf@de.fiat-gm-pwt.com

[2] Centre for Mathematical Sciences, Munich University of Technology
Boltzmannstr. 3, D–85748 Garching, Germany
simeon@mathematik.tu-muenchen.de

received 22 Jul 2002 — accepted 9 Oct 2002

Abstract. Spatial discretization of the initial-boundary value problem of elasto-viscoplasticity leads to a large coupled system where differential equations model the evolution of the material and nonlinear equations stand for the balance of momentum. The present paper classifies this coupled system as differential-algebraic equation and shows how its numerical solution can be improved by using the second order time integration method BDF-2. Moreover, the structure of the nonlinear system in each time step is analysed and an alternative to the standard two-level Newton process is presented that requires the same derivative information as the consistent tangent operator. Simulation examples demonstrate the possible gain in accuracy and efficiency.

1 Introduction

Computational inelasticity is mainly based on the Finite Element Method (FEM) and corresponding software. For various material laws, however, not only the space but also the time dimension plays an important role and the time discretization must be chosen with care. In particular in elasto-viscoplasticity, there is a growing interest in the Differential-Algebraic Equations (DAEs) that arise after space discretization by the FEM. The goal is to apply sophisticated numerical techniques for DAEs in order to get more accurate results and better performance of the time integration.

The present paper demonstrates how this goal can be achieved. Emphasis is placed on two aspects. On the one hand, we introduce a variable stepsize integration scheme based on the BDF-2 method and on the other hand we show how at each time step the structure of the nonlinear system can be exploited in Newton's method. Combined with the FEM in space, this time integration scheme proves to be much more efficient than standard methods.

Figure 1 displays a measurement diagram that was obtained from a tensile test with the austhenitic steel SS 316. Observe the typical irreversible behavior and the quite smooth transition from the elastic to the plastic phase. The

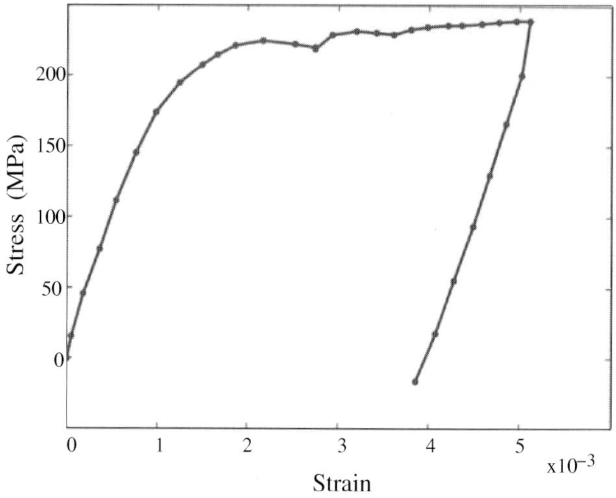

Fig. 1. Measurement of tensile test, steel SS 316

measurements are part of a parameter identification where the model of Chan et al. [3] was validated, see Kollmann et al. [8]. In the simulation runs below, we take these data as an example for testing elasto-viscoplastic behavior.

The aforementioned model is part of the large class of unified viscoplastic materials due to Lubliner [9]. As main feature, such unified models describe both elastic and plastic phases by the same set of evolution equations, i.e., there is no yield surface that separates the two phases. Unified models are mostly used in high temperature environments, e.g., when simulating wear effects in turbine blades. Other applications comprise water networks in power plants and micromechanical devices.

In mathematical terms, a unified model consists of two differential equations

$$\dot{\varepsilon}^p = \boldsymbol{\Phi}(\boldsymbol{\sigma}, \boldsymbol{\alpha}), \qquad \dot{\boldsymbol{\alpha}} = \boldsymbol{\Psi}(\boldsymbol{\sigma}, \boldsymbol{\alpha})$$

for the plastic strain tensor ε^p and the vector of internal variables $\boldsymbol{\alpha}$. The right-hand side $\boldsymbol{\Phi}$ stands for the flow rule and $\boldsymbol{\Psi}$ for the hardening rule. Both rules depend on the internal variables $\boldsymbol{\alpha}$ and the stress tensor $\boldsymbol{\sigma}$.

Two further equations relate stress and strain. Assuming small strains, the total strain ε is additively decomposed into

$$\varepsilon = \varepsilon^e + \varepsilon^p.$$

Finally, Hooke's law holds for the elastic strain tensor ε^e

$$\boldsymbol{\sigma} = \mathsf{C} : \varepsilon^e$$

with elasticity tensor C. The equations given so far describe the material behavior in one point of the body under consideration and are called constitutive equations. If we add the balance of momentum and a kinematic relation between strain and displacement, the initial-boundary value problem of elasto-viscoplasticity is completely specified.

Details of the mathematical model will be discussed below. For the moment, we point out that such unified materials have been extensively studied by Alber [1] and Chelminsky [4]. Existence and uniqueness but also well-posedness results have been obtained for the class of so-called monotone constitutive equations. Thus, there is a sound mathematical theory available as basis for the numerical simulation. We also mention that related work on DAE techniques for the time integration has been recently published by Ellsiepen and Hartmann [5].

The outline of the paper is as follows: In Sect. 2, we introduce the equations of small-deformation elasto-viscoplasticity in the so-called strain space formulation. Next, Sect. 3 discusses the application of the FEM from a global point of view and analyses the structure of the resulting semi-discretized equations in time. The numerical treatment as DAE and corresponding time discretizations are presented in Sect. 4. There, we also concentrate on the BDF-2 method as variable stepsize algorithm and on the solution of the nonlinear system at each time step. Finally, Sect. 5 demonstrates the efficiency of the proposed methods for two simulation examples.

2 Mathematical Model

Kinematics, dynamics, and material law fully describe the elasto-viscoplastic deformation process. More precisely, the kinematic equation determines the geometric properties of the deformation, the dynamic equation expresses the balance of momentum and the constitutive equations characterize the material under consideration. In this section, we state these three ingredients of the mathematical model and summarize some important properties. We refer the interested reader to Alber [1] and Lubliner [9] for more details.

2.1 Kinematic Relations and Balance Law

Let the undeformed elastoplastic body occupy the domain $\Omega \subset \mathrm{R}^3$. The deformation maps each material point $x \in \overline{\Omega}$ at time t to its current position $x + \boldsymbol{u}(x,t)$ with displacement field $\boldsymbol{u}$. With respect to the displacement, the kinematic or Dirichlet boundary conditions are expressed as

$$\boldsymbol{u}_{|\Gamma_0} = \boldsymbol{u}_0$$

on some part Γ_0 of the body boundary Γ with given function $\boldsymbol{u}_0$. Here and in the following we often drop the arguments x and t for convenience.

Strain measures changes of local length elements in the deformed body. We consider small strains only and consequently the (total) strain tensor ε is given by

$$\varepsilon = \mathcal{L}\boldsymbol{u} := \tfrac{1}{2}\left(\nabla \boldsymbol{u} + \nabla \boldsymbol{u}^T\right) \tag{1}$$

where $\mathcal{L}$ stands for a linear operator. By a tensor we mean here a second order tensor in matrix representation, i.e., $\varepsilon = (\varepsilon_{ij})_{3\times 3}$. Stress, on the other hand, is defined as the density of surface force per area element. Like the strain tensor ε, the stress tensor $\boldsymbol{\sigma}$ also depends on time t and is symmetric, i.e., $\boldsymbol{\sigma} = \boldsymbol{\sigma}(x,t)$ and $\boldsymbol{\sigma} = \boldsymbol{\sigma}^T$.

Material laws or constitutive equations relate stress and strain. We will touch upon this point in the next subsection. Before, however, we state the balance of momentum for the quasi-static case,

$$0 = \mathbf{div}\,\boldsymbol{\sigma} + \boldsymbol{\beta} \qquad \text{in } \Omega, \tag{2}$$

where $\boldsymbol{\beta} = \boldsymbol{\beta}(x,t)$ denotes some given volume force. In terms of the stress tensor $\boldsymbol{\sigma}$, we also have the natural or Neumann boundary condition

$$(\boldsymbol{\sigma n})_{|\Gamma_1} = \boldsymbol{t}$$

on some part Γ_1 of the boundary with normal vector $\boldsymbol{n}$ and given surface stress $\boldsymbol{t}$. Note that $\Gamma = \Gamma_0 \cup \Gamma_1$ and $\Gamma_0 \cap \Gamma_1 = \emptyset$.

2.2 Constitutive Equations

The constitutive equations of an elasto-viscoplastic material are based on the assumption that the total strain tensor ε of (1) splits additively into an elastic part ε^e and a plastic part ε^p,

$$\varepsilon = \varepsilon^e + \varepsilon^p. \tag{3}$$

Using the plastic strain ε^p and a vector of l internal variables $\boldsymbol{\alpha} = (\alpha_1, \ldots, \alpha_l)$, the state of a material point at a certain instant of time is modelled by the so-called generalized strain $(\varepsilon^p, \boldsymbol{\alpha})$. This generalized strain satisfies a system of ordinary differential equations in each material point, i.e.,

$$\dot{\varepsilon}^p(x,t) = \boldsymbol{\Phi}(\boldsymbol{\sigma}(x,t), \boldsymbol{\alpha}(x,t)), \tag{4}$$
$$\dot{\boldsymbol{\alpha}}(x,t) = \boldsymbol{\Psi}(\boldsymbol{\sigma}(x,t), \boldsymbol{\alpha}(x,t)). \tag{5}$$

Flow rule (4) and hardening rule (5) depend very much on the particular material and corresponding parameters. Finally, Hooke's law $\boldsymbol{\sigma} = \mathsf{C} : \varepsilon^e$ holds for the elastic strain tensor ε^e and the elasticity tensor C.

Example. The model due to Chan et al. [3] features plastic strain ε^p, isotropic hardening Z (scalar) and directional hardening γ (tensor) as variables. Including some additional relations, the equations read

$$\begin{aligned}
\dot{\varepsilon}^p &= \varphi \operatorname{dev} \boldsymbol{\sigma} / \|\operatorname{dev} \boldsymbol{\sigma}\|, \\
\varphi &= \sqrt{2} D_0 \exp\left[-\tfrac{1}{2}\left(\sqrt{\tfrac{2}{3}} \frac{Z}{\|\operatorname{dev} \boldsymbol{\sigma}\|}\right)^{2n}\right], \\
\dot{Z}^{(i)} &= m_1 (Z_1 - Z^{(i)}) w_p, \\
w_p &= \boldsymbol{\sigma} : \dot{\varepsilon}^p, \\
\dot{\gamma} &= m_2 (Z_3 \cdot \boldsymbol{p} - \gamma) w_p, \\
\boldsymbol{p} &= \boldsymbol{\sigma} / \|\boldsymbol{\sigma}\|, \\
Z &= Z^{(i)} + \gamma : \boldsymbol{p}.
\end{aligned}$$

Note that we abbreviated the scalar product of tensors by a ":". Moreover, the stress deviator is defined by

$$\operatorname{dev} \boldsymbol{\sigma} = \left(\sigma_{ij} - \frac{\sigma_{11} + \sigma_{22} + \sigma_{33}}{3} \delta_{ij}\right).$$

In the model above, the parameters $D_0, n, Z_0, Z_1, Z_3, m_1, m_2$ need be identified for the particular material at hand. E.g., for the steel SS 316 these data have been provided by Seibert [15].

2.3 Initial-Boundary Value Problem

For convenience, we summarize the mathematical model including kinematic, dynamic and constitutive equations.

1.) Linear (small strain) theory

$$\varepsilon(x,t) = \tfrac{1}{2}\left(\nabla \boldsymbol{u}(x,t) + \nabla \boldsymbol{u}(x,t)^T\right).$$

2.) Balance of momentum

$$0 = \mathbf{div}\, \boldsymbol{\sigma}(x,t) + \boldsymbol{\beta}(x,t).$$

3) Hooke's law

$$\boldsymbol{\sigma}(x,t) = \mathsf{C} : (\varepsilon(x,t) - \varepsilon^p(x,t)).$$

4) Evolution equations

$$\begin{aligned}
\dot{\varepsilon}^p(x,t) &= \boldsymbol{\Phi}(\boldsymbol{\sigma}(x,t), \boldsymbol{\alpha}(x,t)) && \text{flow rule,} \\
\dot{\boldsymbol{\alpha}}(x,t) &= \boldsymbol{\Psi}(\boldsymbol{\sigma}(x,t), \boldsymbol{\alpha}(x,t)) && \text{hardening rule.}
\end{aligned}$$

5) Initial values $\varepsilon^p(\cdot, t_0) = \varepsilon_0^p$, $\boldsymbol{\alpha}(\cdot, t_0) = \boldsymbol{\alpha}_0$.
Boundary conditions $\boldsymbol{u}_{|\Gamma_0} = \boldsymbol{u}_0$, $(\boldsymbol{\sigma n})_{|\Gamma_1} = \boldsymbol{t}$.

In total, we have a coupled system of partial differential equations (PDEs) and ordinary differential equations (ODEs). We observe that there are infinitely many ODEs since flow and hardening rules must hold in each material point. Furthermore, there is only a weak interaction between the evolution at two different spatial points. More precisely, this interaction depends merely on the stress tensor $\boldsymbol{\sigma}$, a fact that turns out to be very important for the numerical simulation below.

As mentioned in the Introduction, such unified elasto-viscoplastic materials have been extensively studied by Alber [1] and Chelminsky [4]. Existence and uniqueness as well as well-posedness results have been obtained for the class of so-called monotone constitutive equations. This monotonicity is the analogue of the dissipativity that is typical for elastoplasticity. There, the principle of maximum plastic dissipation is the basis of the evolution equations and guarantees a contractivity property of the elastoplastic flow [16]. For the problem class treated in this paper, however, such a variational principle is in general not available.

3 Space Discretization by the FEM

This section discusses the space discretization by the FEM and analyses the structure of the resulting equations in time. Emphasis is placed on a global point of view, in contrast to the element-wise point of view that prevails in most implementations. Moreover, we strictly separate space and time discretization and present the latter in the next section.

3.1 Weak Form of Balance Equations

Let $V := \{\boldsymbol{v} \in H^1(\Omega)^3 : \boldsymbol{v}_{|\Gamma_0} = 0\}$ be the function space of test functions. Green's theorem implies for $\boldsymbol{v} \in V$

$$\int_\Omega \boldsymbol{v}^T \mathbf{div}\boldsymbol{\sigma}\, dx = \int_\Omega \boldsymbol{v}^T \mathbf{div}(\mathsf{C} : \varepsilon)\, dx - \int_\Omega \boldsymbol{v}^T \mathbf{div}(\mathsf{C} : \varepsilon^p)\, dx$$

$$= -\int_\Omega \mathcal{L}\boldsymbol{u} : \mathsf{C} : \mathcal{L}\boldsymbol{v}\, dx + \int_\Omega \varepsilon^p : \mathsf{C} : \mathcal{L}\boldsymbol{v}\, dx + \int_{\Gamma_1} \boldsymbol{v}^T \boldsymbol{t}\, ds.$$

Therefore, if we apply a test function (or virtual displacement) $\boldsymbol{v} \in V$ to the balance of momentum (2) and integrate the resulting expression over the body, we obtain the weak form as follows:
Find $\boldsymbol{u}(\cdot,t) \in H^1(\Omega)^3$, $\boldsymbol{u}(\cdot,t)_{|\Gamma_0} = \boldsymbol{u}_0(\cdot,t)$, such that at each instant of time

$$0 = \int_\Omega \mathcal{L}\boldsymbol{u} : \mathsf{C} : \mathcal{L}\boldsymbol{v}\, dx - \int_\Omega \boldsymbol{v}^T \boldsymbol{\beta}\, dx - \int_{\Gamma_1} \boldsymbol{v}^T \boldsymbol{t}\, ds - \int_\Omega \varepsilon^p : \mathsf{C} : \mathcal{L}\boldsymbol{v}\, dx \quad \forall \boldsymbol{v} \in V.$$

In a more formal setting, the first integral of the weak form is viewed as a bilinear form

$$a(\boldsymbol{u}, \boldsymbol{v}) := \int_\Omega \mathcal{L}\boldsymbol{u} : \mathsf{C} : \mathcal{L}\boldsymbol{v}\, dx,$$

which is defined on $H^1(\Omega)^3 \times H^1(\Omega)^3$ and which is coercive on $V \times V$. The second and third integrals are written as linear functionals

$$\langle \boldsymbol{\beta}, \boldsymbol{v} \rangle := \int_\Omega \boldsymbol{v}^T \boldsymbol{\beta}\, dx, \qquad \langle \boldsymbol{t}, \boldsymbol{v} \rangle_{\Gamma_1} := \int_{\Gamma_1} \boldsymbol{v}^T \boldsymbol{t}\, ds.$$

Summing up, we write the balance equations as

$$0 = a(\boldsymbol{u}, \boldsymbol{v}) - \langle \boldsymbol{\beta}, \boldsymbol{v} \rangle - \langle \boldsymbol{t}, \boldsymbol{v} \rangle_{\Gamma_1} - \int_\Omega \varepsilon^p : \mathsf{C} : \mathcal{L}\boldsymbol{v}\, dx \qquad \forall \boldsymbol{v} \in V. \tag{6}$$

If we omit the last integral, we recover the standard model of linear elasticity.

3.2 Ritz-Galerkin Approximation

Let now $S \subset V$ be a finite dimensional subspace spanned by basis functions $\{\boldsymbol{\phi}_i : 1 \leq i \leq n\}$ that define, e.g., a finite element method. We approximate the solution $\boldsymbol{u}(x,t)$ of the weak form (6) by

$$\boldsymbol{u}_S(x,t) = \sum_{i=1}^{n} \boldsymbol{\phi}_i(x) q_i(t) + \overline{\boldsymbol{u}}_0(x,t)$$

with unknown coefficients q_i and a given function $\overline{\boldsymbol{u}}_0$ that approximates $\boldsymbol{u}_0$ on the boundary Γ_0. By projection of the weak form (6) onto the subspace S, we obtain

$$0 = a(\boldsymbol{u}_S, \boldsymbol{\phi}_i) - \langle \boldsymbol{\beta}, \boldsymbol{\phi}_i \rangle - \langle \boldsymbol{t}, \boldsymbol{\phi}_i \rangle_{\Gamma_1} - \int_\Omega \varepsilon^p : \mathsf{C} : \mathcal{L}\boldsymbol{\phi}_i\, dx, \qquad i = 1, \ldots, n$$

and due to the properties of the bilinear form a

$$0 = \sum_{j=1}^{n} a(\boldsymbol{\phi}_j, \boldsymbol{\phi}_i) q_j + a(\overline{\boldsymbol{u}}_0, \boldsymbol{\phi}_i) - \langle \boldsymbol{\beta}, \boldsymbol{\phi}_i \rangle - \langle \boldsymbol{t}, \boldsymbol{\phi}_i \rangle_{\Gamma_1} - \int_\Omega \varepsilon^p : \mathsf{C} : \mathcal{L}\boldsymbol{\phi}_i\, dx$$

for $i = 1, \ldots, n$.

At this point, we define the stiffness matrix

$$\boldsymbol{A} := \left(a(\boldsymbol{\phi}_j, \boldsymbol{\phi}_i)\right)_{i,j=1:n}$$

and the force vector

$$\boldsymbol{b}(t) := -\left(a(\overline{\boldsymbol{u}}_0, \boldsymbol{\phi}_i) - \langle \boldsymbol{\beta}, \boldsymbol{\phi}_i \rangle - \langle \boldsymbol{t}, \boldsymbol{\phi}_i \rangle_{\Gamma_1}\right)_{i=1:n}.$$

The remaining integral requires more care. The basis functions $\boldsymbol{\phi}_i$ are known but the plastic strain ε^p is not, since it satisfies a nonlinear evolution equation. Consequently, the integral needs to be evaluated by a quadrature rule

$$\int_\Omega \varepsilon^p : \mathsf{C} : \mathcal{L}\boldsymbol{\phi}_i\, dx \doteq \sum_{k=1}^{m} \gamma_k \varepsilon^p(\xi_k, t) : \mathsf{C} : \mathcal{L}\boldsymbol{\phi}_i(\xi_k)$$

with weights γ_k and nodes or Gauss points ξ_k, $k = 1, \ldots, m$.

In the sequel, we will employ the usual vector representation for the plastic strain tensor ε^p, i.e., we write

$$\varepsilon^p = (\varepsilon^p_{11}, \varepsilon^p_{22}, \varepsilon^p_{33}, 2\varepsilon^p_{12}, 2\varepsilon^p_{23}, 2\varepsilon^p_{13})^T.$$

Thus, the quadrature rule may be written as matrix-vector product

$$\left(\sum_{k=1}^{m} \gamma_k \varepsilon^p(\xi_k, t) : \mathsf{C} : \mathcal{L}\phi_i(\xi_k) \right)_{i=1:n} = \boldsymbol{Q}(\boldsymbol{\xi}) \varepsilon^p(\boldsymbol{\xi}, t)$$

where $\boldsymbol{\xi} = (\xi_1^T, \ldots, \xi_m^T)^T$ and where the matrix $\boldsymbol{Q}$ as well as the vector

$$\varepsilon^p(\boldsymbol{\xi}, t) = (\varepsilon^p(\xi_1, t)^T, \ldots, \varepsilon^p(\xi_m, t)^T)^T.$$

depend on the quadrature nodes $\boldsymbol{\xi}$.

The Ritz-Galerkin approximation of the balance equations finally reads

$$0 = \boldsymbol{A}\boldsymbol{q}(t) - \boldsymbol{b}(t) - \boldsymbol{Q}(\boldsymbol{\xi})\varepsilon^p(\boldsymbol{\xi}, t). \tag{7}$$

Both, the nodal variables $\boldsymbol{q} = (q_1, \ldots, q_n)^T$ and the plastic strain ε^p are unknown quantities. While the first are, assuming a positive definite stiffness matrix $\boldsymbol{A}$, uniquely given by (7), the latter satisfies an additional differential equation.

3.3 Initial Value Problem

Remarkably, the Ritz-Galerkin approximation defines not only a projection of the balance equations but also of the evolution equations (4)and (5). Since the plastic strain ε^p is evaluated in the Gauss points $\boldsymbol{\xi}$ only, we need flow rule (4) and hardening rule (5) in these points but nowhere else. In other words, the infinite dimensional system of evolution equations has been simultaneously reduced to a finite dimensional system of ODEs.

In each Gauss point, we have the differential equations

$$\dot{\varepsilon}^p(\xi_k, t) = \boldsymbol{\Phi}(\boldsymbol{\sigma}(\xi_k, t), \boldsymbol{\alpha}(\xi_k, t)), \tag{8}$$

$$\dot{\boldsymbol{\alpha}}(\xi_k, t) = \boldsymbol{\Psi}(\boldsymbol{\sigma}(\xi_k, t), \boldsymbol{\alpha}(\xi_k, t)) \tag{9}$$

and the relation

$$\boldsymbol{\sigma}(\xi_k, t) = \boldsymbol{C}(\boldsymbol{B}(\xi_k)\boldsymbol{q}(t) - \varepsilon^p(\xi_k, t)) \tag{10}$$

between stress $\boldsymbol{\sigma}$, nodal variables $\boldsymbol{q}$ and plastic strain ε^p. Note that (10) is the discrete analogue of Hooke's law and the decomposition (3). The 6×6 matrix $\boldsymbol{C}$ represents the elasticity tensor while the $6 \times n$ matrix $\boldsymbol{B}(\xi_k)$ is a difference operator that evaluates the total strain from the displacement vector.

At the end of this section, we summarize the equations that result from the Ritz-Galerkin approximation. Anticipating ideas of the next section on DAEs, we write the differential equations in the first place, followed by the discrete balance law

$$\begin{aligned}
\dot{\varepsilon}^p(\xi_k, t) &= \boldsymbol{\Phi}(\boldsymbol{\sigma}(\xi_k, t), \boldsymbol{\alpha}(\xi_k, t)), \\
\dot{\boldsymbol{\alpha}}(\xi_k, t) &= \boldsymbol{\Psi}(\boldsymbol{\sigma}(\xi_k, t), \boldsymbol{\alpha}(\xi_k, t)), \qquad k = 1, \ldots, m \\
\boldsymbol{\sigma}(\xi_k, t) &= \boldsymbol{C}(\boldsymbol{B}(\xi_k)\boldsymbol{q}(t) - \varepsilon^p(\xi_k, t)), \\
0 &= \boldsymbol{A}\boldsymbol{q}(t) - \boldsymbol{b}(t) - \boldsymbol{Q}(\xi)\varepsilon^p(\xi, t).
\end{aligned}$$

Initial values for the variables ε^p and $\boldsymbol{\alpha}$ complete these equations. Looking at the dimensions m (Gauss points) and n (FE basis), we observe that m is proportional to the the number of elements and often $m \geq n$. Moreover, there are $(6 + l) \cdot m$ differential equations in the 3-dimensional case since the flow rule consists of 6 and the hardening rule of l equations ($l < 20$ in general).

4 Time Integration Algorithms – the DAE View Point

This section discusses the Differential-Algebraic Equations (DAEs) that arise after space discretization by the FEM. The goal is to apply sophisticated numerical techniques for DAEs in order to obtain more accurate results and better performance of the time integration.

4.1 The Index

For the following, we rewrite the above initial value problem as semi-explicit DAE

$$\dot{\boldsymbol{y}} = \boldsymbol{f}(\boldsymbol{y}, \boldsymbol{q}), \qquad \boldsymbol{0} = \boldsymbol{g}(\boldsymbol{y}, \boldsymbol{q}, t) \tag{11}$$

where the differential variables $\boldsymbol{y} = (\varepsilon^p, \boldsymbol{\alpha})$ comprise both plastic strain and internal variables. The nodal variables $\boldsymbol{q}$, on the other hand, are called algebraic variables since they appear without time derivative in (11). Note that Hooke's law has been inserted in (11), i.e., we assume that the stress $\boldsymbol{\sigma} = \boldsymbol{\sigma}(\boldsymbol{q}, \varepsilon^p)$ is a given function depending on nodal variables and plastic strain.

First, we investigate the DAE (11) and determine its so-called *index*. The semi-explicit structure with differential equations for $\boldsymbol{y}$ and nonlinear constraints in terms of $\boldsymbol{y}$ and $\boldsymbol{q}$ simplifies the analysis. Under the assumption that the stiffness matrix $\boldsymbol{A} = \partial \boldsymbol{g}/\partial \boldsymbol{q}$ is positive definite, the algebraic or constraint equation of (11) can be formally solved for the algebraic variables $\boldsymbol{q} = \boldsymbol{q}(\boldsymbol{y}, t)$ due to the Implicit Function Theorem. Inserting $\boldsymbol{q}$ in the differential equation yields

$$\dot{\boldsymbol{y}} = \boldsymbol{f}(\boldsymbol{y}, \boldsymbol{q}(\boldsymbol{y}, t)) \tag{12}$$

which is an Ordinary Differential Equation (ODE) for $\boldsymbol{y}$. In DAE terminology, (11) is thus a *system of index 1* [6]. If (12) satisfies the usual Lipschitz condition on the right-hand side, existence and uniqueness of the solution for initial values $\boldsymbol{y}(t_0) = \boldsymbol{y}_0$ are guaranteed. Note that corresponding *consistent initial values* $\boldsymbol{q}(t_0) = \boldsymbol{q}_0$ follow directly from the above relation $\boldsymbol{q}_0 = \boldsymbol{q}(\boldsymbol{y}_0, t_0)$.

We remark that a couple of unified models contain discontinuous right-hand sides that violate the above Lipschitz condition. This is a severe drawback to both the mathematical analysis and the reliable and efficient numerical solution.

A closer look at the structure of the DAE (11) is provided by the Jacobian of the right-hand side. Basically, we have

$$\left(\begin{array}{c|c} \partial \boldsymbol{f}/\partial \boldsymbol{y} & \partial \boldsymbol{f}/\partial \boldsymbol{q} \\ \hline \partial \boldsymbol{g}/\partial \boldsymbol{y} & \partial \boldsymbol{g}/\partial \boldsymbol{q} \end{array}\right) = \left(\begin{array}{cccc|c} * & & & & * \\ & * & & & * \\ & & \ddots & & \vdots \\ & & & * & * \\ \hline * & * & \dots & * & \boldsymbol{A} \end{array}\right) \tag{13}$$

where an asterisk $*$ indicates a block different from zero. Since the evolution in different Gauss points is only coupled via the displacements $\boldsymbol{q}$, we observe, after a suitable reordering of elements, a block diagonal structure of $\partial \boldsymbol{f}/\partial \boldsymbol{y}$. Exploiting this structure will turn out to be crucial for an efficient time integration.

4.2 Implicit Euler and Consistent Tangent

Before focussing on higher order time integration, we briefly discuss the standard approach, i.e., the implicit Euler method. The discretization of the ODE (12) by the implicit Euler method is, up to round off and truncation errors, equivalent to the discretization of the DAE (11) by

$$\begin{aligned} \boldsymbol{y}_{n+1} - \boldsymbol{y}_n &= h\boldsymbol{f}(\boldsymbol{y}_{n+1}, \boldsymbol{q}_{n+1}), \\ 0 &= \boldsymbol{g}(\boldsymbol{y}_{n+1}, \boldsymbol{q}_{n+1}, t_{n+1}) \,. \end{aligned} \tag{14}$$

The discrete approximation $\boldsymbol{y}_{n+1}, \boldsymbol{q}_{n+1}$ at time t_{n+1} is thus given as the solution of the nonlinear system (14). If the stepsize $h = t_{n+1} - t_n$ is sufficiently small and the above assumptions on the index 1 system hold, it can be shown that this solution exists and is unique [6]. Moreover, the overall algorithm is globally convergent of order one in $\boldsymbol{y}$ and $\boldsymbol{q}$, i.e., after n steps we have

$$\boldsymbol{y}(t_n) - \boldsymbol{y}_n = O(h), \qquad \boldsymbol{q}(t_n) - \boldsymbol{q}_n = O(h)$$

for the difference of exact solution and numerical approximation.

The implicit Euler method is a natural choice for the time integration of the equations of viscoplasticity. Moreover, its A-stability is an important property in case of numerical stiffness of the evolution equations. Nevertheless, this method has no adaptive stepsize control, and we will show below that a variable stepsize method can be clearly superior.

The solution of the nonlinear system (14) represents the main computational challenge of the implicit Euler method. Different techniques from DAE and FEM algorithms can be applied here. Standard DAE codes like DASSL [2], which includes the implicit Euler method, treat the nonlinear system directly by a chord-Newton or simplified Newton iteration with fixed Jacobian matrix. Since Jacobian evaluations are costly, the codes try to reuse the Jacobian for several steps. The structure of the equations, cf. example (13), is exploited at the linear algebra level by employing sparse or iterative solvers.

In contrast, the algorithm due to Simo/Hughes [16] used in most FE codes is based on a two-level Newton process. It distinguishes between evolution and equilibrium equations and exploits the structure of (11) and (13) to a high degree. The main idea is to use the discretized evolution equation

$$0 = \boldsymbol{y}_{n+1} - \boldsymbol{y}_n - h\boldsymbol{f}(\boldsymbol{y}_{n+1}, \boldsymbol{q}_{n+1}) \tag{15}$$

to solve for $\boldsymbol{y}_{n+1} = \boldsymbol{y}(\boldsymbol{q}_{n+1})$ depending on the current displacement $\boldsymbol{q}_{n+1}$. Inserting $\boldsymbol{y}_{n+1}$ in the equilibrium equation, we get

$$0 = \boldsymbol{g}(\boldsymbol{y}(\boldsymbol{q}_{n+1}), \boldsymbol{q}_{n+1}, t_{n+1}) \,. \tag{16}$$

Application of Newton's method to the nonlinear system (16) results in two loops. The outer loop updates $\boldsymbol{q}_{n+1}$, the inner loop computes $\boldsymbol{y}_{n+1} = \boldsymbol{y}(\boldsymbol{q}_{n+1})$.

As the evolution equation (15) decouples into m separate subsystems, the inner Newton loop applied here is inexpensive and offers a lot of parallelism. However, concerning the Jacobian or *tangent stiffness matrix* $\boldsymbol{A}_T$ of (16) required in each outer Newton step, additional derivative terms must be provided in order to guarantee quadratic convergence. By the Implicit Function Theorem,

$$\begin{aligned}
\boldsymbol{A}_T^{(i)} &:= \frac{d}{d\boldsymbol{q}}\boldsymbol{g}(\boldsymbol{y}(\boldsymbol{q}), \boldsymbol{q}, t)\Big|_{\boldsymbol{q}=\boldsymbol{q}_{n+1}^{(i)}} \\
&= \left[\frac{\partial \boldsymbol{g}}{\partial \boldsymbol{y}}\frac{\partial \boldsymbol{y}}{\partial \boldsymbol{q}} + \frac{\partial \boldsymbol{g}}{\partial \boldsymbol{q}}\right]_{\boldsymbol{q}=\boldsymbol{q}_{n+1}^{(i)}} \\
&= \left[h\frac{\partial \boldsymbol{g}}{\partial \boldsymbol{y}}\left(\mathbf{I} - h\frac{\partial \boldsymbol{f}}{\partial \boldsymbol{y}}\right)^{-1}\frac{\partial \boldsymbol{f}}{\partial \boldsymbol{q}} + \frac{\partial \boldsymbol{g}}{\partial \boldsymbol{q}}\right]_{\boldsymbol{q}=\boldsymbol{q}_{n+1}^{(i)}} .
\end{aligned} \tag{17}$$

In most FE codes, the computation of this so-called *consistent tangent operator* is implemented at the element level. The user has to specify the material model, time discretization and derivative terms of $\boldsymbol{A}_T$ for the chosen element.

Summing the two different approaches for the solution of the nonlinear system (14) up, it is obvious that the DAE algorithms represent an extension of ODE algorithms while the FEM approach generalizes a nonlinear equations solver to include some additional differential equations. In the following, we will try to combine techniques from both fields to obtain a more efficient and robust integration method for viscoplastic problems opening thereby the possibility for adaptive time steps.

Let us give two additional remarks on this approach. First, the initial guess in Newton's method is computed from the assumption of a so-called *frozen plastic flow*, i.e., the internal variables and the plastic strain are taken from the last step while the displacements are updated by assuming linear elasticity. Second, besides the implicit Euler, one could also try to use the semi-explicit Euler and related time integration schemes [6] on the DAE (11). So far, it is not clear whether the suspected numerical stiffness of the evolution equations prevents the application of such methods.

4.3 BDF-2, a Second Order Method

As explained above, the implicit Euler method is a natural choice in viscoplastic problems since it meets several requirements: it can be easily extended to the DAE (11), it is A-stable and B-stable, which are favorable numerical stability properties, and it fits well in the FEM algorithm for the nonlinear system. The last point is very important in terms of implementation. Sophisticated FEM software exploits the structure of the equilibrium equations to a high degree while specific material models are added as modules containing the already discretized evolution equations. If we want to use a time integration scheme inside a FEM code, we must comply with this program structure. A straightforward application of DAE software is not possible as no right-hand side interfaces are provided.

For this reason, we extend now the first order implicit Euler to a second order method, the BDF-2 scheme, that also meets the above requirements and additionally offers a variable stepsize mechanism. Both second order accuracy and adaptivity in time lead to a much more efficient simulation. Integration methods of even higher order (≥ 3) are not expected to yield much better results due to the low smoothness of the evolution equations. Furthermore, additional storage capacity is needed for providing all necessary history information at the quadrature points. For an application of BDF methods to elasto-plasticity see [10].

Assuming first constant stepsize h, the *second order backward difference formula* (BDF-2) for the DAE (11) reads [2,6]

$$\begin{aligned} \tfrac{3}{2}\boldsymbol{y}_{n+1} - 2\boldsymbol{y}_n + \tfrac{1}{2}\boldsymbol{y}_{n-1} &= h\boldsymbol{f}(\boldsymbol{y}_{n+1}, \boldsymbol{q}_{n+1}), \\ 0 &= \boldsymbol{g}(\boldsymbol{y}_{n+1}, \boldsymbol{q}_{n+1}, t_{n+1}). \end{aligned} \tag{18}$$

Here, $\boldsymbol{y}_{n-1}$ and $\boldsymbol{y}_n$ denote the numerical solution at previous time grid points $t_n = t_{n+1} - h$ and $t_{n-1} = t_n - h$. Obviously, (18) is a straightforward extension of the implicit Euler method where only the difference operator in the differential equation was replaced while the constraint equation discretization remains unchanged. The second order method (18) is thus given by a nonlinear system that has, up to some history data, the same structure as in the implicit Euler case and that can be solved in exactly the same way. Note that the implicit Euler method is often referred to as the BDF-1 method since it also belongs to the class of BDF methods [2].

We also want to allow variable time steps, and for this purpose we have to extend the above scheme. It is only necessary to modify the discretization of the differential equation, the algebraic part remains again unchanged. Suppose that the polynomial $\boldsymbol{\omega}(t)$ interpolates $\boldsymbol{y}_{n-1}, \boldsymbol{y}_n, \boldsymbol{y}_{n+1}$ at t_{n-1}, t_n and t_{n+1}. The unknown value of $\boldsymbol{y}_{n+1}$ can be defined by the requirement that the polynomial $\boldsymbol{\omega}(t)$ satisfies the differential equation at time t_{n+1}, i.e.,

$$\dot{\boldsymbol{\omega}}(t_{n+1}) = \boldsymbol{f}(\boldsymbol{y}_{n+1}, \boldsymbol{q}_{n+1}) \, .$$

For constant stepsize, it is easy to show that this method definition is equivalent to (18). But it extends directly to a variable time grid. If $h_n = t_{n+1} - t_n$ is the current stepsize and $\tau_n = h_n / h_{n-1}$ the stepsize ratio, the above interpolation property results in

$$\frac{1 + 2\tau_n}{1 + \tau_n} \boldsymbol{y}_{n+1} - (1 + \tau_n) \, \boldsymbol{y}_n + \frac{\tau_n^2}{1 + \tau_n} \boldsymbol{y}_{n-1} = h_n \, \boldsymbol{f}(\boldsymbol{y}_{n+1}, \boldsymbol{q}_{n+1}) \, . \tag{19}$$

This variable stepsize method is also globally convergent of order two if the stability condition $\tau_n < 1 + \sqrt{2}$ on the stepsize increase of two consecutive steps is satisfied [6].

The stepsize control usually employed in ODE and DAE solvers relies on a *local* error estimator and some heuristics. We consider here only the inner variables $\boldsymbol{y}$ and neglect the additional information of the displacements or algebraic variables $\boldsymbol{q}$. Let $\boldsymbol{y}_{n+1}$ be the current numerical solution and $\boldsymbol{y}(t_{n+1})$ the exact solution. Given absolute and relative tolerances ATOL and RTOL, the local error or error per step is required to satisfy

$$\|\boldsymbol{y}_{n+1} - \boldsymbol{y}(t_{n+1})\| \leq \text{RTOL} \, \|\boldsymbol{y}(t_{n+1})\| + \text{ATOL} \tag{20}$$

where we tacitly assume that the history data $\boldsymbol{y}_{n-1}, \boldsymbol{y}_n$ are exact. Next, it is convenient to introduce a dimensionless measure ERR for the time integration error. By (20), it should satisfy the inequality

$$ERR := \frac{\|\boldsymbol{y}_{n+1} - \boldsymbol{y}(t_{n+1})\|}{\text{RTOL} \, \|\boldsymbol{y}(t_{n+1})\| + \text{ATOL}} \leq 1 \, . \tag{21}$$

For efficient time integration, the measure ERR has to be close to its upper limit 1 which is accomplished for the BDF-2 method by the optimal time step length [2]

$$h^* = h_n \cdot ERR^{-1/3} \, . \tag{22}$$

The exponent 1/3 reflects the asymptotic behavior $\boldsymbol{y}_{n+1} - \boldsymbol{y}(t_{n+1}) = O(h_n^3)$.

So far, we have derived a procedure to compute an optimal stepsize h^* for the next step based on an error estimate and the current stepsize h_n. We still have to replace the unknown exact solution in (21) by some approximation.

The standard approach is a *predictor-corrector scheme.* A second polynomial, the predictor, interpolating the history data $\boldsymbol{y}_{n-2}, \boldsymbol{y}_{n-1}, \boldsymbol{y}_n$ is evaluated at t_{n+1}. This predictor $\boldsymbol{y}_{n+1}^{(0)}$ gives a good starting value for the nonlinear system or corrector equation (19) and yields also an error estimate. According to [2],

$$ERR \doteq \frac{1}{\alpha_{n+1}} \frac{1}{t_{n+1} - t_{n-2}} \frac{\left\| \boldsymbol{y}_{n+1} - \boldsymbol{y}_{n+1}^{(0)} \right\|}{\mathrm{RTOL} \left\| \boldsymbol{y}_{n+1} \right\| + \mathrm{ATOL}} \tag{23}$$

with leading coefficient $\alpha_{n+1} = 1/(t_{n+1} - t_{n-1}) + 1/(t_{n+1} - t_n)$.

Several remarks on the implementation should be given at this point. The integration algorithm must be set up with an implicit Euler step in order to provide the necessary history data for the first BDF-2 step. After this starting procedure, the second order method continues. The computation of h_{new} is actually done by

$$h_{\text{new}} = h_n \cdot \min(\beta_u, \max(\beta_l, \beta_s \cdot \mathrm{ERR}^{-1/3}))$$

with parameters $\beta_u, \beta_l, \beta_s$ that add a safety margin to the error estimation and prevent the step size from sudden changes. For the safety factor and the lower bound, $\beta_s = 0.9$ and $\beta_l = 0.2$ are fairly good values. The upper bound is restricted by the stability requirement $\beta_u \leq 1 + \sqrt{2}$.

4.4 Solving the Nonlinear System

At the end of this section, let us discuss an alternative approach for solving the nonlinear system (14) in case of implicit Euler or (18) in case of BDF-2. We combine both cases and write the equations as

$$\begin{aligned} 0 &= \alpha \boldsymbol{y}_{n+1} - h \boldsymbol{f}(\boldsymbol{y}_{n+1}, \boldsymbol{q}_{n+1}) + \boldsymbol{c}, \\ 0 &= \boldsymbol{g}(\boldsymbol{y}_{n+1}, \boldsymbol{q}_{n+1}, t_{n+1}). \end{aligned} \tag{24}$$

For implicit Euler we have $\alpha = 1$, $\boldsymbol{c} = -\boldsymbol{y}_n$ and for the BDF-2 method it holds $\alpha = (1 + 2\tau_n)/(1 + \tau_n)$, $\boldsymbol{c} = -(1 + \tau_n)\, \boldsymbol{y}_n + (\tau_n^2)/(1 + \tau_n)\, \boldsymbol{y}_{n-1}$.

As aforementioned, in most FE codes a two-level algorithm is employed:
1) Local iteration in all Gauss points: Solve

$$\boldsymbol{0} = \alpha \boldsymbol{y}_{n+1} - h \boldsymbol{f}(\boldsymbol{y}_{n+1}, \boldsymbol{q}_{n+1}^{(i)}) + \boldsymbol{c} \qquad \text{for } \boldsymbol{y}_{n+1}(\boldsymbol{q}_{n+1}^{(i)}).$$

2) Global iteration step for $\boldsymbol{q}_{n+1}^{(i+1)}$:

$$\boldsymbol{q}_{n+1}^{(i+1)} = \boldsymbol{q}_{n+1}^{(i)} + \Delta q^{(i)}, \qquad \boldsymbol{A}_T^{(i)} \Delta q^{(i)} = -\boldsymbol{g}(\boldsymbol{y}_{n+1}(\boldsymbol{q}_{n+1}^{(i)}), \boldsymbol{q}_{n+1}^{(i)}, t_{n+1}).$$

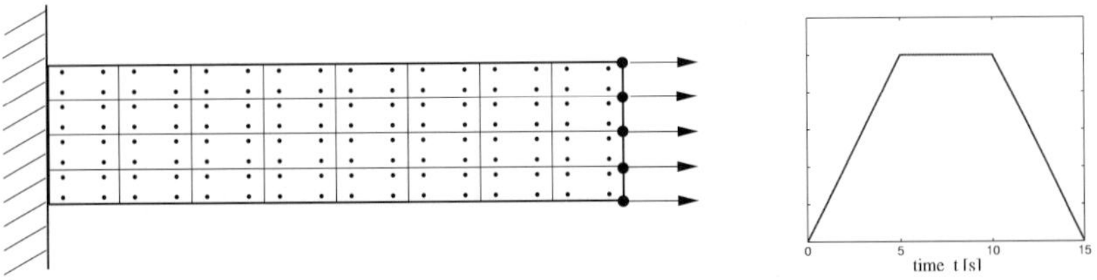

Fig. 2. Tensile Test of a Beam and Loading History.

Aiming at quadratic convergence, this algorithm requires an evaluation and a decomposition of the consistent tangent matrix

$$\boldsymbol{A}_T^{(i)} = \frac{d}{d\boldsymbol{q}} \boldsymbol{g}(\boldsymbol{y}(\boldsymbol{q}), \boldsymbol{q}, t)_{|\boldsymbol{q}=\boldsymbol{q}_{n+1}^{(i)}}$$

in each global iteration step. As shown in (17), the derivative information of blocks $\partial \boldsymbol{f}/\partial \boldsymbol{y}$, $\partial \boldsymbol{f}/\partial \boldsymbol{q}$, $\partial \boldsymbol{g}/\partial \boldsymbol{y}$, and $\partial \boldsymbol{g}/\partial \boldsymbol{q}$ enters the computation of matrix $\boldsymbol{A}_T^{(i)}$.

The alternative that we propose is inspired by DAE techniques used in codes like DASSL [2] or RADAU5 [6]. We also require the derivative information of the consistent tangent operator, but this time the nonlinear system (24) is solved directly, exploiting the block structure of the Jacobian, cf. (13),

$$\left(\begin{array}{c|c} \alpha \boldsymbol{I} - h\partial \boldsymbol{f}/\partial \boldsymbol{y} & -h\partial \boldsymbol{f}/\partial \boldsymbol{q} \\ \hline \partial \boldsymbol{g}/\partial \boldsymbol{y} & \partial \boldsymbol{g}/\partial \boldsymbol{q} \end{array} \right)$$

at the linear algebra level. We have here also a loop over all Gauss points for the evolution equations, but this time only a linear system is solved, in contrast to above where a nonlinear system in each Gauss point is solved.

In some applications, we may even dispense with the requirement of quadratic convergence and apply a chord-Newton process instead, in particular in combination with BDF-2 where a good initial guess $\boldsymbol{y}_{n+1}^{(0)}, \boldsymbol{q}_{n+1}^{(0)}$ is available by extrapolating the predictor polynomial.

5 Simulation Results

The following examples demonstrate that the above DAE techniques improve the numerical simulation in elasto-viscoplasticity both in terms of accuracy and efficiency. We start with the tensile test of a beam and consider then a bolt with a more complex geometry. Both examples are 2D applications under the assumption of plane stress. The model of Chan et al. [3] supplies the constitutive equations where, as pointed out in Subsect. 2.2, the parameters have been identified by Seibert [15] for the steel SS 316. Reference solutions were obtained by computations with very small fixed stepsize and time integration of order 2.

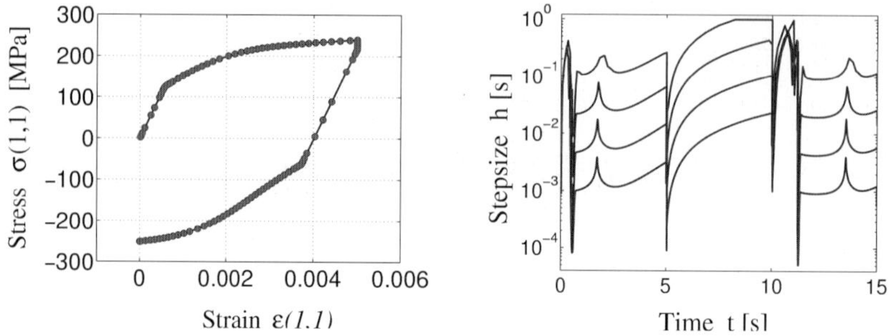

Fig. 3. Stress-Strain Diagram with Discrete Time Steps (left) and Stepsizes at Different Tolerances (right).

5.1 Tensile Test of a Beam

Figure 2 shows the geometry and the spatial discretization of the beam example. We use here 32 elements, each with 8 nodes and 4 Gauss points. On the left built-in end, zero Dirichlet boundary conditions are prescribed in the x_1-direction while on the right free end, a displacement function in form of a ramp models the tensile test. Due to its vertical symmetry, this example shows a one-dimensional behavior and allows a straightforward checking of model and numerical simulation. Strain, stress and also evolution of internal variables are identical in all Gauss points.

In Fig. 3, the stress-strain diagram reflects the typical irreversible behavior. Moreover, it displays also the discrete time steps that were chosen by the BDF-2 method with stepsize control. As one can see, the stepsize varies substantially and its control works reasonably well. On the right, different stepsize sequences are collected for the tolerances ATOL=RTOL$\in \{10^{-3}, 10^{-5}, 10^{-7}, 10^{-9}\}$.

The next Fig. 4 compares the performance of different time integration algorithms for this example. All computations were carried out in a MATLAB environment on an HP B-class workstation. Due to the double logarithmic scaling, one notices clearly the superior order 2 of the BDF-2 time integration, here again in combination with stepsize control. Implicit Euler with fixed stepsize cannot compete with the performance of BDF-2, but it is instructive to compare additionally the variants with regard to the solution of the nonlinear system.

It turns out that direct application of Newton's method to (14) (abbreviated by 'N' in the figure) is more efficient for this example than applying the two-level Newton process with both iterations on the local and on the global level ('A_T'). Furthermore, simplified Newton iterations are sufficient for convergence and lead to a slightly better efficiency than the full iteration ('N_S' and 'A_{TS}').

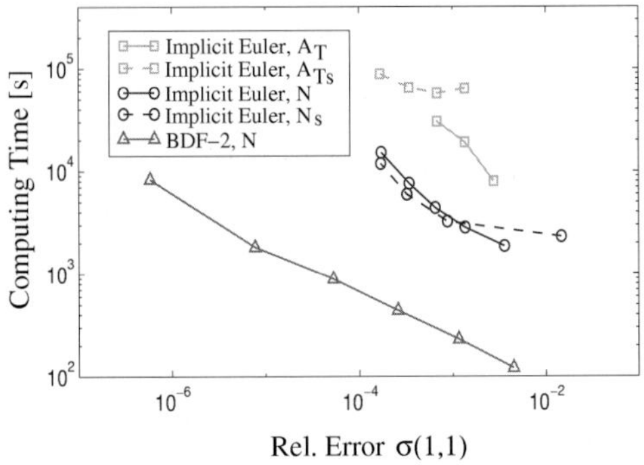

Fig. 4. Performance of Different Algorithms for the Beam Example.

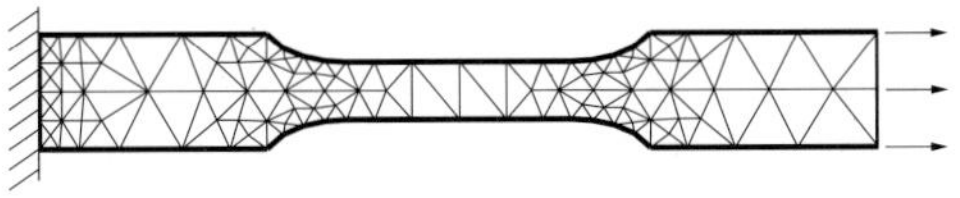

Fig. 5. Tensile Test of a Bolt.

5.2 Bolt Example

The geometry of the second example, the bolt shown in Fig. 5, is inspired by the shape of the specimens examined in [8]. Again, we simulate a tensile test, this time with zero boundary conditions in both x_1 and x_2 on the left and prescribed displacement on the right. The spatial FE discretization consists of 170 triangles, each with 6 nodes and 3 Gauss points. In each Gauss point, 7 evolution equations need be integrated in time, which gives a total of 3570 differential variables $\boldsymbol{y}$ and 113 nodal variables $\boldsymbol{q}$.

As the performance diagram in Fig. 6 demonstrates, the BDF-2 method with stepsize control is once more considerably faster and more precise than implicit Euler. Moreover, direct application of Newton's method proves also to be more efficient than the two-level algorithm. With respect to the stepsize algorithm, we remark that the piecewise definition of the ramp function (see Fig. 2) with its two discontinuities leads in this case to an increased number of rejected steps and small stepsizes. An easy way out of this problem is to stop the integration process exactly at the switching points of the prescribed displacement function and to restart it thereafter.

6 Conclusions

We showed how the DAEs arising in elasto-viscoplasticity after space discretization by the FEM can be efficiently solved. Our emphasis was placed

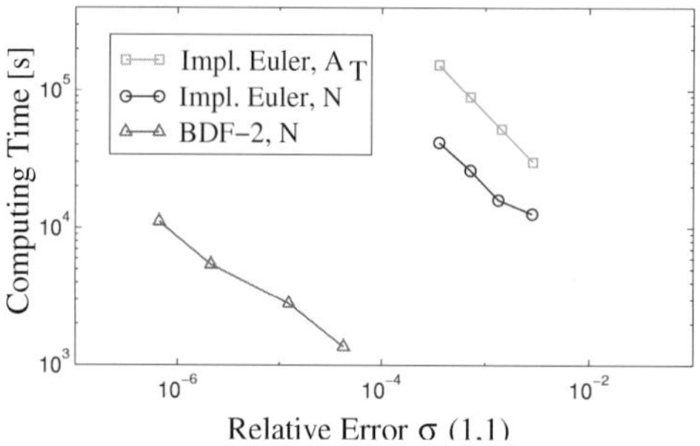

Fig. 6. Performance of Different Algorithms for the Bolt Example.

on two aspects. On the one hand, we have introduced a variable stepsize integration scheme based on the BDF-2 method and, on the other hand, we have analysed how at each time step the structure of the nonlinear system can be exploited in Newton's method.

While second order time integration is definitely promising, even higher order methods seem not appropriate due to a lack of smoothness in the evolution equations and only moderate accuracy requirements. We stress furthermore that the combination of FEM and DAE techniques needs a very detailed tuning in order to obtain efficient algorithms. With respect to this point, the interested reader is referred to [7] where it is shown how the BDF-2 method can be implemented in the widely used FE code FEAP [17].

References

1. Alber, H.D. (1998) Materials with Memory. Springer Lecture Notes in Mathematics 1682. Springer, Berlin Heidelberg
2. Brenan, C., Campbell, S.L., Petzold, L. (1996) The Numerical Solution of Initial Value Problems in Ordinary Differential-Algebraic Equations. 2. Ed., SIAM Philadelphia
3. Chan, K.S., Bodner,S.R., Lindholm, U.S. (1988) Phenomenological Modeling of Hardening and Thermal Recovery in Metals. Journal of Engineering Materials and Technology **110**, 1–8
4. Chelminski, K. (1999) On monotone plastic constitutive equations with polynomial growth condition. Math. Meth. in App. Sci. **22** 547–562
5. Ellsiepen, P., Hartmann, S. (2001) Remarks on the interpretation of current non–linear finite element analyses as differential–algebraic equations, Int. J. Num. Meth. Engng. **51**, 679–707
6. Hairer, E., Wanner, G. (1996) Solving Ordinary Differential Equations II. Springer
7. Kirchner, E., Simeon, B. (1999) A higher order time integration method for viscoplasticity. Comp. Meth. Appl. Mech. Eng. **175**, 1–18
8. Kollmann, F.G., Lehn, J., Harth, T. (2002) Identification of Material Parameters for Inelastic Constitutive Models: Statistical Analysis and Design of Experiments. Lecture Notes in Mechanics, this issue, Springer, Berlin Heidelberg

9. Lubliner, L. (1990) Plasticity Theory. Macmillan Publishing Company
10. Papadopoulos, P., Taylor, R. (1994) On the application of multistep integration methods to inifinitesimal elastoplasticity, Int. Jornal f. Numerical Methods in Engineering **37**, 3169-3184
11. Rentrop, P., Scherf, O., Simeon, B. (1999) Numerical Simulation of Mechanical Multibody Systems with Deformable Components. In Bungartz et al.(Ed.), High Performance Scientific and Engineering Computing. LNCSE 8, Springer, Berlin Heidelberg
12. Rheinboldt, W.C. (1998) Methods for Solving Systems of Nonlinear Equations. SIAM, Philadelphia
13. Scherf, O., Simeon, B. (1999) Viscoplastic deformation from the DAE perspective - a benchmark problem. ZAMM **79**, 17–20
14. Scherf, O. (2000) Numerische Simulation inelastischer Körper. VDI-Verlag, Düsseldorf
15. Seibert,T. (1996) Simulationstechniken zur Untersuchung der Streuung bei der Identifikation der Parameter inelastischer Werkstoffmodelle. Dissertation, TU Darmstadt
16. Simo, J.C., Hughes, T.J.R. (1998) Computational Inelasticity. Springer, New York
17. Zienkiewicz, O.C., Taylor, R.L. (2000) The Finite Element Method, 5th Edition. Butterworth-Heinemann, Oxford

Finite Element Analysis of Anisotropic Structures at Large Inelastic Deformations

Bernhard Eidel and Friedrich Gruttmann

Institute of Materials and Mechanics in Civil Engineering
Darmstadt University of Technology
Alexanderstr. 7, D–64283 Darmstadt, Germany
{eidel, gruttmann}@iwmb.tu-darmstadt.de

received 9 Aug 2002 — accepted 25 Oct 2002

Abstract. A constitutive model for orthotropic elastoplasticity at finite plastic strains is discussed and basic concepts of its numerical implementation are presented. The essential features are the multiplicative decomposition of the deformation gradient in elastic and inelastic parts, the definition of a convex elastic domain in stress space and a representation of the constitutive equations related to the intermediate configuration. The elastic free energy function and the yield function are formulated in an invariant setting by means of the introduction of structural tensors reflecting the privileged directions of the material. The model accounts for kinematic and isotropic hardening. The associative flow rule is integrated using the so–called exponential map which preserves exactly the plastic incompressibility condition. The constitutive equations are implemented in a brick type shell element. Due to special interpolation techniques the element is able to predict a locking-free deformation behaviour even for very thin structures. Representative numerical simulations demonstrate the suitability of the proposed formulations.

1 Introduction

Many elastoplastic materials exhibit anisotropic behavior due to their textured or generally orientation dependent structure. The response of anisotropic materials can be described with scalar-valued functions in terms of several tensor variables, usual deformation or stress tensors and additional structural tensors, which reflect the symmetries of the considered material. Based on representation theorems for tensor functions the general forms can be derived and the type and minimal number of the scalar variables entering the constitutive equations can be given. These forms are automatically invariant under coordinate transformations of elements of the material symmetry group. For an introduction to the invariant formulation of anisotropic constitutive equations based on the concept of structural tensors and their representations as isotropic tensor functions see Betten [5], Boehler [7], Spencer [37].

For a state-of-the-art review of the recent progress in the theory and numerics of anisotropic materials at finite strains we refer to the papers published in a special issue of the International Journal of Solids and Structures Vol.

38 (2001), EUROMECH Colloquium 394, and the references therein. In the following we mention some contributions in the field of anisotropic elastoplasticity being aware that this short overview cannot be complete. A yield criterion which describes the plastic flow of orthotropic metals has been first proposed in the pioneering work of HILL [18]. A numerical study on integration algorithms for Hill's model at small strains is given in DE BORST & FEENSTRA [8]. In [24] MIEHE presented a constitutive framework for the formulation of large strain anisotropic elastoplasticity based on the notion of a plastic metric. In the paper of XIAO et al. [42] a consistent Eulerian type constitutive elastoplasticity theory including isotropic and kinematic hardening was developed, which combines the additive and multiplicative decomposition of the stretch tensor and the deformation gradient. PAPADOPOULOS & LU [27] proposed an anisotropic elastoplasticity model using a family of generalized stress-strain measures. A thermodynamically consistent theory of plastic anisotropy at large deformations taking into account the postulate of Il'iushin is proposed by TSAKMAKIS [39]. To describe the rotation of the underlying substructure evolution equations for the symmetry axes are formulated. In the context of Il'iushin's postulate see also BESDO [2], where constitutive equations of plasticity theory are formulated in strain space.
For some results concerning the mathematical treatment of finite multiplicative elastoplasticity see NEFF [25]. A formulation of multiplicative finite strain elastoplasticity within the framework of generalized standard media was proposed by HACKL in [17].

The essential features of this paper are summarized as follows:
(i) In our formulation the multiplicative decomposition of the deformation gradient in elastic and inelastic parts is assumed to apply. A yield function, related to the intermediate configuration is expressed in terms of the so called Mandel stress tensor and a back stress tensor for kinematic hardening.
(ii) The constitutive equations for elastoplastic orthotropy are formulated in an invariant setting. So-called structural tensors describe the symmetries of the material in the elastic free energy function and the yield condition. The latter is expressed in terms of the invariants of the deviatoric part of the relative Mandel stresses and of the structural tensors.
(iii) The set of constitutive equations is solved by applying a general return method based on an operator split into an elastic predictor and a following corrector step. Plastic incompressibility is fulfilled exactly by means of the exponential map.
(iv) For finite element simulations of engineering problems in structural mechanics we use a formulation of a brick-type shell element, documented in KLINKEL [20], that overcomes artificial stiffening effects, called *locking*, by means of special interpolation techniques. Thus, the element is well suited for the numerical analysis of thin structures.
(v) We investigate four representative numerical examples: The necking of a circular bar and the punching of a conical shell, both for elastoplastic isotropy;

for orthotropic material behaviour we consider the drawing of a circular blank and the bending of a circular plate.
Earlier versions of the authors work on anisotropic elastoplasticity and further numerical results can be found in [11] and [15].

2 Kinematics and Constitutive Framework

The considered body in the reference configuration is denoted by $\mathcal{B} \subset \mathbb{R}^3$. It is parametrized in $\mathbf{X}$ and the current configuration $\mathcal{S} \subset \mathbb{R}^3$ is parametrized in $\mathbf{x}$. The nonlinear deformation map $\varphi_t : \mathcal{B} \to \mathcal{S}$ at time $t \in R_+$ maps points $\mathbf{X} \in \mathcal{B}$ onto points $\mathbf{x} \in \mathcal{S}$. Hence, the deformation gradient $\mathbf{F}$ is defined by $\mathbf{F}(\mathbf{X}) := \mathrm{Grad}\,\varphi_t(\mathbf{X})$ with the Jacobian $J(\mathbf{X}) := \det \mathbf{F}(\mathbf{X}) > 0$. The index notation of $\mathbf{F}$ is $F^a{}_A := \partial x^a / \partial X^A$. Next, the right Cauchy–Green tensor is introduced by $\mathbf{C} = \mathbf{F}^T \mathbf{F}$ with coefficients $C_{AB} = F^a{}_A F^b{}_B g_{ab}$, where g_{ab} denotes the coefficients of the covariant metric tensor in the current configuration.

2.1 Multiplicative Elastoplasticity

Motivated by a micromechanical view of plastic deformations one postulates a multiplicative decomposition of the deformation gradient

$$\mathbf{F}(\mathbf{X}) = \mathbf{F}^e(\mathbf{X})\; \mathbf{F}^p(\mathbf{X})\,, \tag{1}$$

with the elastic and plastic parts $\mathbf{F}^e$ and $\mathbf{F}^p$, respectively. Equation (1) implies a stress-free intermediate configuration, which is in general not compatible. It is well-known that the decomposition is uniquely determined except for a rigid body rotation superposed on the intermediate configuration. Original references dealing with (1) can be found in the textbook by LUBLINER [21]. Furthermore, the plastic incompressibility constraint

$$\det \mathbf{F}^p(\mathbf{X}) = 1 \tag{2}$$

is assumed to hold. The constitutive equations are restricted by the second law of thermodynamics in the form of the Clausius–Duhem-inequality, which reads under the assumption of isothermal deformations with uniform temperature distribution

$$\mathcal{D} = \mathbf{S} : \dot{\mathbf{E}} - \dot{\psi} \geq 0\,. \tag{3}$$

In this local form the dissipation $\mathcal{D}$ denotes the difference between the stress power and the rate of the free energy per unit volume in the reference configuration. $\mathbf{S}$ and $\dot{\mathbf{E}}$ are the Second Piola–Kirchhoff stress tensor and the material time derivative of the Green–Lagrangian strain tensor $\mathbf{E} = \frac{1}{2}(\mathbf{C} - \mathbf{1})$, respectively. Here and in the following $\mathbf{1}$ denotes the second order unit tensor.

Introducing the free energy $\psi = \psi(\mathbf{C}^e, \boldsymbol{\chi})$ as a function of the elastic right Cauchy–Green tensor $\mathbf{C}^e := {\mathbf{F}^e}^T \mathbf{F}^e$ and the internal variables $\boldsymbol{\chi}$ – considered to be in general a set of tensors and scalars and represented as a vector – the associated rate is given by

$$\dot{\psi} = \frac{\partial \psi}{\partial \mathbf{C}^e} : \dot{\mathbf{C}}_e + \frac{\partial \psi}{\partial \boldsymbol{\chi}} \cdot \dot{\boldsymbol{\chi}} \,. \tag{4}$$

The strain rate $\dot{\mathbf{E}} = \frac{1}{2}\,\dot{\mathbf{C}}$ is derived by considering the multiplicative decomposition (1). One obtains, see e.g. [21],

$$\dot{\mathbf{C}} = {\mathbf{F}^p}^T \,[\dot{\mathbf{C}}^e + 2(\mathbf{C}^e\,\mathbf{L}^p)_s]\,\mathbf{F}^p \quad \text{with} \quad \mathbf{L}^p = \dot{\mathbf{F}}^p {\mathbf{F}^p}^{-1} \,, \tag{5}$$

where $\mathbf{L}^p$ denotes the plastic velocity gradient and $()_s$ describes the symmetric part of a tensor. Since inequality (3), considering (4) and (5), must hold for all admissible processes in the material, standard arguments in rational thermodynamics with internal state variables yield the constitutive equations

$$\hat{\mathbf{S}} = 2\frac{\partial \psi}{\partial \mathbf{C}^e} \quad , \quad \boldsymbol{\Xi} = \frac{\partial \psi}{\partial \boldsymbol{\chi}} \,. \tag{6}$$

Here, $\hat{\mathbf{S}} = \mathbf{F}^p\,\mathbf{S}\,{\mathbf{F}^p}^T$ denotes the Second Piola–Kirchhoff stress tensor relative to the intermediate configuration and $\boldsymbol{\Xi}$ the internal stress vector conjugate to $\boldsymbol{\chi}$. Furthermore, one obtains the reduced local dissipation inequality

$$\mathcal{D} = \boldsymbol{\Sigma} : \mathbf{L}^p - \boldsymbol{\Xi} \cdot \dot{\boldsymbol{\chi}} \geq 0 \,, \tag{7}$$

where we call $\boldsymbol{\Sigma} := \mathbf{C}^e\,\hat{\mathbf{S}}$ the Mandel stress tensor, which for anisotropic elasticity is in general nonsymmetric.

The evolution equations for the inelastic strain tensors can be derived by using the principle of maximum plastic dissipation. If the elastic domain E defined by the yield function $\Phi < 0$ is convex, a standard result in convex analysis shows, that, along with the loading-unloading conditions, the following normality rules for the rate equations of inelastic strains must hold

$$\mathbf{L}^p = \lambda \frac{\partial \Phi}{\partial \boldsymbol{\Sigma}} \quad , \quad \dot{\boldsymbol{\chi}} = -\lambda \frac{\partial \Phi}{\partial \boldsymbol{\Xi}} \,. \tag{8}$$

In the following we specify the vector $\boldsymbol{\Xi}$ by introducing the scalar, stress like hardening variable ξ and the back stress tensor $\hat{\boldsymbol{\beta}}$ for isotropic and kinematic hardening, respectively. Furthermore, we assume that the free energy function is additively decoupled in an elastic part ψ^e, a plastic part $\psi^{p,iso}$ due to isotropic hardening and $\psi^{p,kin}$ due to kinematic hardening of Melan–Prager type. The yield criterion Φ is formulated in terms of the relative stresses $\hat{\boldsymbol{\sigma}} := \boldsymbol{\Sigma} - \hat{\boldsymbol{\beta}}$ and the isotropic hardening stress ξ. According to (8) the evolution of the plastic deformation gradient $\mathbf{F}^p$ and of the internal variables $\boldsymbol{\chi}$ are given with Φ as a plastic potential. Here, $\boldsymbol{\chi}$ contains the tensor valued $\boldsymbol{\alpha}$, conjugate to the back stress $\hat{\boldsymbol{\beta}}$ as well as the scalar valued equivalent plastic strain e^p, conjugate to ξ.

2.2 Isotropic Tensor Functions for the Representation of Anisotropic Material Response

In case of anisotropy we introduce a material symmetry group $\mathcal{G}_k$ characterizing the anisotropy class of the material. $\mathcal{G}_k$ is defined with respect to the reference configuration, and we assume that it remains unchanged during plastic deformations. The elements of $\mathcal{G}_k$ are denoted by the unimodular tensors ${}^i\mathbf{Q}|i = 1, ..., n$. Here, the concept of material symmetry will be formulated for an orthotropic elasticity law, which is related to the intermediate configuration and therefore is expressed in terms of the elastic Green strain tensor $\hat{\mathbf{E}}^e = (\mathbf{C}^e - \mathbf{1})/2$. Based on our assumption this concept requires that the elastic material response must be invariant under transformations on the intermediate configuration with elements of the symmetry group $\mathcal{G}_k$

$$\hat{\psi}^e(\mathbf{Q}^T\hat{\mathbf{E}}^e\mathbf{Q}) = \hat{\psi}^e(\hat{\mathbf{E}}^e) \quad \forall\, \mathbf{Q} \in \mathcal{G}_k, \hat{\mathbf{E}}^e \,. \tag{9}$$

We call the function ψ^e a $\mathcal{G}_k$-invariant function. Having only solid materials in mind we set $\mathcal{G}_k \subset \mathrm{SO}(3)$. Based on the mapping $\hat{\mathbf{X}} \to \mathbf{Q}^T\hat{\mathbf{X}}$, applied to the intermediate configuration $\hat{\mathbf{X}}$, for arbitrary rotation tensors $\mathbf{Q} \in \mathrm{SO}(3)$ we have, in view of a coordinate free representation, to fulfill the transformation rule $\mathbf{Q}^T\hat{\mathbf{S}}(\hat{\mathbf{E}}^e, \bullet)\mathbf{Q} = \hat{\mathbf{S}}(\mathbf{Q}^T\hat{\mathbf{E}}^e\mathbf{Q}, \bar{\bullet}) \;\; \forall \mathbf{Q} \in \mathrm{SO}(3)$, where $(\bullet)$ denotes additional tensor arguments, also subject to $\mathcal{G}_k$ and $(\bar{\bullet})$ represents the $\mathbf{Q}$-transformed tensors. In order to construct an isotropic tensor function for the anisotropic constitutive behavior, the $\mathcal{G}_k$-invariant function must be extended in a manner, that it becomes invariant under the special orthogonal group; this is done by the introduction of the so-called structural tensors reflecting the material symmetries. Recall here, that a second order tensor $\mathbf{M}$ is a structural tensor of an anisotropic material characterized by a symmetry group $\mathcal{G}_k$ if $\mathbf{Q}^T\mathbf{M}\mathbf{Q} = \mathbf{M}$ for all $\mathbf{Q} \in \mathcal{G}_k$. Orthotropic materials can be characterized by three symmetry planes, described by three structural tensors ${}^i\mathbf{M}|i = 1, 2, 3$. Thus, the constitutive equation can be expressed as an isotropic scalar-valued tensor function in the arguments $(\hat{\mathbf{E}}^e, {}^1\mathbf{M}, {}^2\mathbf{M}, {}^3\mathbf{M})$ in the form

$$\hat{\psi}^e(\hat{\mathbf{E}}^e, {}^i\mathbf{M}|_{i=1,2,3}) = \hat{\psi}^e(\mathbf{Q}^T\hat{\mathbf{E}}^e\mathbf{Q}, \mathbf{Q}^T\,{}^i\mathbf{M}\mathbf{Q}|_{i=1,2,3}) \; \forall\, \mathbf{Q} \in \mathrm{SO}(3) \,. \tag{10}$$

which fulfills the above postulated transformation rule for the stresses.

2.3 Orthotropic Elastic Free Energy Function

The material symmetry group of the considered orthotropic material is defined by $\mathcal{G}_o := \{\mathbf{I}; \mathbf{S}_1, \mathbf{S}_2, \mathbf{S}_3\}$, where $\mathbf{S}_1, \mathbf{S}_2, \mathbf{S}_3$ are the reflections with respect to the basis planes $({}^2\mathbf{a}, {}^3\mathbf{a})$, $({}^3\mathbf{a}, {}^1\mathbf{a})$ and $({}^1\mathbf{a}, {}^2\mathbf{a})$, respectively. Here, $({}^1\mathbf{a}, {}^2\mathbf{a}, {}^3\mathbf{a})$ represents an orthonormal privileged frame. Based on this, we obtain for this symmetry group the three structural tensors

$${}^1\mathbf{M} := {}^1\mathbf{a} \otimes {}^1\mathbf{a}, \quad {}^2\mathbf{M} := {}^2\mathbf{a} \otimes {}^2\mathbf{a} \quad \text{and} \quad {}^3\mathbf{M} := {}^3\mathbf{a} \otimes {}^3\mathbf{a} \,, \tag{11}$$

which represent the orthotropic material symmetry. Due to the fact that the sum of the three structural tensors yields $\sum_{i=1}^{3} {}^{i}\mathbf{M} = \mathbf{1}$ we may discard ${}^{3}\mathbf{M}$ from the set of structural tensors (11). The integrity basis is given by

$$\mathcal{P} := \{J_1, ..., J_7\} \,. \tag{12}$$

The invariants J_1, J_2, J_3 are defined by the traces of powers of $\hat{\mathbf{E}}^e$, i.e.,

$$J_1 := \mathrm{tr}\hat{\mathbf{E}}^e \,, \quad J_2 := \mathrm{tr}[(\hat{\mathbf{E}}^e)^2] \,, \quad J_3 := \mathrm{tr}[(\hat{\mathbf{E}}^e)^3] \,. \tag{13}$$

The irreducible mixed invariants are given by

$$\left.\begin{array}{ll} J_4 := \mathrm{tr}[{}^1\mathbf{M}\hat{\mathbf{E}}^e] \,, & J_5 := \mathrm{tr}[{}^1\mathbf{M}(\hat{\mathbf{E}}^e)^2] \\ J_6 := \mathrm{tr}[{}^2\mathbf{M}\hat{\mathbf{E}}^e] \,, & J_7 := \mathrm{tr}[{}^2\mathbf{M}(\hat{\mathbf{E}}^e)^2] \end{array}\right\} \,, \tag{14}$$

see e.g. SPENCER [37]. For ψ^e we assume a quadratic form, viz.,

$$\psi^e = \tfrac{1}{2}\lambda J_1^2 + \mu J_2 + \tfrac{1}{2}\alpha_1 J_4^2 + \tfrac{1}{2}\alpha_2 J_6^2 + 2\alpha_3 J_5 + 2\alpha_4 J_7 + \alpha_5 J_4 J_1 + \alpha_6 J_6 J_1 + \alpha_7 J_4 J_6 \,. \tag{15}$$

For the 2nd Piola–Kirchhoff stresses related to the intermediate configuration we have

$$\left.\begin{array}{l} \hat{\mathbf{S}} = \lambda J_1 \mathbf{1} + 2\mu\hat{\mathbf{E}}^e + \alpha_1 J_4 {}^1\mathbf{M} + \alpha_2 J_6 {}^2\mathbf{M} + 2\alpha_3(\hat{\mathbf{E}}^e\,{}^1\mathbf{M} + {}^1\mathbf{M}\hat{\mathbf{E}}^e) \\ \quad +2\alpha_4(\hat{\mathbf{E}}^e\,{}^2\mathbf{M} + {}^2\mathbf{M}\hat{\mathbf{E}}^e) + \alpha_5(J_1 {}^1\mathbf{M} + J_4\mathbf{1}) \\ \quad +\alpha_6(J_1 {}^2\mathbf{M} + J_6\mathbf{1}) + \alpha_7(J_4 {}^2\mathbf{M} + J_6 {}^1\mathbf{M}) \end{array}\right\} \,. \tag{16}$$

In this special case the second derivative of ψ^e yields the constant fourth-order elasticity tensor

$$\left.\begin{array}{l} \mathbb{C}^e = \lambda\mathbf{1}\otimes\mathbf{1} + 2\mu\mathbb{I} + \alpha_1 {}^1\mathbf{M}\otimes{}^1\mathbf{M} + \alpha_2 {}^2\mathbf{M}\otimes{}^2\mathbf{M} \\ \quad +2\alpha_3\mathbb{K}_1 + 2\alpha_4\mathbb{K}_2 + \alpha_5({}^1\mathbf{M}\otimes\mathbf{1} + \mathbf{1}\otimes{}^1\mathbf{M}) \\ \quad +\alpha_6({}^2\mathbf{M}\otimes\mathbf{1} + \mathbf{1}\otimes{}^2\mathbf{M}) + \alpha_7({}^2\mathbf{M}\otimes{}^1\mathbf{M} + {}^1\mathbf{M}\otimes{}^2\mathbf{M}) \end{array}\right\} \tag{17}$$

with $\mathbb{I}_{IJKL} = \delta_{IK}\delta_{JL}$, $\mathbb{K}^1_{IJKL} = \delta_{IK}{}^1M_{JL} + \delta_{JL}{}^1M_{IK}$ and $\mathbb{K}^2_{IJKL} = \delta_{IK}{}^2M_{JL} + \delta_{JL}{}^2M_{IK}$. The elasticity parameters $(\lambda, \mu, \alpha_i | i = 1, \ldots, 7)$ can be identified by using the matrix notation

$$\begin{bmatrix} \hat{S}_{11} \\ \hat{S}_{22} \\ \hat{S}_{33} \\ \hat{S}_{12} \\ \hat{S}_{13} \\ \hat{S}_{23} \end{bmatrix} = \begin{bmatrix} \mathbb{C}_{11} & \mathbb{C}_{12} & \mathbb{C}_{13} & 0 & 0 & 0 \\ \mathbb{C}_{12} & \mathbb{C}_{22} & \mathbb{C}_{23} & 0 & 0 & 0 \\ \mathbb{C}_{13} & \mathbb{C}_{23} & \mathbb{C}_{33} & 0 & 0 & 0 \\ 0 & 0 & 0 & \mathbb{C}_{44} & 0 & 0 \\ 0 & 0 & 0 & 0 & \mathbb{C}_{55} & 0 \\ 0 & 0 & 0 & 0 & 0 & \mathbb{C}_{66} \end{bmatrix} \begin{bmatrix} \hat{E}^e_{11} \\ \hat{E}^e_{22} \\ \hat{E}^e_{33} \\ 2\hat{E}^e_{12} \\ 2\hat{E}^e_{13} \\ 2\hat{E}^e_{23} \end{bmatrix} \,, \tag{18}$$

with the elasticity constants $\mathbb{C}_{ij}$. Choosing the preferred directions as ${}^1\mathbf{a} = (1,0,0)^T$ and ${}^2\mathbf{a} = (0,1,0)^T$ we obtain the material parameters

$$\left.\begin{aligned}
\lambda &= \mathbb{C}_{33} + 2(\mathbb{C}_{44} - \mathbb{C}_{55} - \mathbb{C}_{66})\\
\mu &= \mathbb{C}_{55} + \mathbb{C}_{66} - \mathbb{C}_{44}\\
\alpha_1 &= \mathbb{C}_{11} + \mathbb{C}_{33} - 4\mathbb{C}_{55} - 2\mathbb{C}_{13}\\
\alpha_2 &= \mathbb{C}_{22} + \mathbb{C}_{33} - 4\mathbb{C}_{66} - 2\mathbb{C}_{23}\\
\alpha_3 &= \mathbb{C}_{44} - \mathbb{C}_{66}\\
\alpha_4 &= \mathbb{C}_{44} - \mathbb{C}_{55}\\
\alpha_5 &= \mathbb{C}_{13} - \mathbb{C}_{33} - 2(\mathbb{C}_{44} - \mathbb{C}_{55} - \mathbb{C}_{66})\\
\alpha_6 &= \mathbb{C}_{23} - \mathbb{C}_{33} - 2(\mathbb{C}_{44} - \mathbb{C}_{55} - \mathbb{C}_{66})\\
\alpha_7 &= \mathbb{C}_{12} - \mathbb{C}_{13} - \mathbb{C}_{23} + \mathbb{C}_{33} + 2(\mathbb{C}_{44} - \mathbb{C}_{55} - \mathbb{C}_{66})
\end{aligned}\right\} \qquad (19)$$

in the invariant setting. In case of isotropy the only remaining constants are λ and μ, which can be directly determined from Young's modulus E and Poisson's ratio ν via $\mu = E/2(1+\nu)$, $\lambda = E\nu/(1+\nu)(1-2\nu)$.

2.4 Orthotropic Yield Criterion

In the following, we consider an orthotropic pressure insensitive yield condition using isotropic tensor functions. It is assumed that Φ depends on the symmetric part of the relative stresses $\hat{\boldsymbol{\sigma}}_s := (\boldsymbol{\Sigma} - \hat{\boldsymbol{\beta}})_s$ only. As a consequence the following relations hold for the plastic velocity gradients

$$\mathbf{L}^p = \mathbf{L}^{p\,T} \quad , \quad \mathbf{D}^p := \mathrm{sym}(\mathbf{L}^p) = \mathbf{L}^p \quad , \quad \mathbf{W}^p := \mathrm{skew}(\mathbf{L}^p) = \mathbf{0}\,. \qquad (20)$$

This assumption and its implications will be discussed below.
The integrity basis in terms of the deviatoric part of the relative stresses $\mathrm{dev}\,\hat{\boldsymbol{\sigma}}_s$ and the structural tensors ${}^1\mathbf{M}$ and ${}^2\mathbf{M}$ is given by

$$\left.\begin{aligned}
I_1 := \mathrm{tr}\left[(\mathrm{dev}\hat{\boldsymbol{\sigma}}_s)^2\right],\; I_2 := \mathrm{tr}\left[{}^1\mathbf{M}(\mathrm{dev}\hat{\boldsymbol{\sigma}}_s)^2\right],\; I_3 := \mathrm{tr}\left[{}^2\mathbf{M}(\mathrm{dev}\hat{\boldsymbol{\sigma}}_s)^2\right]\\
I_4 := \mathrm{tr}\left[{}^1\mathbf{M}\mathrm{dev}\hat{\boldsymbol{\sigma}}_s\right],\; I_5 := \mathrm{tr}\left[{}^2\mathbf{M}\mathrm{dev}\hat{\boldsymbol{\sigma}}_s\right],\; I_6 := \mathrm{tr}\left[(\mathrm{dev}\hat{\boldsymbol{\sigma}}_s)^3\right]
\end{aligned}\right\}\,. \qquad (21)$$

The orthotropic flow criterion is formulated as an isotropic tensor function

$$\hat{\Phi}(\mathrm{dev}\hat{\boldsymbol{\sigma}}_s, {}^1\mathbf{M}, {}^2\mathbf{M}) = \hat{\Phi}(\mathbf{Q}^T\mathrm{dev}\hat{\boldsymbol{\sigma}}_s\mathbf{Q}, \mathbf{Q}^{T\,1}\mathbf{M}\mathbf{Q}, \mathbf{Q}^{T\,2}\mathbf{M}\mathbf{Q})\ \forall\ \mathbf{Q} \in SO(3)\,. \qquad (22)$$

Discarding the cubic invariant I_6 in Φ we arrive at a quadratic form in terms of the invariants and six independent material parameters $\eta_i | i = 1, ..., 6$, respectively

$$\Phi = \eta_1\, I_1 + \eta_2\, I_2 + \eta_3\, I_3 + \eta_4\, I_4^2 + \eta_5\, I_5^2 + \eta_6\, I_4 I_5 - \left(1 + \frac{\hat{\xi}(e^p)}{Y_{11}^0}\right)^2\,. \qquad (23)$$

Remark: It can be shown, see CASEY & NAGHDI [9], [10], GREEN & NAGHDI[13] or TSAKMAKIS [39], that under rigid body rotations $\mathbf{Q}$ superposed on the current configuration and – simultanuously – rigid body rotations $\bar{\mathbf{Q}}$ on the intermediate configuration the following transformation rules apply

$$\begin{aligned} \mathbf{F} &\rightarrow \mathbf{F}^* = \mathbf{Q}\,\mathbf{F} = \mathbf{Q}\,\mathbf{F}^e\,\bar{\mathbf{Q}}^T\,\bar{\mathbf{Q}}\,\mathbf{F}^p = \mathbf{Q}\,\bar{\mathbf{F}}^e\,\bar{\mathbf{F}}^p\,, \\ () &\rightarrow ()^* = \bar{\mathbf{Q}}\,()\,\bar{\mathbf{Q}}^T \quad \text{for} \quad \mathbf{C}^e\,,\,\hat{\mathbf{S}}\,,\,\boldsymbol{\Sigma}\,,\,\mathbf{D}^p\,, \\ \mathbf{L}^p &\rightarrow \mathbf{L}^{p\,*} = \bar{\mathbf{Q}}\,\mathbf{L}^p\,\bar{\mathbf{Q}}^T + \dot{\bar{\mathbf{Q}}}\,\bar{\mathbf{Q}}^T\,, \end{aligned} \tag{24}$$

where we have restricted ourselves to tensorial quantities of the intermediate configuration playing an eminent role in the present formulation. Invariance of constitutive equations under rigid body rotations superposed on the current configuration is generally required by the principle of material frame indifference, the latter invariance requirement is due to the well known fact, that the multiplicative decomposition is uniquely defined except for a rigid body rotation superposed on the intermediate configuration; the identity $\bar{\mathbf{Q}}^T\,\bar{\mathbf{Q}}$ can always be inserted, in between $\mathbf{F}^e$ and $\mathbf{F}^p$, see $(24)_1$. As a consequence of the constitutive assumption, that only the symmetric part $\hat{\boldsymbol{\sigma}}_s$ enters the yield function, the flow rule reads $\mathbf{D}^p = \lambda\,\partial_{\Sigma_s}\Phi$, which is, see $(24)_2$, invariant with respect to the arbitrary choice of $\bar{\mathbf{Q}}$, whereas this is not true for the plastic velocity gradient $\mathbf{L}^p$ due to the expression $\dot{\bar{\mathbf{Q}}}\,\bar{\mathbf{Q}}^T$ in $(24)_3$.
As a further consequence of the yield function in terms of $\hat{\boldsymbol{\sigma}}_s$ the six independent material parameters $\eta_i|i = 1,...,6$ can be experimentally identified by three tension tests and three shear tests, respectively, which are independent of each other.

Assume the tests are performed relative to the fixed orientation of the specimen ${}^1\mathbf{a} = (1,0,0)^T$ and ${}^2\mathbf{a} = (0,1,0)^T$. Let Y^0_{ij} be the yield stress in ij-direction, with respect to ${}^i\mathbf{a}$ and ${}^j\mathbf{a}$. The tests with $\hat{\boldsymbol{\beta}} = \mathbf{0}$ are:

1. uniaxial tension in ${}^1\mathbf{a}$-direction:

$$\boldsymbol{\Sigma}_s = \begin{pmatrix} Y^0_{11} & & \\ & 0 & \\ & & 0 \end{pmatrix} \quad \left.\begin{aligned} I_1 &= 2/3(Y^0_{11})^2\,,\ I_4 &&= 2/3Y^0_{11} \\ I_2 &= 4/9(Y^0_{11})^2\,,\ I_5 &&= -1/3\,Y^0_{11} \\ I_3 &= 1/9(Y^0_{11})^2\,,\ I_4\,I_5 &&= -2/9\,(Y^0_{11})^2 \end{aligned}\right\} \tag{25}$$

2. shear test in ${}^1\mathbf{a}$-${}^2\mathbf{a}$ plane:

$$\boldsymbol{\Sigma}_s = \begin{pmatrix} 0 & Y^0_{12} & \\ Y^0_{12} & 0 & \\ & & 0 \end{pmatrix} \quad \left.\begin{aligned} I_1 &= 2(Y^0_{12})^2\,,\ I_4 &&= 0 \\ I_2 &= \ \ (Y^0_{12})^2\,,\ I_5 &&= 0 \\ I_3 &= \ \ (Y^0_{12})^2\,,\ I_4\,I_5 &&= 0 \end{aligned}\right\} \tag{26}$$

We omit an explicit description of the two uniaxial tension tests in ${}^2\mathbf{a}$- and ${}^3\mathbf{a}$-directions and the two shear tests in ${}^1\mathbf{a}$-${}^3\mathbf{a}$ and ${}^2\mathbf{a}$-${}^3\mathbf{a}$ planes, respectively;

analogously each of them leads to a new set of values for the invariants. Evaluating the flow criterion for all six distinct tests yields the linear equations

$$\begin{bmatrix} \frac{2}{3}(Y_{11}^0)^2 & \frac{4}{9}(Y_{11}^0)^2 & \frac{1}{9}(Y_{11}^0)^2 & \frac{4}{9}(Y_{11}^0)^2 & \frac{1}{9}(Y_{11}^0)^2 & -\frac{2}{9}(Y_{11}^0)^2 \\ \frac{2}{3}(Y_{22}^0)^2 & \frac{1}{9}(Y_{22}^0)^2 & \frac{4}{9}(Y_{22}^0)^2 & \frac{1}{9}(Y_{22}^0)^2 & \frac{4}{9}(Y_{22}^0)^2 & -\frac{2}{9}(Y_{22}^0)^2 \\ \frac{2}{3}(Y_{33}^0)^2 & \frac{1}{9}(Y_{33}^0)^2 & \frac{1}{9}(Y_{33}^0)^2 & \frac{1}{9}(Y_{33}^0)^2 & \frac{1}{9}(Y_{33}^0)^2 & \frac{1}{9}(Y_{33}^0)^2 \\ 2(Y_{12}^0)^2 & (Y_{12}^0)^2 & (Y_{12}^0)^2 & 0 & 0 & 0 \\ 2(Y_{13}^0)^2 & (Y_{13}^0)^2 & 0 & 0 & 0 & 0 \\ 2(Y_{23}^0)^2 & 0 & (Y_{23}^0)^2 & 0 & 0 & 0 \end{bmatrix} \begin{bmatrix} \eta_1 \\ \eta_2 \\ \eta_3 \\ \eta_4 \\ \eta_5 \\ \eta_6 \end{bmatrix} = \begin{bmatrix} 1 \\ 1 \\ 1 \\ 1 \\ 1 \\ 1 \end{bmatrix} \tag{27}$$

with the solution

$$\begin{aligned} \eta_1 &= \frac{1}{2}\left(-\frac{1}{(Y_{12}^0)^2} + \frac{1}{(Y_{13}^0)^2} + \frac{1}{(Y_{23}^0)^2}\right), \\ \eta_2 &= \frac{1}{(Y_{12}^0)^2} - \frac{1}{(Y_{23}^0)^2}, \\ \eta_3 &= \frac{1}{(Y_{12}^0)^2} - \frac{1}{(Y_{13}^0)^2}, \\ \eta_4 &= \frac{2}{(Y_{11}^0)^2} - \frac{1}{(Y_{22}^0)^2} + \frac{2}{(Y_{33}^0)^2} - \frac{1}{(Y_{13}^0)^2}, \\ \eta_5 &= -\frac{1}{(Y_{11}^0)^2} + \frac{2}{(Y_{22}^0)^2} + \frac{2}{(Y_{33}^0)^2} - \frac{1}{(Y_{23}^0)^2}, \\ \eta_6 &= -\frac{1}{(Y_{11}^0)^2} - \frac{1}{(Y_{22}^0)^2} + \frac{5}{(Y_{33}^0)^2} + \frac{1}{(Y_{12}^0)^2} - \frac{1}{(Y_{13}^0)^2} - \frac{1}{(Y_{23}^0)^2}. \end{aligned} \tag{28}$$

Now we consider pure isotropy as a special case of orthotropy. For isotropic elasticity $\boldsymbol{\Sigma} = \boldsymbol{\Sigma}^T$ holds and therefore $\mathbf{L}^p = \mathbf{D}^p$. In this case our constitutive assumption for Φ being a function merely of the symmetric part of $\boldsymbol{\Sigma}$ is fulfilled by the elasticity law itself. If we set for the yield normal stresses $Y_{ii}^0 = Y^0$ for $i = 1,2,3$ and for the yield shear stresses $Y_{ij}^0 = Y^0/\sqrt{3}$ for $i \neq j$ with $i,j = 1,2,3$ we arrive at the isotropic von Mises yield criterion

$$\hat{\Phi}(\mathrm{dev}\hat{\boldsymbol{\sigma}}, \xi) = \frac{3}{2}\left(\frac{||\mathrm{dev}\hat{\boldsymbol{\sigma}}||}{Y^0}\right)^2 - \left(1 + \frac{\hat{\xi}(e^p)}{Y^0}\right)^2 \leq 0\,. \tag{29}$$

In the computational benchmark problems in Sect. 6 we apply a nonlinear isotropic hardening function well suited for a fitting of experimental data

$$\hat{\xi}(e^p) = he^p + (Y^\infty - Y^0)(1 - \exp(-\delta e^p))\,, \tag{30}$$

where the expression in terms of Y^∞, Y^0 and δ in (30) is of saturation type.

The constitutive equations are now summarized as follows.

$$
\begin{array}{lll}
\text{elastic strains} & \mathbf{C}^e & = \mathbf{F}^{p\,T-1}\,\mathbf{C}\,\mathbf{F}^{p\,-1} \\
\text{free energy} & \psi & = \hat{\psi}^e(J_1, ..., J_6) + \hat{\psi}^{p,iso}(e^p) + \hat{\psi}^{p,kin}(\boldsymbol{\alpha}) \\
\text{stresses} & \hat{\mathbf{S}} & = 2\,\partial_{C^e}\psi^e \quad , \quad \boldsymbol{\Sigma} = \mathbf{C}^e\,\hat{\mathbf{S}} \\
\text{back stresses} & \hat{\beta} & = \partial_\alpha \psi^{p,kin} \\
\text{isotropic hardening} & \xi & = \partial_{e^p}\psi^{p,iso} \\
\text{relative stresses} & \hat{\boldsymbol{\sigma}}_s & = \boldsymbol{\Sigma}_s - \hat{\boldsymbol{\beta}}_s \\
\text{yield function} & \Phi & = \hat{\Phi}(I_1, ..., I_5, \xi) \\
\text{associative flow rule} & \mathbf{D}^p & = \lambda\,\partial_{\boldsymbol{\Sigma}_s}\Phi \\
\text{evolution of } \boldsymbol{\alpha} & \dot{\boldsymbol{\alpha}} & = -\lambda\,\partial_{\hat{\beta}_s}\Phi \\
\text{evolution of } e^p & \dot{e}^p & = \sqrt{\tfrac{2}{3}}||\mathbf{D}^p|| \\
\text{optimization conditions} & \multicolumn{2}{l}{\lambda \geq 0,\ \Phi \leq 0,\ \ \lambda\Phi = 0}
\end{array}
\tag{31}
$$

3 Integration Algorithm for the Constitutive Equations

To solve the set of constitutive equations at a local level, a so-called operator split along with a general return mapping is applied; for time integration a backward Euler scheme with an exponential map is used. For the solution of the nonlinear finite-element equations on a global level a Newton iteration scheme is used, which requires the consistent tangent matrix. For this reason a simple numerical differentiation technique is applied.

3.1 General Return Algorithm

Based on the definition $(5)_2$ of $\mathbf{L}^p$, and taking $\mathbf{L}^p = \mathbf{D}^p$ into account, we write the flow rule $(31)_8$ for $\mathbf{D}^p$ as

$$
\dot{\mathbf{F}}^p = \mathbf{D}^p\mathbf{F}^p = \lambda\mathbf{N}\,\mathbf{F}^p \quad \text{with} \quad \mathbf{N} := \frac{\partial\Phi}{\partial\boldsymbol{\Sigma}_s}\ . \tag{32}
$$

Within a typical time step $[t_n\,,\,t_{n+1}]$ with time increment $\Delta t := t_{n+1} - t_n$ we integrate (32) by the implicit backward Euler algorithm along with an exponential shift

$$
\mathbf{F}^p_{n+1} = \exp[\gamma\,\partial_{\Sigma_s}\Phi_{n+1}]\ \mathbf{F}^p_n\,, \tag{33}
$$

where $\gamma := \Delta t\,\lambda_{n+1}$ denotes the consistency parameter. For $\mathbf{N}$ we use the corresponding tensor of the trial step defined below. For deviatoric $\mathbf{N}$, i.e. $\mathrm{tr}[\mathbf{N}] = 0$, and applying the identity $\det(\exp[\mathbf{N}]) = \exp[\mathrm{tr}(\mathbf{N})]$, it is obvious that the exponential map (33) preserves plastic incompressibility in the current time step, given that $\det\mathbf{F}^p_n = 1$ holds for the previous step. The rate equations for $\boldsymbol{\alpha}$ and e^p are integrated using a standard backward Euler algorithm. Thus, the procedure for time integration is first order accurate and unconditionally stable. Considering the multiplicative decomposition we obtain for the update of the elastic Cauchy–Green tensor

$$\begin{aligned} \mathbf{C}^e_{n+1} &= \mathbf{F}^{p\,T-1}_{n+1}\,\mathbf{C}_{n+1}\,\mathbf{F}^{p\,-1}_{n+1} \\ &= \exp^T[-\gamma_{n+1}\,\mathbf{N}^{trial}_{n+1}]\,\mathbf{C}^{e\,trial}_{n+1}\,\exp[-\gamma_{n+1}\,\mathbf{N}^{trial}_{n+1}]\,, \end{aligned} \tag{34}$$

where we have introduced by definition $\mathbf{C}^{e\,trial}_{n+1} := \mathbf{F}^{p\,T-1}_{n}\,\mathbf{C}_{n+1}\,\mathbf{F}^{p\,-1}_{n}$ as elastic trial strains. It is well known that for the case of isotropic elasticity $\mathbf{N}_{n+1}$ and $\mathbf{C}^e_{n+1}$ commute, i.e. they have the same principal directions, which allows for a stress update formula that is identical to the classical return mapping algorithm of the geometrically linear theory, see MIEHE & STEIN [22] and SIMO [32].
With the trial values for the Mandel stresses

$$\boldsymbol{\Sigma}^{trial}_{n+1} = 2\,\mathbf{C}^{e\,trial}_{n+1}\,\partial_{\mathbf{C}^{e\,trial}_{n+1}}\psi^e\,, \tag{35}$$

and for the internal variables, i.e. the back stresses $\hat{\boldsymbol{\beta}}^{trial}_{n+1} = \partial_{\boldsymbol{\alpha}_n}\,\psi^{p,kin}$ and the equivalent plastic strain $e^{p\,trial}_{n+1} = \partial_{\xi_n}\,\psi^{p,iso}$ we obtain a trial value for the yield criterion in terms of the deviatoric part of the symmetric relative stresses $\mathrm{dev}\hat{\boldsymbol{\sigma}}_s$ as follows:

$$\Phi^{trial}_{n+1} = \hat{\Phi}(\mathrm{dev}\hat{\boldsymbol{\sigma}}^{trial}_s\,,\,\hat{\mathbf{M}}, e^p_n)^{trial}_{n+1}\,. \tag{36}$$

The time discrete consistency condition reads in the case of plastic loading $\Phi_{n+1} = 0$, which can be solved for γ_{n+1} by applying a Newton solution scheme. At the end of each local iteration the intermediate configuration, described by $\mathbf{F}^p_{n+1}$, and the internal variables $\boldsymbol{\alpha}_{n+1}$ and e^p_{n+1} have to be updated. A summary of the general return mapping algorithm is given in (40) below.

3.2 Algorithmic Elastoplastic Moduli

As we use an Lagrangian formulation of the weak form, which is outlined in Sect. 4, the Second Piola–Kirchhoff stress tensor must be determined by pull back transformation $\mathbf{S} = \mathbf{F}^{p\,-1}\hat{\mathbf{S}}\mathbf{F}^{p\,T-1}$. The nonlinear finite element equations are solved by using a Newton iteration scheme. For this purpose the so-called consistent tangent matrix

$$\mathbb{C}_{ep} = 2\partial\mathbf{S}/\partial\mathbf{C} = \mathbb{C}^{ABCD}\mathbf{e}_A\otimes\mathbf{e}_B\otimes\mathbf{e}_C\otimes\mathbf{e}_D \tag{37}$$

is approximated by numerical differentiation. To this end a simple perturbation technique is applied using the forward difference formula

$$\mathbb{C}^{ABCD} \approx \frac{2}{\epsilon}\left[S^{AB}(\mathbf{C}^{\epsilon}_{(CD)}) - S^{AB}\right]\,. \tag{38}$$

The perturbed Cauchy–Green tensor is computed as

$$\mathbf{C}^{\epsilon}_{(CD)} := \mathbf{C} + \Delta\mathbf{C}^{\epsilon}_{(CD)} \quad \text{with} \quad \Delta\mathbf{C}^{\epsilon}_{(CD)} = \frac{\epsilon}{2}\left(\mathbf{e}_C \otimes \mathbf{e}_D + \mathbf{e}_D \otimes \mathbf{e}_C\right) , \tag{39}$$

where $\mathbf{e}_I$, $I = A, B, C, D$ denotes a fixed Cartesian basis. Computations have shown, that $\epsilon = 10^{-7}$ provides a good choice for the perturbation parameter. As the numerical differentiation requires six additional stress computations, it costs more CPU time than the analytical computation of the moduli. Nevertheless, the numerical determination of consistent algorithmic moduli is advantageous for its simplicity, robustness and for being independent of the material model. It serves as an interface for implementing complicated constitutive models without tedious analytical derivations of tangent operators.

---

1. `Trial step - elastic predictor`

$$\mathbf{C}^{e\,trial}_{n+1} = \mathbf{F}^{p\,T-1}_{n}\,\mathbf{C}_{n+1}\,\mathbf{F}^{p\,-1}_{n}$$

$$\boldsymbol{\Sigma}^{trial}_{n+1} = 2\,\mathbf{C}^{e\,trial}_{n+1}\frac{\partial\psi^e}{\partial\mathbf{C}^{e\,trial}_{n+1}} \;,\quad \hat{\boldsymbol{\beta}}^{trial}_{n+1} = \frac{\partial\psi^{p,kin}}{\partial\boldsymbol{\alpha}_n}$$

$$\hat{\boldsymbol{\sigma}}^{trial}_{n+1} := \boldsymbol{\Sigma}^{trial}_{n+1} - \hat{\boldsymbol{\beta}}^{trial}_{n+1} \;,\quad \mathbf{N}^{trial}_{n+1} = \frac{\partial\Phi}{\partial\boldsymbol{\Sigma}^{trial}_{s\,n+1}}$$

2. `Check yield condition`

$$\texttt{if}\quad \hat{\Phi}(\mathrm{dev}\hat{\boldsymbol{\sigma}}^{trial}_{s\,n+1}, \hat{\mathbf{M}}, e^p_n) > 0 \quad \texttt{go to 3.} \quad \texttt{else exit}$$

3. `Return mapping - corrector step`

$$\texttt{set}\;\; \gamma^{(0)}_{n+1} = 0 \;,\quad e^{p(0)}_{n+1} = e^p_n \;,\quad \boldsymbol{\alpha}^{(0)}_{n+1} = \boldsymbol{\alpha}_n$$

$$\texttt{a)}\quad \mathbf{C}^{e(l)}_{n+1} = \exp^T[-\gamma^{(l)}_{n+1}\,\mathbf{N}^{trial}_{n+1}]\,\mathbf{C}^{e\,trial}_{n+1}\,\exp[-\gamma^{(l)}_{n+1}\,\mathbf{N}^{trial}_{n+1}] \tag{40}$$

$$e^{p(l)}_{n+1} = e^p_n + \gamma^{(l)}_{n+1}\sqrt{\frac{2}{3}}||\mathbf{N}^{trial}_{n+1}|| \;,\quad \boldsymbol{\alpha}^{(l)}_{n+1} = \boldsymbol{\alpha}_n + \gamma^{(l)}_{n+1}\,\mathbf{N}^{trial}_{n+1}$$

$$\texttt{b)}\quad \boldsymbol{\Sigma}^{(l)}_{n+1} = 2\,\mathbf{C}^{e(l)}_{n+1}\frac{\partial\psi^e}{\partial\mathbf{C}^{e(l)}_{n+1}} \;,\quad \hat{\boldsymbol{\beta}}^{(l)}_{n+1} = \frac{\partial\psi^{p,kin}}{\partial\boldsymbol{\alpha}^{(l)}_{n+1}} \;,\quad \xi^{(l)}_{n+1} = \frac{\partial\psi^{p,iso}}{\partial e^{p(l)}{}_{n+1}}$$

$$\texttt{c)}\quad \Phi^{(l)}_{n+1} = \Phi(\gamma^{(l)}_{n+1}) \;,\quad \Phi'^{(l)}_{n+1} \approx [\Phi(\gamma^{(l)}_{n+1} + \epsilon) - \Phi(\gamma^{(l)}_{n+1})]/\epsilon$$

$$\texttt{if}\quad |\Phi^{(l)}_{n+1}| \leq \texttt{tol} \qquad \texttt{go to 4.}$$

$$\texttt{d)}\quad \gamma^{(l+1)}_{n+1} = \gamma^{(l)}_{n+1} - \Phi^{(l)}_{n+1}/\Phi'^{(l)}_{n+1} \qquad \texttt{go to a)}$$

4. `Update intermediate configuration and internal variables`

$$\mathbf{F}^p_{n+1} = \exp[\gamma^{(l)}_{n+1}\mathbf{N}^{trial}_{n+1}]\,\mathbf{F}^p_n \;,\quad e^p_{n+1} = e^{p(l)}_{n+1} \;,\quad \boldsymbol{\alpha}_{n+1} = \boldsymbol{\alpha}^{(l)}_{n+1}$$

3.3 Efficient Calculation of the Exponential Map

For an effective computation of the exponential function of a second order tensor $\boldsymbol{\beta}$ which is not necessarily symmetric we use an recursive algorithm, see MIEHE [23], SANSOUR & KOLLMANN[30]. Firstly, we recall that $\exp[\boldsymbol{\beta}]$ is defined by the power-series

$$\exp[\boldsymbol{\beta}] = \mathbf{1} + \boldsymbol{\beta} + \frac{1}{2!}\boldsymbol{\beta}^2 + \frac{1}{3!}\boldsymbol{\beta}^3 + \ldots + \frac{1}{n!}\boldsymbol{\beta}^n + \ldots \,. \tag{41}$$

Let I_1,I_2,I_3 denote the principal invariants of $\boldsymbol{\beta}$

$$I_1 = \mathrm{tr}[\boldsymbol{\beta}] \;,\; I_2 = \tfrac{1}{2}(I_1^2 - \mathrm{tr}[\boldsymbol{\beta}^2]) \;,\; I_3 = \det[\boldsymbol{\beta}] \,. \tag{42}$$

The Cayley–Hamilton theorem for $\boldsymbol{\beta}$ reads

$$\boldsymbol{\beta}^3 = I_3\mathbf{1} - I_2\boldsymbol{\beta} + I_1\boldsymbol{\beta}^2 \,. \tag{43}$$

Since $\exp[\boldsymbol{\beta}]$ is an isotropic tensor-function of $\boldsymbol{\beta}$, the representation theorem of Rivlin–Ericksen gives

$$\exp[\boldsymbol{\beta}] = \alpha_0(I_1,I_2,I_3)\mathbf{1} + \alpha_1(I_1,I_2,I_3)\boldsymbol{\beta} + \alpha_2(I_1,I_2,I_3)\boldsymbol{\beta}^2 \,. \tag{44}$$

The goal, to calculate $\exp[\boldsymbol{\beta}]$ in an efficient way is achieved by a successive application of the Cayley–Hamilton theorem (43), which allows for a representation of the third and higher powers of $\boldsymbol{\beta}$ in (41) in terms of $\mathbf{1}, \boldsymbol{\beta}, \boldsymbol{\beta}^2$. Thus we achieve the power–series of $\exp[\boldsymbol{\beta}]$ (41) in the representation of (44),

$$\boldsymbol{\beta}^n = \gamma_0^{(n)}\mathbf{1} + \gamma_1^{(n)}\boldsymbol{\beta} + \gamma_2^{(n)}\boldsymbol{\beta}^2 \,, \tag{45}$$

with $\gamma_0^{(n)}$,$\gamma_1^{(n)}$,$\gamma_2^{(n)}$ as functions of the invariants of $\boldsymbol{\beta}$. It can be easily shown by induction, that for arbitrary n the coefficients $\gamma_0^{(n)}$,$\gamma_1^{(n)}$,$\gamma_2^{(n)}$ can be calculated as functions of $\gamma_0^{(n-1)}$,$\gamma_1^{(n-1)}$,$\gamma_2^{(n-1)}$ according to the following recursion formula

$$\left.\begin{aligned} \gamma_0^{(n)} &= \phantom{\gamma_0^{(n-1)} -{}} I_3\gamma_0^{(n-1)} \\ \gamma_1^{(n)} &= \gamma_0^{(n-1)} - I_2\gamma_2^{(n-1)} \\ \gamma_2^{(n)} &= \gamma_1^{(n-1)} + I_1\gamma_2^{(n-1)} \end{aligned}\right\} \,. \tag{46}$$

Comparison of the coefficients in (46) and (44) yields for α_0, α_1, α_2

$$\left.\begin{aligned} &\alpha_0 = 1 + \frac{1}{3!}I_3 + \sum_{n=4}^{N}\frac{1}{n!}\gamma_0^{(n)} \quad , \quad \alpha_1 = 1 - \frac{1}{3!}I_2 + \sum_{n=4}^{N}\frac{1}{n!}\gamma_1^{(n)} \\ &\alpha_2 = \frac{1}{2} + \frac{1}{3!}I_1 + \sum_{n=4}^{N}\frac{1}{n!}\gamma_2^{(n)} \end{aligned}\right\} \,, \tag{47}$$

or with $\gamma_0^{(1)} = 1\,,\, \gamma_1^{(1)} = 1\,,\, \gamma_2^{(1)} = 0\,,\, \gamma_0^{(2)} = 0\,,\, \gamma_1^{(2)} = 0\,,\, \gamma_2^{(2)} = 1\,,\, \gamma_0^{(3)} = I_3\,,\, \gamma_1^{(3)} = -I_2\,,\, \gamma_2^{(3)} = I_1$, we obtain the parameters $\alpha_0, \alpha_1, \alpha_2$ in (44) as

$$\alpha_0 = \sum_{n=1}^{N} \frac{1}{n!}\gamma_0^{(n)} \,,\; \alpha_1 = \sum_{n=1}^{N} \frac{1}{n!}\gamma_1^{(n)} \,,\; \alpha_2 = \sum_{n=1}^{N} \frac{1}{n!}\gamma_2^{(n)} \,. \tag{48}$$

The number N depends on the desired accuracy up to which the exponential series is evaluated. With a chosen tolerance `tol` we define the stop criterion as

$$\frac{\gamma_i^{(n)}}{n!} < \texttt{tol} \,,\; i = 1, 2, 3\,. \tag{49}$$

An alternative method to fulfill the plastic incompressibility condition, which is based on a simple post-processing step, was proposed by the authors in [12], [14] within a formulation of isotropic elastoplasticity using Almansi strain tensors.

4 Variational Formulation

Let $\mathcal{B}$ be the reference body of interest which is bounded by the surface $\partial\mathcal{B}$. The surface is partitioned into two disjoint parts $\partial\mathcal{B} = \partial\mathcal{B}_u \bigcup \partial\mathcal{B}_t$ with $\partial\mathcal{B}_u \bigcap \partial\mathcal{B}_t = \emptyset$. The equation of balance of linear momentum for the static case is governed by the First Piola–Kirchhoff stresses $\mathbf{P} = \mathbf{FS}$ and the body force $\hat{\mathbf{b}}$ in the reference configuration

$$\mathrm{Div}[\mathbf{FS}] + \hat{\mathbf{b}} = \mathbf{0}\,. \tag{50}$$

The Dirichlet and Neumann boundary conditions are given by $\mathbf{u} = \bar{\mathbf{u}}$ on $\partial\mathcal{B}_u$ and $\mathbf{t} = \hat{\mathbf{t}} = \mathbf{PN}$ on $\partial\mathcal{B}_t$, respectively. Here $\hat{\mathbf{t}}$ denotes surface loads, $\mathbf{N}$ represents the exterior unit normal to the boundary surface $\partial\mathcal{B}_t$. With standard arguments of variational calculus we arrive at the variational problem

$$G(\mathbf{u}, \delta\mathbf{u}) = \int_{\mathcal{B}} \mathbf{S} : \delta\mathbf{E}\, dV + G^{ext}\,, \quad \text{with} \tag{51}$$

$$G^{ext}(\delta\mathbf{u}) := -\int_{\mathcal{B}} \hat{\mathbf{b}}\,\delta\mathbf{u}\, dV - \int_{\partial\mathcal{B}_t} \hat{\mathbf{t}}\,\delta\mathbf{u}\, dA\,, \tag{52}$$

where $\delta\mathbf{E} := \frac{1}{2}(\delta\mathbf{F}^T\mathbf{F} + \mathbf{F}^T\delta\mathbf{F})$ characterizes the virtual Green–Lagrangian strain tensor in terms of the virtual deformation gradient $\delta\mathbf{F} := \mathrm{Grad}\delta\mathbf{u}$. The equation of principle of virtual work (51) for a static equilibrium state of the considered body requires $G = 0$. For the solution of this nonlinear equation a standard Newton iteration scheme is applied, which requires the consistent linearization of (51) in order to guarantee quadratic convergence rate near the solution. Since the stress tensor $\mathbf{S}$ is symmetric, the linear increment of G denoted by ΔG is given by

$$\Delta G(\mathbf{u}, \delta\mathbf{u}, \Delta\mathbf{u}) := \int_{\mathcal{B}} (\delta\mathbf{E} : \Delta\mathbf{S} + \Delta\delta\mathbf{E} : \mathbf{S}) \, dV \,, \tag{53}$$

where $\Delta\delta\mathbf{E} := \frac{1}{2}(\Delta\mathbf{F}^T\delta\mathbf{F} + \delta\mathbf{F}^T\Delta\mathbf{F})$ denotes the linearized virtual Green–Lagrange strain tensor as a function of the incremental deformation gradient $\Delta\mathbf{F} := \mathrm{Grad}\Delta\mathbf{u}$. The incremental Second Piola–Kirchhoff stress tensor $\Delta\mathbf{S}$ can be derived as $\Delta\mathbf{S} = \mathbb{C}_{ep}\Delta\mathbf{E}$ with $\Delta\mathbf{E} := \frac{1}{2}(\Delta\mathbf{F}^T\mathbf{F} + \mathbf{F}^T\Delta\mathbf{F})$ and the consistent tangent matrix $\mathbb{C}_{ep}$.

5 Finite Element Discretization

For numerical analyses of thin structures with finite elements it is crucial to avoid *locking*, i.e. artificial stiffening effects. Such effects occur when simple low-order, displacement type elements are used. For an overview with detailed explanations of the sources of different *locking* effects we refer to Klinkel [20]. In this section we describe different effective remedies against this undesired stiffening.
We firstly introduce the formulation of a standard displacement type element. Hence certain modifications are necessary to reduce the locking effects. To avoid *shear locking* the transverse shear strains are approximated by using the interpolation functions of Bathe & Dvorkin [1]. Artificial thickness strains can be avoided by using the interpolation functions of Betsch & Stein [4].

5.1 Displacement Type Formulation

According to the isoparametric concept we use the standard tri-linear shape functions for an eight-node ($nel = 8$) solid element to interpolate the geometry of the initial and the current configurations

$$\mathbf{X}^h = \sum_{I=1}^{nel} N_I(\xi^1, \xi^2, \xi^3)\mathbf{X}_I \quad , \quad \mathbf{x}^h = \sum_{I=1}^{nel} N_I(\xi^1, \xi^2, \xi^3)\mathbf{x}_I \quad \text{with} \tag{54}$$

$$N_I(\xi^1, \xi^2, \xi^3) = \frac{1}{8}(1 + \xi^1_I\xi^1)(1 + \xi^2_I\xi^2)(1 + \xi^3_I\xi^3) \,. \tag{55}$$

The index h denotes the finite element discretization. The convective base vectors of the initial and the current configurations are

$$\mathbf{G}^h_i = \sum_{I=1}^{nel} N_{I,i}\,\mathbf{X}_I \quad , \quad \mathbf{g}^h_i = \sum_{I=1}^{nel} N_{I,i}\,\mathbf{x}_I \,, \tag{56}$$

and the approximation of the virtual strains is given by

$$\delta \mathbf{E}^h = \sum_{I=1}^{nel} \mathbf{B}_I \delta \mathbf{v}_I \quad , \quad \mathbf{B}_I = [\mathbf{B}_I^m, \mathbf{B}_I^s]^T \quad . \tag{57}$$

The components of the virtual nodal displacement vector $\delta \mathbf{v}_I$ are given with respect to the fixed Cartesian basis system. The matrices $\mathbf{B}_I^m$ and $\mathbf{B}_I^s$ are specified below. The expression $\mathbf{S} : \Delta\delta\mathbf{E}^h$ leads to the geometrical matrix $\mathbf{G}_{IJ}$ where the linearized virtual strains $\Delta\delta\mathbf{E}$ were explicitly given above,

$$\mathbf{S} : \Delta\delta\mathbf{E}^h = \sum_{I=1}^{nel} \sum_{J=1}^{nel} \delta \mathbf{v}_I^T \, \mathbf{G}_{IJ} \, \Delta \mathbf{v}_J \quad , \quad \mathbf{G}_{IJ} = \mathrm{diag}\,[S_{IJ}, S_{IJ}, S_{IJ}] \, . \tag{58}$$

The expression $S_{IJ} = S_{IJ}^m + S_{IJ}^s$ results from two parts specified below.

5.2 Shear Stiffness Part

According to Fig. 5.1, four collocation points $M = A, B, C, D$ with given coordinates ξ^i are defined.

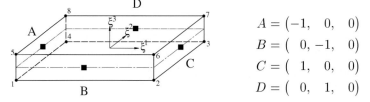

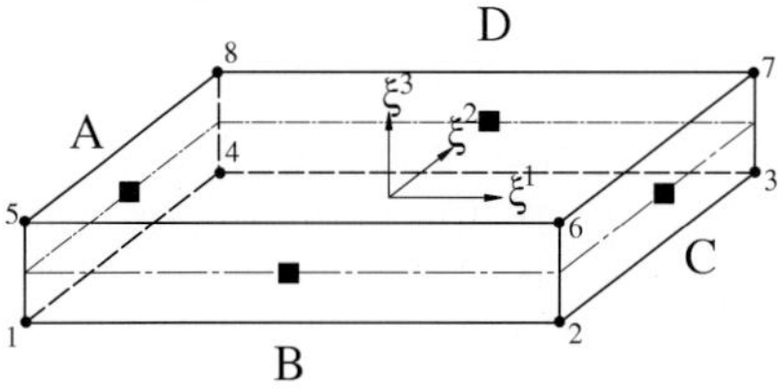

$$\begin{aligned} A &= (-1, \ 0, \ 0) \\ B &= (\ 0, -1, \ 0) \\ C &= (\ 1, \ 0, \ 0) \\ D &= (\ 0, \ 1, \ 0) \end{aligned}$$

Fig. 5.1 Collocation points of the shear strain interpolation

At these points, the shear strains E_{13}^M, E_{23}^M of the Green–Lagrangian strain tensor are evaluated. To overcome *shear locking*, the transverse shear strains E_{13} and E_{23} are obtained via the interpolation in the ξ^1-ξ^2-plane,

$$\begin{bmatrix} 2E_{13}^h \\ 2E_{23}^h \end{bmatrix} = \begin{bmatrix} (1-\xi^2)E_{13}^B + (1+\xi^2)E_{13}^D \\ (1-\xi^1)E_{23}^A + (1+\xi^1)E_{23}^C \end{bmatrix} \quad . \tag{59}$$

According to (59) the transverse shear strains are assumed to be constant in thickness direction within the considered element. Numerical tests have shown that this approximation is sufficient for thin structures. The alternative with two planes and eight collocation points within the element does not lead to significant differences. Hence, the variation of the transverse shear strains can be expressed as

$$\begin{bmatrix} 2\delta E_{13}^h \\ 2\delta E_{23}^h \end{bmatrix} = \sum_{I=1}^{nel} \mathbf{B}_I^s \, \delta \mathbf{v}_I \quad , \quad \text{with} \tag{60}$$

$$\mathbf{B}_I^s = \frac{1}{2}\begin{bmatrix} (1-\xi^2)(\mathbf{g}_3^{B^T} N_{I,1}^B + \mathbf{g}_1^{B^T} N_{I,3}^B) + (1+\xi^2)(\mathbf{g}_3^{D^T} N_{I,1}^D + \mathbf{g}_1^{D^T} N_{I,3}^D) \\ (1-\xi^1)(\mathbf{g}_3^{A^T} N_{I,2}^A + \mathbf{g}_2^{A^T} N_{I,3}^A) + (1+\xi^1)(\mathbf{g}_3^{C^T} N_{I,2}^C + \mathbf{g}_2^{C^T} N_{I,3}^C) \end{bmatrix}. \tag{61}$$

The shape function N_I^M and the current base vectors $\mathbf{g}_i^M$ are obtained by exploitation of the corresponding equation at the collocation points M. The above defined quantity S_{IJ}^s reads

$$S_{IJ}^s = \tfrac{1}{2}[(1-\xi^2)(N_{I,1}^B\, N_{J,3}^B + N_{I,3}^B\, N_{J,1}^B) + (1+\xi^2)(N_{I,1}^D\, N_{J,3}^D + N_{I,3}^D\, N_{J,1}^D)]\; S^{13} + \tfrac{1}{2}[(1-\xi^1)(N_{I,2}^A\, N_{J,3}^A + N_{I,3}^A\, N_{J,2}^A) + (1+\xi^1)(N_{I,2}^C\, N_{J,3}^C + N_{I,3}^C\, N_{J,2}^C)]\; S^{23}\,. \tag{62}$$

5.3 Approximation of the Thickness Strain

For thin shell structures with bending dominated loading a *locking* effect due to artificial thickness strains has been observed by RAMM et al. in [29] when using a direct interpolation of the director vector. To overcome this type of *locking* an *assumed natural strain* (ANS)-interpolation of the thickness strain E_{33} using bi-linear shape functions for four-node shell elements was proposed by BETSCH & STEIN in [4] and by BISCHOFF & RAMM in [6]. Here, we adapt this procedure to the eight-node brick element. According to Fig. 5.2 four collocation points $L = A_1, A_2, A_3, A_4$ are defined in the reference surface with $\xi^3 = 0$.

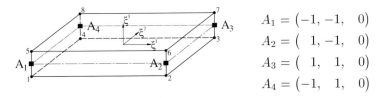

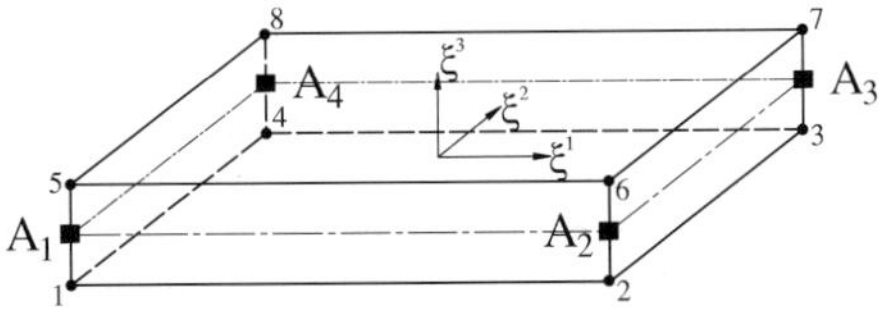

$$\begin{aligned} A_1 &= (-1,\ -1,\ \ 0) \\ A_2 &= (\ \ 1,\ -1,\ \ 0) \\ A_3 &= (\ \ 1,\ \ \ 1,\ \ 0) \\ A_4 &= (-1,\ \ \ 1,\ \ 0) \end{aligned}$$

Fig. 5.2 Collocation points for the thickness strain interpolation

The approximation of E_{33} reads

$$E_{33}^h = \sum_{L=1}^{4} \frac{1}{4}(1+\xi_L^1\xi^1)(1+\xi_L^2\xi^2)\; E_{33}^L \quad , \quad L = A_1,\, A_2,\, A_3,\, A_4\,, \tag{63}$$

where E_{33}^L denotes the thickness strains at the above defined points L. Thus with (63) it is assumed, that within the considered element E_{33} is constant in the ξ^3-direction. This assumption holds for thin structures.
The variation of the thickness strains and the membrane strains are obtained

from

$$\begin{pmatrix} \delta E_{11}^h \\ \delta E_{22}^h \\ \delta E_{33}^h \\ 2\delta E_{12}^h \end{pmatrix} = \sum_{I=1}^{nel} \underbrace{\begin{pmatrix} \mathbf{g}_1^T N_{I,1} \\ \mathbf{g}_2^T N_{I,2} \\ \sum_{L=1}^{4} \frac{1}{4}(1+\xi_L^1\xi^1)(1+\xi_L^2\xi^2)\ (\mathbf{g}_3^L)^T N_{I,3}^L \\ \mathbf{g}_2^T N_{I,1} + \mathbf{g}_1^T N_{I,2} \end{pmatrix}}_{\mathbf{B}_I^m} \delta\mathbf{v}_I \,. \tag{64}$$

Furthermore, the above defined quantity S_{IJ}^m yields

$$\begin{aligned} S_{IJ}^m = \; & S^{11}\ N_{I,1}N_{I,1} + S^{22}\ N_{I,2}N_{I,2} + S^{12}\ (N_{I,1}N_{I,2} + N_{I,2}N_{I,1}) \\ & + S^{33} \sum_{L=1}^{4} \frac{1}{4}(1+\xi_L^1\xi^1)(1+\xi_L^2\xi^2)\ N_{I,3}^L\ N_{J,3}^L \,. \end{aligned} \tag{65}$$

Next, the following stiffness matrices, associated with the element nodes I, J, are introduced

$$\mathbf{K}_{eIJ} = \int_{\mathcal{B}_0^e} (\mathbf{B}_I^T \mathbb{C} \mathbf{B}_J + \mathbf{G}_{IJ})\ \mathrm{d}V \quad , \tag{66}$$

and the vectors

$$\mathbf{f}_{eI}^{int} = \int_{\mathcal{B}_e^0} \mathbf{B}_I^T\ \mathbf{S}\ \mathrm{d}V \quad , \quad \mathbf{f}_{eI}^{ext} = \int_{\mathcal{B}_e^0} N_I\ \rho_0\ \hat{\mathbf{b}}\ \mathrm{d}V + \int_{\partial\mathcal{B}_e^0} N_I \hat{\mathbf{t}}\ \mathrm{d}A \tag{67}$$

are defined. Here, $\mathbf{S}$ denotes the vector of the Second Piola–Kirchhoff stresses obtained by the pull back transformation $\mathbf{S} = \mathbf{F}^{p\,-1}\hat{\mathbf{S}}\mathbf{F}^{p\,T-1}$. Hence, the discretized linearized weak form yields the following system of equations on element level

$$[\mathbf{K}_e]\,[\Delta\mathbf{v}^e] \;=\; [\mathbf{f}_e^{ext} \;-\; \mathbf{f}_e^{int}]\,. \tag{68}$$

Here, $\mathbf{K}_e$, $\mathbf{f}_e^{int}$ and $\mathbf{f}_e^{ext}$ contain the submatrices $\mathbf{K}_{eIJ}$, $\mathbf{f}_{eI}^{int}$ and $\mathbf{f}_{eI}^{ext}$ according to the order of the nodes I and J. Furthermore, $\Delta\mathbf{v}^e$ denotes the vector of the incremental element displacements. The spatial discretization of the considered body leads after assembly $\mathcal{B} \approx \bigcup_{e=1}^{n_{ele}} \mathcal{B}^e$ with n_{ele} finite elements $\mathcal{B}^e$ to a set of algebraic equations which can be solved for the unknown nodal displacements.

6 Numerical Examples

The algorithmic formulation of the orthotropic constitutive model derived and analyzed in the previous sections is implemented in an extended version of FEAP, a general nonlinear finite element code documented in [38]. Four sets of simulations are conducted to test the behavior of the proposed orthotropic model as well as the robustness of the numerical methods. The computational simulations are checked to capture qualitatively the *earing* phenomenon owing to the anisotropic nature of rolled sheet metal. Furthermore, we compare our results quantitatively with numerical simulations of other references and – as far as they are available – with experimental data, the latter for isotropy as a special case of orthotropy. All simulations were run with the 8-node brick type shell element using the ANS -method and a 5-parameter EAS concept, respectively.

6.1 Necking of a Circular Bar

The necking of a circular bar is an example widely investigated in the literature, see e.g. SIMO & ARMERO [33] or KLINKEL [20]. The geometrical data are $R = 6.413\,\text{mm}$, $R_b = 0.982R$ and $L = 26.667\,\text{mm}$. To initialize the necking process we use the reduced radius R_b at $z = 26.667\,\text{mm}$ as a geometrical imperfection. The material data for isotropic elasticity and the isotropic von Mises yield condition (29) with nonlinear isotropic hardening (30) are given as follows.

Elasticity constants:
$E = 206.9\,\text{GPa}\,,\ \nu = 0.29$

Yield parameter:
$Y^0 = 0.45\,\text{GPa}\,,\ Y^\infty = 0.715\,\text{GPa}$
$h = 0.12924\,\text{GPa}\,,\ \delta = 16.93$

Hardening function:
$\hat{\xi}(e^p) = he^p + (Y^\infty - Y^0)(1 - \exp(-\delta e^p))$

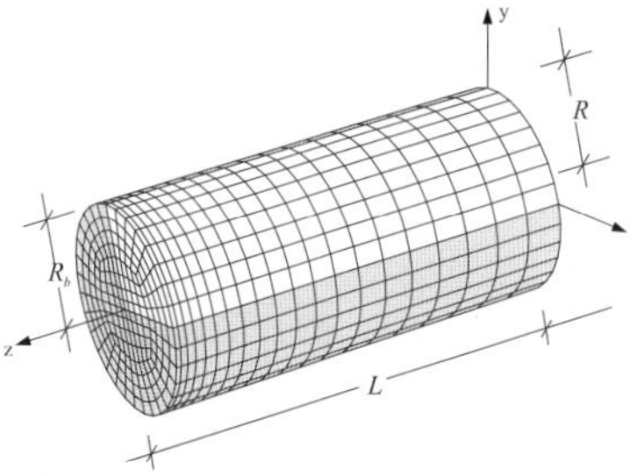

Fig. 6.1. Necking of a circular bar. Elastoplastic material data, geometry and finite element mesh

The finite element discretization of half the bar is depicted in Fig. 6.1. At $z = L$ we impose the symmetry boundary conditions $w = 0\,\text{mm}$, whereas in a displacement controlled computation the axial elongation $w(z = 0\,\text{mm})$ is prescribed. Furthermore, we consider symmetry conditions in the cross-section of the plane. Thus, one eighth of the entire bar of total length $2L$ is discretized with 960 elements, where the thickness direction of the shell elements corresponds to the global z-axis. Figure 6.2 displays the deformed structure at $w = 7\,\mathrm{mm}$ and the equivalent plastic strain, which concentrates in the necking zone. The results are in very good agreement with the computational reference solutions of SIMO & ARMERO [33] and that of KLINKEL [20], see Fig. 6.3. Before the onset of necking the result of our finite element analysis is in accurate accordance with the experimental results reported in NORRIS et al. [26]; it captures pretty well the load bearing capacity of the bar of $79.2\,\mathrm{kN}$, whereas for elongations $w > 4\,\mathrm{mm}$ it is somewhat too weak.

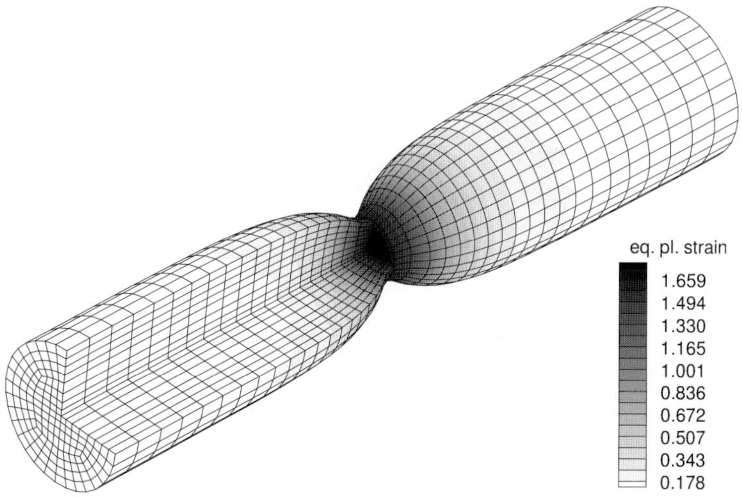

Fig. 6.2. Necking of a circular bar. Equivalent plastic strain at $w = 7\,\mathrm{mm}$

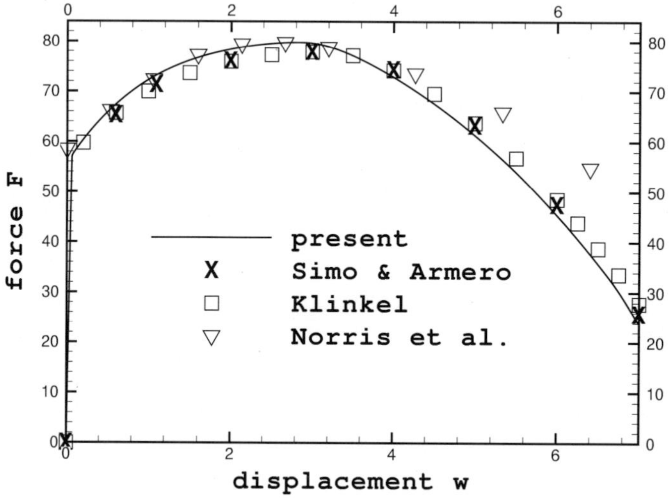

Fig. 6.3. Necking of a circular bar. Computational and experimental results of applied force F [kN] versus axial elongation w [mm]

6.2 Punching of a Conical Shell

In the second example of isotropic elastoplasticity we consider a conical shell subject to a constant ring load $\bar{\lambda}\, p$ with $p = 1\,\text{GPa}\,\text{cm}$, see KLINKEL [20] and references therein. The material data as well as the system with the finite element discretization are depicted in Fig. 6.4. One quarter of the shell is discretized with $8 \times 8 \times 1$ elements. We use the von Mises yield criterion (29) along with isotropic hardening $\hat{\xi}(e^p) = he^p + (Y^\infty - Y^0)(1 - \exp(-\delta e^p))$.

In our computations we apply an arc-length method, where the vertical displacement w of the upper edge is controlled. Note that the shell exhibits two different stability points during the deformation process, namely at $w = 0.02\,\text{cm}$ at the onset of a local rolling of the upper rim, whereas at $w = 1.21\,\text{cm}$ a global snap through of the entire structure is observed, see Figs. 6.5 and 6.6. To avoid *zero-energy-modes* which can occur in applications of the EAS method, we use for the integration of the element stiffness matrix 9 instead of 8 Gauss points, which was firstly proposed by SIMO et al. [34]. The difference can be observed in Fig. 6.6, where the load-displacement curves considerably deviate from one another. Remarkably, the results of the proposed multiplicative model and those obtained by a theory based on generalized stress-strain measures for parameter $m = 0$, which leads to logarithmic stresses and strains, are in very good agreement. For details of the latter formulation, we refer to SCHRÖDER et al. [31]. In this context see also XIAO et al. [42]. The correspondence of the results might shed new light from the numerical side onto a controversial discussion about different competing plasticity models: those, based on the multiplicative decomposition, and others based on the additive framework.

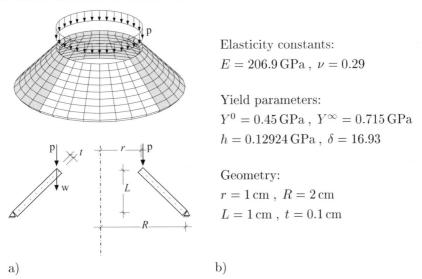

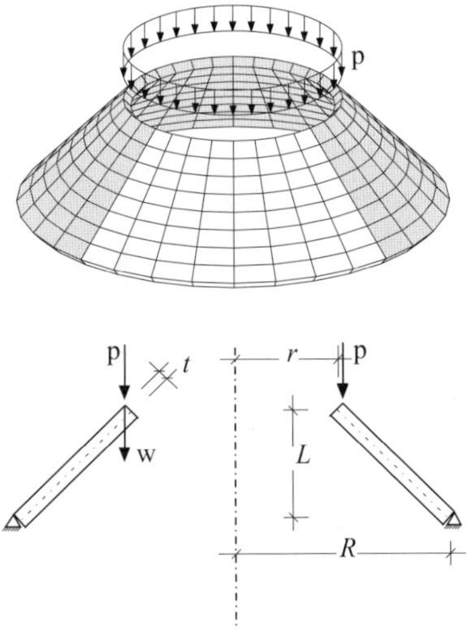

Elasticity constants:

$E = 206.9\,\text{GPa}\,,\ \nu = 0.29$

Yield parameters:

$Y^0 = 0.45\,\text{GPa}\,,\ Y^\infty = 0.715\,\text{GPa}$

$h = 0.12924\,\text{GPa}\,,\ \delta = 16.93$

Geometry:

$r = 1\,\text{cm}\,,\ R = 2\,\text{cm}$

$L = 1\,\text{cm}\,,\ t = 0.1\,\text{cm}$

a) b)

Fig. 6.4. Punching of a conical shell. (**a**) Geometry and finite element mesh. (**b**) Elastoplastic material and geometrical data

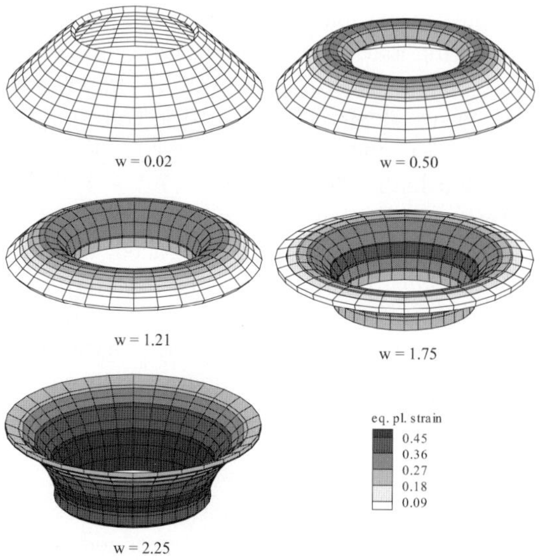

Fig. 6.5. Punching of a conical shell. Equivalent plastic strain on deformed structures at different stages of the punching process

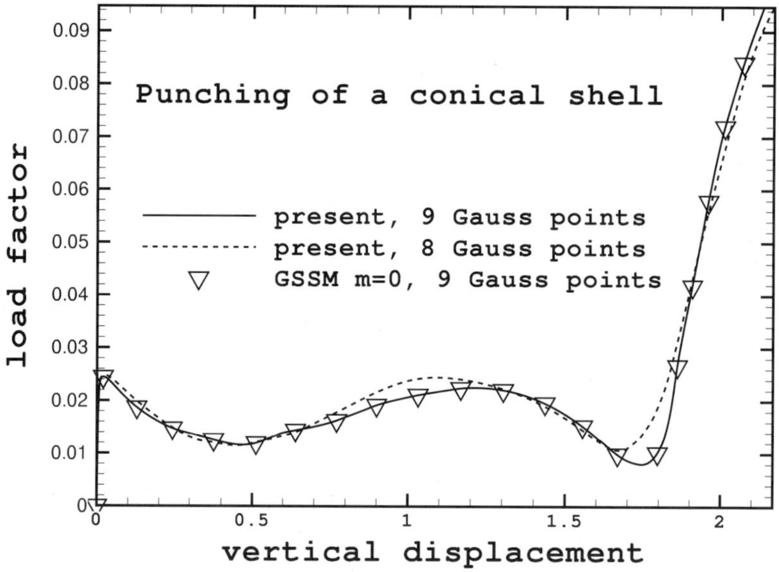

Fig. 6.6. Punching of a conical shell. Comparison of load-deflection curves. Multiplicative model (present) with 8 and 9 Gauss points versus generalized stress-strain measures (GSSM) for $m = 0$ with 9 Gauss points, [31]

6.3 Drawing of a Circular Blank

A circular blank of radius $R = 40\,\text{cm}$ and thickness $t = 1\,\text{cm}$ with a concentric hole of radius $R_i = 20\,\text{cm}$ is deep-drawn into a cup. In order to simulate the drawing process without using contact elements, the inner rim is uniformly pulled inwards in radial direction up to a maximum displacement of $\Delta u_r = 10\,\text{cm}$, while the outer rim is free. We choose plane stress conditions for our simplified model. The material is assumed to be isotropic in elasticity ($\mu = 80.19$ GPa and $\lambda = 110.74$ GPa) but orthotropic in its yield properties. The x- and y-axes of the coordinate system in Fig. 6.7 coincide with the axes of orthotropy. Two different materials are considered for orthotropic yielding. For material A the shear stresses dominate in the yield criterion, we set $Y_{xy} = 0.5 \cdot Y_{xx}/\sqrt{3}$. In contrast to A, for material B the normal stresses are predominant in yielding: for the shear yield stress we choose $Y_{xy} = 2.0 \cdot Y_{xx}/\sqrt{3}$ which is twice the isotropic value. For both materials $Y_{xx} = Y_{yy} = 0.45$ GPa holds.

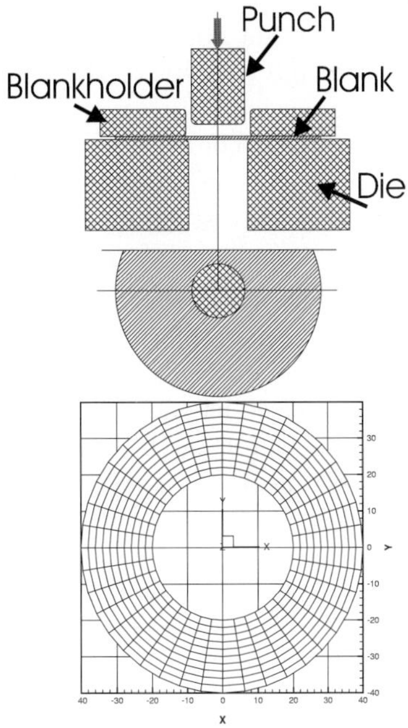

Fig.6.7. Real problem and its 2-dimensional model

As observed in deep-drawing experiments of rolled sheet metal the outer rim of the sheet exhibits waviness called *earing* owing to the anisotropy of the plastic material behaviour. As expected the plastic strain concentrates for material A at a 45° angle in the (x,y)-plane and for material B along the x- and y-axes, see Fig. 6.8 a), b).

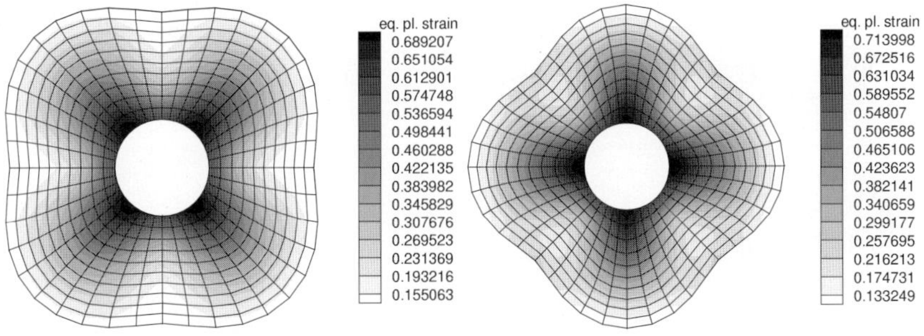

a) b)

Fig.6.8. Drawing of a circular blank. Modelling according to PAPADOPOULOS & LU [27]. Equivalent plastic strain on deformed structures at $\Delta u_r = 10\,\text{cm}$. (**a**) Material A: $Y_{xy} = 0.5 \cdot Y_{xy|isotropic}$. (**b**) Material B: $Y_{xy} = 2.0 \cdot Y_{xy|isotropic}$.

6.4 Simply Supported Circular Plate with Uniform Load

We now consider the elastoplastic deformation of a circular plate under dead load. The plate is simply supported in the z-direction at the bottom of the edges so that horizontal displacements and rotations at the edges may occur. Figure 6.9 a) depicts the geometry of the problem and its finite-element discretization. With respect to symmetry only one quarter of the plate is discretized. The mesh is chosen with one element through the thickness and 192 elements in plane for each quadrant. Again, two different materials, A and B are investigated, which both coincide in isotropic elasticity, but differ in the parameters of orthotropic plastic yielding, see Fig. 6.9 b).

Figure 6.10 depicts the load deflection curves where the load factor $\bar{\lambda}$ in $p_z(\bar{\lambda}) = \bar{\lambda}\, p_{z0}$ is plotted as a function of the vertical displacement of the center point of the circular plate. Again, we note a good agreement between the proposed model and that of the generalized stress-strain measures for $m = 0$. The deflection of the plate at the load levels $\bar{\lambda} = 400$ for material A and $\bar{\lambda} = 600$ for material B is shown in Fig. 6.11. As expected the plastic strains concentrate for material A at a 45° angle in the (x,y)-plane and for material B along the x- and y-axes. For a qualitative comparison see the deep-drawn cup in the centre of Fig. 6.11, taken from RAABE [28]. Our finite element simulation renders physically correct results concerning loci and number of the ears.

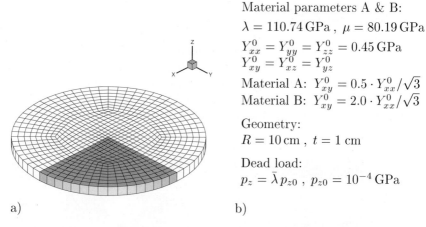

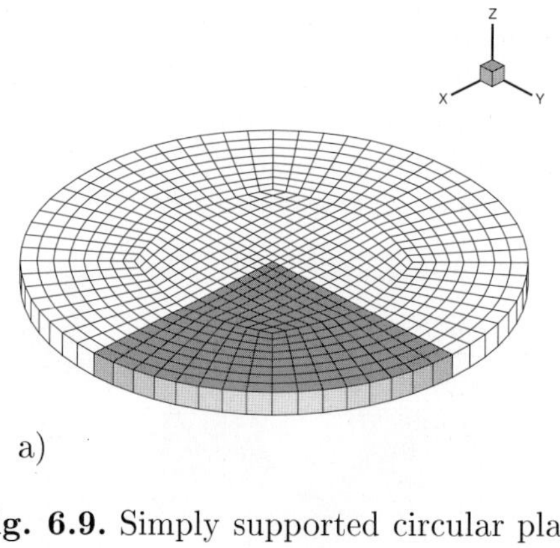

a)

Material parameters A & B:

$\lambda = 110.74\,\text{GPa}\,,\ \mu = 80.19\,\text{GPa}$

$Y^0_{xx} = Y^0_{yy} = Y^0_{zz} = 0.45\,\text{GPa}$

$Y^0_{xy} = Y^0_{xz} = Y^0_{yz}$

Material A: $Y^0_{xy} = 0.5 \cdot Y^0_{xx}/\sqrt{3}$

Material B: $Y^0_{xy} = 2.0 \cdot Y^0_{xx}/\sqrt{3}$

Geometry:

$R = 10\,\text{cm}\,,\ t = 1\,\text{cm}$

Dead load:

$p_z = \bar{\lambda}\, p_{z0}\,,\ p_{z0} = 10^{-4}\,\text{GPa}$

b)

Fig. 6.9. Simply supported circular plate with uniform dead load. (**a**) Geometry and finite element mesh. (**b**) Material, geometrical and load data.

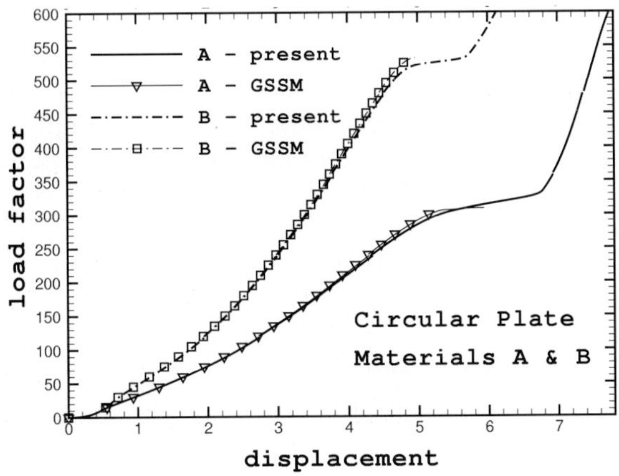

Fig. 6.10. Simply supported circular plate. Load displacement curves for materials A and B. Muliplicative model (present) versus generalized stress strain measures for $m = 0$ (GSSM)

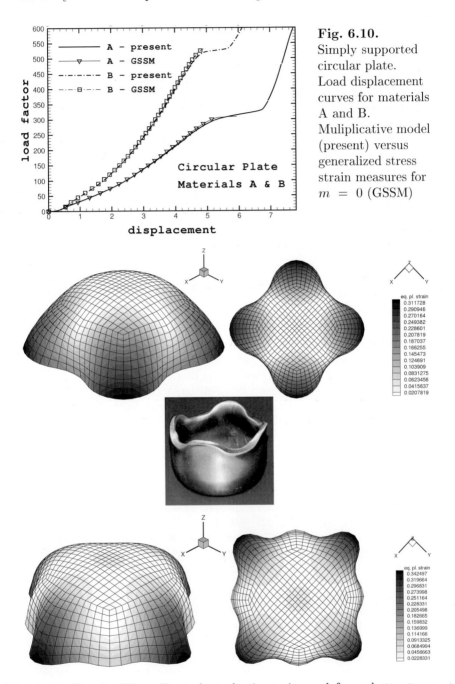

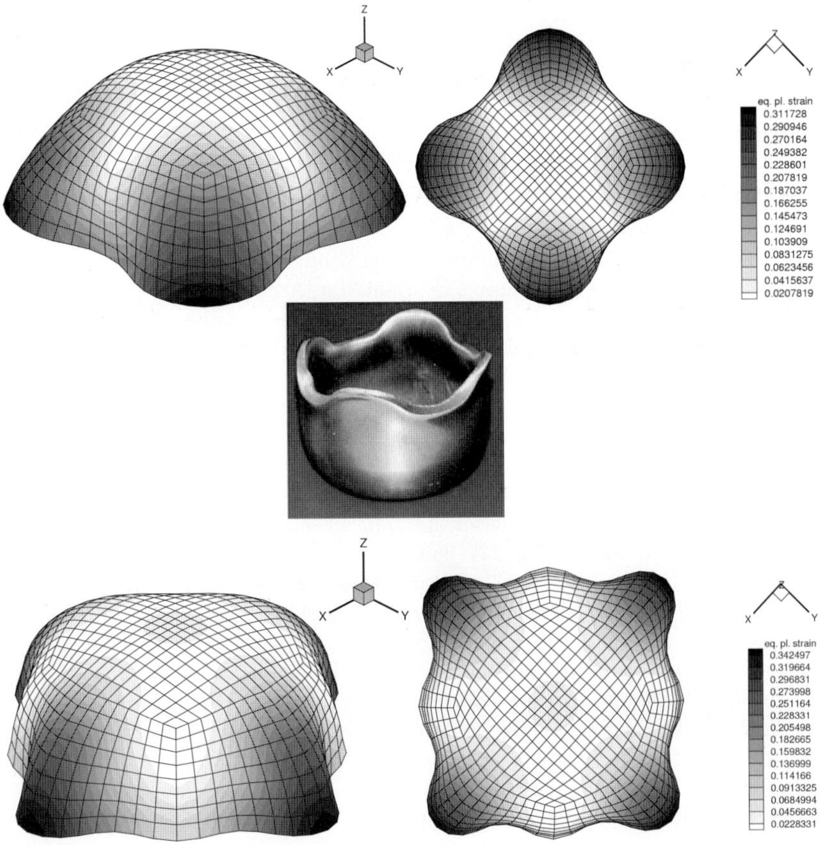

Fig. 6.11. Circular Plate. Equivalent plastic strain on deformed structures from perspective and bird's-eye view for material A at $\bar{\lambda} = 400$ (above), for material B at $\bar{\lambda} = 600$ (below). *Earing* at a deep-drawn cup (centre), [28]

7 Conclusions

In this paper a multiplicative formulation of orthotropic elastoplasticity at finite inelastic strains is presented and aspects of its finite element implementation are addressed. The governing constitutive equations, formulated in an invariant setting by the introduction of structural tensors, are formulated relative to the intermediate configuration. The yield function is expressed in terms of the symmetric part of the Mandel stresses and the associated back stresses. Kinematic as well as isotropic hardening are considered. A general return algorithm along with an exponential map is applied, the latter fulfills plastic incompressibility exactly. Representative numerical examples demonstrate the robustness of our solution algorithms and the predictive capacity of our finite element simulations to capture anisotropic phenomena such as *'earing'*. For the case of isotropy good agreement with computational results from the literature is achieved. Future research will concentrate on the evolution of anisotropy due to plastic deformations.

References

1. Bathe K.-J., Dvorkin E.N. (1984) A continuum mechanics based four node shell element for general nonlinear analysis. Eng Comput 1:77–88
2. Besdo D. (1981) Zur Formulierung von Stoffgesetzen der Plastomechanik im Dehnungsraum nach Ilyushin's Postulat. Ing-Arch 51:1–8
3. Betsch P., Gruttmann F., Stein E. (1996) A 4-node finite shell element for the implementation of general hyperelastic 3D-elasticity at finite strains. Comput Methods Appl Mech Eng 30:57–79
4. Betsch P., Stein E. (1996) An assumed strain approach avoiding artificial thickness straining for a nonlinear 4-node shell element. Commun Numer Methods Eng 11:899–910
5. Betten J. (1987) Formulation of anisotropic constitutive equations. In: Boehler J.P.(Ed.) Applications of tensor functions in solid mechanics, CISM Course No. 292. Springer, Heidelberg, 227–250
6. Bischoff M., Ramm E. (1997) Shear deformable shell elements for large strains and rotations. Int J Numer Methods Eng 40:4427–4449
7. Boehler J.P. (1987) Introduction to the invariant formulation of anisotropic constitutive equations. In: Boehler J. P.(Ed.) Applications of Tensor Functions in Solid Mechanics, CISM Course No. 292. Springer, Heidelberg, 13–30
8. de Borst R., Feenstra P.H. (1990) Studies in anisotropic plasticity with reference to the Hill criterion. Int J Numer Methods Eng 29:315–336
9. Casey J., Naghdi P. (1980) A remark on the use of the decomposition $\mathbf{F} = \mathbf{F}_e\mathbf{F}_p$ in plasticity. J Appl Mech 47:672–675
10. Casey J., Naghdi P. (1981) A correct definition of elastic and plastic deformation and its computational significance. J Appl Mech 48:983–985
11. Eidel B., Gruttmann F. (2002) On the formulation and finite element implementation of anisotropic multiplicative finite strain elastoplasticity. submitted to: PAMM, Proc Appl Math Mech

12. Eidel B., Gruttmann F. (2001) Finite strain inelasticity for isotropy, a simple and efficient finite element formulation. PAMM, Proc Appl Math Mech 1/1:185–186
13. Green A.E., Naghdi P.M. (1971) Some remarks on elastic-plastic deformations at finite strains. Int J Engin Sci 9:1219–1229
14. Gruttmann F., Eidel B. (2001) On the implementation of finite elastoplasticity using Almansi strain tensors and exact fulfillment of plastic incompressibility. IfS Preprint [J01-01].
15. Gruttmann F., Eidel, B. (2002) On the implementation of anisotropic finite strain plasticity. In: Mang H.A., Rammerstorfer F.G., Eberhardsteiner J. (Eds.) Proceedings of the Fifth World Congress on Computational Mechanics (WCCM V), July 7-12, 2002. Vienna University of Technology, Vienna, ISBN 3-9501554-0-6, `http://wccm.tuwien.ac.at`
16. Gruttmann F., Klinkel S., Wagner W. (1995) A finite rotation shell theory with application to composite structures. Rev Eur Élém Finis 4:597–631
17. Hackl K. (1997) Generalized standard media and variational principles in classical and finite strain elastoplasticity. J Mech Phys Solids 5:667–688
18. Hill R. (1948) A theory of the yielding and plastic flow of anisotropic metals. Proc R Soc Lond, Ser A 193:281–297
19. Klinkel S., Wagner W. (1997) An effective geometrically nonlinear brick element based on the EAS-method. Int J Numer Methods Eng 40:4529–4545
20. Klinkel S. (2000) Theorie und Numerik eines Volumen–Schalen–Elementes bei finiten elastischen und plastischen Verzerrungen. PhD Thesis, Universität Karlsruhe
21. Lubliner J. (1990) Plasticity theory, 1st edn. Macmillan Publishing Company, New York
22. Miehe C., Stein E. (1992) A canonical model of multiplicative elasto-plasticity: formulation and aspects of the numerical implementation. Eur J Mech, A/Solids 11:25–43
23. Miehe C. (1996) Exponential map algorithm for stress updates in anisotropic multiplicative elastoplasticity for single crystals. Int J Numer Methods Eng 39:3367–3390
24. Miehe C. (1998) A constitutive frame of elastoplasticity at large strains based on the notion of a plastic metric. Int J Solids Struct 35:3859–3897
25. Neff, P. (2002) Some results concerning the mathematical treatment of finite multiplicative elasto-plasticity. This volume
26. Norris D. M., Moran B., Scudder J. K., Quiñones D. F. (1978) A computer simulation of the tension test. J Mech Phys Solids 26:1–19
27. Papadopoulos P., Lu J. (2001) On the formulation and numerical solution of problems in anisotropic finite plasticity. Comput Methods Appl Mech Eng 190:4889–4910
28. Raabe R.D. (2002) Simulation und Experiment eines Näpfchenziehversuchs. `http://www2.mpie-duesseldorf.mpg.de/msu-web/Mitarbeiter/raabe/3`
29. Ramm E., Bischoff M., Braun M. (1994) Higher order nonlinear shell formulations – a step back into three dimensions. In: Bell K. (Ed.) From finite elements to the troll platform, Dept. of Structural Engineering, Norwegian Institute of Technology, Trondheim, 65–88
30. Sansour C., Kollmann F.G. (1998) Large viscoplastic deformations of shells. Theory and finite element formulation. Comput Mech 21:512–525

31. Schröder J., Gruttmann F., Löblein J. (2002) A simple orthotropic finite elasto-plasticity model based on generalized stress-strain measures. Comput Mech, accepted for publication
32. Simó J.C. (1992) Algorithms for static and dynamic multiplicative plasticity that preserve the classical return mapping schemes of the infinitesimal theory. Comput Methods Appl Mech Eng 99:61–112
33. Simó J.C., Armero F. (1992) Geometrically non-linear enhanced strain mixed methods and the method of incompatible modes. Int J Numer Methods Eng 33:1413–1449
34. Simó J.C., Armero F., Taylor R.L. (1993) Improved version of assumed enhanced strain trilinear elements for 3D finite deformation problems. Comput Methods Appl Mech Eng 110:359–386
35. Simó J.C., Hughes T.J.R. (1998) Computational Inelasticity, 1st edn. Springer, New York
36. Smith G.F., Smith M.M., Rivlin R.S. (1963) Integrity basis for a symmetric tensor and a vector. The crystal classes. Arch Ration Mech Anal 12:93–133
37. Spencer A.J.M. (1971) Theory of invariants. In: Eringen A.C. (Ed.) Continuum Physics Vol. 1, Academic Press, New York. 239–353
38. Taylor R.L. (2000) FEAP - A Finite Element Analysis Program : Users Manual, University of California, Berkeley, http://www.ce.berkeley.edu/rlt
39. Tsakmakis Ch. (2002) Description of plastic anisotropy effects at large deformations. Part I: Restrictions imposed by the second law and the postulate of Il'iushin. Int J Plast, accepted for publication
40. Wagner W., Klinkel S., Gruttmann F. (2002) Elastic and plastic analysis of thin-walled structures using improved hexahedral elements. Comput Struct 80:857-869
41. Wriggers P., Eberlein R., Reese S. (1996) A comparison of three-dimensional continuum and shell elements for finite plasticity. Int J Solids Struct 33:3309–3326
42. Xiao H., Bruhns O.T., Meyers A. (2000) A consistent finite elastoplasticity theory combining additive and multiplicative decomposition of the stretching and the deformation gradient. Int J Plast 16:143–177

Use of the Elastic Predictor–Plastic Corrector Method for Integrating Finite Deformation Plasticity Laws

Charalampos Tsakmakis and Adrian Willuweit

Institute of Mechanics, Darmstadt University of Technology
Hochschulstrasse 1, D–64289, Darmstadt, Germany
tsakmakis@mechanik.tu-darmstadt.de

received 7 Oct 2002 — accepted 21 Nov 2002

Abstract. Classical plasticity theories are formulated by means of ordinary differential equations coupled with algebraic equations, so that the whole system of equations governing the material response is highly nonlinear. To integrate these equations, particular algorithms have been developed, the method of elastic predictor and plastic corrector being often used. This method has turned out to be a very efficient tool, when small deformation plasticity is considered. Especially, the usual constraint condition of plastic incompressibility is preserved exactly. However, when nonlinear geometry is involved, there are several possibilities to employ the method of elastic predictor and plastic corrector. Moreover, some effort has to be made, in order to ensure plastic incompressibility. The exponential map approach is one possibility to overcome the difficulties, but this approach is not suitable when deformation induced anisotropy is considered. The present work addresses the numerical integration of finite deformation plasticity models exhibiting both, isotropic and kinematic hardening. The integration of the evolution equations is based on the elastic predictor and plastic corrector procedure, appropriately adjusted to the structure of the adopted constitutive theory. Plastic incompressibility is preserved by introducing a further unknown into the system of equations to be solved numerically.

1 Introduction - Preliminaries

When integrating small deformation plasticity models the elastic predictor and plastic corrector method is often used. This method has been introduced by KRIEG and KRIEG [12] and WILKINS [31] and was developed further by SIMO and TAYLOR [25]. For nonlinear kinematic hardening of the Armstrong-Frederick type [2] an effective integration scheme has also been proposed by HARTMANN and HAUPT [9]. In either case, plastic incompressibility is automatically preserved.

Incremental finite deformation plasticity was traditionally formulated by using a hypoelastic law to model elasticity. Commonly, the objective time derivative in the hypoelasticity laws has been chosen to be the Jaumann time derivative. An appropriate scheme for integrating such equations was developed by HUGHES and WINGET [11] (see also HUGHES [10]). It has been shown by WEBER [28], that this integration algorithm satisfies the so-called incremental objectivity only in the middle of the time increment. Some interesting issues concerning incremental objectivity are debated also in RUBINSTEIN and ATLURI [19], REED and ATLURI [17], [18] as well as WEBER et al. [30]. As mentioned by SIMO [20] and SIMO and ORTIZ [24], problems arising from incrementally objective integration algorithms may be avoided by using hyperelastic constitutive models instead of hypoelastic ones. Moreover, it has been shown in [20], [21], [22], [23] and [24], how the elastic predictor and plastic corrector integration algorithm may be extended from small to finite deformations.

The present paper is concerned with elastic plastic constitutive models at finite deformations, which exhibit nonlinear isotropic and nonlinear kinematic hardening. Characteristic features in the theory are the multiplicative decomposition of the deformation gradient tensor into elastic and plastic parts, as well as the formulation of the constitutive equations for plasticity relative to the so-called plastic intermediate configuration. Moreover, the yield function is expressed in terms of the Mandel stress tensor. In conformity with this, the back stress tensor is defined to possess the mathematical structure of a Mandel stress tensor as well. The evolution equations governing the hardening response are derived as sufficient conditions for the validity of the dissipation inequality in every admissible process. This way, the obtained plasticity model is thermodynamically consistent, as already demonstrated in [6] and [26]. The aim of this article is, firstly, to show how the elastic predictor and plastic corrector integration scheme may be adjusted to the structure of our constitutive theory. Secondly, two proposals are made for preserving plastic incompressibility during numerical integration. In both cases the set of variables is augmented by a further unknown, an approach, which goes back to LÜHRS, HARTMANN and HAUPT [14]. Finally, it is indicated how the developed integrating algorithm may be implemented in the ABAQUS finite element program.

Throughout the article an explicit reference to space will be dropped. We consider isothermal deformations with uniform distribution of temperature and write $\dot{\varphi}(t)$ for the material time derivative of a function $\varphi(t)$, where t is the time. For real x, $\langle x \rangle$ denotes the function

$$\langle x \rangle = \begin{cases} x & \text{if} \quad x \geq 0 \\ 0 & \text{if} \quad x < 0 \end{cases} . \tag{1}$$

Second-order tensors and vectors, are denoted by bold-face letters, whereas fourth-order tensors are denoted by bold-face calligraphic letters. For second-order tensors $\mathbf{A}$ and $\mathbf{B}$, we write $\operatorname{tr}\mathbf{A}$, $\det\mathbf{A}$ and $\mathbf{A}^T$ for the trace, the determinant and the transpose of $\mathbf{A}$, respectively, while $\mathbf{A}\cdot\mathbf{B} = \operatorname{tr}(\mathbf{A}\mathbf{B}^T)$ is the inner product between $\mathbf{A}$ and $\mathbf{B}$ and $||\mathbf{A}|| = \sqrt{\mathbf{A}\cdot\mathbf{A}}$ is the Euclidean norm of $\mathbf{A}$. Let $\mathbf{a}, \mathbf{b}$ be two vectors. Then, $\mathbf{a} \otimes \mathbf{b}$ denotes the tensor product of these vectors. The identity tensor of second-order is given by

$$\mathbf{1} = \delta_{ij}\, \mathbf{e}_i \otimes \mathbf{e}_j \quad , \tag{2}$$

$i, j = 1, 2, 3$, where δ_{ij} is the Kronecker-delta and $\{\mathbf{e}_i\}$ is an orthonormal basis in the three-dimensional Euclidean vector space. We use the notations $\mathbf{A}^D = \mathbf{A} - \frac{1}{3}(\mathrm{tr}\,\mathbf{A})\mathbf{1}$ for the deviator of $\mathbf{A}$, $\mathbf{A}^{-1}$ for the inverse of $\mathbf{A}$, provided $\mathbf{A}^{-1}$ exists, and $\mathbf{A}^{T-1} = (\mathbf{A}^{-1})^T$. If $\mathbf{A}$ is a symmetric and positive definite second-order tensor, having eigenvalues λ_i and corresponding eigenvectors $\mathbf{a}_i$, then the spectral decomposition $\mathbf{A} = \sum_{i=1}^{3} \lambda_i\, \mathbf{a}_i \otimes \mathbf{a}_i$ applies. In this case, we denote by $\mathbf{A}^m$ the second-order tensor $\mathbf{A}^m = \sum_{i=1}^{3} \lambda_i^m\, \mathbf{a}_i \otimes \mathbf{a}_i$, where m is a real number. Finally, we will denote by A_{ij} or $(\mathbf{A})_{ij}$ the components of a second-order tensor $\mathbf{A}$ with respect to the orthonormal basis $\{\mathbf{e}_i\}$. Similar notations will also be used for fourth-order tensors.

2 Constitutive Theory

2.1 Plasticity with Isotropic and Kinematic Hardening

In [6], [7] and [26] finite deformation plasticity laws, exhibiting nonlinear kinematic and nonlinear isotropic hardening effects, have been proposed. The kinematic hardening rules were introduced in a formal way, by using the internal variable concept of irreversible thermodynamics. It has been proved later in [27], that the kinematic hardening rules of [6], [7], [14], [26] and [29] may alternatively be established in the context of a multiplicative decomposition of the plastic part of the deformation gradient tensor. In the following, we shall present briefly the plasticity theory proposed in [6], [7], [26] and [27]. For the purposes of our work, it suffices to confine ourselves to kinematic hardening rule, which in [27] is referred to as Model C. Thus, the plasticity theory we deal with here is defined by the following equations:

Kinematical relations:

$$\mathbf{F} = \mathbf{F}_e \mathbf{F}_p \quad , \tag{3}$$

$$\mathbf{F}_e = \mathbf{R}_e \hat{\mathbf{U}}_e = \mathbf{V}_e \mathbf{R}_e \quad , \quad \mathbf{R}_e^T = \mathbf{R}_e^{-1} \quad , \tag{4}$$

$$\mathbf{F}_p = \mathbf{R}_p \mathbf{U}_p = \hat{\mathbf{V}}_p \mathbf{R}_p \quad , \quad \mathbf{R}_p^T = \mathbf{R}_p^{-1} \quad , \tag{5}$$

$$\mathbf{L} = \dot{\mathbf{F}}\mathbf{F}^{-1} \quad , \quad \mathbf{D} = \frac{1}{2}(\mathbf{L} + \mathbf{L}^T) \quad , \quad \mathbf{W} = \frac{1}{2}(\mathbf{L} - \mathbf{L}^T) \quad , \tag{6}$$

$$\hat{\mathbf{L}}_p = \dot{\mathbf{F}}_p \mathbf{F}_p^{-1} \quad , \quad \hat{\mathbf{D}}_p = \frac{1}{2}(\hat{\mathbf{L}}_p + \hat{\mathbf{L}}_p^T) \quad , \quad \hat{\mathbf{W}}_p = \frac{1}{2}(\hat{\mathbf{L}}_p - \hat{\mathbf{L}}_p^T) \quad , \tag{7}$$

$$\hat{\mathbf{C}}_e = \mathbf{F}_e^T \mathbf{F}_e = \hat{\mathbf{U}}_e^2 \quad , \quad \hat{\boldsymbol{\Gamma}}_e = \frac{1}{2}(\hat{\mathbf{C}}_e - \mathbf{1}) \quad , \quad \mathbf{B}_e = \mathbf{F}_e \mathbf{F}_e^T = \mathbf{V}_e \quad , \tag{8}$$

$$\hat{\mathbf{B}}_p = \mathbf{F}_p \mathbf{F}_p^T = \hat{\mathbf{V}}_p^2 \quad , \quad \hat{\boldsymbol{\Gamma}}_p = \frac{1}{2}(\mathbf{1} - \hat{\mathbf{B}}_p^{-1}) \quad , \tag{9}$$

$$\hat{\boldsymbol{\Gamma}} = \hat{\boldsymbol{\Gamma}}_e + \hat{\boldsymbol{\Gamma}}_p \quad . \tag{10}$$

Specific free energy:

$$\Psi = \Psi_e + \Psi_p^{(is)} + \Psi_p^{(kin)} \quad . \tag{11}$$

Elasticity law:

$$\mathbf{S} = (\det \mathbf{F})\mathbf{T} = \mathbf{F}_e \hat{\mathbf{T}} \mathbf{F}_e^T \tag{12}$$

$$\hat{\mathbf{T}} = \rho_0 \frac{\partial \hat{\Psi}_e(\hat{\boldsymbol{\Gamma}}_e)}{\partial \hat{\boldsymbol{\Gamma}}_e} \quad , \tag{13}$$

$$\Psi_e = \hat{\Psi}_e(\hat{\boldsymbol{\Gamma}}_e) : \text{isotropic tensor function.} \tag{14}$$

Yield function:

$$\hat{\mathbf{P}} = (\mathbf{1} + 2\hat{\boldsymbol{\Gamma}}_e)\hat{\mathbf{T}} \quad , \tag{15}$$

$$f = \hat{f}(\hat{\mathbf{P}}, \hat{\boldsymbol{\xi}}) := \sqrt{\frac{3}{2}(\hat{\mathbf{P}} - \hat{\boldsymbol{\xi}})^D \cdot (\hat{\mathbf{P}} - \hat{\boldsymbol{\xi}})^D} \quad , \tag{16}$$

$$F(t) = \hat{F}(\hat{\mathbf{P}}, \hat{\boldsymbol{\xi}}, k) = \hat{f}(\hat{\mathbf{P}}, \hat{\boldsymbol{\xi}}) - R - k_0 \quad , \tag{17}$$

$$\hat{\boldsymbol{\xi}} : \text{back-stress tensor,} \tag{18}$$

$$R : \text{isotropic hardening} \quad . \tag{19}$$

Flow rule:

$$\hat{\mathbf{D}}_p = \dot{s} \frac{\partial \hat{F}}{\partial \hat{\mathbf{P}}} = \dot{s} \sqrt{\frac{3}{2}} \frac{(\hat{\mathbf{P}} - \hat{\boldsymbol{\xi}})^D}{\|(\hat{\mathbf{P}} - \hat{\boldsymbol{\xi}})^D\|} \quad , \tag{20}$$

$$\dot{s} = \sqrt{\frac{2}{3} \hat{\mathbf{D}}_p \cdot \hat{\mathbf{D}}_p} \quad . \tag{21}$$

Plasticity:

$$F = 0 : \text{ yield condition} \quad , \tag{22}$$

$$\dot{s} \begin{cases} > 0 & \text{for} \quad F = 0 \;\&\; (\dot{F})_{\mathbf{F}_p = \text{const}} > 0 \\ = 0 & \text{otherwise} \end{cases} \quad . \tag{23}$$

Viscoplasticity:

$$F : \text{ overstress,} \tag{24}$$

$$\dot{s} = \frac{\langle F \rangle^m}{\eta} \geq 0 \quad . \tag{25}$$

Isotropic hardening law:

$$\Psi_p^{(is)} = \hat{\Psi}_p^{(is)}(r) = \frac{\gamma}{2} r^2 \quad , \quad R = \rho_0 \frac{\partial \Psi_p^{(is)}}{\partial r} = \gamma r \quad , \tag{26}$$

$$\dot{r} = (1 - \beta r)\dot{s} \quad \Longleftrightarrow \quad \dot{R} = (\gamma - \beta R)\dot{s} \quad . \tag{27}$$

Kinematic hardening law:

$$\boldsymbol{\xi} = (\mathbf{1} + 2\hat{\mathbf{Y}})\hat{\mathbf{Z}} \quad , \tag{28}$$

$$\Psi_p^{(kin)} = \hat{\Psi}_p^{(kin)}(\hat{\mathbf{Y}}) = \frac{c}{2} \hat{\mathbf{Y}} \cdot \hat{\mathbf{Y}} \quad , \quad \hat{\mathbf{Z}} = \rho_0 \frac{\partial \hat{\Psi}_p^{(kin)}}{\partial \hat{\mathbf{Y}}} = c\hat{\mathbf{Y}} \quad , \tag{29}$$

$$\overset{\triangledown}{\hat{\mathbf{Y}}} = \dot{\hat{\mathbf{Y}}} - \hat{\mathbf{L}}_p \hat{\mathbf{Y}} - \hat{\mathbf{Y}} \hat{\mathbf{L}}_p^T = \hat{\mathbf{D}}_p - \dot{s} b \hat{\mathbf{Y}} \quad \Longleftrightarrow \tag{30}$$

$$\overset{\triangledown}{\hat{\mathbf{Z}}} = \dot{\hat{\mathbf{Z}}} - \hat{\mathbf{L}}_p \hat{\mathbf{Z}} - \hat{\mathbf{Z}} \hat{\mathbf{L}}_p^T = c\hat{\mathbf{D}}_p - \dot{s}\, c\, b\, \hat{\mathbf{Z}} \quad . \tag{31}$$

In these relations, $\mathbf{F}$ is the deformation gradient tensor ($\det \mathbf{F} > 0$), $\mathbf{F} = \partial \bar{\mathbf{x}} / \partial \mathbf{X} = \mathrm{GRAD}\, \bar{\mathbf{x}}$, with $\mathbf{x} = \bar{\mathbf{x}}(\mathbf{X}, t)$ describing the motion of the material body $\mathcal{B}$ in the three-dimensional Euclidean point space E. $\mathbf{X}$ and $\mathbf{x}$ are the position vectors to the places occupied by one and the same particle in the reference configuration (at time $t = 0$) and the current configuration (at time t), respectively. The body occupies the region R_R in the reference configuration and the region R_t in the current configuration. As usually, we shall utilize R_R, R_t in order to denote the regions in E and at the same time the corresponding configuration maps.

Equation (3) represents the multiplicative decomposition of $\mathbf{F}$ into elastic and visco-plastic parts. (References dealing with the multiplicative decomposition of $\mathbf{F}$ can be found e.g. in Lubliner [13] or Maugin [15]). Since it is assumed that $\det \mathbf{F} > 0$, and because of the plastic incompressibility $\det \mathbf{F}_p = 1$, postulated by $\mathrm{tr}\, \hat{\mathbf{D}}_p = 0$ in (20), we have $\det \mathbf{F}_e > 0$, so that the polar decompositions (4), (5) apply. It is important to notice, that the transformation $\mathbf{F}_p$ introduces a so-called plastic intermediate configuration denoted by $\hat{\mathsf{R}}_t$. Thus, $\hat{\boldsymbol{\Gamma}}_e$ defined by (8) and $\hat{\mathbf{T}}$ defined by (13) are strain and stress tensors, respectively, acting in tangent spaces of material points in the plastic intermediate configuration $\hat{\mathsf{R}}_t$. In (12) and (15), $\mathbf{T}$, $\mathbf{S}$ and $\hat{\mathbf{P}}$ are the Cauchy, the weighted Cauchy and the Mandel stress tensors, respectively. Since an isotropic elasticity law is assumed to hold (see (13), (14)), the commutativity relation $\hat{\boldsymbol{\Gamma}}_e \hat{\mathbf{T}} = \hat{\mathbf{T}} \hat{\boldsymbol{\Gamma}}_e$ applies and hence the stress tensor $\hat{\mathbf{P}}$ is symmetric. Eq. (17) defines a v. Mises yield function with kinematic and isotropic hardening, expressed in terms of Mandel stress tensors. That means, the back-stress tensor $\hat{\boldsymbol{\xi}}$ is postulated to have the mathematical structure of the Mandel stress. (More details about back-stress tensors of Mandel type may be found in [27]). Equation (20) represents the flow rule (associated normality rule), formulated relative to $\hat{\mathsf{R}}_t$. The yield function may be employed either to describe rate-independent (plastic) or rate-dependent (viscoplastic) material behaviour. In the first case, F serves to formulate the yield condition, as well as to determine $\dot{s}$ from the consistency condition $\dot{F} = 0$. In the second case, F represents an overstress, which is used to define $\dot{s}$ directly by means of the constitutive relation (25). Isotropic hardening is introduced by (26), (27). The kinematic hardening law, composed of equations (28)-(31), has been proposed in [6], [7], [26] and [27] as a possible generalization of the Armstrong-Frederick hardening law [2] from small to finite deformations. For arbitrary second-order tensors $\hat{\mathbf{X}}$ with respect to $\hat{\mathsf{R}}_t$, the definitions

$$\overset{\triangledown}{\hat{\mathbf{X}}} := \dot{\hat{\mathbf{X}}} - \hat{\mathbf{L}}_p\hat{\mathbf{X}} - \hat{\mathbf{X}}\hat{\mathbf{L}}_p^T \quad , \tag{32}$$

$$\overset{\vartriangle}{\hat{\mathbf{X}}} := \dot{\hat{\mathbf{X}}} + \hat{\mathbf{L}}_p^T\hat{\mathbf{X}} + \hat{\mathbf{X}}\hat{\mathbf{L}}_p \quad , \tag{33}$$

represent Oldroyd derivatives relative to $\hat{\mathsf{R}}_t$. Finally, the quantities k_0, m, β, γ, c, b are nonnegative material parameters.

It is perhaps of interest to mention that the hardening laws (26)-(31) have been introduced in [6], [7], [26] and [27] as sufficient conditions for the dissipation inequality. The whole model is generally established in a thermodynamically consistent way, i.e., the second law of thermodynamics is fulfilled in every admissible process. Additionally, the constitutive theory is formulated with respect to the plastic intermediate configuration $\hat{\mathsf{R}}_t$ and satisfies the invariance requirements under superposed rigid body motions demanded by GREEN and NAGHDI [8] (see also CASEY and NAGHDI [3]) for both the plastic intermediate configuration and the current configuration. Plastic spin concepts are not necessary here, since isotropy is assumed to apply for the yield function F and the specific free energy functions $\hat{\Psi}_e$, $\hat{\Psi}_p^{(is)}$, $\hat{\Psi}_p^{(kin)}$. As shown in [3], plastic spin concepts are needed only when anisotropy effects, other than the Bauschinger effect, are present in the constitutive behaviour. (For the notion of plastic spin see, e.g., DAFALIAS [4] and DAFALIAS and AIFANTIS [5]).

2.2 Eulerian Form of the Constitutive Equations

To implement the constitutive model (3)-(31) into finite element codes like ABAQUS we have to transform all equations appropriately into the current configuration. It is important to note that all physical aspects are already incorporated in (3)-(31), which are formulated relative to $\hat{\mathsf{R}}_t$. Therefore, the transformation formulas with respect to R_t will have formal character only and may be defined on the basis of reasons of convenience. To elaborate, we introduce the Almansi strain tensor $\mathbf{A}$ with respect to R_t,

$$\mathbf{A} = \frac{1}{2}(\mathbf{1} - \mathbf{F}^{T-1}\mathbf{F}^{-1}) = \mathbf{F}_e^{T-1}\hat{\boldsymbol{\Gamma}}\mathbf{F}_e^{-1} \quad , \tag{34}$$

so that

$$\mathbf{A} = \mathbf{A}_e + \mathbf{A}_p \quad , \tag{35}$$

$$\mathbf{A}_e = \frac{1}{2}(\mathbf{1} - \mathbf{B}_e^{-1}) = \mathbf{F}_e^{T-1}\hat{\boldsymbol{\Gamma}}_e\mathbf{F}_e^{-1} \quad , \tag{36}$$

$$\mathbf{A}_p = \mathbf{F}_e^{T-1}\hat{\boldsymbol{\Gamma}}_p\mathbf{F}_e^{-1} \quad . \tag{37}$$

Equation (35) is the Eulerian counterpart of (10) we will deal with. Because of the assumed elastic isotropy, we have

$$\hat{\Psi}_e(\hat{\boldsymbol{\Gamma}}_e) = \bar{\Psi}(\mathbf{B}_e) \quad . \tag{38}$$

After standard calculations, the elasticity law (13) may be rewritten, with respect to R_t, as

$$\mathbf{S} = 2\,\rho_0\,\mathbf{B}_e\frac{\partial\bar{\Psi}}{\partial\mathbf{B}_e} \quad . \tag{39}$$

Similar to (32), (33), we define for an arbitrary Eulerian second-order tensor **X** the Oldroyd derivatives

$$\overset{\triangledown}{\mathbf{X}} := \dot{\mathbf{X}} - \mathbf{L}\mathbf{X} - \mathbf{X}\mathbf{L}^T \quad , \tag{40}$$

$$\overset{\triangle}{\mathbf{X}} := \dot{\mathbf{X}} + \mathbf{L}^T + \mathbf{X}\mathbf{L} \quad , \tag{41}$$

with respect to R_t. Then, after some algebraic manipulations we may gain the results

$$\overset{\triangle}{\mathbf{A}} = \mathbf{D} = \mathbf{F}_e^{T-1}\overset{\triangle}{\hat{\boldsymbol{\Gamma}}}\mathbf{F}_e^{-1} = \overset{\triangle}{\mathbf{A}}_e + \overset{\triangle}{\mathbf{A}}_p \quad , \tag{42}$$

$$\overset{\triangle}{\mathbf{A}}_e = \mathbf{F}_e^{T-1}\overset{\triangle}{\hat{\boldsymbol{\Gamma}}}_e\mathbf{F}_e^{-1} \quad , \tag{43}$$

$$\overset{\triangle}{\mathbf{A}}_p = \mathbf{F}_e^{T-1}\overset{\triangle}{\hat{\boldsymbol{\Gamma}}}_p\mathbf{F}_e^{-1} \quad , \quad \overset{\triangle}{\hat{\boldsymbol{\Gamma}}}_p = \hat{\mathbf{D}}_p \quad . \tag{44}$$

¿From (15), we see that

$$\mathbf{B}_e\mathbf{S} = \mathbf{F}_e\hat{\mathbf{P}}\mathbf{F}_e^T \quad . \tag{45}$$

We regard $\mathbf{B}_e\mathbf{S}$ as an Eulerian counterpart of $\hat{\mathbf{P}}$. If one denotes by $\boldsymbol{\xi}$ an Eulerian counterpart of $\hat{\boldsymbol{\xi}}$, which transforms according to (45), then

$$\boldsymbol{\xi} = \mathbf{F}_e\hat{\boldsymbol{\xi}}\mathbf{F}_e^T \quad , \tag{46}$$

and the yield-function (17) may be recast as

$$F = f - R - k_0 \quad , \tag{47}$$

$$f = \sqrt{\frac{3}{2}(\mathbf{S} - \mathbf{B}_e^{-1}\boldsymbol{\xi})^D \cdot (\mathbf{S} - \mathbf{B}_e^{-1}\boldsymbol{\xi})^D} \quad . \tag{48}$$

Furthermore, the flow rule (20) becomes

$$\overset{\triangle}{\mathbf{A}}_p = \frac{3\dot{s}}{2f}(\mathbf{S} - \mathbf{B}_e^{-1}\boldsymbol{\xi})^D \mathbf{B}_e^{-1} \quad . \tag{49}$$

Numerical calculations indicate that the iterative established solutions based on the hardening law (31), which is an evolution equation for $\hat{\mathbf{Z}}$, converge faster than iterative established solutions related to the evolution equation (30) for $\hat{\mathbf{Y}}$. Therefore, we refer the analysis on (31) and define the tensors

$$\mathbf{Z} = \mathbf{F}_e \hat{\mathbf{Z}} \mathbf{F}_e^T \quad , \quad \mathbf{Y} = \mathbf{F}_e^{T-1} \hat{\mathbf{Y}} \mathbf{F}_e^{-1} \quad , \tag{50}$$

relative to R_t, so that

$$\overset{\triangledown}{\mathbf{Z}} = \mathbf{F}_e \overset{\triangledown}{\hat{\mathbf{Z}}} \mathbf{F}_e^T \quad , \quad \boldsymbol{\xi} = (\mathbf{1} + 2\mathbf{B}_e \mathbf{Y})\mathbf{Z} \quad . \tag{51}$$

It follows then from (29), (30) that

$$\mathbf{Z} = c\mathbf{B}_e \mathbf{Y} \mathbf{B}_e \quad , \tag{52}$$

$$\overset{\triangledown}{\mathbf{Z}} = c\mathbf{B}_e \overset{\triangle}{\mathbf{A}}_p \mathbf{B}_e - \dot{s}bc\mathbf{Z} \quad . \tag{53}$$

Summarizing, the equations to be solved are

$$\mathbf{A} = \frac{1}{2}(\mathbf{1} - \mathbf{F}^{T-1}\mathbf{F}^{-1}) = \mathbf{A}_e + \mathbf{A}_p \quad , \tag{54}$$

$$\mathbf{S} = 2\rho_0 \mathbf{B}_e \frac{\partial \bar{\Psi}}{\partial \mathbf{B}_e} \quad , \tag{55}$$

$$F = f - R - k_0 \quad , \tag{56}$$

$$f = \sqrt{\frac{3}{2}(\mathbf{S} - \mathbf{B}_e^{-1}\boldsymbol{\xi})^D \cdot (\mathbf{S} - \mathbf{B}_e^{-1}\boldsymbol{\xi})^D} \quad , \tag{57}$$

$$\overset{\triangle}{\mathbf{A}}_p = \dot{\mathbf{A}}_p + \mathbf{L}^T \mathbf{A}_p + \mathbf{A}\mathbf{L} = \frac{3\dot{s}}{2f}(\mathbf{S} - \mathbf{B}_e^{-1}\boldsymbol{\xi})^D \mathbf{B}_e^{-1} \quad , \tag{58}$$

$$\boldsymbol{\xi} = (\mathbf{1} + 2\mathbf{B}_e \mathbf{Y})\mathbf{Z} \quad , \tag{59}$$

$$\mathbf{Y} = \frac{1}{c}\mathbf{B}_e^{-1} \mathbf{Z} \mathbf{B}_e^{-1} \quad , \tag{60}$$

$$\dot{R} = (\gamma - \beta R)\dot{s} \quad , \tag{61}$$

$$\overset{\triangledown}{\mathbf{Z}} = \dot{\mathbf{Z}} - \mathbf{L}\mathbf{Z} - \mathbf{Z}\mathbf{L}^T = c\mathbf{B}_e \overset{\triangle}{\mathbf{A}}_p \mathbf{B}_e - \dot{s}bc\mathbf{Z} \quad , \tag{62}$$

Plasticity:

$\dot{s}$ has to be determined from $\dot{F} = 0$ for plastic loading (63)

Viscoplasticity:

$$\dot{s} = \frac{\langle F \rangle^m}{\eta} \quad . \tag{64}$$

It is worth noticing, that the elastic deformations, and therefore the elastic strains as well, are not subjected to some restrictions, i.e., they may be arbitrarily large. Also, any hyperelasticity law for the stress tensor can be used in (12). However, for definiteness, in the sequel we shall be concerned with the relation

$$\mathbf{S} = \mu(\mathbf{B}_e - \mathbf{1}) + \frac{\lambda}{2}\{\ln(\det \mathbf{B}_e)\}\mathbf{1} \quad , \tag{65}$$

where μ and λ are elasticity constants. Eq. (65) represents a compressible Neo-Hookean type hyperelasticity law, which has been proposed by Simo and Ortiz [24].

3 Time Integration

An operator split algorithm according to [24] will be employed. From (54)-(65), the pertinent differential equations to be solved are (58), (61) and (62), i.e.,

$$\dot{\mathbf{A}}_p = \underbrace{-\mathbf{L}^T\mathbf{A}_p - \mathbf{A}_p\mathbf{L}}_{:={}^I\mathbf{G}_p} + \underbrace{\dot{s}\left\{\frac{3}{2f}(\mathbf{S} - \mathbf{B}_e^{-1}\boldsymbol{\xi})\mathbf{B}_e^{-1}\right\}}_{:={}^{II}\mathbf{G}_p} = {}^I\mathbf{G}_p + {}^{II}\mathbf{G}_p \quad , \tag{66}$$

$$\dot{R} = \underbrace{0}_{{}^I G_R} + \underbrace{\dot{s}(\gamma - \beta R)}_{{}^{II}G_R} = {}^I G_R + {}^{II} G_R \quad , \tag{67}$$

$$\dot{\mathbf{Z}} = \underbrace{\mathbf{L}\mathbf{Z} + \mathbf{Z}\mathbf{L}^T}_{:={}^I\mathbf{G}_Z} + \underbrace{\dot{s}\left\{\frac{3c}{2f}\mathbf{B}_e(\mathbf{S} - \mathbf{B}_e^{-1}\boldsymbol{\xi})^D - bc\mathbf{Z}\right\}}_{{}^{II}\mathbf{G}_Z} = {}^I\mathbf{G}_Z + {}^{II}\mathbf{G}_Z \quad . \tag{68}$$

Finite element programs (like ABAQUS), which rely on the principle of virtual work, require to solve a deformation controlled constitutive problem (see [10], [29] and [30]). In other words, restricting attention to quasi-static problems, we assume that at time t the material state is known and that the body $\mathcal{B}$ occupies the region R_t in E. The known values of quantities at time t are indicated by an index (0). This way, we denote by $\mathbf{u}^{(0)} = \mathbf{u}(\mathbf{X}, t) = \mathbf{x}^{(0)} - \mathbf{X}$ the displacement vector of the material point with position vector $\mathbf{X}$ in R_R. For this material point, the finite element procedure generates in a time increment Δt the displacement increment $\Delta\mathbf{u}$. Now, the problem consists in finding the material state (i.e. the value of all variables appearing in the constitutive equations) at time $t+\Delta t$. This means, a so-called two-point, deformation driven problem must be solved. The region in E occupied by the material body $\mathcal{B}$ at time $t+\Delta t$ is denoted by $\mathsf{R}_{t+\Delta t}$. Values of quantities at time $t+\Delta t$ will be indicated by an index (1). Hence, for the material point with position vector $\mathbf{X}$ in R_R, the relation $\Delta\mathbf{u} = \mathbf{x}^{(1)} - \mathbf{x}^{(0)}$ applies. Moreover, we use the notations

$$\mathbf{F}^{(0)} := \frac{\partial \mathbf{x}^{(0)}}{\partial \mathbf{X}} \quad , \quad \mathbf{F}^{(1)} := \frac{\partial \mathbf{x}^{(1)}}{\partial \mathbf{X}} \quad , \quad \Delta\mathbf{F} := \frac{\partial \mathbf{x}^{(1)}}{\partial \mathbf{x}^{(0)}} = \frac{\partial \Delta\mathbf{u}}{\partial \mathbf{x}^{(0)}} + \mathbf{1} \quad , \tag{69}$$

so that

$$\mathbf{F}^{(1)} = (\Delta\mathbf{F})\mathbf{F}^{(0)}. \tag{70}$$

These tensors are provided by the ABAQUS UMAT-subroutine, so they can be considered to be known at every time. Let $\zeta \in [t, t+\Delta t]$ and assume the deformation path between $\mathbf{x}^{(0)}$ and $\mathbf{x}^{(1)}$ to be given by $\mathbf{x} = \tilde{\mathbf{x}}(\mathbf{x}^{(0)}, \zeta)$. Evidently, $\mathbf{x}^{(0)} = \tilde{\mathbf{x}}(\mathbf{x}^{(0)}, t)$ and $\mathbf{x}^{(1)} = \tilde{\mathbf{x}}(\mathbf{x}^{(0)}, t+\Delta t)$. We set

$$\tilde{\mathbf{F}}(\mathbf{x}^{(0)}, \zeta) := \frac{\partial \tilde{\mathbf{x}}(\mathbf{x}^{(0)}, \zeta)}{\partial \mathbf{x}^{(0)}} \quad , \tag{71}$$

$$\mathbf{v} = \tilde{\mathbf{v}}(\mathbf{x}^{(0)}, \zeta) = \frac{d}{d\zeta}\tilde{\mathbf{x}}(\mathbf{x}^{(0)}, \zeta) \quad , \tag{72}$$

$$\mathbf{L} = \tilde{\mathbf{L}}(\mathbf{x}^{(0)}, \zeta) = \frac{\partial \tilde{\mathbf{v}}(\mathbf{x}^{(0)}, \zeta)}{\partial \mathbf{x}} = \left(\frac{d}{d\zeta}\tilde{\mathbf{F}}(\mathbf{x}^{(0)}, \zeta)\right)\left(\tilde{\mathbf{F}}(\mathbf{x}^{(0)}, \zeta)\right)^{-1} \quad , \tag{73}$$

so that $\tilde{\mathbf{F}}(\mathbf{x}^{(0)}, t) = \mathbf{1}$ and $\tilde{\mathbf{F}}(\mathbf{x}^{(0)}, t+\Delta t) = \Delta\mathbf{F}$. Next, we set $\mathbf{x}^{(0)}$ fixed and express all field quantities as functions of time ζ only. Preparatory to the integration approach to be presented below, we define the bar-transformed tensors $\bar{\mathbf{A}}(\zeta)$ and $\bar{\mathbf{Z}}(\zeta)$ by

$$\bar{\mathbf{A}}_p(\zeta) = \tilde{\mathbf{F}}^T(\zeta)\mathbf{A}_p(\zeta)\tilde{\mathbf{F}}(\zeta) \quad , \tag{74}$$

$$\bar{\mathbf{Z}}(\zeta) = \tilde{\mathbf{F}}^{-1}(\zeta)\mathbf{Z}(\zeta)\tilde{\mathbf{F}}^{T-1}(\zeta) \quad , \tag{75}$$

which are counterparts relative to R_t of the tensors $\bar{\mathbf{A}}(\zeta)$ and $\bar{\mathbf{Z}}(\zeta)$. The differential equations (66)-(68) can be integrated by making use of a two-part operator split algorithm according to [24]. This means, we split the right-hand sides of $\dot{\mathbf{A}}$, $\dot{R}$ and $\dot{\mathbf{Z}}$ additively as indicated in (66), (67) and (68).

Then, we integrate the differential equations (part I) $\dot{\mathbf{A}}_p =^I \mathbf{G}_p$, $\dot{R} = 0$ and $\dot{\mathbf{Z}} =^I \mathbf{G}_Z$ by using the given initial conditions. The results of this integration serve as initial conditions for part II of the algorithm, which consists of of the differential equations $\dot{\mathbf{A}}_p =^{II} \mathbf{G}_p$, $\dot{R} =^{II} G_R$ and $\dot{\mathbf{Z}} =^{II} \mathbf{G}_Z$.

The results of the second integration are the solution of the whole problem. To be more specific, in part I we have to solve the differential equations

$$\frac{\mathrm{d}\mathbf{A}_p}{\mathrm{d}\zeta} =^I \mathbf{G}_p = -\mathbf{L}^T\mathbf{A}_p - \mathbf{A}_p\mathbf{L} \quad , \tag{76}$$

$$\frac{\mathrm{d}R}{\mathrm{d}\zeta} = {}^I G_R = 0 \quad , \tag{77}$$

$$\frac{\mathrm{d}\mathbf{Z}}{\mathrm{d}\zeta} =^I \mathbf{G}_Z = \mathbf{L}\mathbf{Z} + \mathbf{Z}\mathbf{L}^T \quad . \tag{78}$$

Obviously, $\dot{s} = 0$ holds during part I. This is why part I is called elastic predictor or trial elastic part. The differential equations to be solved in to part II are

$$\frac{\mathrm{d}\mathbf{A}_p}{\mathrm{d}\zeta} =^{II} \mathbf{G}_p = \dot{s}\frac{3}{2f}(\mathbf{S} - \mathbf{B}_e^{-1}\boldsymbol{\xi})^D\mathbf{B}_e^{-1} \quad , \tag{79}$$

$$\frac{\mathrm{d}R}{\mathrm{d}\zeta} =^{II} G_R = \dot{s}(\gamma - \beta R) \quad , \tag{80}$$

$$\frac{\mathrm{d}\mathbf{Z}}{\mathrm{d}\zeta} =^{II} \mathbf{G}_Z = \dot{s}\left\{\frac{3c}{2f}\mathbf{B}_e(\mathbf{S} - \mathbf{B}_e^{-1}\boldsymbol{\xi})^D - bc\mathbf{Z}\right\} \quad , \tag{81}$$

and are called the plastic part or the plastic corrector.

3.1 Part I: Elastic Predictor

¿From (76)-(78), the equations

$$\frac{\mathrm{d}\mathbf{A}_p}{\mathrm{d}\zeta} + \mathbf{L}^T\mathbf{A}_p + \mathbf{A}_p\mathbf{L} = \mathbf{0} \quad , \tag{82}$$

$$\frac{\mathrm{d}R}{\mathrm{d}\zeta} = 0 \quad , \tag{83}$$

$$\frac{\mathrm{d}\mathbf{Z}}{\mathrm{d}\zeta} - \mathbf{L}\mathbf{Z} - \mathbf{Z}\mathbf{L}^T = \mathbf{0} \tag{84}$$

have to be solved in part I, which, after transforming them back to the configuration R_t with the aid of (74), (75), are equivalent to

$$\frac{\mathrm{d}}{\mathrm{d}\zeta}\bar{\mathbf{A}}_p(\zeta) = \mathbf{0} \quad , \quad \frac{\mathrm{d}}{\mathrm{d}\zeta}R(\zeta) = 0 \quad , \quad \frac{\mathrm{d}}{\mathrm{d}\zeta}\bar{\mathbf{Z}}(\zeta) = \mathbf{0} \quad . \tag{85}$$

The initial conditions for these equations are

$$\bar{\mathbf{A}}_p(\zeta = t) = \mathbf{A}_p^{(0)} \quad , \quad R(\zeta = t) = R^{(0)} \quad , \quad \bar{\mathbf{Z}}(\zeta = t) = \mathbf{Z}^{(0)} \quad . \tag{86}$$

For $\zeta = t + \Delta t$, it follows from (85), (86) that

$$\bar{\mathbf{A}}_p^{(1)} = \mathbf{A}_p^{(0)} \quad , \quad R^{(1)} = R^{(0)} \quad , \quad \bar{\mathbf{Z}}^{(1)} = \mathbf{Z}^{(0)} \quad . \tag{87}$$

If we take into account that

$$\bar{\mathbf{A}}_p^{(1)} = (\Delta\mathbf{F})^T \mathbf{A}_p^{(1)} (\Delta\mathbf{F}) \quad , \quad \bar{\mathbf{Z}}^{(1)} = (\Delta\mathbf{F})^{-1} \mathbf{Z}^{(1)} (\Delta\mathbf{F})^{T-1} \quad , \tag{88}$$

(cf. (74), (75)), then we conclude that

$$\begin{aligned} {}^I\mathbf{A}_p{}^{(1)} &= (\Delta\mathbf{F})^{T-1} \mathbf{A}_p^{(0)} (\Delta\mathbf{F})^{-1} \quad , \\ {}^I R^{(1)} &= R^{(0)} \quad , \\ {}^I\mathbf{Z}^{(1)} &= (\Delta\mathbf{F}) \mathbf{Z}^{(0)} (\Delta\mathbf{F})^T \quad , \end{aligned} \tag{89}$$

where ${}^I(\cdot)$ denotes the quantity $(\cdot)$ as evaluated in part I.

The next step is to calculate the values ${}^I\mathbf{S}^{(1)}$ and ${}^I\boldsymbol{\xi}^{(1)}$. To this end we have to determine the value of ${}^I\mathbf{B}_e{}^{(1)}$. Obviously (34)-(36) imply that

$$\mathbf{B}_e^{-1}(\zeta) = \mathbf{1} - 2(\mathbf{A}(\zeta) - \mathbf{A}_p(\zeta)) = \mathbf{F}^{T-1}(\zeta)\mathbf{F}^{-1}(\zeta) + 2\mathbf{A}_p(\zeta) \quad , \tag{90}$$

and hence

$$\left(\mathbf{B}_e^{(0)}\right)^{-1} = \left(\mathbf{F}^{(0)}\right)^{T-1} \left(\mathbf{F}^{(0)}\right)^{-1} + 2\mathbf{A}_p^{(0)} \quad , \tag{91}$$

$$\left(\mathbf{B}_e^{(1)}\right)^{-1} = \left(\mathbf{F}^{(1)}\right)^{T-1} \left(\mathbf{F}^{(1)}\right)^{-1} + 2\mathbf{A}_p^{(1)} \quad . \tag{92}$$

Keeping in mind (70) and $(89)_1$, we deduce from (92)

$$\left(\mathbf{B}_e^{(1)}\right)^{-1} = (\Delta\mathbf{F})^{T-1} \left\{ \left(\mathbf{F}^{(0)}\right)^{T-1} \left(\mathbf{F}^{(0)}\right)^{-1} + 2\mathbf{A}_p^{(0)} \right\} (\Delta\mathbf{F})^{-1} \quad . \tag{93}$$

The latter, together with (91), yields the value of $\mathbf{B}_e$ in part I, at time $t + \Delta t$:

$$\left({}^I\mathbf{B}_e{}^{(1)}\right)^{-1} = (\Delta\mathbf{F})^{T-1} \left(\mathbf{B}_e^{(0)}\right)^{-1} (\Delta\mathbf{F})^{-1} \quad . \tag{94}$$

Alternatively, and equivalently, (94) may be derived from (44), (82). Actually, these imply $\overset{\triangle}{\mathbf{A}}_p = \mathbf{F}_e^{T-1} \hat{\mathbf{D}}_p \mathbf{F}_e^{-1} = \mathbf{0}$ in part I, and therefore

$$\hat{\mathbf{D}}_p(\zeta) = \frac{1}{2} \left(\hat{\mathbf{L}}_p(\zeta) + \hat{\mathbf{L}}_p^T(\zeta) \right) = \mathbf{0} \quad \Longleftrightarrow \quad \frac{\mathrm{d}}{\mathrm{d}\zeta} s(\zeta) = 0 \quad , \tag{95}$$

where (21) has been taken into account. Eq. $(95)_2$ furnishes for part I

$${}^I s^{(1)} = s^{(0)} = \text{const} \quad . \tag{96}$$

We may use (7), (5) to obtain

$$\hat{\mathbf{L}}_p(\zeta) = \left(\frac{\mathrm{d}}{\mathrm{d}\zeta}\mathbf{R}_p(\zeta)\right)\mathbf{R}_p(\zeta) + \mathbf{R}_p(\zeta)\left(\frac{\mathrm{d}}{\mathrm{d}\zeta}\mathbf{U}_p(\zeta)\right)\mathbf{U}_p^{-1}(\zeta)\mathbf{R}_p^T(\zeta) \quad . \tag{97}$$

Since $\left(\frac{\mathrm{d}}{\mathrm{d}\zeta}\mathbf{R}_p(\zeta)\right)\mathbf{R}_p(\zeta)$ is skew symmetric, (97) together with (95) furnish $\frac{\mathrm{d}}{\mathrm{d}\zeta}\mathbf{U}_p(\zeta) = \mathbf{0}$, or $\mathbf{U}_p(\zeta) = \mathbf{U}_p^{(0)} = \text{const}$. Thus, we have

$$\mathbf{F}(\zeta) = \mathbf{F}_e(\zeta)\mathbf{F}_p(\zeta) = \mathbf{F}_e(\zeta)\mathbf{R}_p(\zeta)\mathbf{U}_p^{(0)} \quad . \tag{98}$$

This equation states that $\mathbf{R}_p(\zeta)$, and therefore $\mathbf{F}_p(\zeta)$ too, are not necessarily constant in part I, which differs from other approaches that assume $\mathbf{F}_p =$ const in part I.

We see from the definition $(69)_1$ for $\mathbf{F}^{(0)}$ and the definition (71) for $\tilde{\mathbf{F}}(\zeta)$ that

$$\mathbf{F}(\zeta) = \tilde{\mathbf{F}}(\zeta)\mathbf{F}^{(0)} \quad , \tag{99}$$

and hence

$$\mathbf{F}(\zeta) = \tilde{\mathbf{F}}(\zeta)\mathbf{F}_e^{(0)}\mathbf{F}_p^{(0)} = \tilde{\mathbf{F}}(\zeta)\mathbf{F}_e^{(0)}\mathbf{R}_p^{(0)}\mathbf{U}_p^{(0)} \quad . \tag{100}$$

Combining (100) with (98) implies

$$\mathbf{F}_e(\zeta) = \tilde{\mathbf{F}}(\zeta)\mathbf{F}_e^{(0)}\mathbf{R}_p^{(0)}\mathbf{R}_p^T(\zeta) \quad . \tag{101}$$

For $\zeta = t + \Delta t$ we have $\tilde{\mathbf{F}}(t+\Delta t) = \Delta\mathbf{F}$, and (101) reduces to

$$\mathbf{F}_e^{(1)} = \Delta\mathbf{F}\mathbf{F}_e^{(0)}\left\{\mathbf{R}_p^{(0)}(\mathbf{R}_p^{(1)})^T\right\} \quad . \tag{102}$$

Eq. (102) leads to (94), if one takes into account $(8)_3$ and that $\mathbf{R}_p^{(0)}(\mathbf{R}_p^{(1)})^T$ is an orthogonal second-order tensor.

Having established ${}^I\mathbf{B}_e{}^{(1)}$, the value of ${}^I\mathbf{S}^{(1)}$ may be calculated from (65):

$${}^I\mathbf{S}^{(1)} = \mu\left({}^I\mathbf{B}_e{}^{(1)} - \mathbf{1}\right) + \frac{\lambda}{2}\left\{\ln\left(\det{}^I\mathbf{B}_e{}^{(1)}\right)\right\}\mathbf{1} \quad . \tag{103}$$

The value of ${}^I\boldsymbol{\xi}^{(1)}$ follows from (59),

$${}^I\boldsymbol{\xi}^{(1)} = \left\{\mathbf{1} + 2\left({}^I\mathbf{B}_e{}^{(1)}\right)\left({}^I\mathbf{Y}^{(1)}\right)\right\}{}^I\mathbf{Z}^{(1)} \quad , \tag{104}$$

where ${}^I\mathbf{Z}^{(1)}$ is given by $(89)_3$, and ${}^I\mathbf{Y}^{(1)}$ is given by (60),

$${}^I\mathbf{Y}^{(1)} = \frac{1}{c}\left({}^I\mathbf{B}_e{}^{(1)}\right)^{-1}\left({}^I\mathbf{Z}^{(1)}\right)\left({}^I\mathbf{B}_e{}^{(1)}\right)^{-1} \quad . \tag{105}$$

To complete part I, it still remains to examine wether the yield condition is satisfied at $t + \Delta t$. In other words, we have to examine if

$$ {}^{I}F^{(1)} = {}^{I}f^{(1)} - {}^{I}R^{(1)} - k_0 > 0 \quad , \tag{106} $$

$$ {}^{I}f^{(1)} = \sqrt{\frac{3}{2}\left\{{}^{I}\mathbf{S}^{(1)} - \left({}^{I}\mathbf{B}_e{}^{(1)}\right)^{-1}\left({}^{I}\boldsymbol{\xi}^{(1)}\right)\right\}^{D} \cdot \left\{{}^{I}\mathbf{S}^{(1)} - \left({}^{I}\mathbf{B}_e{}^{(1)}\right)^{-1}\left({}^{I}\boldsymbol{\xi}^{(1)}\right)\right\}^{D}} $$

is satisfied. If (106) is not fulfilled, then only elastic strains are produced and the results within part I furnish the final solutions of the whole problem at time $t + \Delta t$. If (106) is satisfied, then plastic strains are produced during the increment $\Delta\mathbf{F}$ and the final solutions of the problem will be obtained from part II of the algorithm. Notice, that this definition for the generation of plastic strains is the algorithmic version of the classical condition for plastic loading in the case of rate-independent plasticity (cf. (22), (23)).

3.2 Part II: Plastic Corrector

Let ${}^{I}F^{(1)} > 0$. First, consider the differential equations (79)-(81), which have to be integrated numerically. Here, we shall employ the Eulerian backward scheme to deduce, after rearranging terms,

$$ {}^{II}\mathbf{A}_p{}^{(1)} = {}^{II}\mathbf{A}_p{}^{(0)} + \left(\Delta t\dot{s}^{(1)}\right)\frac{3}{2\left({}^{II}f^{(1)}\right)}\left\{{}^{II}\mathbf{S}^{(1)} - \left({}^{II}\mathbf{B}_e{}^{(1)}\right)^{-1}\left({}^{II}\boldsymbol{\xi}^{(1)}\right)\right\}^{D}\left({}^{II}\mathbf{B}_e{}^{(1)}\right)^{-1} , \tag{107} $$

$$ {}^{II}R^{(1)} = {}^{II}R^{(0)} + \left(\Delta t\dot{s}^{(1)}\right)\left[\gamma - \beta\left({}^{II}R^{(1)}\right)\right] \quad , \tag{108} $$

$$ {}^{II}\mathbf{Z}^{(1)} = {}^{II}\mathbf{Z}^{(0)} + \left(\Delta t\dot{s}^{(1)}\right)\left[\frac{3c}{2\left({}^{II}f^{(1)}\right)}\left({}^{II}\mathbf{B}_e{}^{(1)}\right)\left\{{}^{II}\mathbf{S}^{(1)} - \left({}^{II}\mathbf{B}_e{}^{(1)}\right)^{-1}\left({}^{II}\boldsymbol{\xi}^{(1)}\right)\right\}^{D} - bc\left({}^{II}\mathbf{Z}^{(1)}\right)\right] . \tag{109} $$

Here, ${}^{II}(\cdot)$ denotes the quantity $(\cdot)$ evaluated in part II. To rewrite (107)-(109), we define

$$ \Theta^{(1)} := \Delta t\dot{s}^{(1)} \quad , \tag{110} $$

and recall that ${}^{II}\mathbf{A}_p{}^{(0)} = {}^{I}\mathbf{A}_p{}^{(1)}$, ${}^{II}R^{(0)} = {}^{I}R^{(1)} = R^{(0)}$, ${}^{II}\mathbf{Z}^{(0)} = {}^{I}\mathbf{Z}^{(1)}$, these values being known from part I. Then,

$$ {}^{II}\mathbf{A}_p{}^{(1)} = {}^{I}\mathbf{A}_p{}^{(1)} + \frac{3\Theta^{(1)}}{2({}^{II}f^{(1)})} \left\{ {}^{II}\mathbf{S}^{(1)} - \left({}^{II}\mathbf{B}_e{}^{(1)}\right)^{-1} \left({}^{II}\boldsymbol{\xi}^{(1)}\right) \right\}^D \left({}^{II}\mathbf{B}_e{}^{(1)}\right)^{-1}, \tag{111} $$

$$ {}^{II}R^{(1)} = R^{(0)} + \Theta^{(1)} \left[\gamma - \beta \left({}^{II}R^{(1)} \right) \right] \quad , \tag{112} $$

$$ {}^{II}\mathbf{Z}^{(1)} = {}^{I}\mathbf{Z}^{(1)} + \Theta^{(1)} \left[\frac{3c}{2({}^{II}f^{(1)})} \left({}^{II}\mathbf{B}_e{}^{(1)}\right) \left\{ {}^{II}\mathbf{S}^{(1)} - \left({}^{II}\mathbf{B}_e{}^{(1)}\right)^{-1} \left({}^{II}\boldsymbol{\xi}^{(1)}\right) \right\}^D - bc \left({}^{II}\mathbf{Z}^{(1)}\right) \right]. \tag{113} $$

${}^{II}\mathbf{B}_e{}^{(1)}$ may be calculated as in (92),

$$ \left({}^{II}\mathbf{B}_e{}^{(1)}\right)^{-1} = \left(\mathbf{F}^{(1)}\right)^{T-1} \left(\mathbf{F}^{(1)}\right)^{-1} + 2\left({}^{II}\mathbf{A}_p{}^{(1)}\right) \quad , \tag{114} $$

${}^{II}f^{(1)}$ is defined by

$$ {}^{II}f^{(1)} = \left\{ \frac{3}{2} \left[{}^{II}\mathbf{S}^{(1)} - \left({}^{II}\mathbf{B}_e{}^{(1)}\right)^{-1} \left({}^{II}\boldsymbol{\xi}^{(1)}\right) \right]^D \cdot \left[{}^{II}\mathbf{S}^{(1)} - \left({}^{II}\mathbf{B}_e{}^{(1)}\right)^{-1} \left({}^{II}\boldsymbol{\xi}^{(1)}\right) \right]^D \right\}^{1/2} \quad , \tag{115} $$

${}^{II}\mathbf{S}^{(1)}$ follows from (65),

$$ {}^{II}\mathbf{S}^{(1)} = \mu \left({}^{II}\mathbf{B}_e{}^{(1)} - \mathbf{1} \right) + \frac{\lambda}{2} \left\{ \ln \left(\det {}^{II}\mathbf{B}_e{}^{(1)} \right) \right\} \mathbf{1} \quad , \tag{116} $$

while (59) yields ${}^{II}\boldsymbol{\xi}^{(1)}$,

$$ {}^{II}\boldsymbol{\xi}^{(1)} = \left\{ \mathbf{1} + 2 \left({}^{II}\mathbf{B}_e{}^{(1)}\right) \left({}^{II}\mathbf{Y}^{(1)}\right) \right\} {}^{II}\mathbf{Z}^{(1)} \quad , \tag{117} $$

where

$$ {}^{II}\mathbf{Y}^{(1)} = \frac{1}{c} \left({}^{II}\mathbf{B}_e{}^{(1)}\right)^{-1} \left({}^{II}\mathbf{Z}^{(1)}\right) \left({}^{II}\mathbf{B}_e{}^{(1)}\right)^{-1} \quad . \tag{118} $$

We see that 38 equations, namely equations (111)-(118), are available to solve for the 39 unknowns ${}^{II}\mathbf{S}^{(1)}$, ${}^{II}\boldsymbol{\xi}^{(1)}$, ${}^{II}\mathbf{Z}^{(1)}$, ${}^{II}R^{(1)}$, ${}^{II}{\mathbf{A}_p}^{(1)}$, ${}^{II}{\mathbf{B}_e}^{(1)}$, ${}^{II}\mathbf{Y}^{(1)}$, ${}^{II}f^{(1)}$, $\Theta^{(1)}$. We close the set of governing equations by adding the relation

$$ {}^{II}f^{(1)} - {}^{II}R^{(1)} - k_0 = 0 \tag{119} $$

for plasticity, or the relation

$$ \Theta^{(1)} = \frac{\langle {}^{II}f^{(1)} - {}^{II}R^{(1)} - k_0 \rangle^m}{\eta} \Delta t \tag{120} $$

for viscoplasticity. Eq. (119) represents the numerical counterpart of the consistency condition given in (63), while (120) follows from (64) at time $t + \Delta t$.

This way, part II of the algorithm reduces to solving equations (111)-(118) together with (119) or (120). This nonlinear implicit set of equations is solved with a Newton-Raphson iteration scheme. The required Jacobian is approximated numerically with a forward difference-quotient matrix. As mentioned before, the solutions of this system are the final solutions of the problem. Some examples are discussed in Sect. 4 below.

3.3 Plastic Incompressibility

Plastic incompressibility requires to satisfy the equality $\det \mathbf{F}_p = 1$, which is equivalent to $\det \hat{\mathbf{B}}_p^{-1} = 1$. By using (37), (9), we may express $\hat{\mathbf{B}}_p^{-1}$ in terms of $\mathbf{A}_p$, $\mathbf{B}_e$:

$$ \hat{\mathbf{B}}_p^{-1} = \mathbf{1} - 2\mathbf{F}_e^T \mathbf{A}_p \mathbf{F}_e = \mathbf{F}_e^T (\mathbf{B}_e^{-1} - 2\mathbf{A}_p) \mathbf{F}_e \quad . \tag{121} $$

On taking the determinant of (121), the condition of plastic incompressibility becomes

$$ \det(\mathbf{1} - 2\mathbf{B}_e \mathbf{A}_p) = 1 \quad . \tag{122} $$

Evidently, because of numerical errors, the solutions according to part II will not satisfy (122). This is the reason why some authors prefer to apply so called exponential map schemes in part II (see e.g. [23] and [29]). Such integration algorithms may preserve plastic incompressibility exactly. Alternatively, plastic incompressibility may be preserved by using some correction terms. Such terms have been introduced for the first time by LÜHRS ET AL [14] in the context of small elastic strains. Two possibilities will be pursued here. The first possibility is obtained by adding in (111) a correction term of the form $\Lambda^{(1)}\mathbf{1}$, with $\Lambda^{(1)}$ being scalar valued,

$$ \begin{aligned} {}^{II}{\mathbf{A}_p}^{(1)} = {}^{I}{\mathbf{A}_p}^{(1)} + & \\ & \frac{3\Theta^{(1)}}{2({}^{II}f^{(1)})} \left\{ {}^{II}\mathbf{S}^{(1)} - \left({}^{II}{\mathbf{B}_e}^{(1)}\right)^{-1} \left({}^{II}\boldsymbol{\xi}^{(1)}\right) \right\}^D \left({}^{II}{\mathbf{B}_e}^{(1)}\right)^{-1} \\ & + \Lambda^{(1)}\mathbf{1} \quad , \end{aligned} \tag{123} $$

and by requiring (122) to hold at time $t + \Delta t$,

$$\det\left\{\mathbf{1} - 2\left({}^{II}\mathbf{B}_e{}^{(1)}\right)\left({}^{II}\mathbf{A}_p{}^{(1)}\right)\right\} = 1 \quad . \tag{124}$$

The second possibility arises by using the correction term $\Lambda^{(1)}\mathbf{1}$ to modify (111) as follows,

$$\begin{aligned} {}^{II}\mathbf{A}_p{}^{(1)} = {}^{I}\mathbf{A}_p{}^{(1)} + & \\ & \frac{3\Theta^{(1)}}{2({}^{II}f^{(1)})}\left[\left\{{}^{II}\mathbf{S}^{(1)} - \left({}^{II}\mathbf{B}_e{}^{(1)}\right)^{-1}\left({}^{II}\boldsymbol{\xi}^{(1)}\right)\right\}^{D} + \Lambda^{(1)}\mathbf{1}\right] \\ & \left({}^{II}\mathbf{B}_e{}^{(1)}\right)^{-1} \quad , \end{aligned} \tag{125}$$

where (124) is again required to apply.

Again, the set of equations (111)-(118) together with (119) or (120) are solved numerically. But now, in order to preserve plastic incompressibility, (111) is replaced by (123) or (125), combined with the constraint (124). 40 equations are available to solve for the 40 unknowns ${}^{II}\mathbf{S}^{(1)}$, ${}^{II}\boldsymbol{\xi}^{(1)}$, ${}^{II}\mathbf{Z}^{(1)}$, ${}^{II}R^{(1)}$, ${}^{II}\mathbf{A}_p{}^{(1)}$, ${}^{II}\mathbf{B}_e{}^{(1)}$, ${}^{II}\mathbf{Y}^{(1)}$, ${}^{II}f^{(1)}$, $\Theta^{(1)}$ and $\Lambda^{(1)}$. The capabilities of both methods to preserve plastic incompressibility are illustrated in Sect. 4, where some examples of deformation processes are discussed.

3.4 Implementation into the ABAQUS Finite-Element-Program

For the implementation of new constitutive models, the finite element program ABAQUS provides the possibility of writing a user material subroutine (UMAT). Inputs provided to and outputs expected from UMAT are discussed in [1], [29] and [30]. The essential output expected from the UMAT-subroutine is a fourth-order tensor $\boldsymbol{\mathcal{M}}$, representing the constitutive contribution to the Jacobian matrix used in the Newton procedure for global equilibrium, and the Cauchy stress tensor $\mathbf{T}^{(1)}$.

Proceeding to determine $\boldsymbol{\mathcal{M}}$, we interpolate the velocity gradient tensor $\mathbf{L}$ during the time increment Δt by a constant value, as indicated by HUGHES and WINGET [11],

$$\mathbf{L} = \text{const} = \frac{2}{\Delta t}\left(\Delta\mathbf{F} - \mathbf{1}\right)\left(\Delta\mathbf{F} + \mathbf{1}\right)^{-1} \quad . \tag{126}$$

Consequently,

$$\mathbf{D} = \frac{1}{2}(\mathbf{L} + \mathbf{L}^T) = \text{const} \quad , \quad \mathbf{W} = \frac{1}{2}(\mathbf{L} - \mathbf{L}^T) = \text{const} \quad , \tag{127}$$

during the time step Δt. ABAQUS requires the stress tensor $\mathbf{T}^{(1)}$ to be given as a function of the strain increment $\Delta\varepsilon$ and the rotation $\mathbf{Q}^{(1)}$,

$$T_{ij}^{(1)} = \sigma_{ij}(\Delta\varepsilon_{mn}, Q_{rs}^{(1)}) \quad , \tag{128}$$

where $\Delta\varepsilon := \Delta t\mathbf{D}$, and $\mathbf{Q}^{(1)}$ is the solution of the initial value problem

$$\frac{\mathrm{d}}{\mathrm{d}\zeta}\mathbf{Q}(\zeta) = \mathbf{W}(\zeta)\mathbf{Q}(\zeta) \quad , \quad \mathbf{Q}(t) = \mathbf{1} \quad . \tag{129}$$

It has been shown by HUGHES and WINGET [11] (see also WEBER [28]) that $\mathbf{Q}^{(1)}$ has the form

$$\mathbf{Q}^{(1)} = \left(\mathbf{1} - \frac{\Delta t}{2}\mathbf{W}\right)^{-1}\left(\mathbf{1} + \frac{\Delta t}{2}\mathbf{W}\right) \quad . \tag{130}$$

The tensor $\boldsymbol{\mathcal{M}}$ is then defined by (see also DIEGELE et al. [7])

$$\mathcal{M}_{ijkl} = \frac{\partial\sigma_{ij}(\Delta\varepsilon_{pq}, Q^{(1)}_{rs})}{\partial\Delta\varepsilon_{kl}} \quad . \tag{131}$$

We shall now show how $\boldsymbol{\mathcal{M}}$ may be approximated numerically. Let

$$\Delta\bar{\varepsilon}^{(kl)} = \frac{\chi}{2}\left(\mathbf{e}_k \otimes \mathbf{e}_l + \mathbf{e}_l \otimes \mathbf{e}_k\right) \tag{132}$$

be a small perturbation of $\Delta\varepsilon$, where χ is a small real value. The components of $\Delta\bar{\varepsilon}^{(kl)}$ are

$$\Delta\bar{\varepsilon}^{(kl)}_{pq} = \mathbf{e}_p \cdot \Delta\bar{\varepsilon}^{(kl)}\mathbf{e}_q = \frac{\chi}{2}\left(\delta_{pk}\delta_{ql} + \delta_{pl}\delta_{qk}\right) \quad . \tag{133}$$

Then, $\boldsymbol{\mathcal{M}}$ may be approximated by the formula

$$\mathcal{M}_{ijkl} \approx \frac{\sigma_{ij}(\Delta\tilde{\varepsilon}^{(kl)}_{pq}, Q^{(1)}_{rs}) - \sigma_{ij}(\Delta\varepsilon_{pq}, Q^{(1)}_{rs})}{\chi} \quad , \tag{134}$$

where

$$\Delta\tilde{\varepsilon}^{(kl)} = \Delta\varepsilon + \Delta\bar{\varepsilon}^{(kl)} \quad . \tag{135}$$

On the other hand, our time integration algorithm will provide the value of $\mathbf{T}^{(1)}$ as a function of $\Delta\mathbf{F}$, i.e., $T^{(1)}_{ij} = T^{(1)}_{ij}(\Delta F_{mn})$, so that $T^{(1)}_{ij}(\Delta F_{mn}) = \sigma_{ij}(\Delta\varepsilon_{pq}, Q^{(1)}_{rs})$. This means, for determining $\boldsymbol{\mathcal{M}}$ in (134), we must find the deformation gradient tensor $\Delta\tilde{\mathbf{F}}^{(kl)}$ which produces the strain increment $\Delta\tilde{\varepsilon}^{(kl)}$. We remark that $\Delta\tilde{\varepsilon}^{(kl)}$ is related to the velocity gradient tensor $\tilde{\mathbf{L}}^{(kl)}$ by

$$\tilde{\mathbf{L}}^{(kl)} = \mathbf{L} + \frac{1}{\Delta t}\Delta\tilde{\varepsilon}^{(kl)} \quad , \tag{136}$$

where $\mathbf{L}$ is the velocity gradient caused by the deformation gradient $\Delta\mathbf{F}$ and the time increment Δt (see (126)). The deformation gradient tensor $\Delta\tilde{\mathbf{F}}^{(kl)}$, which produces $\Delta\tilde{\mathbf{L}}^{(kl)}$ during the time step Δt, follows then from (126),

$$\Delta\tilde{\mathbf{F}}^{(kl)} = \left(\mathbf{1} - \frac{\Delta t}{2}\Delta\tilde{\mathbf{L}}^{(kl)}\right)^{-1}\left(\mathbf{1} + \frac{\Delta t}{2}\Delta\tilde{\mathbf{L}}^{(kl)}\right) \quad . \tag{137}$$

This way, the tensor $\boldsymbol{\mathcal{M}}$ may be calculated from

$$\mathcal{M}_{ijkl} \approx \frac{T_{ij}^{(1)}(\Delta\tilde{F}_{mn}^{(kl)}) - T_{ij}^{(1)}(\Delta F_{mn})}{\chi} \quad , \tag{138}$$

where $\chi \ll 1$ (see MIEHE [16]) and $\Delta\tilde{\mathbf{F}}^{(kl)}$ is given by (133) and (135)-(137).

4 Examples - Concluding Remarks

The examples in this section aim to demonstrate what consequences may arise when integrating with algorithms which do preserve plastic incompressibility or not. In the ensuing analysis, the material parameters are chosen to be

$$\mu = 76923\,\text{MPa} \quad , \tag{139}$$

$$\lambda = 115384\,\text{MPa} \quad , \tag{140}$$

$$k_0 = 200\,\text{MPa} \quad , \tag{141}$$

$$\beta = \gamma = 0 \quad , \tag{142}$$

$$c = 20000\,\text{MPa} \quad , \tag{143}$$

$$b = 50 \quad . \tag{144}$$

That means, for small deformations μ and λ in (139) and (140) correspond to values of 200000 MPa and $0,3$ for the Young's-modulus and the Poisson ratio, respectively.

First, we consider a rectangular specimen with a circular hole subjected to displacement controlled tensile loading in Y-direction. The specimen geometry and the finite element mesh used are illustrated in Fig. 1 and Fig. 2. At the beginning of the deformation process, R_0 is the radius of the hole, B_0, L_0 and T_0 are the the width, the length and the thickness of the specimen, while X, Y, Z is the Cartesian coordinate system we refer to. The origin of the coordinate system coincides with the center of the circular hole in the X-Y-plane, and is located in the middle of the plate in the Z-direction. Because of various symmetry conditions, only 1/8 of the specimen has been meshed with 169 twenty-node solid elements (C3D20 in the ABAQUS element library). Fig. 3 shows the deviation of $\det \mathbf{F}_p$ from the value 1, when that numerical algorithm is used which does not preserve plastic incompressibility. For this result, the overall strain (global strain) of the specimen is $e^* = (L - L_0)/L_0 = 0.1$, where L is the current length of the inhomogeneously deformed specimen. Distributions of the associated stress components T_{xx}, T_{yy}, T_{zz} and T_{xy} are plotted in Figs 4-7. Moreover, Figs 4-7 illustrate corresponding stress distributions resulting from an integration algorithm preserving plastic incompressibility. The stress components T_{xz}, T_{yz} vanish everywhere. For all calculations the same step size has been chosen.

It is important to note, that the two possibilities to preserve plastic incompressibility, proposed in Sect. 3, imply identical results for all cases studied. Therefore, no explicit reference to the integration algorithm preserving plastic incompressibility is made in Figs. 4-7. Also, the same is true for Figs. 10-13 below.

As can be seen from Fig. 3, noticeable deviations of $\det \mathbf{F}_p$ from the value 1 may become visible, if the integration algorithm is not augmented to preserve plastic incompressibility. On the other hand, the graphs in Figs. 4-7 reveal, that the predicted stress responses according to the two integration algorithms do not exhibit significant differences in the stress components for the considered example. However, from these results it should not be concluded, that the two algorithms predict stress behaviour in every deformation process, which is only slightly different from one another. To demonstrate this, we consider the deformation process of simple shear, for which the deformation gradient tensor $\mathbf{F}$ is given by

$$\mathrm{F}_{ij}(t) = \begin{pmatrix} 1 & \gamma(t) & 0 \\ 0 & 1 & 0 \\ 0 & 0 & 1 \end{pmatrix} \quad . \tag{145}$$

The amount of shear $\gamma(t)$ satisfies the relation $\gamma(t) = \tan \Theta(t)$, where $\Theta(t)$ is the angle of shear in the X-Y-coordinate plane (see Fig 8). We focus attention on the special deformation process, where the shear strain is increasing at the beginning until the value 0.5 is reached (loading), and then decreases to the value -0.3 (unloading-reloading). As can be seen from Fig. 9, when integrating with the algorithm which does not preserve plastic incompressibility, the values of $\det \mathbf{F}_p$ deviate only marginally from the value 1 during simple shear. Nevertheless, significant deviations exist between the correct values of the stress components and the predicted ones. This may be recognized from Figs. 10-13, which show calculated stress responses according to the two integration algorithms. It has to be noted that the responses according to the integration algorithm preserving plastic incompressibility may be regarded to represent the correct solutions. In fact, these are identical to the corresponding solutions obtained analytically. To be more specific, simple shear is a homogenous deformation process, so the stress components may be calculated from the constitutive equations only. The latter may be directly integrated by using e.g. an explicit Runge-Kutta algorithm with automatic step size control. Solutions obtained in this way are regarded to be equivalent to solutions gained analytically.

To conclude, the numerical examples studied suggest that significant solution errors may occur when integrating without preserving plastic incompressibility. The latter may be satisfied easily by using the methods proposed in the article. It is perhaps of interest to mention that, for the examples discussed here, the CPU-time required from the algorithm preserving plastic incompressibility is only about 10% longer than for the algorithm not preserving plastic incompressibility.

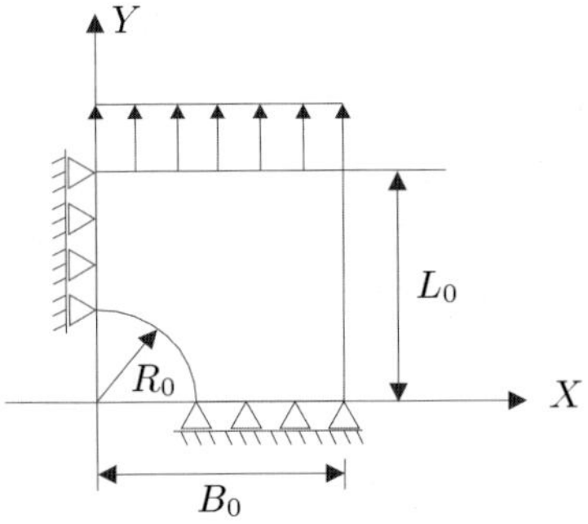

Fig. 1. Quarter of a rectangular specimen with a circular hole: $R_0 = 40$ cm, $B_0 = 100$ cm, $L_0 = 100$ cm, $T_0 = 10$ cm.

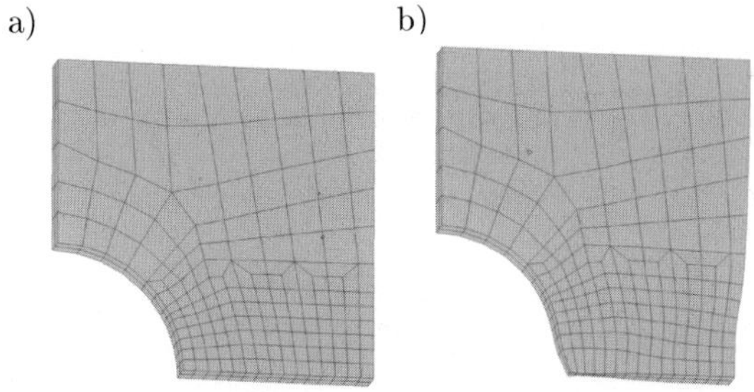

Fig. 2. a) Undeformed and b) deformed finite element model.

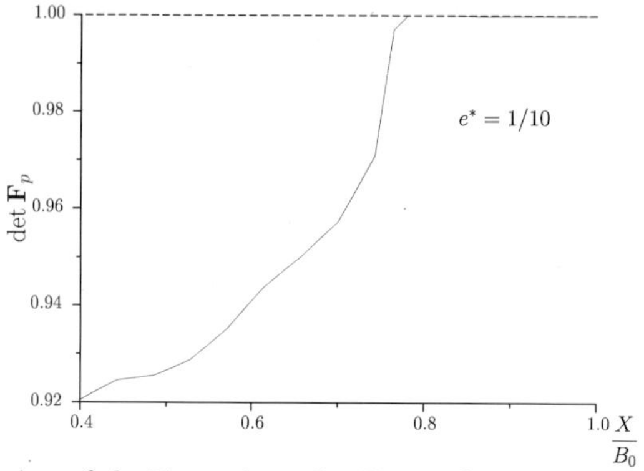

Fig. 3. Distribution of $\det \mathbf{F}_p$ against the X-axis for $Z = Y = 0$. Plastic incompressibility is not preserved by the numerical algorithm employed.

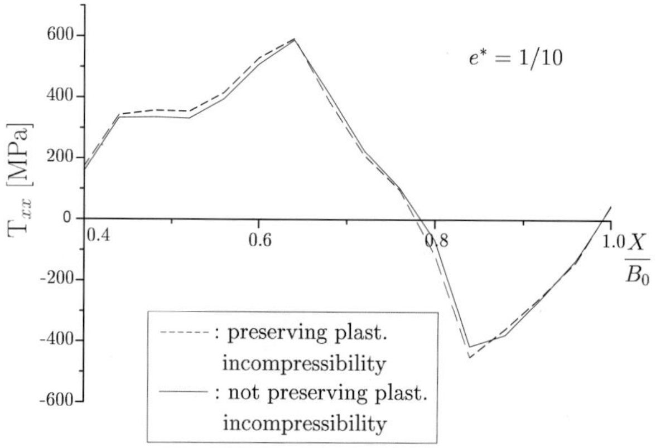

Fig. 4. Distribution of stress component T_{xx} against the X-axis for $Z = Y = 0$.

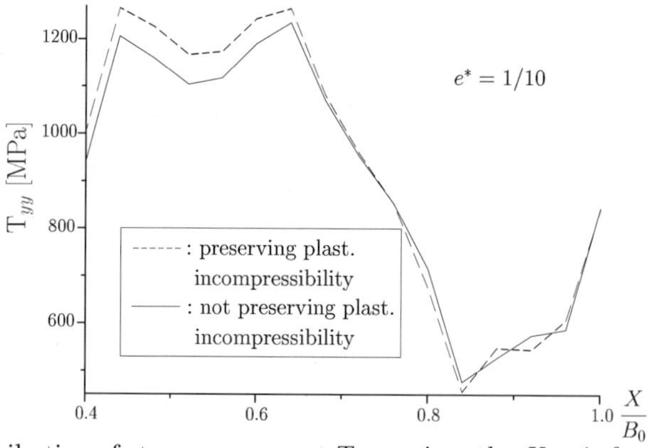

Fig. 5. Distribution of stress component T_{yy} against the X-axis for $Z = Y = 0$.

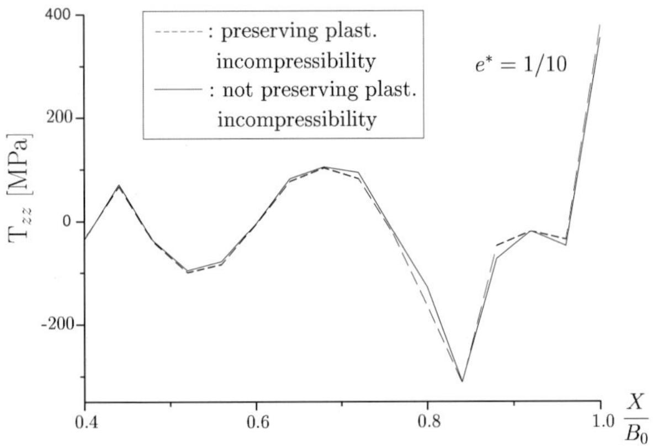

Fig. 6. Distribution of stress component T_{zz} against the X-axis for $Z = Y = 0$.

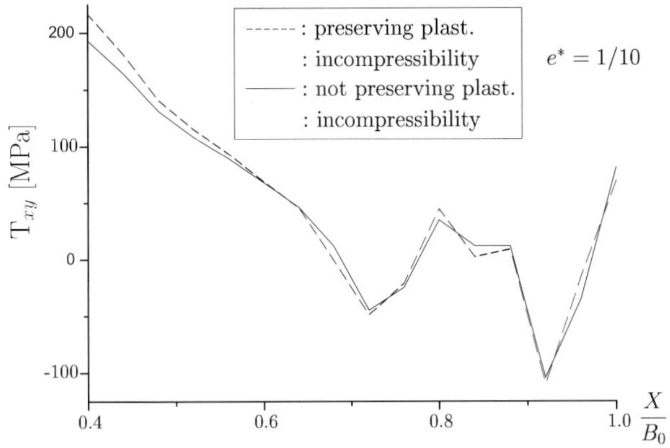

Fig. 7. Distribution of stress component T_{xy} against the X-axis for $Z = Y = 0$.

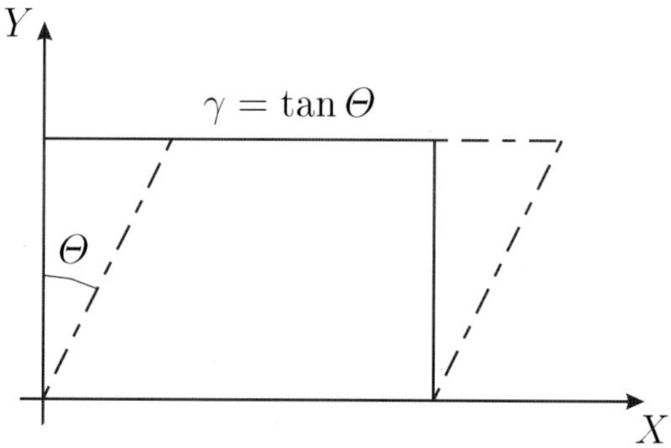

Fig. 8. Simple shear deformation.

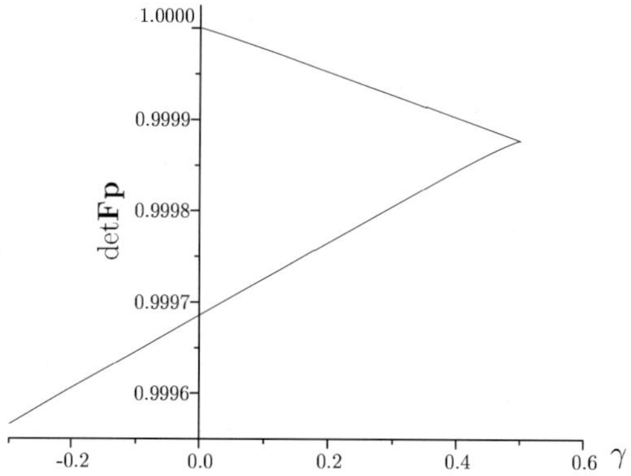

Fig. 9. Prdicted values of det $\mathbf{F}_p$ against the shear strain γ. Plastic incompressibility is not preserved by the numerical algorithm employed.

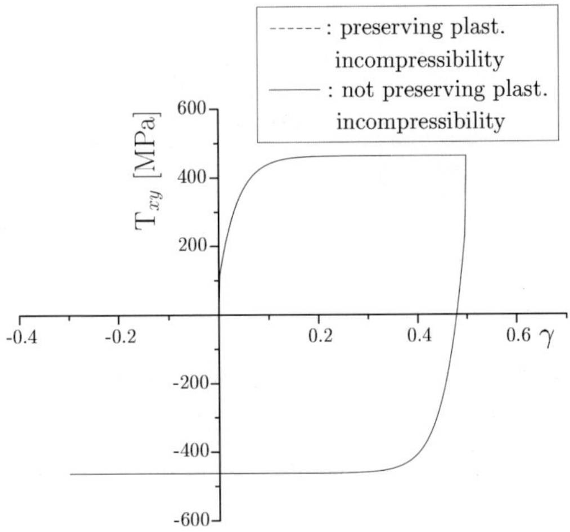

Fig. 10. Predicted values of the shear stress $\mathbf{T}_{xy}$

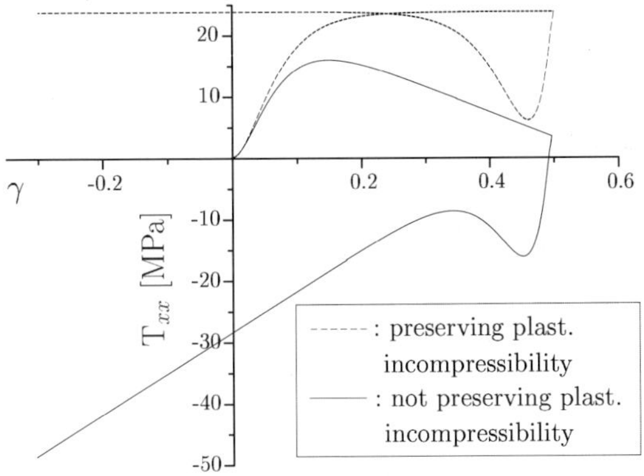

Fig. 11. Predicted values of the normal stress $\mathbf{T}_{xx}$

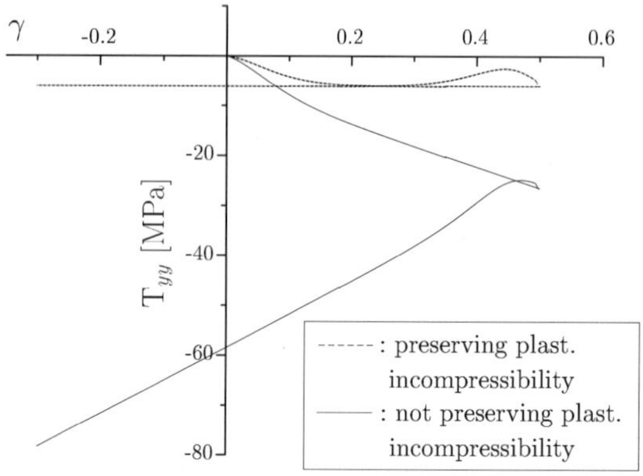

Fig. 12. Predicted values of the normal stress $\mathbf{T}_{yy}$

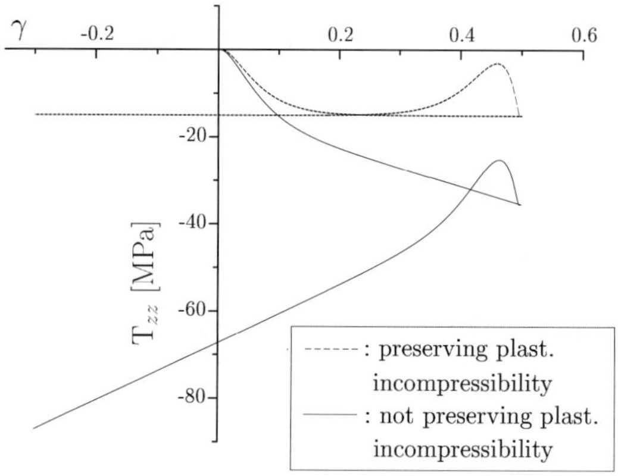

Fig. 13. Predicted values of the normal stress $\mathbf{T}_{zz}$

References

1. ABAQUS/Standard-Version 5.8. User's Manual. Hibbitt, Karlsson & Sorensen, Inc.
2. Armstrong P.J., Frederick C.O. (1966) A mathematical representation of the multiaxial Bauschinger effect. General Electricity Generating Board, Report No. RD/B/N731, Berceley Nuclear Laboratories
3. Casey J., Naghdi P. (1981) A remark on the use of the decomposition $\mathbf{F} = \mathbf{F}_e\mathbf{F}_p$ in plasticity. J. Appl. Mech. **48** 983–985
4. Dafalias Y. (1990) The plastic spin in viscoplasticity. Int. J. of Solids and Structures. **26** 149–163
5. Dafalias Y., Aifantis E. (1990) On the microscopic origin of the plastic spin. Acta Mechanica. **82** 31–48
6. Diegele E., Jansohn W., Tsakmakis C. (1995) Viscoplasticity and dual Variables. Proceedings of the ASME Materials Division. **1** 449–467
7. Diegele E., Jansohn W., Tsakmakis Ch. (2000) Finite deformation plasticity and viscoplasticity laws exhibiting nonlinear hardening rules Part I: Constitutive theory and numerical integration. Computational Mechanics. **25** 1–12
8. Green A., Naghdi P. (1971) Some remarks on elastic-plastic deformations at finite strains. Int. J. Eng. Sci. **9** 1219–1229
9. Hartmann S., Haupt P. (1993) Stress computation and consistent tangent operator using non-linear kinematic hardening models. International Journal for Numerical Methods In Engineering **36** 3801–3814
10. Hughes T. (1984) Numerical implementation of constitutive models: Rate-independent deviatoric plasticity. In: A Theoretical Foundation for Large Scale Computation of Nonlinear Material Behaviour, Dordrecht, Martinus Nijhoff Publisher
11. Hughes T., Winget J. (1980) Finite rotation effects in numerical integration of rate constitutive equations in large-deformation analysis. International Journal for Numerical Methods In Engineering **15** 1862–1867
12. Krieg R.D., Krieg D.B. (1977) Accuracies of numerical solution methods for the elastic-perfectly plastic model. J. Pressure Vessel Tech. ASME 99
13. Lubliner L. (1990) Plasticity theory. Macmillian Publishing Company, New York
14. Lührs G., Hartmann S., Haupt P. (1996) On the numerical treatment of finite deformations in elastoviscoplasticity. Computer Methods in Applied Mechanics and Engineering **144** 1–21
15. Maugin G. (1992) The thermodynamics of plasticity and fracture. Cambridge Univesity Press, New York etc.
16. Miehe C. (1996) Numerical computation of algorithmic (consistent) tangent moduli in large-strain computational inelasticity. Comput. Methods Appl. Mech. Eng. **134** 223–240
17. Reed K., Atluri S. (1983) Analysis of large quasistatic deformations of inelastic bodies by a new hybrid-stress finite element allgorithm. Comp. Meth. Appl. Mech. Eng. **39** 245–295
18. Reed K., Atluri S. (1985) Constitutive modelling and computational implementation for finite strain plasticity. Int. J. Plasticity **1** 63–87
19. Rubinstein R., Atluri S. (1983) Objectivity of incremental constitutive relations over finite time steps in computational finite deformation analyses. Comp. Meth. Appl. Mech. Eng. **36** 277–290

20. Simo J.C. (1985) On the computational significance of the intermediate configuration and hyperelastic stress relations in finite deformation elastoplasticity. Mechanics of Materials **4** 439–451
21. Simo J.C. (1988) A framework for finite strain elasoplasticity based on maximum plastic dissipation and the multiplicative decomposition. Part 1: Continuum formulation. Comp. Meth. Appl. Mech. Eng. **66** 199–219
22. Simo J.C. (1988) A framework for finite strain elasoplasticity based on maximum plastic dissipation and the multiplicative decomposition. Part 2: Computational aspects. Comp. Meth. Appl. Mech. Eng. **68** 1-31
23. Simo J.C. (1992) Algorithms for static and dynamic multiplicative plasticity that preserve the classical return mapping schemes of the infinitesimal theory. Comp. Meth. Appl. Mech. Eng. **99** 61-112
24. Simo J.C., Ortiz M. (1985) A unified approach to finite deformation elastoplastic analysis based on the use of hyperelastic constitutive equations. Comp. Meth. Appl. Mech. Eng. **49** 221–245
25. Simo J.C., Taylor R.L. (1985) Consistent tangent operators for rate-independent elastoplasticity. Computer Methods In Applied Mechanics And Engineering **48** 101–118
26. Tsakmakis C. (1996) Kinematic hardening rules in finite plasticity, Part 1: A constitutive approach. Continuum Mech. Thermodyn. **8** 215–231
27. Tsakmakis Ch., Willuweit A. A comparative study of kinematic hardening rules at finite deformations. Int. J. Nonl. Mech., in press
28. Weber G. (1988) Computational procedures for a new class of finite deformation elastic-plastic constitutive equations. Ph.D. thesis, Massachusetts Institute of Technology
29. Weber G., Anand L. (1990) Finite deformation constitutive equations and a time integration Procedure for Isotropic, Hyperelastic-Viscoplastic Solids. Comp. Meth. Appl. Mech. Eng. **79** 173–202
30. Weber G.G., Lush A.M., Zavaliangos A., Anand L. (1990) An objective time-integration procedure for isotropic rate-independent and rate-dependent elastic-plastic constitutive equations. Int. J. Plasticity **6** 701–744
31. Wilkins M.L. (1964) Calculation of elasic-plasic flow. Methods of Computational Physics, Vol. 3

A Model of Finite Strain Viscoplasticity with an Anisotropic Elastic Constitutive Law

Carlo Sansour[1], Franz Gustav Kollmann[2], and Jozef Bocko[3]

[1] School of Petroleum Engineering, University of Adelaide
Adelaide SA 5005, Australia
carlo.sansour@adelaide.edu.au
[2] Mauerkircherstr. 15, D–81679 München
fg.kollmann@t-online.de
[3] Department of Applied Mechanics, Technical University of Košice
Letná 9, 04187 Košice, Slovakia

received 29 Jul 2002 — accepted 19 Nov 2002

Abstract. We deal with the anisotropic formulation for the stress tensor within theories of viscoplasticity and a multiplicative framework. As the formulation of an anisotropic strain energy function has not been yet established in the literature, we first discuss the theory and give a justification for it from a purely theoretical point of view. The inelastic behaviour is assumed to be governed by evolution equations of the unified type. As the anisotropic formulation considerably complicates the numerical approach, the numerical treatment is developed in full detail. Various numerical examples with applications to shells are presented.

1 Introduction

Anisotropy is a phenomenon of great importance in almost all fields of mechanics. It features a high level of complexity in issues of modelling and simulation. The anisotropy of elastic material behaviour can perhaps be considered well understood. The crystallographic groups were specified at an early stage, and issues of representation theorems, developed over the years, have arrived at a high level of development which is reflected in the monograph of Zheng [26], where a whole body of literature can be found. In fact, following a basic result of the representation theory an anisotropic formulation of the free energy function, or any other physical quantity of interest, can be achieved by considering it as an isotropic function of certain mechanical agencies (like the strain tensor) and of the so-called structural tensors formulated as the dyadic products of the privileged direction.

In contrast to the purely elastic behaviour, the issues of anisotropy in conjunction with the inelastic material response are still far from being considered well understood. This statement is especially true for large strain formulations, specifically those based on the multiplicative decomposition of the deformation gradient: $\mathbf{F} = \mathbf{F}_e\mathbf{F}_p$. The increasing activity in the field is reflected in the very recent work of Boehlke and Bertram [4], Gasser and Holzapfel [7], Haupt and Kersten [8], Menzel and Steinmann [12], Reese [14], Sansour and Kollmann [17,18], Svendsen [22], Tsakmakis [24] (much of it is yet to be published).

The term 'anisotropic inelasticity' encompasses a wide range of anisotropic phenomena which manifest themselves in different aspects of the constitutive behaviour, the full description of which necessitates that two sets of constitutive equations are to be specified: those for the thermodynamical forces, specifically the law for the stress tensor falls within this category, and those for the evolution equations of the internal variables. Anisotropy can be incorporated in both categories. Hill's criteria for an orthotropic flow rule (see [9]) falls within the latter category and can be considered as the best known formulation of anisotropic inelastic behaviour. In addition, kinematical hardening is considered to be an anisotropic effect. Extensions of small strain formulations hereof to the case of large strains are discussed e.g. in Ekh and Runesson [6], Tsakmakis [23], and Xiao et al. [25].

The formulation of an anisotropic constitutive law for the stress tensor (as a thermodynamical force) in the frame of large deformations has only recently been addressed in the literature. The issues play a dominant role not only from a physical but also from a numerical point of view as they dictate, to a large extent, the solution schemes to be adopted in numerical analysis. As far as additive structures are concerned, there are no conceptual difficulties in extending formulations of linear elasticity to the case of inelastic material behaviour. Formulations along this line can be found in Papadopoulous and Lu [13]. On the other hand, when the inelastic formulation is based on the multiplicative decomposition of the deformation gradient, the issue of anisotropy of the stress tensor is much more complicated at both the theoretical and the numerical level. From a theoretical point of view, the question arises whether the privileged directions of the material are affected by the inelastic part of the deformation gradient, and if so, what kind of transformation may be considered.

This paper is intended at a formulation of finite strain viscoplasticity, based on the multiplicative decomposition of the deformation gradient, including anisotropic constitutive law for the stress tensor. We focus on the anisotropic formulation of the free energy function, discuss and motivate the transformation of structural tensors as mixed-variant tensors, and elaborate on the thermodynamical consequences of such a transformation. The viscoplastic model itself is of the Bodner-Partom-type.

We note that the internal structure of the material may undergo changes due to inelastic deformation. The accumulation of changes at the micro scale can be so high that they are remarkable also at a macro scale. Accordingly, the directions of anisotropy may undergo changes. This kind of induced isotropy will not be considered here. We assume instead that the directions of anisotropy are fixed with respect to a certain reference configuration.

Since anisotropy in general, and that of the stress tensor in particular, complicates the numerical schemes considerably, we discuss the numerical algorithms adopted in full. The inelastic theory is applied to a shell formulation which allows for the application of fully three-dimensional constitutive laws. As the shell theory and the corresponding finite element formulation have already been presented in previous publications (see [16,19]), we will restrict our attention to the generally three-dimensional developments of the numerical schemes.

The paper is organized as follows. In Sect. 2 the theoretical basis for an anisotropic multiplicative inelastic formulation is provided. First the definition of structural tensors, which contain the information about the privileged directions of the material, is introduced. Based on theoretical arguments, the structural tensors are first transformed in a mixed-variant fashion before including them in a formulation of the free energy function. The evaluation of the dissipation inequality provides the thermodynamically active stress tensors which prove to differ from those appearing in a purely isotropic formulation. In Sect. 3 details of the time integration are given. The integration of the evolution equations in the presence of an anisotropic free energy function is considered in detail. It is shown that the anisotropy of the elastic constitutive law significantly complicates the numerical schemes at the local iteration level as well as at the level of the tangent operator. Sect. 4 provides various examples of computations carried out for thin bodies.

2 Anisotropic Inelasticity

In this section the theoretical grounds are laid down for a theory of viscoplasticity on the basis of the multiplicative decomposition of the deformation gradient. The anisotropy of the elastic constitutive law will be addressed in detail.

2.1 Kinematics and Basic Relations

Let $\mathbf{F}$ be the deformation gradient. We start by employing its multiplicative decomposition in an elastic part $\mathbf{F}_\mathrm{e}$ and an inelastic part $\mathbf{F}_\mathrm{p}$ according to ([1,10,11])

$$\mathbf{F} = \mathbf{F}_\mathrm{e}\mathbf{F}_\mathrm{p}. \tag{1}$$

For metals, the inelastic part of the above decomposition is accompanied with the assumption of incompressibility: $\mathbf{F}_\mathrm{p} \in SL^+(3,\mathbb{R})$. Here, $SL^+(3,\mathbb{R})$ is the special linear group with determinant equal to one: $\det(\mathbf{F}_\mathrm{p}) = 1$.

On the basis of the above decomposition we define the following Cauchy-Green-type deformation tensors

$$\mathbf{C} := \mathbf{F}^{\mathrm{T}}\mathbf{F}\,,\ \mathbf{C}_{\mathrm{e}} := \mathbf{F}_{\mathrm{e}}^{\mathrm{T}}\mathbf{F}_{\mathrm{e}}\,,\ \mathbf{C}_{\mathrm{p}} := \mathbf{F}_{\mathrm{p}}^{\mathrm{T}}\mathbf{F}_{\mathrm{p}}\,, \tag{2}$$

$$\mathbf{b} := \mathbf{F}\mathbf{F}^{\mathrm{T}}\,,\ \mathbf{b}_{\mathrm{e}} := \mathbf{F}_{\mathrm{e}}\mathbf{F}_{\mathrm{e}}^{\mathrm{T}}\,. \tag{3}$$

Whereas $\mathbf{b}$ and $\mathbf{b}_{\mathrm{e}}$ are defined with respect to the actual configuration (spatial tensors), $\mathbf{C}$ and $\mathbf{C}_{\mathrm{p}}$ constitute material tensors, and $\mathbf{C}_{\mathrm{e}}$ is given with respect to the so-called intermediate configuration.

The time derivative of the deformation gradient can be formulated as

$$\dot{\mathbf{F}} = \mathbf{l}\mathbf{F} \quad , \quad \dot{\mathbf{F}} = \mathbf{F}\mathbf{L}\,, \tag{4}$$

where $\mathbf{l}$ and $\mathbf{L}$ are the left and right rates, respectively. Both rates are mixed tensors and are related according to

$$\mathbf{L} = \mathbf{F}^{-1}\mathbf{l}\mathbf{F}\,. \tag{5}$$

Due to the fact that $\mathbf{F}_{\mathrm{p}} \in SL^{+}(3, \mathrm{I\!R})$ we can define a right rate of the inelastic deformations according to

$$\dot{\mathbf{F}}_{\mathrm{p}} = \mathbf{F}_{\mathrm{p}}\mathbf{L}_{\mathrm{p}}\,, \tag{6}$$

which, once pushed-forward under the action of $\mathbf{F}$, leads to the spatial inelastic rate

$$\mathbf{l}_{\mathrm{p}} = \mathbf{F}\mathbf{L}_{\mathrm{p}}\mathbf{F}^{-1}\,. \tag{7}$$

The quantities will be of use later on.

The internal power can be formulated in terms of the Kirchhoff stress tensor or its material counter part according to

$$\mathcal{W} = \boldsymbol{\tau} : \mathbf{l} = \boldsymbol{\Xi} : \mathbf{L}\,. \tag{8}$$

Here, $\boldsymbol{\tau}$ is the Kirchhoff stress tensor and $\mathbf{l}, \mathbf{L}$ are defined in (4). The double dot product of two second order tensors $\mathbf{a}, \mathbf{b}$ is equivalent to $\mathbf{a} : \mathbf{b} = \mathrm{tr}\, \mathbf{a}\mathbf{b}^{\mathrm{T}}$, where tr is the trace operation. Eq. (8), together with (5), reveals the relation

$$\boldsymbol{\Xi} = \mathbf{F}^{\mathrm{T}}\boldsymbol{\tau}\mathbf{F}^{-\mathrm{T}}\,. \tag{9}$$

The stress tensor $\boldsymbol{\Xi}$ is, accordingly, the mixed variant pull-back of the Kirchhoff tensor. Up to a spherical part and a sign, it coincides with Eshelby's stress tensor. Accordingly, we refer to it as Eshelby-like tensor.

A common property of unified inelastic constitutive models is the use of further internal variables in addition to $\mathbf{F}_{\mathrm{p}}$. These describe some properties of the material, and their presence in the theory requires additional constitutive relations. We denote a typical set of such internal variables by $\mathbf{Z}$. The free energy function ψ can be stated either as $\psi = \psi(\mathbf{C}_{\mathrm{e}}, \mathbf{Z})$ or, alternatively, as

$\psi = \psi(\mathbf{b}_e, \mathbf{Z})$ depending on the chosen elastic strain measure which is defined either in the intermediate or in the spatial configuration.

Given the form of ψ, the localized form of the dissipation inequality for an isothermal process reads

$$\mathcal{D} = \boldsymbol{\tau} : \mathbf{l} - \rho_{\text{ref}}\dot{\psi} = \boldsymbol{\Xi} : \mathbf{L} - \rho_{\text{ref}}\dot{\psi} \geq 0 , \tag{10}$$

where ρ_{ref} is the density in the reference configuration. The evaluation of (10) leads to the definition of the thermodynamical forces and to a reduced dissipation inequality. The elaboration depends on the chosen stress and strain measures as well as on certain assumptions concerning the anisotropy to be employed in the definition of the free energy function.

2.2 Structural Tensors and Free Energy Functions

The elastic case The description of elastic anisotropy is carried out by using the method of structural tensors as developed by Rivlin, Spencer, Smith, Wang, and Boehler (see [3,20,21,26]). To be specific the developments are carried out for the case of orthotropy. Similar results hold for any kind of anisotropy described by structural tensors.

A structural tensor is defined as the tensor product

$$\mathbf{M} = \mathbf{v} \otimes \mathbf{v} , \tag{11}$$

where $\mathbf{v}$ is a privileged direction of the material in the reference configuration. In case of more than one privileged direction, correspondingly more tensors are introduced. The case of orthotropy is described by means of three structural tensor

$$\mathbf{M}_i = \mathbf{v}_i \otimes \mathbf{v}_i , \quad i = 1, 2, 3 , \tag{12}$$

with

$$\mathbf{v}_i \cdot \mathbf{v}_j = \delta_{ij} , \tag{13}$$

where δ_{ij} is Kronecker's delta.

Let us assume a purely elastic deformation. The set of invariants

$$\begin{aligned} J_1 &= \text{tr}(\mathbf{M}_1\mathbf{C}) , \; J_2 = \text{tr}(\mathbf{M}_2\mathbf{C}) , \; J_3 = \text{tr}(\mathbf{M}_3\mathbf{C}) , \; J_4 = \text{tr}(\mathbf{M}_1\mathbf{C}^2) , \\ J_5 &= \text{tr}(\mathbf{M}_2\mathbf{C}^2) , \; J_6 = \text{tr}(\mathbf{M}_3\mathbf{C}^2) , \; J_7 = \text{tr}(\mathbf{C}^3) . \end{aligned} \tag{14}$$

constitute the integrity basis for a material description and the formulation is, accordingly, complete. A free energy function ψ is then to be formulated in terms of these invariants: $\psi = \psi(J_1, ..., J_7)$.

The physically equivalent formulation in terms of the spatial tensor $\mathbf{b}$ (which shares with $\mathbf{C}$ the same invariants) can be established using the modified structural tensors defined in the following way

$$\mathbf{m}_i = \mathbf{F}\mathbf{M}_i\mathbf{F}^{-1}, \quad i = 1, 2, 3. \tag{15}$$

In fact, we can immediately see that the relations

$$\mathrm{tr}(\mathbf{m}_i\mathbf{b}) = \mathrm{tr}(\mathbf{M}_i\mathbf{C}), \quad i = 1, 2, 3 \tag{16}$$

hold. The same holds true for all other invariants as well. Accordingly, the formulation of a free energy function in terms of the material structural tensors and $\mathbf{C}$, $\psi = \psi(\mathbf{C}, \mathbf{M}_i)$, is equivalent to that in terms of the modified structural tensors and $\mathbf{b}$, $\psi = \psi(\mathbf{b}, \mathbf{m}_i)$.

Having established the case of pure elasticity we turn our attention to the inelastic case.

The inelastic case According to the experience in the purely elastic case described above, the formulation of the free energy function will be carried out in terms of $\mathbf{b}_\mathrm{e}$ together with the modified spatial structural tensors $\mathbf{m}_i$. Corresponding invariants can be formulated along the same lines described in the preceding section. They read

$$\begin{aligned} I_1 &= \mathrm{tr}(\mathbf{m}_1\mathbf{b}_\mathrm{e}), \; I_2 = \mathrm{tr}(\mathbf{m}_2\mathbf{b}_\mathrm{e}), \; I_3 = \mathrm{tr}(\mathbf{m}_3\mathbf{b}_\mathrm{e}), \; I_4 = \mathrm{tr}(\mathbf{m}_1\mathbf{b}_\mathrm{e}^2), \\ I_5 &= \mathrm{tr}(\mathbf{m}_2\mathbf{b}_\mathrm{e}^2), \; I_6 = \mathrm{tr}(\mathbf{m}_3\mathbf{b}_\mathrm{e}^2), \; I_7 = \mathrm{tr}(\mathbf{b}_\mathrm{e}^3). \end{aligned} \tag{17}$$

After having established the formulation in terms of spatial quantities we ask for an equivalent formulation in terms of quantities defined with respect to the so-called intermediate configuration. First we note that the elastic spatial deformation tensor $\mathbf{b}_\mathrm{e}$ (Eq. (3_2)) is equivalent to $\mathbf{C}_\mathrm{e}$ by the fact that they share the same invariants. While the latter is defined in the intermediate configuration, the former is given in the actual configuration. To define invariants which are equivalent to (17) but formulated in terms of $\mathbf{C}_\mathrm{e}$ the material structural tensors have to be modified. It is straightforward to show that the invariants defined in (17) are equivalent to corresponding invariants incorporating $\mathbf{C}_\mathrm{e}$ only when the corresponding structural tensors are defined according to

$$\mathbf{M}_i^\star = \mathbf{F}_\mathrm{p}\mathbf{M}_i\mathbf{F}_\mathrm{p}^{-1}, \quad i = 1, 2, 3. \tag{18}$$

In analogy with (17) we consider the following set of invariants

$$\begin{aligned} I_1 &= \mathrm{tr}(\mathbf{M}_1^\star\mathbf{C}_\mathrm{e}), \; I_2 = \mathrm{tr}(\mathbf{M}_2^\star\mathbf{C}_\mathrm{e}), \; I_3 = \mathrm{tr}(\mathbf{M}_3^\star\mathbf{C}_\mathrm{e}), \; I_4 = \mathrm{tr}(\mathbf{M}_1^\star\mathbf{C}_\mathrm{e}^2), \\ I_5 &= \mathrm{tr}(\mathbf{M}_2^\star\mathbf{C}_\mathrm{e}^2), \; I_6 = \mathrm{tr}(\mathbf{M}_3^\star\mathbf{C}_\mathrm{e}^2), \; I_7 = \mathrm{tr}(\mathbf{C}_\mathrm{e}^3). \end{aligned} \tag{19}$$

Remarkably, the invariants can alternatively be written as

$$\begin{aligned} I_i &= \mathrm{tr}(\mathbf{M}_i \mathbf{C}_\mathrm{p}^{-1}\mathbf{C})\,, \; I_{i+3} = \mathrm{tr}(\mathbf{M}_i (\mathbf{C}_\mathrm{p}^{-1}\mathbf{C})^2)\,, \\ I_7 &= \mathrm{tr}((\mathbf{C}_\mathrm{p}^{-1}\mathbf{C})^3)\,, \quad i = 1,2,3\,. \end{aligned} \tag{20}$$

The following remarks are pertinent. 1) The transformation (18) does not change the invariants of the structure tensors. 2) All three sets of invariants given in (17), (19), and (20) are equivalent. 3) The structural tensors and not the privileged directions themselves are transformed. The strong theoretical argument motivating and justifying this transformation is the physical equivalence of all formulations whether material, intermediate, or spatial. The validity of a symmetric, and more classical, transformation of the type $\mathbf{F}_\mathrm{p}\mathbf{M}_i\mathbf{F}_\mathrm{p}^T$ is questionable as the following simple calculation shows: $\mathrm{tr}[(\mathbf{F}_\mathrm{p}\mathbf{M}_i\mathbf{F}_\mathrm{p}^T)\mathbf{C}_\mathrm{e}] = \mathrm{tr}(\mathbf{M}_i\mathbf{C})$. That is, the inelastic effect is eliminated from terms linear in $\mathbf{C}_\mathrm{e}$. Accordingly, the anisotropy is reduced to a nonlinear (higher order) effect which in fact is questionable. Nevertheless, transformation of the symmetric type are used in [14] and [22]. 4) Contrasting the set of invariants (14), the present set of invariants defined in (20) does not constitute an integrity basis, the same holds true for the equivalent sets defined in (17) or (19). This is due to the fact that the tensor $\mathbf{C}_\mathrm{p}^{-1}\mathbf{C}$ is not symmetric. Although it is straightforward to construct such a set by splitting $\mathbf{C}_\mathrm{p}^{-1}\mathbf{C}$ into symmetric and skew-symmetric parts, this is hardly necessary as in practical applications only a few number of invariants is selected to include in the formulation of the free energy function. Accordingly, we restrict ourselves to the invariants formulated in (20). The complete set of the integrity basis is reported in [18].

2.3 Evaluation of the Dissipation Inequality

We start with the dissipation inequality (10) which is a general thermodynamical statement and as such independent of the kind of anisotropy under consideration. With $(4)_2$ and (6), the evaluation of the new form of the free energy function leads first to

$$\begin{aligned} \mathcal{D} = &\left(\boldsymbol{\Xi} - \rho_\mathrm{ref}\mathbf{F}^\mathrm{T}\frac{\partial\psi(\mathbf{C}_\mathrm{e},\,\mathbf{Z})}{\partial\mathbf{F}}\right) : \mathbf{L} \\ &-\rho_\mathrm{ref}\mathbf{F}_\mathrm{p}^\mathrm{T}\frac{\partial\psi(\mathbf{C}_\mathrm{e},\,\mathbf{Z})}{\partial\mathbf{F}_\mathrm{p}} : \mathbf{L}_\mathrm{p} - \rho_\mathrm{ref}\frac{\partial\psi(\mathbf{C}_\mathrm{e},\,\mathbf{Z})}{\partial\mathbf{Z}}\cdot\dot{\mathbf{Z}} \geq 0\,, \end{aligned} \tag{21}$$

from which we first deduce

$$\boldsymbol{\Xi} = \rho_\mathrm{ref}\mathbf{F}^T\frac{\partial\psi}{\partial\mathbf{F}}\,. \tag{22}$$

With the definitions

$$\bar{\boldsymbol{\Xi}} = -\rho_\mathrm{ref}\mathbf{F}_\mathrm{p}^\mathrm{T}\frac{\partial\psi(\mathbf{C}_\mathrm{e},\,\mathbf{Z})}{\partial\mathbf{F}_\mathrm{p}}\,, \qquad \mathbf{Y} := -\rho_\mathrm{ref}\frac{\partial\psi(\mathbf{C}_\mathrm{e},\,\mathbf{Z})}{\partial\mathbf{Z}}\,, \tag{23}$$

the reduced dissipation inequality takes the form

$$\mathcal{D}_{\mathrm{p}} := \bar{\boldsymbol{\Xi}} : \mathbf{L}_{\mathrm{p}} + \mathbf{Y} \cdot \dot{\mathbf{Z}} \geq 0. \tag{24}$$

At this stage we confine ourselves to the following form of the free energy function, quadratic in the invariants:

$$\begin{aligned}\psi = \alpha_1 I_1 + \alpha_2 I_1^2 + \alpha_3 I_2 + \alpha_4 I_2^2 + \alpha_5 I_3 + \alpha_6 I_3^2 + \alpha_7 I_1 I_2 + \alpha_8 I_1 I_3 \\ +\alpha_9 I_2 I_3 + \alpha_{10} I_4 + \alpha_{11} I_5 + \alpha_{12} I_6 \,,\end{aligned} \tag{25}$$

where $\alpha_1, \ldots, \alpha_{12}$ are the material constants of which only nine are linearly independent. Equation (25) implicitly ensures linear behavior at the elastic range.

Explicitly, for the anisotropy under consideration and with (20),(22), and (25) we may elaborate

$$\begin{aligned}\boldsymbol{\Xi} = \sum_{i=1}^{3} \Big[\frac{\partial \psi}{\partial I_i} (\mathbf{C}\mathbf{M}_i \mathbf{C}_{\mathrm{p}}^{-1} + \mathbf{C}\mathbf{C}_{\mathrm{p}}^{-1}\mathbf{M}_i) \\ + \frac{\partial \psi}{\partial I_{i+3}} (\mathbf{C}\mathbf{M}_i \mathbf{C}_{\mathrm{p}}^{-1} \mathbf{C}\mathbf{C}_{\mathrm{p}}^{-1} + \mathbf{C}\mathbf{C}_{\mathrm{p}}^{-1}\mathbf{M}_i \mathbf{C}\mathbf{C}_{\mathrm{p}}^{-1} \\ + \mathbf{C}\mathbf{C}_{\mathrm{p}}^{-1}\mathbf{C}\mathbf{M}_i\mathbf{C}_{\mathrm{p}}^{-1} + \mathbf{C}\mathbf{C}_{\mathrm{p}}^{-1}\mathbf{C}\mathbf{C}_{\mathrm{p}}^{-1}\mathbf{M}_i)\Big] \,.\end{aligned} \tag{26}$$

Further, the material is assumed to be stress free at the reference configuration. Accordingly, the above expression is valid under the side condition $\boldsymbol{\Xi} = \mathbf{0}$ for $\mathbf{C} = \mathbf{1}$ and $\mathbf{C}_{\mathrm{p}}^{-1} = \mathbf{1}$. Using (26) we arrive at the relation

$$\begin{aligned}\boldsymbol{\Xi}|_{(\mathbf{C}=\mathbf{1};\mathbf{C}_{\mathrm{p}}^{-1}=\mathbf{1})} = \Big[2\frac{\partial \psi}{\partial I_1}\mathbf{M}_1 + 2\frac{\partial \psi}{\partial I_2}\mathbf{M}_2 + 2\frac{\partial \psi}{\partial I_3}\mathbf{M}_3 + \\ 4\frac{\partial \psi}{\partial I_4}\mathbf{M}_1 + 4\frac{\partial \psi}{\partial I_5}\mathbf{M}_2 + 4\frac{\partial \psi}{\partial I_6}\mathbf{M}_3 \Big]_{(\mathbf{C}=\mathbf{1};\mathbf{C}_{\mathrm{p}}^{-1}=\mathbf{1})} = \mathbf{0}\,.\end{aligned} \tag{27}$$

Since the structural tensors are independent of each other, the above equation splits into the equivalent three equations:

$$2\frac{\partial \psi}{\partial I_1}|_{(\mathbf{C}=\mathbf{1};\mathbf{C}_{\mathrm{p}}^{-1}=\mathbf{1})} + 4\frac{\partial \psi}{\partial I_4}|_{(\mathbf{C}=\mathbf{1};\mathbf{C}_{\mathrm{p}}^{-1}=\mathbf{1})} = 0\,, \tag{28}$$

$$2\frac{\partial \psi}{\partial I_2}|_{(\mathbf{C}=\mathbf{1};\mathbf{C}_{\mathrm{p}}^{-1}=\mathbf{1})} + 4\frac{\partial \psi}{\partial I_5}|_{(\mathbf{C}=\mathbf{1};\mathbf{C}_{\mathrm{p}}^{-1}=\mathbf{1})} = 0\,, \tag{29}$$

$$2\frac{\partial \psi}{\partial I_3}|_{(\mathbf{C}=\mathbf{1};\mathbf{C}_{\mathrm{p}}^{-1}=\mathbf{1})} + 4\frac{\partial \psi}{\partial I_6}|_{(\mathbf{C}=\mathbf{1};\mathbf{C}_{\mathrm{p}}^{-1}=\mathbf{1})} = 0\,. \tag{30}$$

From (25) we have

$$\frac{\partial \psi}{\partial I_1} = \alpha_1 + 2\alpha_2 I_1 + \alpha_7 I_2 + \alpha_8 I_3\,, \tag{31}$$

$$\frac{\partial \psi}{\partial I_2} = \alpha_3 + 2\alpha_4 I_2 + \alpha_7 I_1 + \alpha_9 I_3\,, \tag{32}$$

$$\frac{\partial \psi}{\partial I_3} = \alpha_5 + 2\alpha_6 I_3 + \alpha_8 I_1 + \alpha_9 I_2\,, \tag{33}$$

$$\frac{\partial \psi}{\partial I_4} = \alpha_{10}\,, \tag{34}$$

$$\frac{\partial \psi}{\partial I_5} = \alpha_{11}\,, \tag{35}$$

$$\frac{\partial \psi}{\partial I_6} = \alpha_{12}\,. \tag{36}$$

Finally, with equations (28) - (30) at hand, the set of equations (31) - (36) can be rewritten as

$$\frac{\partial \psi}{\partial I_1} = \alpha_1 + 2\alpha_2 I_1 + \alpha_7 I_2 + \alpha_8 I_3\,, \tag{37}$$

$$\frac{\partial \psi}{\partial I_2} = \alpha_3 + 2\alpha_4 I_2 + \alpha_7 I_1 + \alpha_9 I_3\,, \tag{38}$$

$$\frac{\partial \psi}{\partial I_3} = \alpha_5 + 2\alpha_6 I_3 + \alpha_8 I_1 + \alpha_9 I_2\,, \tag{39}$$

$$\frac{\partial \psi}{\partial I_4} = \alpha_{10} = -\frac{1}{2}(\alpha_1 + 2\alpha_2 + \alpha_7 + \alpha_8\,, \tag{40}$$

$$\frac{\partial \psi}{\partial I_5} = \alpha_{11} = -\frac{1}{2}(\alpha_3 + 2\alpha_4 + \alpha_7 + \alpha_9)\,, \tag{41}$$

$$\frac{\partial \psi}{\partial I_6} = \alpha_{12} = -\frac{1}{2}(\alpha_5 + 2\alpha_6 + \alpha_8 + \alpha_9)\,. \tag{42}$$

As is evident from the side conditions, only nine constants are independent.

Having established the expression for the material stress tensor $\boldsymbol{\Xi}$, we focus now our attention to the formulation of the reduced dissipation inequality. In fact, due to the dependency of the structural tensors on the inelastic deformation, the reduced dissipation inequality is modified. It is the modified stress tensor $\bar{\boldsymbol{\Xi}}$ which acts as thermodynamical conjugate to the inelastic rate $\mathbf{L}_\mathrm{p}$. The modified stress tensor $\bar{\boldsymbol{\Xi}}$ is defined in $(23)_1$ which, with (20) and (25), leads to the expression

$$\bar{\boldsymbol{\Xi}} = \boldsymbol{\Xi} + \sum_{i=1}^{3} \left[\frac{\partial \psi_\mathrm{e}}{\partial I_i} \left(\mathbf{M}_i \mathbf{C}\mathbf{C}_\mathrm{p}^{-1} - \mathbf{C}\mathbf{C}_\mathrm{p}^{-1}\mathbf{M}_i \right) + \frac{\partial \psi_\mathrm{e}}{\partial I_{i+3}} \left(\mathbf{M}_i \mathbf{C}\mathbf{C}_\mathrm{p}^{-1}\mathbf{C}\mathbf{C}_\mathrm{p}^{-1} - \mathbf{C}\mathbf{C}_\mathrm{p}^{-1}\mathbf{C}\mathbf{C}_\mathrm{p}^{-1}\mathbf{M}_i \right) \right]. \tag{43}$$

The expression is a result of rather involved but otherwise elementary algebraic operations the details of which are omitted for the sake of brevity. We note also that the difference between $\bar{\Xi}$ and Ξ constitutes of terms which vanish in the case of isotropy. The terms reflect the non-commutativity of the structural tensors with $\mathbf{C}\mathbf{C}_{\mathrm{p}}^{-1}$.

2.4 Inelastic Constitutive Model

The base for further developments is the inelastic model suggested by Bodner and Partom [2] as modified to accommodate large strains in [15,19]. According to (24) the stress tensor $\bar{\Xi}$ and the plastic rate tensor $\mathbf{L}_{\mathrm{p}}$ are conjugate variables. By that the stress tensor $\bar{\Xi}$ is to be considered as the driving force for the inelastic deformation which motivates the formulation of the evolution equation for $\mathbf{L}_{\mathrm{p}}$ in terms of exactly the same stress quantity. The following set of evolution equations is considered.

$$\mathbf{L}_{\mathrm{p}} = \dot{\phi}\bar{\boldsymbol{\nu}}^{T}\,, \tag{44}$$

$$\dot{Z} = \frac{m}{Z_0}(Z_1 - Z)\dot{W}_{\mathrm{p}}\,, \tag{45}$$

$$\dot{W}_{\mathrm{p}} = \Pi\dot{\phi}(\Pi, Z)\,, \tag{46}$$

$$\Pi = \sqrt{\frac{3}{2} dev\bar{\Xi} : dev\bar{\Xi}}\,, \tag{47}$$

$$\dot{\phi} = \frac{2}{\sqrt{3}} D_0 \exp\left[-\frac{1}{2}\frac{N+1}{N}\left(\frac{Z}{\Pi}\right)^{2N}\right], \tag{48}$$

$$\bar{\boldsymbol{\nu}} = \frac{3}{2}\frac{dev\bar{\Xi}}{\Pi}\,, \tag{49}$$

$$W_{\mathrm{p}}(0) = 0\,, \tag{50}$$

$$Z(0) = Z_0\,. \tag{51}$$

Here, Z_0, Z_1, D_0, N, m are material parameters.

The following remarks are relevant.

1. The flow rule defines in a natural form a plastic spin. Whereas the push-forward of Ξ will provide us with the symmetric Kirchhoff stress tensor, the same is not true for the modified material stress tensor $\bar{\Xi}$. Its push-forward to the actual configuration results necessarily in a nonsymmetric quantity. Accordingly the corresponding push-forward of the inelastic rate $\mathbf{L}_{\mathrm{p}}$ is nonsymmetric as well. The skew-symmetric part can then be viewed as defining the plastic spin, or at least contributing to whatever definition one may employ for the plastic spin.
2. The plastic spin, or the above contribution to it, is directly related to the anisotropy in the elastic constitutive law. In the case of isotropy, the difference between both tensors $\bar{\Xi}$ and Ξ vanishes and we end up with a symmetric pushed-forward inelastic rate.

3. Evidently, the flow rule as chosen in (44) is isotropic. As we are considering elastic anisotropy it is very likely that the flow itself will be anisotropic as well. The kind of anisotropy of the flow rule does not necessarily coincide with that of the elastic law. May this be as it is, from a theoretical point of view, the formulation of an anisotropic inelastic flow rule does not impose any difficulties. Inspite of this, in all computations to be considered in the next sections we will assume an isotropic flow rule and concentrate on the implications of the elastic anisotropy on the computational schemes. In fact, the extension of these schemes to a possible anisotropic flow rule is straightforward, once an elastic anisotropy is defined.

3 Numerical Implementation

In this section, formulational and computational aspects of the integration algorithm as well as the algorithmic tangent operator are described.

In the solution of the nonlinear evolution problem for the viscoplastic model described in the previous sections, a numerical time integration procedure is used, which provides an approximate solution at discrete times for the evolution equations in (44)-(51).

3.1 Time Integration and Local Iteration

Let $\Delta t = t_{n+1} - t_n$ be the time increment, t_n and t_{n+1}two successive times. Since the tensor $\mathbf{F}_\mathrm{p}$ is an element of a Lie group and the tensor $\mathbf{L}_\mathrm{p}$ is an element of the corresponding Lie algebra the following equation can be formulated for the time integration

$$\mathbf{F}_{\mathrm{p}|n+1} = \mathbf{F}_{\mathrm{p}|n} \exp[\Delta t \mathbf{L}_\mathrm{p}] \,, \tag{52}$$

for some $\mathbf{L}_\mathrm{p}$ in the interval Δt the choice of which is defined by means of the integration procedure. This algorithm preserves the condition of plastic incompressibility exactly. The inverse follows immediately as

$$\mathbf{F}_\mathrm{p}^{-1}|_{n+1} = \exp[-\Delta t \mathbf{L}_\mathrm{p}] \mathbf{F}_\mathrm{p}^{-1}|_n = \exp[\overline{\boldsymbol{\beta}}] \mathbf{F}_\mathrm{p}^{-1}|_n \,. \tag{53}$$

Here, the substitution

$$\overline{\boldsymbol{\beta}} = -\Delta t \mathbf{L}_\mathrm{p} = -\Delta t \dot{\phi} \overline{\boldsymbol{\nu}}^\mathrm{T} \tag{54}$$

is used for a certain value of $\dot{\phi}$ and $\overline{\boldsymbol{\nu}}^\mathrm{T}$. For all scalar quantities we apply the mid-point rule according to which we have

$$\Pi = \frac{1}{2} \left(\Pi_{n+1} + \Pi_n \right) \,, \tag{55}$$

$$Z = \frac{1}{2} \left(Z_{n+1} + Z_n \right) \,, \tag{56}$$

$$\dot{Z} = \frac{Z_{n+1} - Z_n}{\Delta t} \,. \tag{57}$$

Using (45), (46) as well as (56) and (57), we obtain for the internal variable Z the equation

$$Z = \frac{m\Delta t\Pi\dot{\phi}Z_1 + 2Z_0 Z_n}{m\Delta t\Pi\dot{\phi} + 2Z_0}, \tag{58}$$

which depends explicitly on $\dot{\phi}$. Substituting (58) into (48) we finally obtain the nonlinear equation for $\dot{\phi}$ in the form

$$\dot{\phi} = \frac{2}{\sqrt{3}} D_0 \exp\left[-\frac{1}{2}\frac{N+1}{N}\left(\frac{Z(\Pi,\dot{\phi})}{\Pi}\right)^{2N}\right]. \tag{59}$$

Eq. (59) is solved iteratively by the Newton method. The iteration necessitates the elaboration of certain derivatives which involve the definition of the anisotropic stress tensor and makes the operations quite involved. In what follows we provide the basic quantities needed in the iteration process. The increment $\Delta\dot{\phi}$ at every iteration step is given by

$$\Delta\dot{\phi} = \frac{\overline{a}}{\overline{B}}, \tag{60}$$

where

$$\overline{a} = \dot{\phi} - \frac{2}{\sqrt{3}} D_0 \exp\left[-\frac{1}{2}\frac{N+1}{N}\left(\frac{Z(\Pi,\dot{\phi})}{\Pi}\right)^{2N}\right], \tag{61}$$

$$\begin{aligned}\overline{B} = {} & \frac{2}{\sqrt{3}} D_0 \exp\left[-\frac{1}{2}\frac{N+1}{N}\left(\frac{Z(\Pi,\dot{\phi})}{\Pi}\right)^{2N}\right] \times \\ & \times \left[-(N+1)\left(\frac{Z(\Pi,\dot{\phi})}{\Pi}\right)^{2N-1}\right] \times \\ & \times \left[\frac{1}{\Pi}\frac{\partial Z(\Pi,\dot{\phi})}{\partial\dot{\phi}} + \left(\frac{1}{2\Pi}\frac{\partial Z(\Pi,\dot{\phi})}{\partial\Pi} - \frac{Z(\Pi,\dot{\phi})}{2\Pi^2}\right)\frac{\partial\Pi_{|n+1}}{\partial\dot{\phi}}\right] - 1\,.\end{aligned} \tag{62}$$

Derivatives of the variable Z are computed using Eq. (58), they read

$$d_1 = \frac{\partial Z}{\partial\dot{\phi}} = \frac{2Z_0\,(Z_1 - Z_n)\, m\Delta t\Pi}{\left(m\Delta t\Pi\dot{\phi} + 2Z_0\right)^2}, \tag{63}$$

$$2d_2 = \frac{\partial Z}{\partial\Pi} = \frac{2Z_0\,(Z_1 - Z_n)\, m\Delta t\dot{\phi}}{\left(m\Delta t\Pi\dot{\phi} + 2Z_0\right)^2}. \tag{64}$$

Using the abbreviations

$$f_1 = \frac{2}{\sqrt{3}} D_0 \exp\left[-\frac{1}{2}\frac{N+1}{N}\left(\frac{Z(\Pi,\dot{\phi})}{\Pi}\right)^{2N}\right] , \tag{65}$$

$$f_2 = -(N+1)\left(\frac{Z(\Pi,\dot{\phi})}{\Pi}\right)^{2N-1}\frac{1}{\Pi} , \tag{66}$$

$$f_3 = (N+1)\left(\frac{Z(\Pi,\dot{\phi})}{\Pi}\right)^{2N-1}\left(\frac{Z}{2\Pi^2}\right) , \tag{67}$$

together with (63) and (64), one has for (62)

$$\overline{B} = f_1\left[f_2 d_1 + (f_2 d_2 + f_3)\frac{\partial \Pi_{n+1}}{\partial \dot{\phi}}\right] - 1 . \tag{68}$$

The last unknown term in (68) is the derivative $\partial \Pi_{n+1}/\partial\dot{\phi}$. Its computation is carried out according to

$$\frac{\partial \Pi_{n+1}}{\partial \dot{\phi}} = \frac{\partial \Pi_{n+1}}{\partial \overline{\Xi}}\frac{\partial \overline{\Xi}}{\partial \dot{\phi}} = \overline{\boldsymbol{\nu}}^{\mathrm{T}}_{n+1} : \frac{\partial \overline{\Xi}}{\partial \dot{\phi}} , \tag{69}$$

where the relation

$$\frac{\partial \Pi_{|n+1}}{\partial \overline{\Xi}} = \overline{\boldsymbol{\nu}}^{\mathrm{T}}_{|n+1} , \tag{70}$$

which follows from (47), has been used. Again the term $\partial\overline{\Xi}/\partial\dot{\phi}$ is unknown and has to be determined.

At this stage an expression is needed for the derivative of the exponential map with respect to its argument. Detailed discussion of this issue has been recently given in [19], where the reader can find a derivation of the formula

$$\frac{\partial \mathbf{F}_{\mathrm{p}}^{-1}{}_{|n+1}}{\partial \dot{\phi}} = \frac{\partial\left(\exp[\overline{\beta}]\mathbf{F}_{\mathrm{p}}^{-1}{}_{|n}\right)}{\partial \dot{\phi}} = \frac{\partial \overline{\beta}}{\partial \dot{\phi}}\left(\exp[\overline{\beta}]\mathbf{F}_{\mathrm{p}}^{-1}{}_{|n}\right) . \tag{71}$$

The formula makes use of the multiplicative nature of the updating formula for $\mathbf{F}_{\mathrm{p}}^{-1}$. Alternatively, it is possible to develop a more classical expression (based on an additive interpretation of the updating) for the above derivative which is included in the Appendix. Whatever formula one chooses the procedure as such remains un-effected. To be specific, we will use the multiplicative interpretation documented in the above formula in our further developments.

In what follows, and for the sake of clarity, we resort to a notation in indices in deriving explicitly all necessary expressions. First, Eq. (53) leads to

$$\begin{aligned}\frac{\partial \overline{\Xi}_{mn}}{\partial \dot{\phi}} &= \frac{\partial \overline{\Xi}_{mn}}{\partial (\mathbf{F}_{\mathrm{p}}^{-1})_{ij}}\frac{\partial \overline{\beta}_{io}}{\partial \dot{\phi}}(\mathbf{F}_{\mathrm{p}}^{-1})_{oj}|_{(n+l)} \\ &\quad + \frac{\partial \overline{\Xi}_{mn}}{\partial (\mathbf{F}_{\mathrm{p}}^{-1})_{ij}}\frac{\partial \overline{\beta}_{io}}{\partial \overline{\Xi}_{ab}}(\mathbf{F}_{\mathrm{p}}^{-1})_{oj}\frac{\partial \overline{\Xi}_{ab}}{\partial \dot{\phi}}|_{(n+l)} .\end{aligned} \tag{72}$$

The index $_{(n+l)}$ should emphasize the fact that $\mathbf{F}_\mathrm{p}^{-1}$ is taken from the actual l-th local iteration. This equation can be reformulated to give

$$\frac{\partial \overline{\Xi}_{ab}}{\partial \dot{\phi}} = \left[\delta_{ma}\delta_{nb} - \frac{\partial \overline{\Xi}_{mn}}{\partial (\mathbf{F}_\mathrm{p}^{-1})_{ij}} \frac{\partial \overline{\beta}_{io}}{\partial \overline{\Xi}_{ab}} (\mathbf{F}_\mathrm{p}^{-1})_{oj}|_{(n+l)} \right]^{-1} \times \\ \times \frac{\partial \overline{\Xi}_{mn}}{\partial (\mathbf{F}_\mathrm{p}^{-1})_{ij}} \frac{\partial \overline{\beta}_{io}}{\partial \dot{\phi}} (\mathbf{F}_\mathrm{p}^{-1})_{oj}|_{(n+l)} \,. \tag{73}$$

For the elaboration of this expression we need the relations

$$\frac{\partial \overline{\beta}_{io}}{\partial \dot{\phi}} = -\Delta t\, \overline{\nu}_{oi} \,, \tag{74}$$

$$\frac{\partial \overline{\beta}_{io}}{\partial \overline{\Xi}_{ab}} = \frac{\partial \overline{\beta}_{io}}{\partial \overline{\nu}_{pq}} \frac{\partial \overline{\nu}_{pq}}{\partial \overline{\Xi}_{ab}} \tag{75}$$

and

$$\frac{\partial \overline{\beta}_{io}}{\partial \overline{\nu}_{pq}} = -\Delta t \dot{\phi} \delta_{iq} \delta_{op} \,, \tag{76}$$

which follow from (54). In addition one has

$$\frac{\partial \overline{\nu}_{pq}}{\partial \overline{\Xi}_{ab}} = \frac{1}{\mathit{\Pi}} \left[-\overline{\nu}_{pq}\overline{\nu}_{ab} + \frac{3}{2} \left(\delta_{pa}\delta_{qb} - \frac{1}{3}\delta_{ab}\delta_{pq} \right) \right] , \tag{77}$$

which results from (47) and (49).

To derive $\partial \overline{\Xi}_{mn} / \partial (\mathbf{F}_\mathrm{p}^{-1})_{ij}$, as it appears in Eq. (73), we recall first the relations

$$\mathbf{C}_\mathrm{p}^{-1} = \mathbf{F}_\mathrm{p}^{-1} \mathbf{F}_\mathrm{p}^{-T} \,, \tag{78}$$

and

$$\frac{\partial (\mathbf{C}_\mathrm{p}^{-1})_{rs}}{\partial (\mathbf{F}_\mathrm{p}^{-1})_{ij}} = \delta_{ri} (\mathbf{F}_\mathrm{p}^{-1})_{sj} + \delta_{si} (\mathbf{F}_\mathrm{p}^{-1})_{rj} \,. \tag{79}$$

Accordingly, one has

$$\frac{\partial \overline{\Xi}_{mn}}{\partial (\mathbf{F}_\mathrm{p}^{-1})_{ij}} = \frac{\partial \overline{\Xi}_{mn}}{\partial (\mathbf{C}_p^{-1})_{rs}} \frac{\partial (\mathbf{C}_\mathrm{p}^{-1})_{rs}}{\partial (\mathbf{F}_\mathrm{p}^{-1})_{ij}} \,. \tag{80}$$

Finally, (43) and (26) give

$$\begin{aligned}
\frac{\partial \overline{\Xi}_{mn}}{\partial (\mathbf{C}_{\mathrm{p}}^{-1})_{rs}} &= \sum_{i=1}^{3} \left[\frac{\partial \psi}{\partial I_i} \left(C_{mo} (\mathbf{M}_i)_{or} \delta_{ns} + (\mathbf{M}_i)_{mo} C_{or} \delta_{ns} \right) \right. \\
&+ \frac{\partial \psi}{\partial I_{i+3}} \big(C_{mo} (\mathbf{M}_i)_{or} C_{sb} (\mathbf{C}_{\mathrm{p}}^{-1})_{bn} + C_{mo} (\mathbf{M}_i)_{ot} (\mathbf{C}_{\mathrm{p}}^{-1})_{tq} C_{qr} \delta_{ns} \\
&+ C_{mr} (\mathbf{M}_i)_{sq} C_{qb} (\mathbf{C}_{\mathrm{p}}^{-1})_{bn} + C_{mo} (\mathbf{C}_{\mathrm{p}}^{-1})_{ot} (\mathbf{M}_i)_{tq} C_{qr} \delta_{ns} \\
&+ C_{mr} C_{sq} (\mathbf{M}_i)_{qb} (\mathbf{C}_{\mathrm{p}}^{-1})_{bn} + C_{mo} (\mathbf{C}_{\mathrm{p}}^{-1})_{ot} C_{tq} (\mathbf{M}_i)_{qr} \delta_{ns} \\
&+ \left. (\mathbf{M}_i)_{mo} C_{or} C_{sb} (\mathbf{C}_{\mathrm{p}}^{-1})_{bn} + (\mathbf{M}_i)_{mo} C_{ot} (\mathbf{C}_{\mathrm{p}}^{-1})_{tq} C_{qr} \delta_{ns} \big) \right] \\
&+ \left(2\alpha_2 (\mathbf{M}_1)_{rt} C_{ts} + \alpha_7 (\mathbf{M}_2)_{rt} C_{ts} + \alpha_8 (\mathbf{M}_3)_{rt} C_{ts} \right) \\
&\quad \left(C_{mo} (\mathbf{M}_1)_{ot} (\mathbf{C}_{\mathrm{p}}^{-1})_{tn} + (\mathbf{M}_1)_{mo} C_{ot} (\mathbf{C}_{\mathrm{p}}^{-1})_{tn} \right) \\
&+ \left(2\alpha_4 (\mathbf{M}_2)_{rt} C_{ts} + \alpha_7 (\mathbf{M}_1)_{rt} C_{ts} + \alpha_9 (\mathbf{M}_3)_{rt} C_{ts} \right) \\
&\quad \left(C_{mo} (\mathbf{M}_2)_{ot} (\mathbf{C}_{\mathrm{p}}^{-1})_{tn} + (\mathbf{M}_2)_{mo} C_{ot} (\mathbf{C}_{\mathrm{p}}^{-1})_{tn} \right) \\
&+ \left(2\alpha_6 (\mathbf{M}_3)_{rt} C_{ts} + \alpha_8 (\mathbf{M}_1)_{rt} C_{ts} + \alpha_9 (\mathbf{M}_2)_{rt} C_{ts} \right) \\
&\quad \left(C_{mo} (\mathbf{M}_3)_{ot} (\mathbf{C}_{\mathrm{p}}^{-1})_{tn} + (\mathbf{M}_3)_{mo} C_{ot} (\mathbf{C}_{\mathrm{p}}^{-1})_{tn} \right) , \qquad (81)
\end{aligned}$$

which completes the local iteration.

From the preceding developments it is evident that anisotropy extremely complicates the numerical approach. For the sake of clarity we summarize in box 1 the basic equations.

The increment of the inelastic rate within a local iteration step is given by

$$\Delta\dot{\phi} = \frac{\overline{a}}{\overline{\overline{B}}} \qquad \text{with} \qquad \overline{a} = \dot{\phi} - \frac{2}{\sqrt{3}} D_0 \exp\left[-\frac{1}{2}\frac{N+1}{N}\left(\frac{Z(\Pi,\dot{\phi})}{\Pi}\right)^{2N}\right]$$

$$\overline{B} = f_1\left[f_2 d_1 + (f_2 d_2 + f_3)\frac{\partial \Pi_{n+1}}{\partial \dot{\phi}}\right] - 1$$

$$d_1 = \frac{\partial Z}{\partial \dot{\phi}} = \frac{2Z_0 (Z_1 - Z_n)\, m \Delta t \Pi}{\left(m \Delta t \Pi \dot{\phi} + 2Z_0\right)^2}$$

$$2d_2 = \frac{\partial Z}{\partial \Pi} = \frac{2Z_0 (Z_1 - Z_n)\, m \Delta t \dot{\phi}}{\left(m \Delta t \Pi \dot{\phi} + 2Z_0\right)^2}$$

$$f_1 = \frac{2}{\sqrt{3}} D_0 \exp\left[-\frac{1}{2}\frac{N+1}{N}\left(\frac{Z(\Pi,\dot{\phi})}{\Pi}\right)^{2N}\right]$$

$$f_2 = -(N+1)\left(\frac{Z(\Pi,\dot{\phi})}{\Pi}\right)^{2N-1}\frac{1}{\Pi}$$

$$f_3 = (N+1)\left(\frac{Z(\Pi,\dot{\phi})}{\Pi}\right)^{2N-1}\left(\frac{Z}{2\Pi^2}\right)$$

$$\frac{\partial \Pi_{n+1}}{\partial \dot{\phi}} = -\Delta T\, \overline{\boldsymbol{\nu}}^{\mathrm{T}}_{n+1} : (\boldsymbol{H}^{-1}\boldsymbol{T})$$

$$(\boldsymbol{H})_{abmn} = [\delta_{ma}\delta_{nb} - (\boldsymbol{G})_{abmn}]$$

$$(\boldsymbol{G})_{abmn} = \frac{\Delta T \dot{\phi}}{\Pi}\frac{\partial \overline{\Xi}_{mn}}{\partial (\mathbf{F}_{\mathrm{p}}^{-1})_{ij}}\Big[\overline{\nu}_{ab}\overline{\nu}_{oi}(\mathbf{F}_{\mathrm{p}}^{-1})_{oj} - \frac{3}{2}\left((\mathbf{F}_{\mathrm{p}}^{-1})_{aj}\delta_{ib} - \frac{1}{3}\delta_{ab}(\mathbf{F}_{\mathrm{p}}^{-1})_{ij}\right)\Big]$$

$$(\boldsymbol{T})_{mn} = \frac{\partial \overline{\Xi}_{mn}}{\partial (\mathbf{F}_{\mathrm{p}}^{-1})_{ij}}\overline{\nu}_{oi}(\mathbf{F}_{\mathrm{p}}^{-1})_{oj}$$

The term $\partial\overline{\Xi}_{mn}/\partial(\mathbf{F}_{\mathrm{p}}^{-1})_{ij}$ is formulated in Eqs. (78)-(81).

Box 1: The local iteration

3.2 The Algorithmic Tangent Operator

The algorithmic tangent operator is obtained by linearization of the second Piola-Kirchhoff stress tensor $\mathbf{S}$ with respect to the right Cauchy-Green deformation tensor. The stress tensor $\mathbf{S}$ can be written in the form

$$\mathbf{S} = \mathbf{C}^{-1}\boldsymbol{\Xi}\,. \tag{82}$$

The mentioned derivative is then

$$\frac{\partial \mathbf{S}}{\partial \mathbf{C}} = \frac{\partial \mathbf{C}^{-1}}{\partial \mathbf{C}}\boldsymbol{\Xi} + \mathbf{C}^{-1}\frac{\partial \boldsymbol{\Xi}}{\partial \mathbf{C}}\,. \tag{83}$$

Using indices the equation can be rewritten as

$$\frac{\partial S_{ib}}{\partial C_{rs}} = -C^{-1}_{ir}C^{-1}_{sa}\Xi_{ab} + C^{-1}_{ia}\frac{\partial \Xi_{ab}}{\partial C_{rs}}\,. \tag{84}$$

The computation of the term $\partial \Xi_{mn}/\partial C_{rs}$ is carried out very much in the spirit of the preceding section. First we have to notice that $\boldsymbol{\Xi}$ depends on $\mathbf{C}$ in two manners. On the one hand explicitly, and on the other hand implicitly through its dependence on $\mathbf{F}_{\mathrm{p}}^{-1}$ and the dependence of the latter on $\mathbf{C}$. Taking these facts into account one can write

$$\frac{\partial \Xi_{mn}}{\partial C_{rs}} = \frac{\partial \Xi_{mn}}{\partial C_{rs}}|_{expl} + \frac{\partial \Xi_{mn}}{\partial (\mathbf{F}_{\mathrm{p}}^{-1})_{ij}}\left(\frac{\partial \overline{\beta}_{io}}{\partial \dot{\phi}}\frac{\partial \dot{\phi}}{\partial \Xi_{ab}} + \frac{\partial \overline{\beta}_{io}}{\Xi_{ab}}\right)(\mathbf{F}_{\mathrm{p}}^{-1})_{oj}\frac{\partial \Xi_{ab}}{\partial C_{rs}}\,. \tag{85}$$

Here, $\partial \Xi_{mn}/\partial C_{rs}|_{expl}$ is the explicit derivative of Ξ_{mn} with respect to C_{rs} which, with the help of (20) and (26) is given as

$$
\begin{aligned}
\frac{\partial \Xi_{mn}}{\partial C_{rs}}|_{expl} = & \sum_{i=1}^{3} \left[\frac{\partial \psi}{\partial I_i} \left(\delta_{mr} (\mathbf{M}_i)_{st} (\mathbf{C}_{\mathrm{p}}^{-1})_{tn} + \delta_{mr} (\mathbf{C}_{\mathrm{p}}^{-1})_{st} (\mathbf{M}_i)_{tn} \right) \right. \\
& + \frac{\partial \psi}{\partial I_{i+3}} \left(\delta_{mr} (\mathbf{M}_i)_{st} (\mathbf{C}_{\mathrm{p}}^{-1})_{tq} C_{qb} (\mathbf{C}_{\mathrm{p}}^{-1})_{bn} \right. \\
& + C_{mo} (\mathbf{M}_i)_{ot} (\mathbf{C}_{\mathrm{p}}^{-1})_{tr} (\mathbf{C}_{\mathrm{p}}^{-1})_{sn} \\
& + \delta_{mr} (\mathbf{C}_{\mathrm{p}}^{-1})_{st} (\mathbf{M}_i)_{tq} C_{qb} (\mathbf{C}_{\mathrm{p}}^{-1})_{bn} \\
& + C_{mo} (\mathbf{C}_{\mathrm{p}}^{-1})_{ot} (\mathbf{M}_i)_{tr} (\mathbf{C}_{\mathrm{p}}^{-1})_{sn} \\
& + \delta_{mr} (\mathbf{C}_{\mathrm{p}}^{-1})_{st} C_{tq} (\mathbf{M}_i)_{qb} (\mathbf{C}_{\mathrm{p}}^{-1})_{bn} \\
& + C_{mo} (\mathbf{C}_{\mathrm{p}}^{-1})_{or} (\mathbf{M}_i)_{sb} (\mathbf{C}_{\mathrm{p}}^{-1})_{bn} \\
& + \delta_{mr} (\mathbf{C}_{\mathrm{p}}^{-1})_{st} C_{tq} (\mathbf{C}_{\mathrm{p}}^{-1})_{qb} (\mathbf{M}_i)_{bn} \\
& \left. \left. + C_{mo} (\mathbf{C}_{\mathrm{p}}^{-1})_{or} (\mathbf{C}_{\mathrm{p}}^{-1})_{sb} (\mathbf{M}_i)_{bn} \right) \right] \\
& + \left(2\alpha_2 (\mathbf{C}_{\mathrm{p}}^{-1})_{rt} (\mathbf{M}_1)_{ts} + \alpha_7 (\mathbf{C}_{\mathrm{p}}^{-1})_{rt} (\mathbf{M}_2)_{ts} \right. \\
& \left. + \alpha_8 (\mathbf{C}_{\mathrm{p}}^{-1})_{rt} (\mathbf{M}_3)_{ts} \right) \\
& \quad \left(C_{mo} (\mathbf{M}_1)_{ot} (\mathbf{C}_{\mathrm{p}}^{-1})_{tn} + C_{mo} (\mathbf{C}_{\mathrm{p}}^{-1})_{ot} (\mathbf{M}_1)_{tn} \right) \\
& + \left(2\alpha_4 (\mathbf{C}_{\mathrm{p}}^{-1})_{rt} (\mathbf{M}_2)_{ts} + \alpha_7 (\mathbf{C}_{\mathrm{p}}^{-1})_{rt} (\mathbf{M}_1)_{ts} \right. \\
& \left. + \alpha_9 (\mathbf{C}_{\mathrm{p}}^{-1})_{rt} (\mathbf{M}_3)_{ts} \right) \\
& \quad \left(C_{mo} (\mathbf{M}_2)_{ot} (\mathbf{C}_{\mathrm{p}}^{-1})_{tn} + C_{mo} (\mathbf{C}_{\mathrm{p}}^{-1})_{ot} (\mathbf{M}_2)_{tn} \right) \\
& + \left(2\alpha_6 (\mathbf{C}_{\mathrm{p}}^{-1})_{rt} (\mathbf{M}_3)_{ts} + \alpha_8 (\mathbf{C}_{\mathrm{p}}^{-1})_{rt} (\mathbf{M}_1)_{ts} \right. \\
& \left. + \alpha_9 (\mathbf{C}_{\mathrm{p}}^{-1})_{rt} (\mathbf{M}_2)_{ts} \right) \\
& \quad \left(C_{mo} (\mathbf{M}_3)_{ot} (\mathbf{C}_{\mathrm{p}}^{-1})_{tn} + C_{mo} (\mathbf{C}_{\mathrm{p}}^{-1})_{ot} (\mathbf{M}_3)_{tn} \right) . \qquad (86)
\end{aligned}
$$

With Eq. (85) at hand, the derivative of the material stress tensor with respect to the right Cauchy-Green tensor is given as

$$
\begin{aligned}
\frac{\partial \Xi_{ab}}{\partial C_{rs}} = & \left[\delta_{ma} \delta_{nb} - \left(\frac{\partial \Xi_{mn}}{\partial (\mathbf{F}_{\mathrm{p}}^{-1})_{ij}} \frac{\partial \overline{\beta}_{io}}{\partial \dot{\phi}} (\mathbf{F}_{\mathrm{p}}^{-1})_{oj} \frac{\partial \dot{\phi}}{\partial \Xi_{ab}} \right. \right. \\
& \left. \left. + \frac{\partial \Xi_{mn}}{\partial (\mathbf{F}_{\mathrm{p}}^{-1})_{ij}} \frac{\partial \overline{\beta}_{io}}{\partial \Xi_{ab}} (\mathbf{F}_{\mathrm{p}}^{-1})_{oj} \right) \right]^{-1} \frac{\partial \Xi_{mn}}{\partial C_{rs}}|_{expl} . \qquad (87)
\end{aligned}
$$

From the last local iteration, the quantities $\partial \overline{\beta}_{io}/\partial \dot{\phi}$ and $\partial \overline{\beta}_{io}/\partial \Xi_{ab}$ are known. What we still need are formulas for $\partial \dot{\phi}/\partial \boldsymbol{\Xi}$ and for $\partial \Xi_{mn}/\partial (\mathbf{F}_{\mathrm{p}}^{-1})_{ij}$.

The first is obtained using Eq. (59) which directly gives

$$\begin{aligned}\frac{\partial\dot{\phi}}{\partial\Xi} = & \frac{2}{\sqrt{3}}D_0 \exp\left[-\frac{1}{2}\frac{N+1}{N}\left(\frac{Z(\Pi,\dot{\phi})}{\Pi}\right)^{2N}\right]\times \\ & \times\left[-(N+1)\left(\frac{Z(\Pi,\dot{\phi})}{\Pi}\right)^{2N-1}\right]\times \\ & \left[\frac{1}{2\Pi}\frac{\partial Z}{\partial \Pi} - \frac{Z}{2\Pi^2}\right]\frac{\partial \Pi_{n+1}}{\partial\Xi} \\ & + \frac{2}{\sqrt{3}}D_0 \exp\left[-\frac{1}{2}\frac{N+1}{N}\left(\frac{Z(\Pi,\dot{\phi})}{\Pi}\right)^{2N}\right]\times \\ & \times\frac{1}{\Pi}\left[-(N+1)\left(\frac{Z(\Pi,\dot{\phi})}{\Pi}\right)^{2N-1}\right]\frac{\partial Z}{\partial\dot{\phi}}\frac{\partial\dot{\phi}}{\partial\Xi}\,. \end{aligned} \tag{88}$$

From the last equation we deduce

$$\frac{\partial\dot{\phi}}{\partial\Xi_{rs}} = \frac{f_1\,(0.5 f_2 d_2 + f_3)}{1 - f_1 . f_2 d_1}\frac{\partial \Pi_{n+1}}{\partial \Xi_{rs}}\,, \tag{89}$$

where the abbreviations (63)-(67) are used. The term $\partial\Pi_{n+1}/\partial\Xi_{rs}$ is also known from the last local iteration. As far as $\partial\Xi_{mn}/\partial(\mathbf{F}_\mathrm{p}^{-1})_{ij}$ is concerned, the elaboration goes as follows. First, one has

$$\frac{\partial\Xi_{mn}}{\partial(\mathbf{F}_\mathrm{p}^{-1})_{ij}} = \frac{\partial\Xi_{mn}}{\partial(\mathbf{C}_p^{-1})_{rs}}\frac{\partial(\mathbf{C}_\mathrm{p}^{-1})_{rs}}{\partial(\mathbf{F}_\mathrm{p}^{-1})_{ij}}\,, \tag{90}$$

where Eq. (79) is to be considered. In addition, it is straightforward to derive from Eq. (26) the expression

$$
\begin{aligned}
\frac{\partial \Xi_{mn}}{\partial C^{p-1}_{rs}} = & \sum_{i=1}^{3} \Bigg[\frac{\partial \psi}{\partial I_i} \left(C_{mo}(\mathbf{M}_i)_{or}\delta_{ns} + C_{mr}(\mathbf{M}_i)_{sn}\right) \\
& + \frac{\partial \psi}{\partial I_{i+3}} \Big(C_{mo}(\mathbf{M}_i)_{or} C_{sb} C^{p-1}_{bn} + C_{mo}(\mathbf{M}_i)_{ot} C^{p-1}_{tq} C_{qr}\delta_{ns} \\
& + C_{mr}(\mathbf{M}_i)_{sq} C_{qb} C^{p-1}_{bn} + C_{mo} C^{p-1}_{ot} (\mathbf{M}_i)_{tq} C_{qr}\delta_{ns} \\
& + C_{mr} C_{sq} (\mathbf{M}_i)_{qb} C^{p-1}_{bn} + C_{mo} C^{p-1}_{ot} C_{tq} (\mathbf{M}_i)_{qr}\delta_{ns} \\
& + C_{mr} C_{sq} C^{p-1}_{qb} (\mathbf{M}_i)_{bn} + C_{mo} C^{p-1}_{ot} C_{tr} (\mathbf{M}_i)_{sn} \Big) \Bigg] \\
& + \left(2\alpha_2(\mathbf{M}^1)_{rt} C_{ts} + \alpha_7(\mathbf{M}^2)_{rt} C_{ts} + \alpha_8(\mathbf{M}^3)_{rt} C_{ts}\right) \\
& \quad \left(C_{mo}(\mathbf{M}^1)_{ot} C^{p-1}_{tn} + C_{mo} C^{p-1}_{ot} (\mathbf{M}^1)_{tn}\right) \\
& + \left(2\alpha_4(\mathbf{M}^2)_{rt} C_{ts} + \alpha_7(\mathbf{M}^1)_{rt} C_{ts} + \alpha_9(\mathbf{M}^3)_{rt} C_{ts}\right) \\
& \quad \left(C_{mo}(\mathbf{M}^2)_{ot} C^{p-1}_{tn} + C_{mo} C^{p-1}_{ot} (\mathbf{M}^2)_{tn}\right) \\
& + \left(2\alpha_6(\mathbf{M}^3)_{rt} C_{ts} + \alpha_8(\mathbf{M}^1)_{rt} C_{ts} + \alpha_9(\mathbf{M}^2)_{rt} C_{ts}\right) \\
& \quad \left(C_{mo}(\mathbf{M}^3)_{ot} C^{p-1}_{tn} + C_{mo} C^{p-1}_{ot} (\mathbf{M}^3)_{tn}\right) . \qquad (91)
\end{aligned}
$$

Hence, all terms necessary for the evaluation of the tangent operator are provided. Note that the tangent is non–symmetric in general. The loss of symmetry is not a direct consequence of the anisotropy itself but of the fact that the update formula for the stress tensor can not be reduced to a simple additive form. For the sake of completeness and clarity we summarize the basic steps in box 2.

4 Numerical Examples

In this section we present several numerical examples to illustrate the range of applications and performance capabilities of the described formulation. The formulation is general and as such three-dimensional. In the following the applications are carried out for thin bodies (shells). The used shell theory and the corresponding finite element formulations have been presented in previous publications (see [16], [19]). The shell theory has the main feature of allowing for the application of a three-dimensional constitutive law. That is, no modifications of the constitutive law whatsoever are necessary. We recall that we have restricted ourselves to a free energy function which is quadratic in the strains. Accordingly, the derivative of the free energy function with respect to $\mathbf{C}_\mathrm{e}$ is necessarily linear allowing for the use of a fourth order tensor to write

$$
2\rho_{\mathrm{ref}} \frac{\partial \psi_\mathrm{e}}{\partial \mathbf{C}_\mathrm{e}} = \mathbf{H}\mathbf{E}_\mathrm{e} \,, \qquad \mathbf{E}_\mathrm{e} = \frac{1}{2}(\mathbf{C}_\mathrm{e} - \mathbf{1}) \,. \qquad (92)
$$

$$\frac{\partial S_{ib}}{\partial C_{rs}} = -C_{ir}^{-1} S_{sb} + C_{ia}^{-1} \frac{\partial \Xi_{ab}}{\partial C_{rs}}$$

$$\frac{\partial \Xi_{ab}}{\partial C_{rs}} = \left[\delta_{ma}\delta_{nb} - \left(\frac{\partial \Xi_{mn}}{\partial (\mathbf{F}_{\mathrm{p}}^{-1})_{ij}} \frac{\partial \overline{\beta}_{io}}{\partial \dot{\phi}} (\mathbf{F}_{\mathrm{p}}^{-1})_{oj} \frac{\partial \dot{\phi}}{\partial \Xi_{ab}} \right.\right.$$
$$\left.\left. + \frac{\partial \Xi_{mn}}{\partial (\mathbf{F}_{\mathrm{p}}^{-1})_{ij}} \frac{\partial \overline{\beta}_{io}}{\partial \Xi_{ab}} (\mathbf{F}_{\mathrm{p}}^{-1})_{oj} \right) \right]^{-1} \frac{\partial \Xi_{mn}}{\partial C_{rs}} |_{expl}$$

$$\frac{\partial \dot{\phi}}{\partial \Xi_{rs}} = \frac{f_1\,(0.5 f_2 d_2 + f_3)}{1 - f_1 . f_2 d_1} \frac{\partial \Pi_{n+1}}{\partial \Xi_{rs}}$$

$$\frac{\partial \Xi_{mn}}{\partial (\mathbf{F}_{\mathrm{p}}^{-1})_{ij}} = \frac{\partial \Xi_{mn}}{\partial (\mathbf{C}_p^{-1})_{rs}} \left(\delta_{ri} F_{sj}^{p-1} + \delta_{si} F_{rj}^{p-1} \right)$$

The term $\partial \Xi_{mn} / \partial C_{rs}|_{expl}$ is formulated in Eq. (86), that of $\partial \Xi_{mn} / \partial (\mathbf{C}_p^{-1})_{rs}$ in Eq. (91). The quantities $f_1, f_2, f_3, d_1, d_2, \partial \overline{\beta}_{io} / \partial \dot{\phi}, \partial \overline{\beta}_{io} / \partial \Xi_{ab}$, and $\partial \Pi_{n+1} / \partial \Xi_{rs}$ are known from the last local iteration step (Box 1).

Box 2: The tangent operator

The expression allows for a direct comparison with models of linear elasticity. Alternatively, one can understand $\mathbf{H}$ as a quadratic 6×6 matrix, with $\mathbf{E}_{\mathrm{e}}$ being a vector representation of $\frac{1}{2}(\mathbf{C}_{\mathrm{e}} - \mathbf{1})$. In this case, $\mathbf{H}$ has the classical matrix representation

$$\mathbf{H} = \begin{pmatrix} a & f & e & 0 & 0 & 0 \\ f & b & d & 0 & 0 & 0 \\ e & d & c & 0 & 0 & 0 \\ 0 & 0 & 0 & g & 0 & 0 \\ 0 & 0 & 0 & 0 & h & 0 \\ 0 & 0 & 0 & 0 & 0 & j \end{pmatrix}, \tag{93}$$

where the relations hold

$$\alpha_2 = \frac{1}{4}\left(\frac{a}{2} + g - h - j\right) , \tag{94}$$

$$\alpha_4 = \frac{1}{4}\left(\frac{b}{2} + h - j - g\right) , \tag{95}$$

$$\alpha_6 = \frac{1}{4}\left(\frac{c}{2} + j - g - h\right) , \tag{96}$$

$$\alpha_7 = \frac{f}{4} , \tag{97}$$

$$\alpha_8 = \frac{e}{4} , \tag{98}$$

$$\alpha_9 = \frac{d}{4} , \tag{99}$$

$$\alpha_{10} = \frac{1}{4}(h + j - g) , \tag{100}$$

$$\alpha_{11} = \frac{1}{4}(j + g - h) , \tag{101}$$

$$\alpha_{12} = \frac{1}{4}(g + h - j) . \tag{102}$$

The last equations provide relations between the constants α_i which appear in the definition of the free energy function and those of the elasticity matrix $\mathbf{H}$.

Altogether two sets of material constants have to be provided: that of the elastic constitutive law, the constants $a, b, c, ...$ etc. and that of the evolution equations, the constants D_0, Z_0, Z_1, m, N.

In the presented examples the following material constants are considered

Material I

$$D_0 = 10000\ 1/\text{sec},$$
$$Z_0 = 1150\ \text{N/mm}^2,$$
$$Z_1 = 1540\ \text{N/mm}^2,$$
$$m = 100,$$
$$N = 1,$$

$a = 48968.\ \text{N/mm}^2$, $b = 13588.\ \text{N/mm}^2$, $c = 13414.\ \text{N/mm}^2$,
$d = 4669.2\ \text{N/mm}^2$, $e = 1874.7\ \text{N/mm}^2$, $f = 3469.2\ \text{N/mm}^2$,
$g = 4804.3\ \text{N/mm}^2$, $h = 4804.3\ \text{N/mm}^2$, $j = 4214.3\ \text{N/mm}^2$,

At least in one example a further set of elastic material constants have been examined. The constants read:

Material II

$a = 175770.\ \text{N/mm}^2$, $b = 7498.1\ \text{N/mm}^2$, $c = 7498.1\ \text{N/mm}^2$,

$d = 1898.1\ \mathrm{N/mm^2}$, $e = 2349.\ \mathrm{N/mm^2}$, $f = 2349.\ \mathrm{N/mm^2}$,
$g = 3500.\ \mathrm{N/mm^2}$, $h = 3500.\ \mathrm{N/mm^2}$, $j = 1400.\ \mathrm{N/mm^2}$.

All other constants are taken to be the same as for material I. Implicitly in all examples material I is considered. The use of the material II will be explicitly mentioned. All computations are carried out displacement-controlled.

In the above, and due to the lack of experimental data, each set is chosen to be realistic but both sets are not based experimentally on one and the same material. The viscoplastic parameters are determined by Bodner and Partom using one-dimensional experiments. The anisotropic constants are well known for metals. They fulfill certain restrictions as to provide a positive definite elastic tensor and physically reasonable behaviour. The value of one Young's modulus and one Possion's ratio are comparable with those related to the isotropic viscoplastic model (experiments of Bodner and Partom), which makes them good candidates for the theory presented in the paper.

4.1 Extension of a Bar

The geometry of a bar loaded at its free end by a longitudinal force is shown in Fig. 1. The boundary conditions on the second end of the bar allow for a free contraction of the specimen. Here, only the left corner point is fixed. The specimen is modelled using 10×26 4-node elements. Three computations are carried out for the privileged direction of orthotropy given by the angle $0°, 60°$, and $90°$. The prescribed deformation velocity of the left top corner is 0.035 mm/sec. The load-displacement curves are shown in Fig. 2 and the corresponding deformed configurations in Fig. 3. The results demonstrate clearly that anisotropy in general leads to a non-homogenous deformation which influences significantly the onset of instability.

4.2 Simply Supported Rectangular Plate

The second example considered is that of a simply supported plate under dead load. The plate consists of 16×32 4-node elements and its initial configuration is given in Fig. 4. The deformation velocity of the middle point is $0.1\,\mathrm{mm\,s^{-}1}$. Here, the results of computations for two different sets of material constants are provided, where in each case three different orientations of the privileged directions are considered. The load-displacement curves for material I are given in Fig. 5 and those for material II in Fig. 6. The deformed configuration of the plate related to one of the computations is included as an inset.

4.3 Pinched Cylinder with Rigid Diaphragm

The final problem considers a cylinder with rigid diaphragms at both ends and loaded by a line load acting on a segment of a length of only $5/28 \times \pi R$,

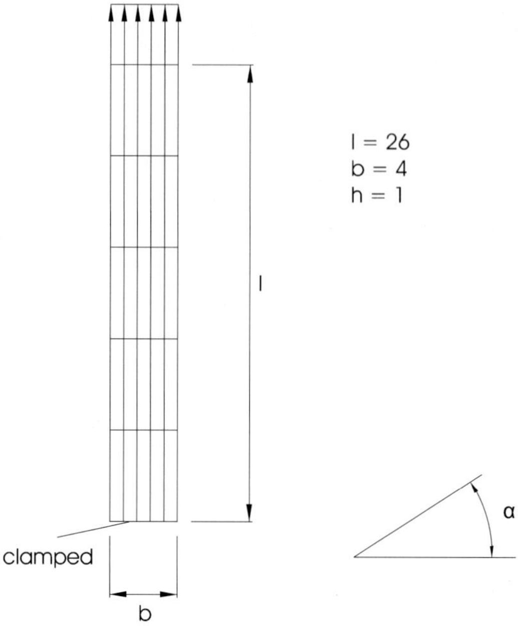

Fig. 1. Strip under tension. Problem definition - 10 x 26 Elements

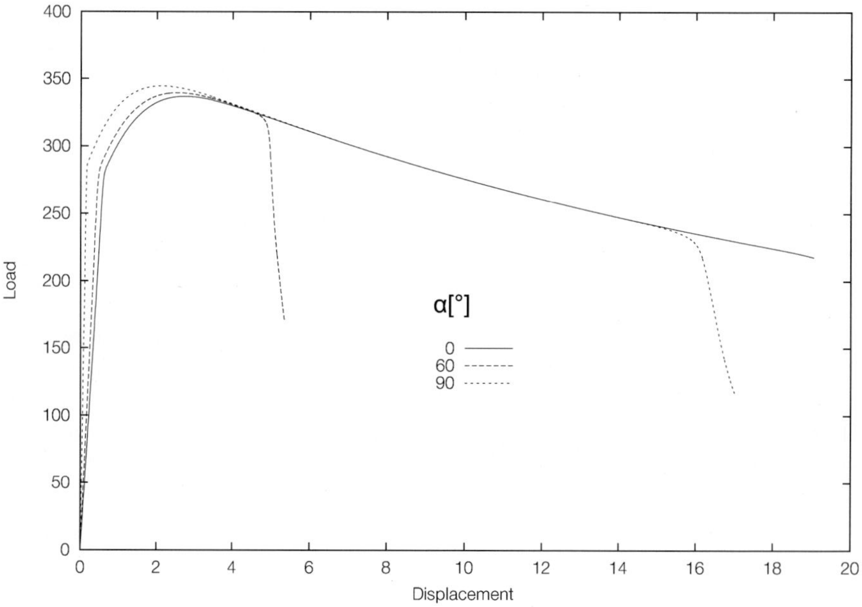

Fig. 2. Load-displacement curve

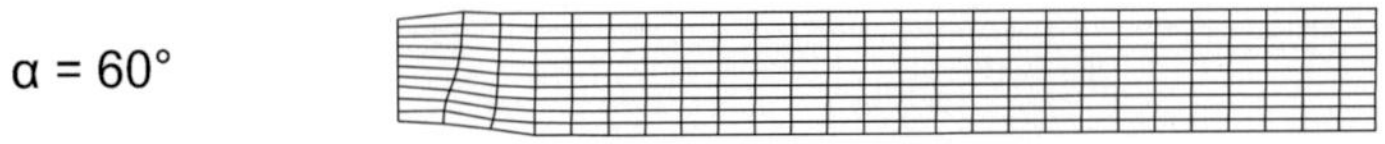

Fig. 3. Deformed configuration

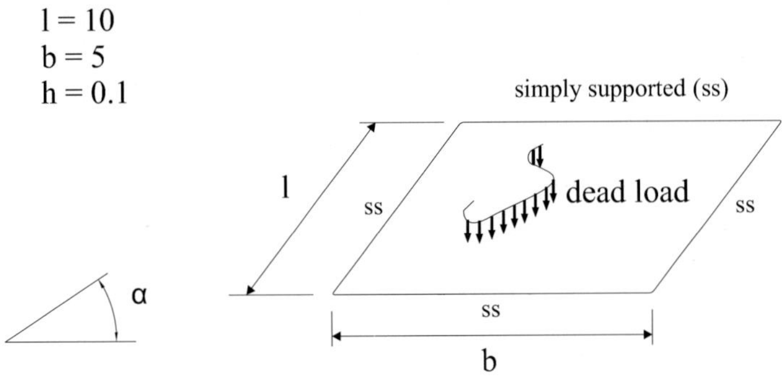

Fig. 4. Plate under transversal dead loading. Problem definition I - 16 x 32 elements

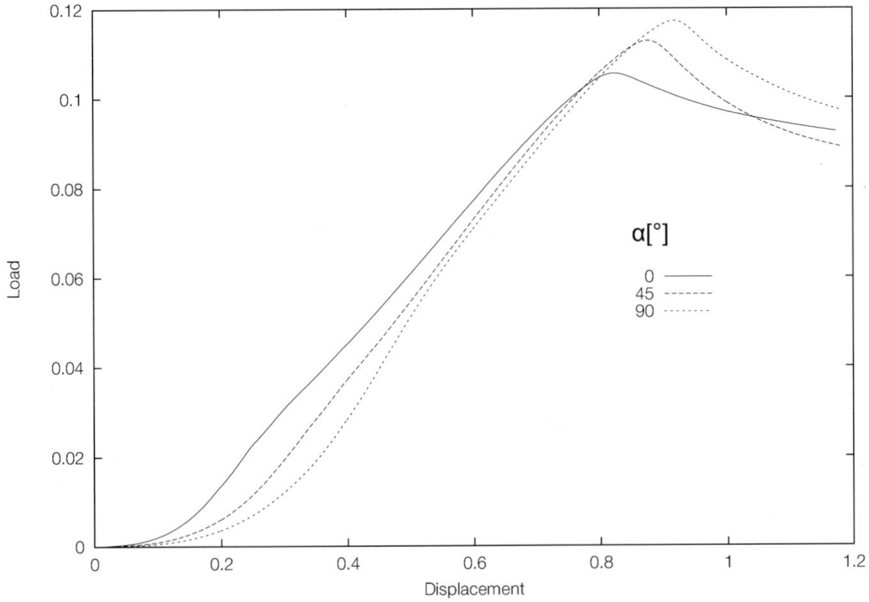

Fig. 5. Load-displacement curve for midpoint (Material I)

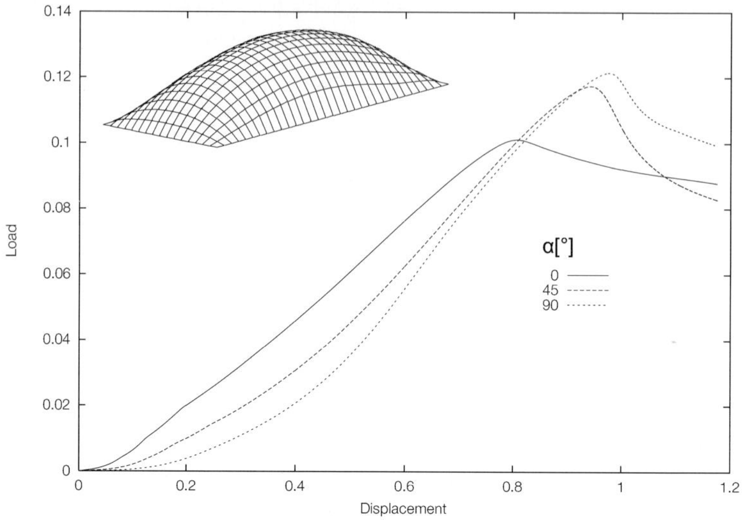

Fig. 6. Load-displacement curve for the plate midpoint (Material II). Inset: Deformed configuration for $\alpha = 45°$

as shown in Fig. 7. Considering symmetry conditions, only one eighth of the cylinder is modelled using 28×28 4-node elements. Accordingly, the whole cylinder is considered to consist of eight parts designed appropriately in order to assure the considered symmetries along the corresponding axes. The deformation velocity of the central point A is $0.32\,\mathrm{mm\,s^{-}1}$. The load-displacement curves, here too for three different orientations of the privileged directions, are given in Fig. 8 for point A at the top of the cylinder as well as for point B at its side. The deformed configuration is shown in Fig. 9.

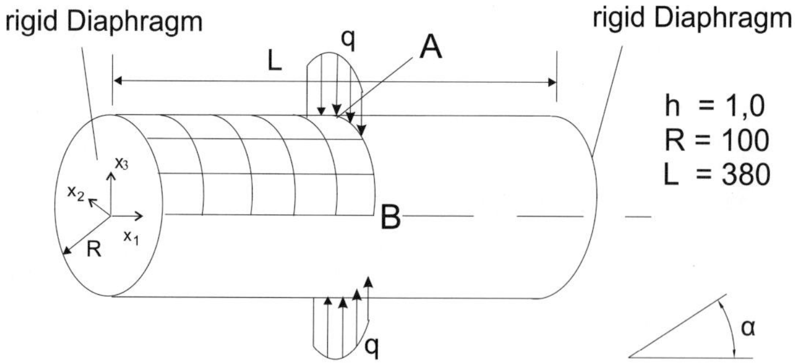

Fig. 7. Cylinder with rigid diaphragms. Problem definition

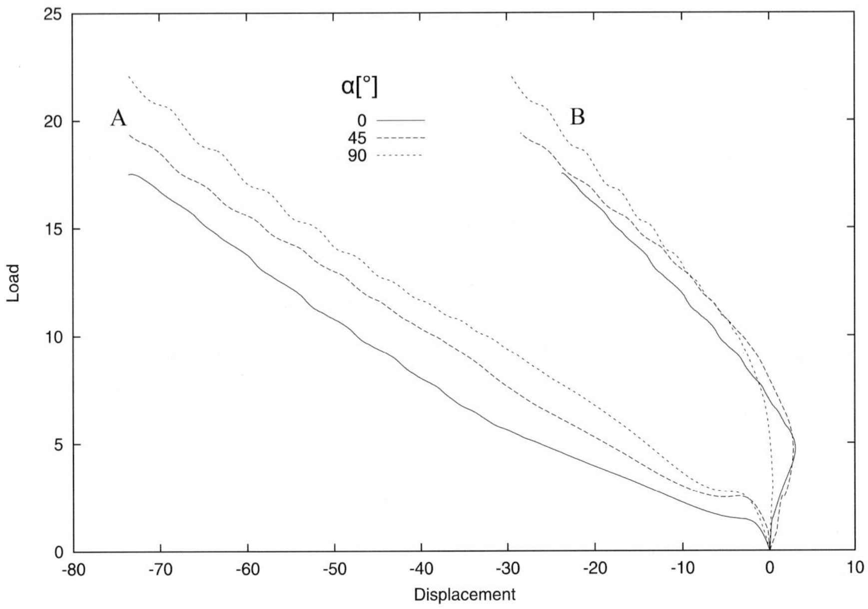

Fig. 8. Load-displacement curves

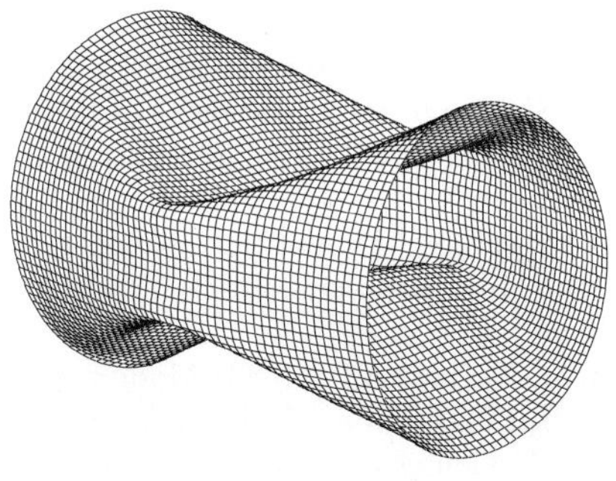

Fig. 9. Deformed configuration for $\alpha = 0°$

5 Conclusion

A theory and numerical algorithms for multiplicative inelasticity with assumed orthotropic constitutive law for the stress tensor have been considered. The viscoplastic unified model used is of Bodner & Partom-type and takes a mechanism of isotropic hardening into account. The orthotropic properties of the material are formulated by appropriate definition of the free energy function as an orthotropic tensor function. Specifically, it is assumed that the structural tensors defining the anisotropy under consideration should be transformed under the action of the inelastic part of the deformation gradient as mixed variant tensors. The integration of evolution equations is carried out using the exponential map which exactly fulfils the condition of incompressibility for inelastic deformation.

The paper focuses on establishing the expressions for the integration of the internal variables (local iteration) as well as on deriving the tangent operator. The latter is non-symmetric in general. The elaborations make clear that anisotropy complicates considerably the numerical procedures. The resulting equations are quite involved making the computational effort quite high.

References

1. Besseling J.F. (1968) A thermodynamic approach to rheology. In: Parkus H., Seddov, L.I. (Eds.) Irreversible Aspects of Continuum Mechanics. Springer, Wien
2. Bodner S. R., Partom Y. (1975) Constitutive equations for elastic-viscoplastic strain-hardening materials. ASME Journal of Applied Mechanics **42**: 385–389
3. Boehler J.P. (Ed.) (1987) Applications of Tensor Functions in Solid Mechanics. Springer, Wien
4. Boehlke T., Bertram A. (2001) The evolution of Hooke's law due to texture development in FCC polycrystals. International Journal of Solids and Structures **38**: 9437–9459
5. Dubrovin B.A., Fomenko A.T., Novikov S.P. (1984) Modern Geometry- Methods and Applications Part I. Springer-Verlag, New York
6. Ekh M., Runesson K. (2001) Modeling and numerical issues in hyperelastic-plasticity with anisotropy. International Journal of Solids and Structures **38**: 9461–9478
7. Gasser T.C., Holzapfel G.A. (2002) A rate-independent elastoplastic constitutive model for (biological) fiber-reinforced composites at finite strains: Continuum basis, algorithmic formulation and finite element implementation. Computational Mechanics, to appear
8. Haupt P., Kersten Th. (2002) On the modelling of anisotropic material behaviour in viscoplasticity. Int. J. Plasticity, submitted
9. Hill, R. (1948) A theory of yielding and plastic flow of anisotropic metals. Proceedings of the Royal Society **A 193**: 281-297.
10. Kröner E. (1960) Allgemeine Kontinuumstheorie der Versetzungen und Eigenspannungen. Archive for Rational Mechanics and Analysis **4**: 273–334

11. Lee E.H. (1969) Elastic-plastic deformation at finite strains. ASME Journal of Applied Mechanics **36**: 1–6
12. Menzel A., Steinmann P. (2002) On spatial formulation of anisotropic multiplicative elasto-plasticity, preprint
13. Papadopoulos P., Lu J. (2001) On the formulation and numerical solution of problems in anisotropic finite plasticty. Comp. Meth. Appl. Mech. Engrg. **190**: 4889-4910
14. Reese S. (2002) Meso-macro modelling of fibre-reinforced rubber-like composites exhibiting large elastoplastic deformations. Int. J. of Solids and Structures, to appear
15. Sansour C., Kollmann F.G. (1997) On theory and numerics of large viscoplastic deformation. Computer Methods in Applied Mechanics and Engineering **146**: 351–369
16. Sansour C., Kollmann F.G. (1998) Large viscoplastic deformations of shells. Theory and finite element formulation. Computational Mechanics **21**: 512–525
17. Sansour C., Kollmann F.G. (2001) Anisotropic formulations for finite strain viscoplasticity. Applications to shells, in: Wall W.A., Bletzinger K.-U., Schweizerhof K. (Eds.) Trends in Computational Structural Mechanics, CIMNE, Barcelona, 198-207
18. Sansour C., Kollmann F.G. (2002) On the formulation of anisotropic elastic laws within multiplicative finite strain viscoplasticity, submitted
19. Sansour C., Wagner W. (2001) A model of finite strain viscoplasticity based on unified constitutive equations. Theoretical and computational considerations with applications to shells. Computer Methods in Applied Mechanics and Engineering **191**: 423–450
20. Smith G.F. (1971) On isotropic functions of symmetric tensors, skew-symmetric tensors and vectors. International Journal of Engineering Science **9**: 899–916
21. Spencer AJM. (1971) Theory of invariants. In: Continuum Physics Vol.I C. Eringen (Ed.). Academic Press, New York 239–353
22. Svendsen B. (2001) On the modelling of anisotropic elastic and inelastic material behaviour at large deformation. International Journal of Solids and Structures **38**: 9579–9599
23. Tsakmakis C. (1996) Kinematic hardening rules in finite plasticity, part i: a constitutive approach. Continuum Mechanics Thermodynam. **8**: 215-231
24. Tsakmakis Ch. (2001) Description of plastic anisotropy effects at large deformations. Part I: Restrictions imposed by the second law and the postulate of Il'iushin, preprint.
25. Xiao H., Bruhns O.T., Meyers A. (2000) A consistent finite elastoplasticity theory combining additive and multiplicative decomposition of the stretching and the deformation gradient. Int. J. Plasticity **16**: 143-177
26. Zheng Q-S. (1994) Theory of representations for tensor functions- A unified invariant approach to constitutive equations. Applied Mechanics Reviews **47**: 545–587

Part II

Constitutive Behaviour — Experiments and Verification

Identification of Material Parameters for Inelastic Constitutive Models: Stochastic Simulation

Tobias Harth[1], Jürgen Lehn[1], and Franz Gustav Kollmann[2]

[1] Department of Mathematics, Darmstadt University of Technology
Schlossgartenstr. 7 D–64289 Darmstadt, Germany
{harth, lehn}@mathematik.tu-darmstadt.de
[2] Mauerkircherstr. 15, D–81679 München
fg.kollmann@t-online.de

received 23 Jul 2002 — accepted 30 Oct 2002

Abstract. The material parameters of a constitutive model have to be identified by minimization of the distance between the model response and the experimental data using an appropriate optimization algorithm. However, measurement failures and differences in the specimens lead to uncertainties in the determined parameters. Since the amount of test data is not sufficient for a statistical analysis, a method of stochastic simulation is introduced in order to generate artificial data with the same behaviour as the experimental data. The stochastic simulations are validated by comparing the results of the parameter fits from artificial data with the results from experimental data. Furthermore, the results of the parameter identification for different inelastic constitutive models are presented and a principal component analysis is performed to study the correlation structure of the estimated material parameters. The presented simulation method is a suitable tool for various kinds of analysis purposes since the artificial data presents an alternative to a high amount of experimental data.

1 Introduction

In order to describe inelastic deformations of metallic materials constitutive models are applied. In the last decades so called unified models have been developed which predict the time dependent and time independent response of a material to a given loading history by one and the same system of constitutive equations. We cite BODNER and PARTOM [2], CHABOCHE [6], KREMPL et al. [16], and STECK [31] as typical examples of unified constitutive models for isotropic materials.

Since all constitutive models contain parameters which characterize a material at a given temperature w. r. t. the model equations, identification experiments have to be performed. The test data available for this work consist of 132 experiments with specimens of stainless steel AINSI SS316 measured at 600°C. The experimental data split up into 11 different kinds of experiments which are given in detail in Sect. 2 together with a brief description of the applied material.

The constitutive models studied in this work are the model of Chan, Bodner, and Lindholm [9], a model of Chaboche [7], and a model of Chaboche, where the kinematic hardening variable proposed by Haupt, Kamlah, and Tsakmakis [13] is applied. The model equations are stated in Sect. 3.

In order to determine the material parameters very efficient numerical methods have to be applied. The constitutive models consist of systems of ODEs which cannot be solved analytically. Since the model parameters are identified simultaneously by minimization of the distance of the model response to the experimental data, the model equations must be integrated numerically thousands of times. Integration methods for viscoplastic models have been studied e. g. in [12] and [28]. Literature about parameter identification can be found e. g. in [5], [14], [17], and [20]. For the optimization of the parameter fits stochastic methods described in Sect. 4 are applied. The results of the parameter fits are stated and the accuracies of the applied models are compared.

Since the amount of test data is statistically insufficient for an analysis of the uncertainties in the identified parameters and other statistical applications, a method of stochastic simulation is introduced in Sect. 5, which is based on time series analysis. This stochastic method enables us to generate artificial data that exhibits the same mechanical behaviour as the experimental data. Many applications of stochastic methods in engineering are given e. g. in [10].

In Sect. 6 the stochastic simulations are validated by a comparison of the parameter fits from experimental data with the parameter fits from artificial data. This comparison is performed with appropriate nonparametric statistical tests.

Finally, in order to study the correlation structure of the identified parameters the method of principal component analysis is applied. In Sect. 7 the results of this study are given.

2 Material and Experimental Data

The applied material in this work is the austhenitic steel SS316, which belongs to the group of stainless steels. The short notation of this material is 1.4404 X2 CrNiMo 18 10. In Table 1 the chemical composition according to DIN 17440 together with a chemical analysis is stated.

The available test data consist of measurements from 132 experiments at a constant temperature of 600°C. These experiments split up into three different types, which are tension-relaxation tests, cyclic tension-compression

Table 1. Chemical composition of steel SS316

Element		Compostion of SS316	
		DIN 17440	Chemical Analysis
Cr	(Chrome)	16.5 to 18.5 %	16.74 %
Ni	(Nickel)	11.0 to 14.0 %	11.1 %
Mo	(Molybdenum)	2.0 to 2.5 %	2.0 %
Mn	(Manganese)	2.0 %	1.36 %
Si	(Silicon)	1.0 %	0.56 %
C	(Carbon)	$\leq$ 0.03 %	0.03 %
P	(Phosphorus)	$\leq$ 0.045 %	0.025 %
S	(Sulphur)	$\leq$ 0.03 %	0.025 %

tests, and creep tests. The first two types are strain controlled, whereas the latter is stress controlled. A detailed description is given below.

2.1 Tension-Relaxation Experiments

A tension experiment is a strain controlled experiment where a specimen is subjected to a constant positive strain rate. In a relaxation period the strain is kept constant and the measured stress shows a decrease in time. The performed experiments in this work are combined tension-relaxation experiments, where a specimen is strained by 1% and then a relaxation period of 15000 s is initiated. This process is repeated four times, such that a total strain of 4% is reached.

The available data of tension-relaxation experiments result from among 36 experiments, which have been performed with the strain rates 10^{-3}s^{-1}, 10^{-4}s^{-1}, and 10^{-5}s^{-1} respectively. Each of the three kinds of tests is performed with 12 specimens. The measured data of all 12 tests at strain rate 10^{-4}s^{-1} are plotted in Fig. 1.

The measurement data are collected at 1000 time instants, 50 equidistant measurements in every tension period and 200 equidistant measurements in every relaxation period.

2.2 Tension-Compression Experiments

The cyclic experiments consist of a sequence of tension periods at constant strain rate followed by compression periods equally at constant strain rate. The available test data result from 72 experiments of 5 tension-compression cycles measured with strain amplitudes of 0.25% and 0.5% at strain rates 10^{-3}s^{-1},10^{-4}s^{-1}, and 10^{-5}s^{-1}, respectively. Experiments for each of the six

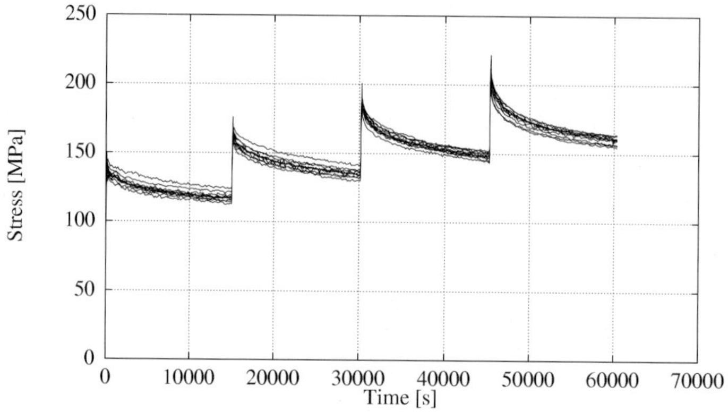

Fig. 1. Measured data of tension-relaxation experiments at strain rate $10^{-4}s^{-1}$

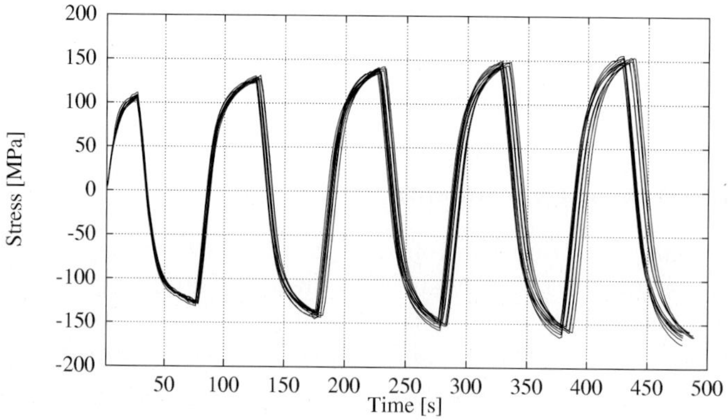

Fig. 2. Measured data of cyclic experiments at strain rate $10^{-4}s^{-1}$ and maximum strain of 0.25%

combinations of strain amplitudes and strain rates have been performed with 12 specimens. In Fig. 2 the experimental data of all 12 experiments at strain rate $10^{-4}s^{-1}$ and strain amplitude 0.25% are plotted.

For the tests with strain amplitude 0.25% measurements at 250 time instants are available and 485 measurements for the tests with 0.5% strain amplitude.

2.3 Creep Experiments

Creep experiments are stress controlled. They are performed with dead weight machines. A weight is fixed on a specimen and acts as a static force on the material, hence the specimen is subjected to a constant hold stress. In this work data from 24 experiments of two different kinds of creep tests are available. Creep tests with a hold stress of 230 MPa and a duration of 100

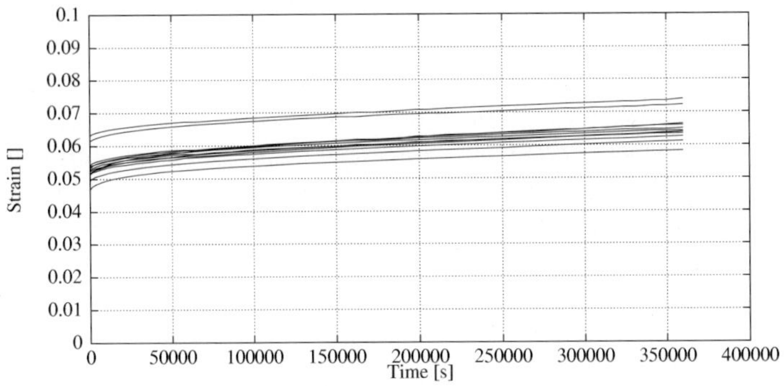

Fig. 3. Measured data of creep experiments with a duration of 100 hours

hours and tests with a hold stress of 160 MPa and a duration of 1000 hours have been performed. Each of these two kinds of creep tests is performed with 12 specimens. Figure 3 presents the data of all 12 creep tests with a duration of 100 hours.

Usually, for the description of creep experiments the inelastic or creep strain is used. This means that from the measured total strain ε the elastic strain ε^{e} is subtracted. Since in creep experiments the stress is constant the elastic strain is also constant and can be computed by HOOKE's law, which e. g. yields $\varepsilon^{\mathrm{e}} = 0.001586$ for the creep tests with a duration of 100 hours and a hold stress of 230 MPa, and a modulus of elasticity of $E = 145\,000$ MPa (see Sect. 4.4). Therefore, the inelastic strain ε^{p} is obtained from the measured total strain by $\varepsilon^{\mathrm{p}} = \varepsilon - \varepsilon^{\mathrm{e}}$. However, the time scale in Fig. 3 can not reveal the evolution of strain in the initial phase. The hold stress is reached after 2 seconds, the first measurement of strain is taken after 36 seconds. Fig. 3 shows the large total strain, which occurs particularly at high temperatures.

However, in our study the measured data is not manipulated, since we want to compare the measured data directly with data obtained from numerical simulation. Therefore, to the simulated inelastic strain the elastic part is added. This procedure has the additional advantage that for comparison with simulations in all three kinds of experiments (tension with intermediate relaxation, creep and tension-compression tests under cyclic straining) described in this paper one and the same strain is used, namely the total strain.

For the creep tests with a hold stress of 230 MPa and a duration of 100 hours 50 equidistant measurements of strain are available and 150 equidistant measurements for the tests with a hold stress of 160 MPa and a duration of 1000 hours.

3 Constitutive Models

A mathematical frame for the prediction of the behaviour of loaded materials is called constitutive model. Its equations describe the relation between the observable variables for stress and strain. Since the response of a material to a given loading is strongly dependent on its mechanical properties, a constitutive model contains free material parameters, which have to be fitted by experimental data. The total strain, ε is additively decomposed into an elastic part, ε^{e} and an inelastic part, ε^{p}. The stress σ is determined by HOOKE's law. Below the three constitutive models applied in this work are presented, where only the uniaxial case is taken into account.

3.1 Chaboche Model

CHABOCHE proposed a yield surface model, which describes viscoplastic behaviour only in the presence of overstress, i. e. when the stress lies outside a boundary determined by kinematic and isotropic hardening variables respectively. The flow rule of this model is defined by

$$\dot{\varepsilon}^{\mathrm{p}} = \left\langle \frac{|\sigma - X| - R - h}{K} \right\rangle^{n} \cdot \operatorname{sign}(\sigma - X), \tag{1}$$

where the variables $X(t)$ and $R(t)$ describe kinematic and isotropic hardening respectively. The constant h denotes the initial yield stress, such that the evolution of the size of the elastic domain is given by $R(t) + h$ with $R(0) = 0$ as initial condition. The brackets $\langle \cdot \rangle$ denote the function given by

$$\langle x \rangle = \max(0, x). \tag{2}$$

In [8] the isotropic hardening variable R is defined by the differential equation

$$\dot{R} = b(q - R)\dot{p}, \tag{3}$$

where p denotes the accumulated plastic strain, i. e. $\dot{p} = |\dot{\varepsilon}^{\mathrm{p}}|$ holds. For the description of recovery effects at high temperatures an additional term

$$-\gamma_r R^{m_r}, \quad \text{so that} \quad \dot{R} = b(q - R)\dot{p} - \gamma_r R^{m_r}, \tag{4}$$

is applied as proposed in [8] and [14], since the experimental data was performed at 600 °C. The parameter b indicates the speed of stabilization, whereas the value of the parameter q is an asymptotic value according to the evolution of the isotropic hardening. For small accumulated plastic strain rates recovery takes place, which is controlled in its intensity by the parameters γ_r and m_r.

In order to obtain a more accurate modelling of kinematic hardening we follow the advice in [7] to consider more than only one variable. A sum of

Table 2. The constitutive model of CHABOCHE

Strain:

$$\varepsilon(t) = \varepsilon^{\mathrm{e}}(t) + \varepsilon^{\mathrm{p}}(t)$$

HOOKE's Law:

$$\sigma(t) = E \cdot \varepsilon^{\mathrm{e}}(t)$$

Flow Rule:

$$\dot{\varepsilon}^{\mathrm{p}}(t) = \left\langle \frac{|\sigma(t) - X(t)| - R(t) - k}{K} \right\rangle^{n} \cdot \operatorname{sign}(\sigma(t) - X(t))$$

Hardening Variables:

$$\begin{aligned} X(t) &= X_1(t) + X_2(t) \\ \dot{X}_i(t) &= c_i \cdot \dot{\varepsilon}^{\mathrm{p}}(t) - a_i \cdot X_i(t) \cdot |\dot{\varepsilon}^{\mathrm{p}}(t)| - \gamma_i \cdot |X_i(t)|^{m_i} \cdot \operatorname{sign}(X_i(t)) \qquad i = 1, 2 \\ \dot{R}(t) &= b \cdot (q - R(t)) \cdot |\dot{\varepsilon}^{\mathrm{p}}(t)| - \gamma_r \cdot R(t)^{m_r} \end{aligned}$$

Initial Conditions:

$$\varepsilon^{\mathrm{p}}(0) = 0, \quad X_1(0) = 0, \quad X_2(0) = 0, \quad R(0) = 0$$

Parameters:

h	Yield strength
K, n	Flow rule
a_1, a_2, c_1, c_2	Kinematic hardening
b, q	Isotropic hardening
$\gamma_1, \gamma_2, \gamma_r, m_1, m_2, m_r$	Recovery

two non-linear kinematic hardening variables $X = X_1 + X_2$ is applied. Their evolution is given by ARMSTRONG and FREDERICK type equations

$$\dot{X}_i = c_i \dot{\varepsilon}^{\mathrm{p}} - a_i X_i \dot{p} \qquad (i = 1, 2), \tag{5}$$

where again the terms

$$-\gamma_i |X_i|^{m_i} \cdot \operatorname{sign}(X_i) \qquad (i = 1, 2) \tag{6}$$

as in [8] are considered for recovery effects. As above (6) is added to the right hand side of (5). The complete model is stated in Table 2. The values of the parameters a_1 and a_2 denote the speed of saturation, and, according to these values, the parameters c_1 and c_2 are asymptotic values of the kinematic hardening variables.

A superposition of two kinematic hardening variables leads to a better description, since each variable covers its own strain range, where it yields a suitable modelling. Hence, one variable describes the hardening for rather large strains, whereas the other models particularly the transition from the elastic to the plastic domain. Hence, one variable is stabilizing rapidely. The model of CHABOCHE in the form presented here has 15 material parameters and its inelastic part consists of a system of four ordinary differential equations.

3.2 Extended Chaboche Model

In order to improve the description of cyclic behaviour more complex variables for kinematic hardening have to be applied. In [13] an extension of the ARMSTRONG and FREDERICK equation is introduced which considers an enhanced dependence on the strain range by an appropriate positive function g dependent on an additional internal variable S. The kinematic hardening equation becomes

$$\dot{X} = a\dot{\varepsilon}^{\mathrm{p}} - g(S)X\dot{p}, \tag{7}$$

where p denotes the accumulated plastic strain. Since one hardening variable in the previously described CHABOCHE model stabilizes rapidly and serves more or less only for modelling purposes, only one of the two hardening variables is exchanged by equation (7) to obtain a new constitutive model. The flow rule (1) and the other equations for the hardening variables stated in (5) for $i = 1$ and in (3) together with the recovery terms in (4) and (6) are not changed. For the sake of convenience we denote the exchanged kinematic hardening variable by X_2, which, according to [13], is defined by

$$\dot{X}_2 = c_2\dot{\varepsilon}^{\mathrm{p}} - \frac{a_2}{1 + vS}X_2\dot{p}, \tag{8}$$

where the evolution equation of the variable S is given by

$$\dot{S} = \frac{p}{p_0}(|X_2| - S) \tag{9}$$

with initial condition $S(0) = 0$.

When $v = 0$ in equation (8) equation (5) is obtained. Again a recovery term of the form

$$-\gamma_2|X_2|^{m_2} \cdot \mathrm{sign}(X_2) \tag{10}$$

is additively included in the evolution law of the variable X_2. The full model is presented in Table 3.

This extension of the model of CHABOCHE has 17 material parameters and its inelastic part consists of a system of five ODE's. It is the most complex model that is studied in this work.

3.3 The Model of Chan, Bodner, and Lindholm

The constitutive model of BODNER and PARTOM [2] is not a yield surface model as that of CHABOCHE. The model is applied exactly in the form as

Table 3. The extended constitutive model of CHABOCHE

Strain:

$$\varepsilon(t) = \varepsilon^{\mathrm{e}}(t) + \varepsilon^{\mathrm{p}}(t)$$

HOOKE's Law:

$$\sigma(t) = E \cdot \varepsilon^{\mathrm{e}}(t)$$

Flow Rule:

$$\dot{\varepsilon}^{\mathrm{p}}(t) = \left\langle \frac{|\sigma(t)-X(t)|-R(t)-k}{K} \right\rangle^{n} \cdot \mathrm{sign}(\sigma(t) - X(t))$$

Hardening Variables:

$$\begin{aligned}
X(t) &= X_1(t) + X_2(t) \\
\dot{X}_1(t) &= c_1 \cdot \dot{\varepsilon}^{\mathrm{p}}(t) - a_1 \cdot X_1(t) \cdot |\dot{\varepsilon}^{\mathrm{p}}(t)| - \gamma_1 \cdot |X_1(t)|^{m_1} \cdot \mathrm{sign}(X_1(t)) \\
\dot{X}_2(t) &= c_2 \cdot \dot{\varepsilon}^{\mathrm{p}}(t) - \frac{a_2}{1+v \cdot S(t)} \cdot X_2 \cdot |\dot{\varepsilon}^{\mathrm{p}}(t)| - \gamma_2 \cdot |X_2(t)|^{m_2} \cdot \mathrm{sign}(X_2(t)) \\
\dot{S}(t) &= \frac{|\dot{\varepsilon}^{\mathrm{p}}(t)|}{p_0} \cdot (|X_2(t)| - S(t)) \\
\dot{R}(t) &= b \cdot (q - R(t)) \cdot |\dot{\varepsilon}^{\mathrm{p}}(t)| - \gamma_r \cdot R(t)^{m_r}
\end{aligned}$$

Initial Conditions:

$$\varepsilon^{\mathrm{p}}(0) = 0, \quad X_1(0) = 0, \quad X_2(0) = 0,$$
$$S(0) = 0, \quad R(0) = 0$$

Parameters:

h	Yield strength
K, n	Flow rule
$a_1, a_2, c_1, c_2, v, p_0$	Kinematic hardening
b, q	Isotropic hardening
$\gamma_1, \gamma_2, \gamma_r, m_1, m_2, m_r$	Recovery

it is described in CHAN et al. [9]. This model also consists of a flow rule dependent on variables for kinematic and isotropic hardening, respectively. Again only the uniaxial formulation is considered. The flow rule is defined by the differential equation

$$\dot{\varepsilon}^{\mathrm{p}} = \sqrt{\frac{4}{3}} D_0 \cdot \mathrm{sign}(\sigma) \cdot \exp\left(-\frac{1}{2}\left(\frac{Z^2}{\sigma^2}\right)^{n}\right). \tag{11}$$

The parameter n indicates the sensitivity of the plastic strain rate for changes in stress σ (rate sensitivity). All equations of the CHAN, BODNER, and LINDHOLM model are stated in Table 4. This model contains 12 material parameters, and its inelastic part consists of a system of three ODEs.

The equation for $\dot{W}_{\mathrm{p}}$ describes the dissipated inelastic work. The hardening variable Z is dependent on a kinematic (directional) and an isotropic hardening variable $Z^{(\mathrm{D})}$ and $Z^{(\mathrm{I})}$, respectively.

Table 4. The constitutive model of CHAN, BODNER, and LINDHOLM

Strain:

$$\varepsilon(t) = \varepsilon^{\mathrm{e}}(t) + \varepsilon^{\mathrm{p}}(t)$$

HOOKE's Law:

$$\sigma(t) = E \cdot \varepsilon^{\mathrm{e}}(t)$$

Flow Rule:

$$\dot{\varepsilon}^{\mathrm{p}}(t) = \sqrt{\tfrac{4}{3}} \cdot D_0 \cdot \operatorname{sign}(\sigma(t)) \cdot \exp\left(-\tfrac{1}{2} \cdot \left(\tfrac{Z^2(t)}{\sigma^2(t)}\right)^n\right)$$

Hardening Variables:

$$\begin{aligned} Z(t) &= Z^{(\mathrm{I})}(t) + Z^{(\mathrm{D})}(t) \\ \dot{Z}^{(\mathrm{I})}(t) &= m_1 \cdot (Z_1 - Z^{(\mathrm{I})}(t)) \cdot \sigma(t) \cdot \dot{\varepsilon}^{\mathrm{p}}(t) - A_1 \cdot Z_1 \cdot \left(\tfrac{Z^{(\mathrm{I})}(t) - Z_2}{Z_1}\right)^{r_1} \\ Z^{(\mathrm{D})}(t) &= \beta(t) \cdot \operatorname{sign}(\sigma(t)) \\ \dot{\beta}(t) &= m_2 \cdot (Z_3 \cdot \operatorname{sign}(\sigma(t)) - \beta(t)) \cdot \sigma(t) \cdot \dot{\varepsilon}^{\mathrm{p}}(t) \\ &\quad - A_2 \cdot Z_1 \cdot \left(\tfrac{|\beta(t)|}{Z_1}\right)^{r_2} \cdot \operatorname{sign}(\beta(t)) \end{aligned}$$

Initial Conditions:

$$\varepsilon^{\mathrm{p}}(0) = 0, \quad Z^{(\mathrm{I})}(0) = Z_0, \quad \beta(0) = 0$$

Parameters:

D_0, n	Flow rule
Z_3, m_2	Kinematic hardening
Z_0, Z_1, m_1	Isotropic hardening
A_1, A_2, r_1, r_2, Z_2	Recovery

The values of m_1 and m_2 control the isotropic and kinematic hardening rates, respectively, whereas the parameters Z_1 and Z_3 are saturation values of the two hardening variables. The parameter D_0 represents the limiting value for the plastic strain rate. In CHAN et al. [9] the number of parameters is reduced by the following choice of values for the recovery parameters

$$A_1 = A_2, Z_0 = Z_2 \quad \text{and} \quad r_1 = r_2. \tag{12}$$

In [30] this set-up for the recovery parameters was also chosen with accurate results. At room temperature, where recovery can be neglected the parameters A_1 and A_2 can be even set to zero. In this case the number of material parameters would reduce to 7. Since the data applied in this work is performed at 600 °C the number can only be reduced by (12) to 9 material parameters.

4 Numerical Methods

4.1 Integration

In order to compute model responses for a given loading history and given material parameters the differential equations of the constitutive model have to be solved. Therefore, the constitutive model is rewritten as a system of ordinary differential equations. Let k be the number of evolution equations of the constitutive model, then the time dependent variables $y_1, \ldots, y_k$ are combined to a vector

$$\boldsymbol{y}(t) := (y_1, \ldots, y_k)^{\mathrm{T}}, \tag{13}$$

where T indicates transposition. This yields the k-dimensional ODE system

$$\dot{\boldsymbol{y}}(t) = (f_1(t, \boldsymbol{y}(t)), \ldots, f_k(t, \boldsymbol{y}(t)))^{\mathrm{T}} \tag{14}$$

with initial conditions

$$\boldsymbol{y}(0) = (y_1^0, \ldots, y_k^0)^{\mathrm{T}}, \tag{15}$$

where the functions $f_1, \ldots, f_k$ are defined by the model equations according to the Tables 2, 3, and 4, respectively.

We refer to [25] and [29] for a description of the numerical methods which can be applied to solve differential equations of this kind very efficiently. An investigation of the efficiency is given in [28] with the result that the generalized RUNGE-KUTTA method is best suited for the equations of this model. However, the explicit EULER method is the fastest algorithm for the integration of the model equations for tension-relaxation tests, but it is not suitable for computations of a high precision.

Since the integration of the model equations must be performed very often to determine the model parameters, a step-size control is implemented to reduce the computation time. In [25] an algorithm for adaptive step-size control is given which is applied to improve the accuracy and to decrease the number of computation steps.

4.2 Distance Function

The material parameters of a constitutive model are determined by minimization of the distance between the model response and the experimental data. Hence, a distance function is required which serves as a target function dependent on the material parameters for the optimization process. In [29] different distance functions are studied, and it is shown that the identification results are hardly affected by the choice of those functions. A distance function is applied which is derived from the L_2-Norm by

$$d(\boldsymbol{\theta}) := \sqrt{\frac{\sum\limits_{i=1}^{n} \left(y_i - \tilde{y}_i(\boldsymbol{\theta})\right)^2}{\sum\limits_{i=1}^{n} y_i^2}}, \tag{16}$$

where $\boldsymbol{\theta}$ is a parameter vector consisting of the parameters of a constitutive model, y_i is the measured value at discrete time t_i, $\tilde{y}_i(\boldsymbol{\theta})$ is the model response obtained from $\boldsymbol{\theta}$ at time t_i and n is the number of all discrete time points.

For simultaneous parameter fits from different kinds of experiments it is important to apply a distance function which yields comparable values for the fits from the different kinds of experiments. This is the reason why the function d is defined by the relative deviation w. r. t. the L_2-Norm.

In order to determine simultaneously the distance between m different experiments and the corresponding model responses dependent on one parameter vector $\boldsymbol{\theta}$ the function G is defined by

$$G(\boldsymbol{\theta}) := \sum_{j=1}^{m} \alpha^{(j)} \cdot d^{(j)}(\boldsymbol{\theta}), \tag{17}$$

where $d^{(j)}$ is the L_2-Distance according to (16) between the data of experiment j and its model response, calculated with weights $\alpha^{(j)} > 0$. A minimization of function (17) yields one parameter vector $\boldsymbol{\theta}$ such that the distances between the data of the m experiments and the model responses according to $\boldsymbol{\theta}$ are minimized with respect to the weights $\alpha^{(j)}$. For an appropriate choice of these weights the number of experiments belonging to different types of experiments should be considered, thus a choice of $\alpha^{(1)} = \ldots = \alpha^{(m)}$ is not always suitable. In this paper the weights are chosen such that $\sum_{j=1}^{m} \alpha^{(j)} = 1$. For every type of tests the same weight is assigned and for all tests belonging to one type of tests the weights are again equally assigned. The distance function in (17) is the target function of the optimization process for the parameter identification.

4.3 Optimization

The aim of parameter identification is to find a parameter vector $\boldsymbol{\theta}$ which minimizes the target function G defined in the previous section. Since the constitutive model consists of highly nonlinear differential equations for the minimization of the target function nonlinear optimization methods have to be applied. In [25] and [28] different optimization strategies are investigated and their computation times are compared. All these methods contain stochastic elements. However, it turned out that these methods require too much computation time because of their specialization on particular tasks of the minimization problem, i. e. the task of determining appropriate start parameters, the global search and the following local search for the minimum. Thus, combined methods yield the best results. The optimization strategy PRINO turned out to be the most efficient strategy in this set-up. This optimization strategy combines the cluster oriented strategy of Price [23] with the local optimization method of Müller, Nollau, and Polovinkin [22]. Again we refer to [29] for a description of the algorithms.

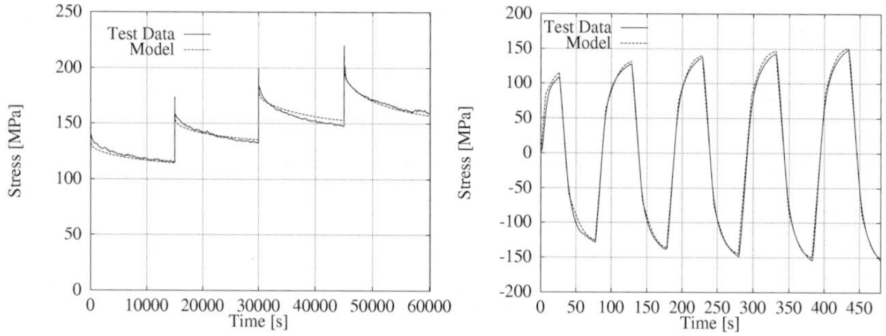

Fig. 4. Fit of the CHAN, BODNER, and LINDHOLM model to (**a**) a tension-relaxation test at strain rate $10^{-4}\mathrm{s}^{-1}$, and (**b**) a cyclic test at strain rate $10^{-4}\mathrm{s}^{-1}$ and strain amplitude of 0.25%

4.4 Identification Results

YOUNG's modulus E describes the linear relationship between stress and elastic strain in the purely elastic regime. Usually, YOUNG's modulus is determined in preliminary experiments and is then considered as a fixed material parameter. Since no such data of preliminary experiments is available for our work, YOUNG's modulus is estimated by linear regression in the elastic domains of the tension and tension-compression experiments. The linear regression of the experiments on SS316 at 600 °C yield a value of $E = 145\,000$ MPa. Hence, throughout this work YOUNG's modulus will be fixed to this value. In Fig. 4 fits of the CHAN, BODNER, and LINDHOLM model to a tension-relaxation test at strain rate $10^{-4}\mathrm{s}^{-1}$ are presented and a cyclic experiment at strain rate $10^{-4}\mathrm{s}^{-1}$ with strain amplitude 0.25% are presented, which show that the applied numerical methods work with high precision.

In the sequel, the results of simultaneous parameter fits to all 132 experiments are presented. The obtained parameter vector is called optimal parameter vector, since these material parameters yield the best description of the data. Thus, the parameters from the optimal parameter vector are considered to be the material parameters which correspond to the material SS316 and the constitutive model that is applied. A computation of an optimal parameter vector with the applied numerical methods takes between 9 and 12 hours on a PC.

In Table 5 the optimal parameter vector of the CHABOCHE model is given. YOUNG's modulus is fixed to a value of 145 000 MPa. For the parameter fit to all experimental data with the CHABOCHE model, the value of 7.72% of the target function is obtained.

The identified parameters from the data of all 132 experiments of the extended CHABOCHE model are given Table 6. The distance value for the fit with this model is 7.03%, which is clearly better than the result from the fit with the CHABOCHE model. This fact can only be due to the kinematic hardening variable of HAUPT, KAMLAH, and TSAKMAKIS.

Table 5. Optimal parameter vector of the CHABOCHE model

a_1	c_1	a_2	c_2	K	n	h	q	b
3547.78	43006.94	314.07	15655.34	17.03	6.38	62.64	265.70	9.69

γ_1	m_1	γ_2	m_2	γ_r	m_r
6.11e-5	2.59	8.34e-9	1.47	3.36e-9	1.79

Table 6. Optimal parameter vector of the extended CHABOCHE model

a_1	c_1	a_2	c_2	v	p_0	K	n	h
3776.96	36920.24	674.75	24979.44	11.36	27.24	17.69	3.26	67.44

q	b	γ_1	m_1	γ_2	m_2	γ_r	m_r
93.99	32.52	3.82e-4	1.57	1.33e-8	1.08	3.01e-9	1.93

Table 7. Optimal parameter vector of the CHAN, BODNER, and LINDHOLM model

D_0	Z_0	Z_1	Z_3	m_1	m_2	n	A_1	r_1
13252.02	356.36	2178.42	182.05	0.0186	1.97	1.45	24.91	19.29

In Table 7 the optimal parameter vector of the CHAN, BODNER, and LINDHOLM model is presented. The accuracy value for the parameter fit with this model is 8.73%. Since in this work YOUNG's modulus is fixed, this optimal parameter vector differs from the one given in [25], where YOUNG's modulus was considered as a free material parameter.

The improvement from the CHABOCHE model to the extended CHABOCHE model results mainly from the better description of cyclic experiments; this is shown in Table 8, where the accuracy values of the parameter fits of the three models to all cyclic experiments, tension-relaxation, and creep experiments are stated. It can be seen that the kinematic hardening variable of HAUPT, KAMLAH, and TSAKMAKIS yields a better description than the ARMSTRONG and FREDERICK variables in the CHABOCHE model. Both kinds of CHABOCHE models describe cyclic experiments much better than the model of CHAN, BODNER, and LINDHOLM.

However, both CHABOCHE models have two kinematic hardening equations, whereas the model of CHAN, BODNER, and LINDHOLM has only one and thus the integration and therefore the process of parameter identification is slower than for the latter model.

Table 8. Accuracy values from fits to the different types of experiments

Constitutive model	Cyclic	Tension-Relaxation	Creep
Model of CHABOCHE	7.73%	3.98%	4.38%
Extended Model of CHABOCHE	6.94%	3.63%	4.18%
Model of CHAN, BODNER, and LINDHOLM	9.52%	4.99%	5.43%

5 Stochastic Simulation

The identification experiments consist of scattered data and thus cause uncertainties for the parameter fits. Since not enough test data is available for a statistical analysis, a method of stochastic simulation is introduced, where artificial data are generated which exhibit the same properties as the experimental data. The experimental data is considered as a stochastic process and by a detailed analysis of the test data the behaviour of the scattered data can be modeled by time series. A detailed description can be found in [25]. The stochastic models of the data are based on first order autoregressive processes (AR[1]-processes).

However, the following remark applies. The artificial data generated by stochastic simulation can only model the stochastic properties of the limited number of experiments. As already mentioned, each such experiment has to be considered as a realization of a stochastic process. To represent this stochastic process in theory an infinite number of its realizations, i. e. experiments is required. If the limited number of performed experiments does not represent the properties of the stochastic process adequately, this deficiency can not be cured by stochastic simulation. Even worse, no methodology exists in the stochastic literature which enables one to decide whether the rather limited number of experiments represents adequately the properties of this underlying stochastic process. Therefore, our stochastic analysis is based on the rather strong assumption that our experiments yield an adequate representation of the properties of the underlying stochastic process. Thus, the accuracy of the stochastic simulations is dependent on the quality and the number of the considered experiments. The stochastic simulations do not add new information, but they give us the possibility to produce a larger variety of input than in a deterministic approach. This enables us to study correlations in the identification process at a higher level of confidence. In [24] and [10] more applications of stochastic simulation in engineering can be found.

5.1 Stochastic Models for Creep Tests

The data of each of the 12 performed creep tests with a duration of 100 hours consist of 50 measurements of strains $x_i^{(j)}$ at time t_i with $i = 1, \ldots, 50$ and

Table 9. Mean values and standard deviations of the coefficients $\hat{b}_1^{(j)}, \ldots, \hat{b}_4^{(j)}$ for the creep tests with a duration of 100 hours

l	1	2	3	4
$\bar{b}_l$	$5.39 \cdot 10^{-2}$	$-4.58 \cdot 10^{-2}$	$3.6 \cdot 10^{-5}$	0.4539
$\hat{\sigma}_l$	$4.7 \cdot 10^{-3}$	$1.443 \cdot 10^{-2}$	$1.14 \cdot 10^{-5}$	$2.34 \cdot 10^{-2}$

$j = 1, \ldots, 12$. For a description of the measured strain curves we choose the design functions $g^{(j)}$ defined by

$$g^{(j)}(t) = \hat{b}_1^{(j)} + \hat{b}_2^{(j)} \cdot \exp(-t) + \hat{b}_3^{(j)} \cdot t^{\hat{b}_4^{(j)}}, \tag{18}$$

where the coefficients $\hat{b}_1^{(j)}, \hat{b}_2^{(j)}, \hat{b}_3^{(j)}$ and $\hat{b}_4^{(j)}$ are estimated for every experiment j by the method of LEVENBERG-MARQUARD, which presents a method of non-linear regression given in [21]. The resulting mean values $\bar{b}_l$ and standard deviations $\hat{\sigma}_l$ with $l = 1, \ldots, 4$ of these coefficients are stated in Table 9.

The distance between the fitted function $g^{(j)}$ and the data of the process $x_i^{(j)}$ is represented by residuals r_i, which are defined by

$$r_i^{(j)} = x_i^{(j)} - g^{(j)}(t_i). \tag{19}$$

A study of their autocorrelations and partial autocorrelations has shown that an AR[1]-process is suitable for a description of this time series. First order autoregressive processes are stochastic processes $(P_i)_i$ with

$$P_i = \alpha \cdot P_{i-1} + Z_i, \tag{20}$$

where $(Z_i)_i$ is a sequence of independent and identically distributed random variables with an expectation of 0 and variance σ^2. The process $(Z_i)_i$ is called white-noise-process. Usually, the random variables Z_i are assumed to be normally distributed. The process $(P_i)_i$ is applied to describe the residuals in the stochastic model.

The parameters α and σ of the AR[1]-process modelling the residuals are estimated by the YULE-WALKER-Equations described in [4], which yield in this situation $\alpha = 0.66$ and $\sigma = 4.49 \cdot 10^{-5}$ such that the process

$$P_i = 0.66 \cdot P_{i-1} + Z_i \tag{21}$$

is obtained, where $(Z_i)_i$ is a sequence of independent and identically normally distributed random variables with an expectation of 0 and a standard deviation of $4.49 \cdot 10^{-5}$.

Furthermore, the assumption is made that the estimated coefficients $\hat{b}_l^{(j)}$ of the functions $g^{(j)}$ are realizations of normally distributed random variables

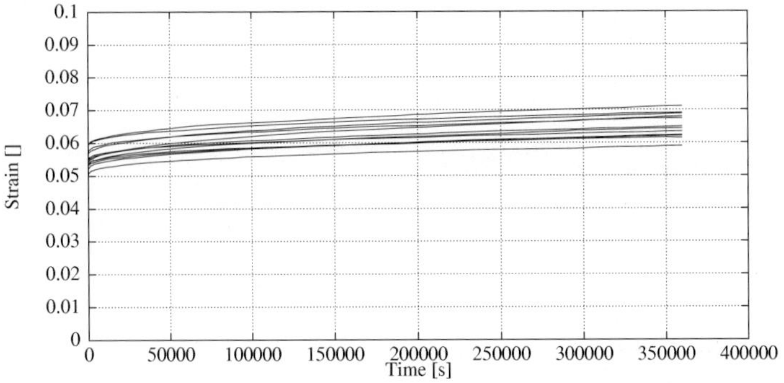

Fig. 5. Generated data for creep tests with a duration of 100 hours

b_l with distribution parameters as in Table 9. This assumption leads to the following definition of a sequence of random variables $(X_i)_i$

$$X_i = b_1 + b_2 \cdot \exp(-t_i) + b_3 \cdot t_i^{b_4} + P_i, \tag{22}$$

which serves as a stochastic model for the creep tests with a duration of 100 hours. In Fig. 5 the measured data and 12 generated curves for this kind of creep tests are given.

The stochastic model for the creep tests with a duration of 1000 hours is derived analogously.

5.2 Tension Tests with Relaxation

In order to develop stochastic models for the tension tests with relaxation basically the same methods as for the creep tests are applied. However, two differences have to be considered. At first, the performed tension tests consist of 4 different periods of tension and 4 periods of relaxation, thus the data of this type of tests has to be divided into 8 parts. Each of these parts is described by a partial model. The model of the total experiment consists therefore of a sequence of 8 partial models. The overall model is continuous with discontinuities of the stress derivative at the linking points. The second difference is that the tension periods do not exhibit equal duration times, which forces us to estimate these durations in the model. This estimator is assumed to be normally distributed. The design of the stochastic models for tension-relaxation tests is given in full detail in [25]. In Fig. 6 the measured data and 12 generated stress curves for tension tests at strain rate 10^{-4}s^{-1} are shown.

5.3 Cyclic Tension-Compression Tests

For the cyclic tension-compression tests the same problems as for the tension-relaxation tests appear. The tension periods have to be separated from the

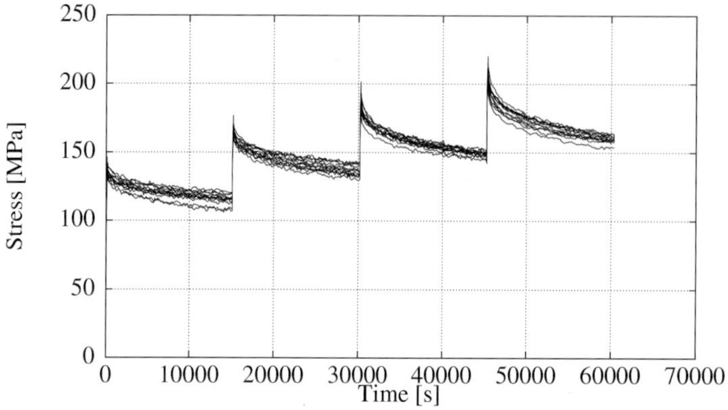

Fig. 6. Generated data for tension-relaxation tests at strain rate $10^{-4}s^{-1}$

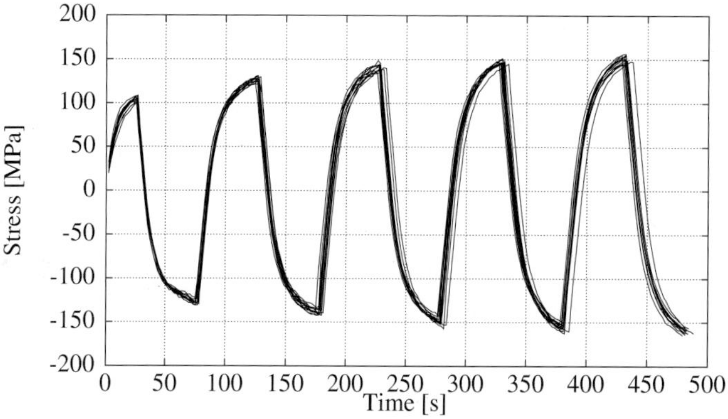

Fig. 7. Generated data for cyclic tests at strain rate $10^{-4}s^{-1}$ and maximum strain of 0.25%

compression periods in order to find model functions which describe the behaviour of the measured data. Since the durations of the single periods of tension and compression differ from each other they have to be estimated in the stochastic model. Again the assumption is made that this estimator is normally distributed. In Fig. 7 the measured data and 12 generated curves for cyclic tension-compression tests at strain rate $10^{-4}s^{-1}$ and maximum strain of 0.25% are given.

6 Validation of the Stochastic Simulations

In order to validate the stochastic simulations the parameter fits from artificial data must be compared with fits from experimental data. Only if the identified parameters from both fits are similar, it can be concluded that the artificial data exhibits the same mechanical behaviour as the experimental data and thus present a permissible alternative to a high amount of expensive tests. For a comparison a statistical analysis of a large number of parameter fits must be performed. Unfortunately, however, this analysis is restricted by the small amount of test data in our data base.

We decided to compare simultaneous fits from one experiment out of each of the eleven kinds of experiments, i. e. two kinds of creep tests, three kinds of tension-relaxation tests, and six kinds of cyclic experiments. This means that from every of the eleven different kinds of experiments one of the twelve available tests is randomly chosen and the parameters are simultaneously identified to these eleven experiments. This method enables us to repeat this process many times, since the number of possible combinations of randomly chosen experiments from our data base is 12^{11}. Another reason why not more experiments are considered for a simultaneous identification of parameters is the computation time. A simultaneous fit from 11 experiments belonging to different kinds of tests takes between 50 and 70 minutes on a PC dependent on the complexity of the applied model. Hence, we decided to repeat these parameter fits 250 times, which takes between 9 and 12 days on a PC.

Exactly the same method is applied for parameter fits to artificial data. Data from one experiment from each of the eleven kinds of experiments are generated by the stochastic simulations and then the material parameters are identified. This process is also repeated 250 times.

Since it was assumed that the parameters of the design functions of the stochastic simulations as e. g. the function defined in (18) are realizations of normally distributed random variables $b_i \sim N(\mu_i, \sigma_i^2)$ with $i = 1, \ldots, 4$, it must be verified that this assumption does not affect the results of the parameter identification. Thus random variables with a significantly different distribution function have to be applied to show the robustness of the stochastic simulations. The random variables b_i are replaced by uniformly distributed random variables $\tilde{b}_i \sim U(\mu_i - (\sqrt{12}/2)\sigma_i, \mu_i + (\sqrt{12}/2)\sigma_i)$. Again 250 parameter fits are performed from data generated by the stochastic simulations dependent on the random variables $\tilde{b}_i$ instead of the variables b_i.

6.1 Robustness of the Simulations

To show the robustness of the stochastic simulations, the results of the parameter identification from artificial data generated with normally distributed design parameters have to be compared with the results obtained from generated data with uniformly distributed design parameters. The 250 fits for every material parameter in both cases can be interpreted as realizations of independent and identically distributed random variables with distribution functions $F_1, \ldots, F_m$ and $G_1, \ldots, G_m$, where m denotes the number of material parameters of the model. For generating random numbers an inversive congruential generator is applied, which has better independence properties

Table 10. Absolute values of the WILCOXON rank-sum statistic for the CHABOCHE model

a_1	c_1	a_2	c_2	K	n	h	q	b
0.141	0.703	1.71	0.762	1.473	0.555	1.04	0.44	0.081

γ_1	m_1	γ_2	m_2	γ_r	m_r
1.131	0.181	0.565	2.192	1.391	0.436

than the widely used linear congruential generators and is suitable for parallel processing. This type of generator has been introduced by EICHENAUER and LEHN [11]. Since we cannot assume that the distributions F_i and G_i are normal distributions nonparametric methods have to be applied.

A nonparametric statistical test which checks equality of two distribution functions is the WILCOXON rank-sum test [18]. In the present situation the null hypothesis $H_0 : F_i = G_i$ is rejected by the two-sided Wilcoxon rank-sum statistic W for $|W| \geq 1.96$. Below the results of this test for every $i = 1, \ldots, m$ are given, where $m = 15$ for the CHABOCHE model, $m = 17$ for the extended CHABOCHE model, and for the CHAN, BODNER, and LINDHOLM model $m = 9$.

In Table 10 the absolute values of the WILCOXON rank-sum statistic for the parameter fits for the CHABOCHE model are presented. The null hypothesis has to be rejected only for the parameter m_2 in the recovery term of one of the two kinematic hardening variables. Since the values of $|W|$ are generally small it is concluded that the stochastic simulations are robust according to these results with the CHABOCHE model. In Fig. 8 the parameter fits for the parameters q and m_2 are presented in boxplots. The box in a boxplot shows the median, i. e. the 50th percentile, as a line and the 25th and 75th percentile are shown as the lower and upper parts of the box, respectively. The whiskers above and below the box represent the largest and smallest observed values.

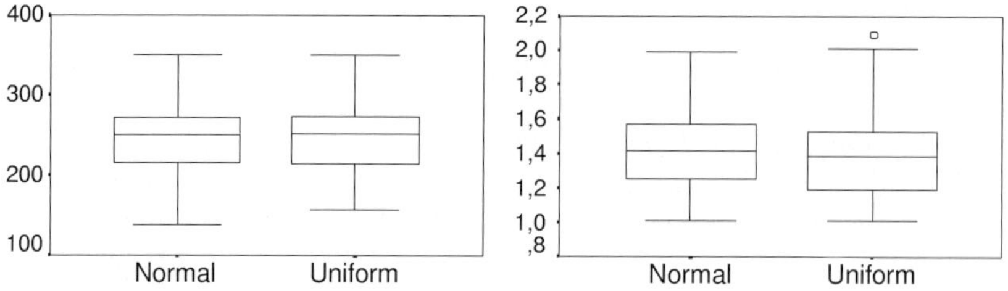

Fig. 8. Parameter fits of the CHABOCHE model for (**a**) parameter q, and (**b**) parameter m_2

Table 11. Absolute values of the WILCOXON rank-sum statistic for the extended CHABOCHE model

a_1	c_1	a_2	c_2	v	p_0	K	n	h
0.567	0.038	1.882	1.259	1.553	0.624	0.221	0.332	0.462

q	b	γ_1	m_1	γ_2	m_2	γ_r	m_r
0.695	0.134	1.721	0.615	1.395	1.111	0.451	0.548

Table 12. Absolute values of the WILCOXON rank-sum statistic for the CHAN, BODNER, and LINDHOLM model

D_0	Z_0	Z_1	Z_3	m_1	m_2	n	A_1	r_1
2.379	1.008	1.803	1.394	1.633	0.294	1.444	1.339	1.002

Similar results are obtained for the extended CHABOCHE model. The results are stated in Table 11. For no material parameter a significant value for the WILCOXON rank-sum statistic is obtained. It can be concluded that there is no significant difference for the parameter identification when normally distributed design parameters or uniformly distributed design parameters are chosen for the simulation.

The values of $|W|$ for the CHAN, BODNER, and LINDHOLM model are presented in Table 12. There is only one rejection of the null hypothesis, namely for the parameter D_0. According to [3] this parameter presents an upper bound of the plastic strain rate and can even be chosen as a fixed value. Hence, again the test shows a good coincidence between the results from the simulations with normally distributed design parameters and from the simulations with uniformly distributed design parameters.

6.2 Validation of the Stochastic Simulations

In the previous section the robustness of the stochastic simulations is verified. However, the parameter fits from experimental data have not been compared yet with the fits from artificial data. The 250 identified parameters from experimental data for each of the m parameters of a constitutive model are interpreted as realizations from independent and identically distributed random variables with distribution functions $H_1, \ldots, H_m$. The results from the fits from experimental data are compared with the fits from artificial data with normally distributed and uniformly distributed design parameters, respectively. Since the WILCOXON rank-sum test is a test for equality of only two samples, it is not applicable to this problem. A nonparametric statistical test for the null hypothesis $H_0 : F_i = G_i = H_i$ is the KRUSKAL-WALLIS test, which can be found in [18]. In this situation the null hypothesis must be

Table 13. Values of the KRUSKAL-WALLIS statistic for the CHABOCHE model

a_1	c_1	a_2	c_2	K	n	h	q	b
5.71	3.97	5.44	2.29	10.45	8.67	1.22	1.96	2.61

γ_1	m_1	γ_2	m_2	γ_r	m_r
6.50	2.95	106.1	5.04	3.30	0.92

Table 14. Values of the KRUSKAL-WALLIS statistic for the extended CHABOCHE model

a_1	c_1	a_2	c_2	v	p_0	K	n	h
2.78	2.43	3.56	7.04	2.60	7.50	0.54	0.18	0.16

q	b	γ_1	m_1	γ_2	m_2	γ_r	m_r
0.46	1.96	4.04	0.57	2.30	4.04	43.95	3.79

rejected if the values of the KRUSKAL-WALLIS statistic K are greater than 5.99. As in the previous section the results of the KRUSKAL-WALLIS test for all parameters of the three constitutive models are given.

The values of the KRUSKAL-WALLIS statistic K for the CHABOCHE model are presented in Table 13. There are rejections of the null hypothesis for the parameters γ_1, γ_2, K and n. Since the recovery parameters are hard to identify, because they are determined by relaxation periods which, in our data base, are only a part of the tension-relaxation tests, it can be concluded that this result shows a good coincidence between the fits from the stochastic simulations and the experimental data.

The results are even better for the extended CHABOCHE model which are stated in Table 14. Only three rejections have to be made, namely for the parameters c_2, p_0, and γ_r. Again this result emphasizes that the stochastic simulation yield good parameter fits in comparison to the fits from experimental data.

For the CHAN, BODNER, and LINDHOLM model the values of the KRUSKAL-WALLIS statistic lead to rejections for almost all the material parameters as presented in Table 15. However, also in Table 12 the values of the WILCOXON rank-sum statistic have been greater than the values of this statistic for the other two models. It can be concluded that the parameters for this model cannot be determined with the same accuracy as for the CHABOCHE models, i. e. the parameter identification seems to be more stable for the two CHABOCHE models. Thus, further investigations are necessary.

In Table 16 the absolute values $|W|$ of the WILCOXON rank-sum statistic W to the parameter fits from experimental data and from artificial data

Table 15. Values of the KRUSKAL-WALLIS statistic for the CHAN, BODNER, and LINDHOLM model

D_0	Z_0	Z_1	Z_3	m_1	m_2	n	A_1	r_1
5.45	145.50	16.22	114.98	8.68	3.93	202.42	7.79	1.27

Table 16. Absolute values of the WILCOXON rank-sum statistic for the Chan, Bodner, and Lindholm model

D_0	Z_0	Z_1	Z_3	m_1	m_2	n	A_1	r_1
1.04	10.86	3.98	9.70	1.55	1.64	12.65	1.57	0.10

with normally distributed design parameters are presented. This test leads to rejections only for the parameters Z_0, Z_1, Z_3, and n. This result emphasizes that in contrast to the two CHABOCHE models all three distributions F_i, G_i, and H_i according to the 9 parameters of the model exhibit small differences.

To gain more knowledge about the identified parameters and in order to find an explanation for the unexpected results for the CHAN, BODNER, and LINDHOLM model, a principal component analysis is performed in the next section.

7 Principal Component Analysis

A principal component analysis reveals the correlation structure of high dimensional data. Let m be the number of parameters of a constitutive model; then the aim of a principal component analysis is to find $r < m$ factors, which by linear combination, give an approximation to the data of identified parameters. The coefficients of the linear combinations corresponding to the r factors are called factor loadings. If r is not too small, then the factor loadings still contain all relevant information about the correlation of the samples of identified material parameters. Thus, the method of principal component analysis is applied for dimension reduction of high dimensional data and it is given in full detail e. g. in [15]. The absolute values of the factor loadings lie in the unit interval.

The factor loadings are determined by a diagonalization of the correlation matrix R of all identified parameter vectors. By neglecting the eigenvectors of the $m - r$ smallest eigenvalues a reduction of dimension is achieved and an approximation of the correlation matrix R is obtained by

$$R \approx L \cdot L^{\mathrm{T}}, \tag{23}$$

where L is the $m \times r$-matrix of factor loadings.

Table 17. Factor loadings for the CHABOCHE model

parameter	1	2	3	4	5	6
a_1	-0.102	0.155		0.942		0.109
c_1						0.985
a_2		0.955		0.135		
c_2		0.973				
K	0.109		0.936	0.137		
n					0.991	
q	0.938			-0.123		
b	-0.986					
h	0.448	-0.280	-0.598	0.385	-0.216	-0.182

The factor loadings corresponding to the parameter fits from the experimental data are now presented for all models. Since the number of parameters and also the number of extracted factors is large, the analysis is restricted to the most important parameters of each model, i. e. the recovery parameters will be neglected. This restriction does not affect the presented results. It should be noted that almost exactly the same results are obtained when the samples of parameter fits from the stochastic simulations are applied. This fact explains again that the artificial data comprise a good simulation of the experimental data.

The rows of the loading matrix contain the coefficients of the r factors in order to describe the sample of parameter fits for one parameter. If there is only one parameter which has a high loading on one factor, then this factor is called the unique factor of this parameter. A factor is called common factor, if two or more parameters have high loadings on this factor. Hence, parameters which have a common factor exhibit a relationship. Loadings with an absolute value smaller than 0.1 are omitted in the following tables since small loadings are irrelevant for this analysis and make the tables harder to read.

In Table 17 the loading matrix of the CHABOCHE model is stated. For the 9 considered parameters $r = 6$ factors are chosen. The parameters show a low complexity, since each of them has a high loading only on one factor. The only exception is the parameter h which describes the yield stress. The high complexity of this parameter is due to its influence on the hardening variables, since they describe the location and size of the yield surface. The parameters a_2 and c_2 as well as the parameters q and b each share a common factor. This is a very plausible result, since the parameters a_2 and c_2 are from the same kinematic hardening variable and the parameters q and b are from the isotropic hardening variable.

Table 18. Factor loadings for the extended CHABOCHE model

parameter	1	2	3	4	5	6	7
a_1							0.990
c_1						0.994	
a_2	0.332	0.863	0.230	0.126		-0.103	
c_2		0.941	-0.247				
v	0.514		0.453	0.635		-0.104	
p_0				0.951			
K			-0.358	0.127	-0.840		
n			-0.494	0.133	0.759		0.126
q	-0.928	-0.125					
b	0.937		-0.258				
h	-0.184		0.887	0.132			

If YOUNG's modulus were not assumed to be fixed but fitted simultaneously together with the inelastic parameters of the model, then it would have a unique factor. This result emphasizes that it is admissible to fix YOUNG's modulus to a reasonable value without affecting the parameter fits in a negative way.

The principal component analysis yields very similar results when the parameter fits for the extended CHABOCHE model are investigated. The loading matrix is given in Table 18. Again the parameters q and b from the isotropic hardening variable and the parameters a_2 and c_2 from one of the two kinematic hardening variables share common factors. For this model the parameter h shows low complexity. However, most of the parameters show a slight dependence on those factors, on which h has its highest load. Again the loading matrix for this model reveals a clear correlation structure from this model.

The parameters of the CHAN, BODNER, and LINDHOLM model neither exhibit high complexity, as can be seen in the loading matrix in Table 19. In contrast to the two CHABOCHE models we consider all variables in the principal component analysis. An extraction of only 5 factors is sufficient for the data of the fits of all 9 material parameters. However, the resulting correlation structure is not as clear as for the CHABOCHE models. The parameters Z_0, Z_3, and n share a common factor as well as the parameters r_1, Z_1, and m_1. The parameter D_0 has a unique factor, which means that the recommendation to fix this parameter to a reasonable value would not affect the parameter fits of this model.

Table 19. Factor loadings for the CHAN, BODNER, and LINDHOLM model

parameter	1	2	3	4	5
D_0					0.997
A_1				0.995	
r_1	0.243	0.839			
Z_0	0.917	-0.125	-0.295		
Z_1	0.494	-0.811		0.112	
Z_3	0.961				
m_1	-0.141	0.933			
m_2	-0.306		0.944		
n	-0.952		0.199		

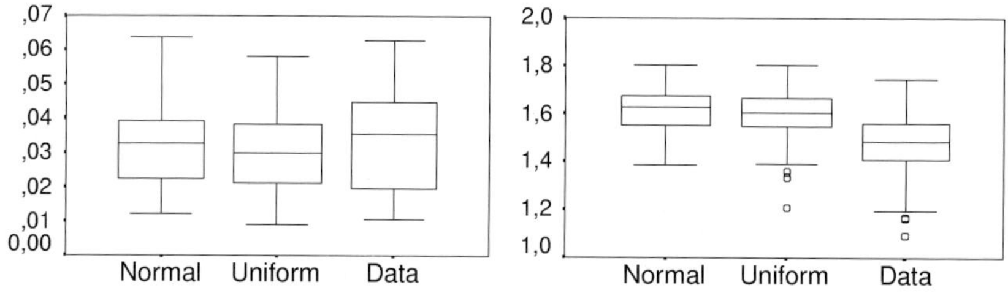

Fig. 9. Parameter fits of the CHAN, BODNER, and LINDHOLM model for (**a**) parameter m_1, and (**b**) parameter n

From Table 16 it could be deduced that the null hypothesis for the equality of distributions $H_0 : F_i = G_i$ is rejected for the parameters Z_0, Z_1, Z_3, and n. Now it can be seen that particularly the parameters Z_0, Z_3, and n share a common factor and are highly correlated. This can explain the results from the previous section. If there are small differences between the artificial and the experimental data, such that a rejection for one parameter is obtained, then it is very likely that also the highly correlated parameters will be rejected. However, from Fig. 9 it can be seen that the parameter fits do not differ too much.

We cannot give the reason why the model of CHAN, BODNER, and LINDHOLM yields worse results than the CHABOCHE models. A reason could be that Z_1 is a parameter of the isotropic hardening variable $Z^{(I)}$ and a parameter of the recovery terms of the kinematic and isotropic hardening variable. Thus, it might be that the model equations are unable to distinguish the different kinds of material behaviour precisely enough.

8 Conclusion

In this work three viscoplastic constitutive models are investigated and their material parameters are determined by stochastic methods. It is shown that these methods work accurately and an optimal parameter vector for each model by a simultaneous parameter fit to all available experiments is identified. The values of the distance function according to the optimal parameter vectors and the test data are compared and it turns out that the accuracy, i. e. the description of the experimental data, is the best for the CHABOCHE type models. In particular, the kinematic hardening equation by HAUPT, KAMLAH, and TSAKMAKIS improves the description of cyclic experiments significantly.

Since there are not enough experimental data available for statistical investigations as e. g. an analysis of the uncertainties in the parameter fits for the constitutive models, a method of stochastic simulation is presented. This method enables us to generate a large amount of artificial data with the same properties as the original experimental data. An analysis is carried out in order to check the robustness of these stochastic methods. This analysis is performed by applying significantly different distributions for the design variables of the stochastic simulations. A comparison of the identified parameters by statistical tests shows the robustness of the applied stochastic simulation method. Furthermore, again by applying statistical tests, it is shown that the results for the parameter fits from experimental data and from artificial data coincide well. A substantial part of the theory and the ideas behind these methods and further topics can be found in [1] and [10].

Moreover, the correlation structure of fitted model parameters is analyzed by the principle component analysis and it turned out that this method is appropriate to give a deeper insight into the dependencies of the parameters.

It can be seen from the applied methods in this work that the stochastic simulations will be a useful tool for analyzing the influence of scatter in the experimental data on the material parameters. In [26], [27], and [29] investigations for the CHAN, BODNER, and LINDHOLM model are performed. Besides, these methods will be essential to perform an experimental design in order to find out the minimum number of experiments which yield an appropriate result for the parameter identification.

References

1. Bard, Y. (1974) Nonlinear Parameter Estimation. Academic Press, Inc., New York
2. Bodner, S.R., Partom, Y. (1975) Constitutive Equations for Elastic-Viscoplastic Strain Hardening Materials. Journal of Applied Mechanics **42**, 385–389
3. Bodner, S.R (2002) Unified Plasticity for Engineering Applications. Mathematical Concepts and Methods in Science and Engineering **47**, Kluwer Academic / Plenum Publishers, New York

4. Brockwell, P.J., Davis, R.A. (1991) Time Series: Theory and Methods, Second Edition, Springer Verlag, New York
5. Bruhns, O., Anding, D.K. (1999) On the Simultaneous Estimation of Model Parameters Used in Constitutive Laws for Inelastic Material Behaviour. Int. J. Plasticity **15** (12), 1311–1340.
6. Chaboche, J.L. (1977) Viscoplastic Constitutive Equations for the Description of Cyclic and Anisotropic Behaviour of Metals. Bulletin de l'Academie Polonaise des Sciences, Série Sc. et Techn. **15** (1), 33–41
7. Chaboche, J.L. (1989) Constitutive Equations for Cyclic Plasticity and Cyclic Viscoplasticity. Int. J. Plasticity **5**, 283–295
8. Chaboche, J.L., Rousselier, G. (1983) On the Plastic and Viscoplastic Constitutive Equations–Part I: Rules Developed with Internal Variable Concept. Journal of Pressure Vessel Technology **105**, 153–158
9. Chan, K.S., Bodner, S.R., Lindholm, U.S. (1988) Phenomenological Modelling of Hardening and Thermal Recovery in Metals. Journal of Engineering Materials and Technology **110**, 1–8
10. Doltsinis, I. (1999) Stochastic Analysis of Multivariate Systems in Computational Mechanics and Engineering. CIMNE, Barcelona
11. Eichenauer, J., Lehn, J. (1986) A Non-linear Congruential Pseudo Random Number Generator. Statistical Papers **27**, 315–326
12. Hartmann, G., Kollmann F.G. (1987) A Computational Comparison of the Inelastic Constitutive Models of Hart and Miller. Acta Mechan. **69**, 139–165
13. Haupt, P., Kamlah, M., Tsakmakis, Ch. (1992) Continuous Representation of Hardening Properties in Cyclic Plasticity. Int. J. Plasticity **8**, 803–817
14. Huber, N. (2000) Anwendung Neuronaler Netze bei nichtlinearen Problemen der Mechanik. Habilitationsschrift, Wissenschaftliche Berichte FZKA 6504
15. Kennedy, W.J.,Jr., Gentle, J.E. (1980) Statistical Computing. Statistics: Textbooks and Monographs **33**, Marcel Dekker, Inc., New York
16. Krempl, E., McMahon, J.J., Yao, D. (1986) Viscoplasticity Based on Overstress with a Differential Growth Law for the Equilibrium Stress. Mechanics of Materials **5**, 35–48
17. Kunkel, R., Kollmann F.G. (1997) Identification of Constants of a Viscoplastic Constitutive Model for a Single Crystal Alloy. Acta Mechan. **124**, 27–45
18. Lehmann, E.L. (1975) Nonparametrics: Statistical Methods Based on Ranks. McGraw-Hill
19. Lemaitre, J., Chaboche, J.L. (1990) Mechanics of Solid Materials. Cambridge University Press
20. Mahnken, R., Stein, E. (1996) Parameter Identification for Viscoplastic Models Based on Analytical Derivatives of a Least-squares Functional and Stability Investigations. Int. J. Plasticity **12** 4, 451–479
21. Moré, J.J. (1977) The Levenberg-Marquard Algorithm: Implementation and Theory. Lecture Notes Math. **630**, Springer Verlag, Berlin - Heidelberg - New York, 105–116
22. Müller, P.H., Nollau, V., Polovinkin, A.I. (1986). Stochastische Suchverfahren. Verlag Harry Deutsch, Thun u. Frankfurt/Main
23. Price, W.L. (1978) A Controlled Random Search Procedure for Global Optimization. In: Dixon, L.C.W., Szegö, G.P., Towards Global Optimization **2**, North-Holland Publishing Company, Amsterdam, pp. 71–84
24. Reuter, R., Hülsmann, J. (2000) Achieving Design Targets through Stochastic Simulation. Madymo User's Conference, Paris

25. Schwan, S. (2000) Identifikation der Parameter inelastischer Werkstoffmodelle: Statistische Analyse und Versuchsplanung. Dissertation, Technische Universität Darmstadt
26. Schwan, S., Lehn, J., Harth, T., Kollmann, F.G. (2002) Identification of Material Parameters for Inelastic Constitutive Models, Part I: Stochastic Methods. Int. J. Plasticity, submitted
27. Schwan, S., Lehn, J., Harth, T., Kollmann, F.G. (2002) Identification of Material Parameters for Inelastic Constitutive Models, Part II: Statistical Analysis and Design of Experiments. Int. J. Plasticity, submitted
28. Seibert, T. (1996) Simulationstechniken zur Untersuchung der Streuungen bei der Identifikation der Parameter inelastischer Werkstoffmodelle. Dissertation, Fachbereich Mathematik der Technischen Hochschule Darmstadt
29. Seibert, T., Lehn, J., Schwan, S., Kollmann, F.G. (2000) Identification of Material Parameters for Inelastic Constitutive Models: Stochastic Simulations for the Analysis of Deviations. Continuum Mechanics and Thermodynamics **12**, 95–120
30. Senchenkov, I.K., Tabieva, G.A. (1996) Determination of the Parameters of the Bodner-Partom Model for Thermoviscoplastic Deformation of Materials. International Applied Mechanics **32** (2), 132–139
31. Steck, E.A. (1985) A Stochastic Model for the High-Temperature Plasticity of Metals. Int. J. Plasticity **1**, 243–258

A Simple Model for Describing Yield Surface Evolution During Plastic Flow

Yannis F. Dafalias[1,2], David Schick[3], and Charalampos Tsakmakis[3]

[1] Dept. of Mechanics, Faculty of Applied Mathematical and Physical Science, National Technical University of Athens
Zographou 15780, Greece
[2] Department of Civil and Environmental Engineering, University of California at Davis
California 95616, USA
yfdafalias@ucdavis.edu
[3] Institute of Mechanics, Darmstadt University of Technology
Hochschulstrasse 1, D–64289, Darmstadt, Germany
{schick, tsakmakis}@mechanik.tu-darmstadt.de

received 7 Oct 2002 — accepted 14 Nov 2002

Abstract. It has been recognized for a long time that inelastic deformations may induce anisotropy in the material response, even if this is initially isotropic. For metallic materials, deformation induced anisotropy is reflected above all by translation, rotation and distortion of the yield surface. This has been confirmed by several experimental investigations independent of the way the yield point is defined. In the present paper a simple, thermodynamically consistent model is proposed, describing the evolving anisotropy of the yield surface. The model is first theoretically established, based on a sufficient condition for the dissipation inequality to be satisfied. Then, it is applied to predict the subsequent yield surfaces, after various prestressings, which have been observed experimentally by ISHIKAWA.

1 Introduction

An important feature in the constitutive theory of rate-independent plasticity and rate-dependent (visco-)plasticity is the assumption of the existence of a yield surface in the stress or strain space, which separates purely elastic states from elastic-plastic states (see e.g. [21], [29]). Closely related to the yield surface are also the so-called loading conditions, which decide whether or not inelastic flow has to be involved. These conditions are satisfied for the case of work hardening plasticity if the actual strain or stress state is on the yield surface and the imposed strain or stress increment points outward from the yield surface (see e.g. [4], [8]). On the other hand, when viscoplasticity is concerned, loading conditions are defined commonly to be fulfilled if a nonvanishing, so-called overstress applies. The notion overstress has been introduced by KREMPL [24] and PERZYNA [30] and is defined as a scalar valued function of a stress state which is outward from the yield surface in stress space (for more details see [40]).

Also, the concept of yield surface plays a crucial role if the plastic strain is supposed to obey an associated normality rule, i.e. if the plastic strain rate is positive proportional to the outer normal at the yield surface. Such evolution equations, termed "flow rules", may often be obtained, at least for isotropic material response, from some overall work postulates (a long list of papers dealing with work postulates in plasticity is given in [39]).

Conformity of the yield surface concept with experimental results has been examined in several works. Among others we mention here the works [13]-[20], [22], [23], [26], [27], [31]-[34], [36], [37], [43], [44]. Generally there are some differences in the approaches employed to measure yield surfaces. For example, the definition of plastic yielding is not unique. Customary, the method of departure from the linearity (proportional limit), the method of backward extrapolation and the stress at a strain offset by a given small amount are utilized to determine the initial yield surface as well as subsequent yield surfaces after preloading. The first method is used e.g. in [26], [27], [32]-[34], the second one e.g. in [22], [36], while the offset criterion has been employed e.g. in [15], [18]-[20], [22], [23], [26], [27], [37], [43], [44]. Further references on experimental determination of yield surfaces can be found in the review papers [16], [17], [31]. As it can be seen from these works, the assumed definition of yielding affects the identified yield surface crucially. Similarly, the form of the measured yield surfaces depends heavily on the loading-unloading-reloading paths chosen. Essentially, after preloadings the subsequent yield surfaces may translate, rotate and distort, even if an initially isotropic yield surface has been recorded. In some cases, when the offset strains are very small, the subsequent yield surfaces have been observed to exhibit a sharpening in the direction of preloading and a flattening on the opposite side. However, when the yield surfaces are measured by partial unloading from the actual stress state to the assumed center of the yield surface, the yield loci referred to plane stress loadings have turned out to form rather ellipses (see [18], [19], [37]).

Some effort has been made to describe theoretically the evolution of yield surfaces during plastic flow (see e.g. the literature given in [42]). Because of their simplicity, yield functions which contain a fourth-order state tensor and are quadratic functions of the stress tensor are very attractive. This is the case e.g. for the constitutive models proposed by Backhaus [2], Baltov and Sawczuk [3], Ishikawa [18], Rees [35], Williams and Svensson [43], [44], Wu et al. [45] and Yoshimura [46]. It is worth noting, that all these works are formulated in a purely mechanical context.

The aim of the present paper is to show how deformation induced anisotropy of the yield surface may be formulated in a thermodynamically consistent manner. This is achieved by establishing sufficient conditions for the satisfaction of the so-called dissipation inequality. For the sake of simplicity, we shall demonstrate the proposed model for yield surfaces which initially are isotropic and the initial yield surface may be approximated with sufficient accuracy by a VON MISES yield function. This refers to e.g. the experiments by ISHIKAWA [18], which will be used in order to discuss the capabilities of the model. To make our work self-contained, a brief summary of the experimental results by ISHIKAWA [18], will be given in the following.

2 Subsequent Yield Surfaces of Stainless Steel Examined by Ishikawa

The specimens used in ISHIKAWA's [18] investigations are drawing tubes of type SUS304 stainless steel, subjected to solution heat treatment. Stress controlled deformation processes, with a constant stress rate of 4.3 MPa/s have been imposed, consisting of axial-torsional loadings. Yielding was defined by a VON MISES effective strain of 50 μm/m, which is small enough to detect yield surfaces in stress space by using a single specimen. The yield surfaces are determined by partially unloading the specimen from the actual stress state to the center of the yield surface, the location of which has been approximated during the preloading part of the stress path. More precisely, the center of the subsequent yield surface in the experiment is simulated by using the constitutive model given in [19]. Figs. 1-7 illustrate the imposed stress paths, and the resulting yield surfaces, with respect to the σ-$\sqrt{3}\tau$-coordinate system (used stress space), where σ, τ are the axial and shear stress components, respectively. For all specimens, the initial yield locus may be approximated well by a VON MISES yield circle. These approximated initial yield circles are denoted by broken lines, the corresponding radii being taken from Table 1 in Ishikwa [18]. All detected subsequent yield loci may be approximated well by ellipses, as shown in the figures by solid lines. During proportional loading the ellipses are translated and compressed in the direction of the prestress. During non-proportional loading the subsequent yield ellipses translate, rotate and change their shape. The approximations of the subsequent yield loci indicated in Figs. 1-7 are constructed from the experimental data by using a fitting procedure. Note that, because of time effects (rate-dependence) in the material behaviour, the subsequent yield loci do generally not contain the prestress points. The experimental results e.g. in Figs. 1, 2 indicate that isotropic hardening is not present. In fact, existence of isotropic hardening (softening) would imply subsequent yield loci, after radial loading, which are broader (smaller) in the direction perpendicular to the preloading path. But such behaviour is not observed (see Figs. 1c, 2c).Therefore, the experimental results in Figs. 1-7 are so interpreted that isotropic hardening is generally absent, the mean value of the constant yield stress being $k_0 = 194$ MPa.

3 Proposed Constitutive Model

3.1 Basic Relations

We confine attention to small elastic-viscoplastic (rate-dependent) deformations and denote by $\mathbf{E}$ and $\mathbf{T}$ the linearized strain tensor and the CAUCHY stress tensor, respectively. Thereby, second-order tensors, like vectors, are presented by bold-face letters and fourth-order tensors are denoted by calligraphic bold-face letters. For two second-order tensors $\mathbf{A}$ and $\mathbf{B}$, we write $\mathrm{tr}\mathbf{A}$ for the trace of $\mathbf{A}$, $\mathbf{A}^T$ for the transpose of $\mathbf{A}$, $\mathbf{A}\cdot\mathbf{B} = \mathrm{tr}(\mathbf{A}\mathbf{B}^T)$ for the inner product between $\mathbf{A}$ and $\mathbf{B}$, and $||\mathbf{A}|| = \sqrt{\mathbf{A}\cdot\mathbf{A}}$ for the EUCLIDEAN norm of $\mathbf{A}$. Since the formulation is not affected by a space dependence, an explicit reference to space will be dropped. Only isothermal deformations with homogeneous temperature distribution will be considered, so that a temperature dependence will be dropped as well. As usually, we assume the strain tensor to satisfy the decomposition

$$\mathbf{E} = \mathbf{E}_e + \mathbf{E}_p \quad , \tag{1}$$

where $\mathbf{E}_e$ and $\mathbf{E}_p$ are the elastic and the plastic strain parts, respectively. Furthermore, we assume the existence of a specific free energy ψ with a corresponding elastic and plastic decomposition as

$$\psi = \psi_e + \psi_p \quad . \tag{2}$$

For simplicity, we suppose ψ_e to be an isotropic function of $\mathbf{E}_e$, of the form

$$\psi_e = \hat{\psi}_e(\mathbf{E}_e) = \frac{1}{2\rho}\mathbf{E}_e\cdot\boldsymbol{\mathcal{C}}[\mathbf{E}_e] \quad , \tag{3}$$

$$\boldsymbol{\mathcal{C}} = 2\mu\,\boldsymbol{\mathcal{E}} + \lambda\,\mathbf{1}\otimes\mathbf{1} \quad . \tag{4}$$

Here, ρ is the mass density and μ, λ are the elasticity parameters. $\boldsymbol{\mathcal{E}}$ is the fourth-order identity tensor operating in the space of all symmetric second-order tensors. The symbol $\mathbf{1}$ denotes the second-order identity tensor. With respect to an orthonormal basis $\{\boldsymbol{e}_i\}$, $i = 1,2,3$ in the three-dimensional EUCLIDEAN vector space, in which the material body under consideration is postulated to move, we have

$$\begin{aligned}\boldsymbol{\mathcal{E}} &= \mathcal{E}_{imjn}\,\mathbf{e}_i\otimes\mathbf{e}_m\otimes\mathbf{e}_j\otimes\mathbf{e}_n \\ &= \tfrac{1}{2}\left(\delta_{ij}\delta_{mn} + \delta_{in}\delta_{mj}\right)\mathbf{e}_i\otimes\mathbf{e}_m\otimes\mathbf{e}_j\otimes\mathbf{e}_n \quad ,\end{aligned} \tag{5}$$

$$\mathbf{1} = \delta_{ij}\,\mathbf{e}_i\otimes\mathbf{e}_j \quad , \tag{6}$$

where $\otimes$ represents the tensor product and δ_{ij} is the KRONECKER delta symbol. Let $\mathcal{C}_{ijkl}$ be the components of the fourth-order tensor $\boldsymbol{\mathcal{C}}$ with respect to the orthonormal basis, $\{\boldsymbol{e}_i\}$. Then $\boldsymbol{\mathcal{C}}$ has the properties

$$\mathcal{C}_{ijkl} = \mathcal{C}_{klij} = \mathcal{C}_{jikl} \quad . \tag{7}$$

Eq. $(7)_1$ states that the tensor $\boldsymbol{\mathcal{C}}$ is symmetric, i.e. $\boldsymbol{\mathcal{C}} = \boldsymbol{\mathcal{C}}^T$.
According to the assumptions made, the second law of thermodynamics, in the form of the CLAUSIUS-DUHEM-inequality, reads

$$\mathcal{D}_{C-D} = \mathbf{T} \cdot \dot{\mathbf{E}} - \rho\,\dot{\psi} = \left(\mathbf{T} - \rho \frac{\partial \hat{\psi}_e}{\partial \mathbf{E}_e} \right) \cdot \dot{\mathbf{E}} - \rho\,\dot{\psi}_p \geq 0 \quad . \tag{8}$$

Using standard arguments, one can show that the relations

$$\mathbf{T} = \rho\,\frac{\partial \hat{\psi}_e}{\partial \mathbf{E}_e} = \boldsymbol{\mathcal{C}}[\mathbf{E}_e] \quad , \tag{9}$$

$$\mathcal{D}_d := \mathbf{T} \cdot \dot{\mathbf{E}}_p - \rho\,\dot{\psi}_p \geq 0 \tag{10}$$

are necessary and sufficient conditions for inequality (8) to be satisfied in the case of rate-dependent (visco-)plasticity we are concerned with in the present paper. Relation (10) is known as the dissipation inequality.

3.2 Yield Function - Flow Rule

We interpret ISHIKAWA's experimental results as follows. Kinematic hardening effects are present, whereas isotropic hardening is absent. All subsequent yield loci may be approximated well by ellipses. This means, after prestressing, the subsequent yield loci translate and distort. For simplicity, rotation of the subsequent yield loci is not assumed explicitly.

To model these phenomena, we assume the existence of a yield function $\hat{F}(\mathbf{T}, \boldsymbol{\xi}, \boldsymbol{\mathcal{H}})$ in terms of a fourth-order tensor $\boldsymbol{\mathcal{H}}$ of the form

$$F = \hat{F}(\mathbf{T}, \boldsymbol{\xi}, \boldsymbol{\mathcal{H}}) = \hat{f}(\mathbf{T}, \boldsymbol{\xi}, \boldsymbol{\mathcal{H}}) - k_0 \quad , \tag{11}$$

$$f = \hat{f}(\mathbf{T}, \boldsymbol{\xi}, \boldsymbol{\mathcal{H}}) := \sqrt{(\mathbf{T} - \boldsymbol{\xi})^D \cdot \boldsymbol{\mathcal{H}}[(\mathbf{T} - \boldsymbol{\xi})^D]} \quad , \tag{12}$$

with

$$\boldsymbol{\mathcal{H}} = \boldsymbol{\mathcal{H}}_0 + \varphi \boldsymbol{\mathcal{A}} \tag{13}$$

and where yield occurs when $F = 0$.

In these equations, $\boldsymbol{\xi}$ is the so-called back stress tensor and k_0 is a material parameter representing constant yield stress. We write $\mathbf{S}^D = \mathbf{S} - \frac{1}{3}(tr\mathbf{S})\mathbf{1}$ for the deviator of a second order tensor $\mathbf{S}$, so that $\mathbf{T}^D$ and $\boldsymbol{\xi}^D$ are the deviators of $\mathbf{T}$ and $\boldsymbol{\xi}$, respectively. $\boldsymbol{\mathcal{H}}_0$ and $\boldsymbol{\mathcal{A}}$ are fourth-order tensors modelling distortion, but not explicitly rotation of the subsequent yield surfaces. In particular, we assume $\boldsymbol{\mathcal{H}}_0$ to be constant, representing the initial value of $\boldsymbol{\mathcal{H}}$, while $\boldsymbol{\mathcal{A}}$ evolves with plastic flow from its initial zero value, representing the anisotropic development, and satisfies homogeneous initial conditions. The scalar φ will be discussed subsequently. The experiments by ISHIKAWA suggest modelling of the initial yield surface by using the VON MISES yield function. Therefore, we set

$$\boldsymbol{\mathcal{H}}_0 = \tfrac{3}{2}\left(\boldsymbol{\mathcal{E}} - \tfrac{1}{3}\,\mathbf{1} \otimes \mathbf{1}\right) \quad . \tag{14}$$

With respect to the orthonormal basis $\{\boldsymbol{e}_i\}$, $\boldsymbol{\mathcal{H}}_0$ exhibits the properties (7) as well as the property

$$(\mathcal{H}_0)_{iikl} = 0 \quad . \tag{15}$$

Accordingly, we also require from $\mathcal{A}_{ijkl}$ the properties (7), (15) as well, where $\mathcal{A}_{ijkl}$ are the components of $\boldsymbol{\mathcal{A}}$ with respect to the orthonormal basis $\{\boldsymbol{e}_i\}$. Notice that (12) together with normality $\dot{\mathbf{E}}_p \sim \partial f/\partial \mathbf{T}$, satisfies the incompressibility condition $tr\dot{\mathbf{E}}_p = 0$ without any restriction on $\boldsymbol{\mathcal{H}}$. But since there are only five independent components of $(\mathbf{T} - \boldsymbol{\xi})^D$, the 21 components of $\boldsymbol{\mathcal{H}}$ must reduce to 15 independent ones. This is in fact achieved by the 6 equations (15) to both $\boldsymbol{\mathcal{H}}_0$ and $\boldsymbol{\mathcal{A}}$, hence, to $\boldsymbol{\mathcal{H}}$ [6].

The experiments illustrated in Figs. 5, 6 show that the yield loci shrink after prestressing and expand after unloading to the state where the back stress tensor is nearly vanishing. To incorporate such phenomena into the constitutive model, we assume φ to be a constitutive function of $\boldsymbol{\xi}$. For the range of experimental results by ISHIKAWA, the assumption

$$\varphi = 1 + \varphi_0\, e^{\,\varphi_1 \|\boldsymbol{\xi}\|} \tag{16}$$

seems to be appropriate for the purposes of the present paper, with φ_0 and φ_1 being material constants.

Now, we postulate inelastic flow to occur, if a positive overstress applies. Consequently, on defining overstress to be given by F, inelastic flow occurs only if $F > 0$. Moreover, for an associated flow rule the yield function serves also as a plastic potential, thus

$$\dot{\mathbf{E}}_p = \dot{s}\sqrt{\frac{3}{2}}\frac{\frac{\partial \hat{f}}{\partial \mathbf{T}}}{\|\frac{\partial \hat{f}}{\partial \mathbf{T}}\|} = \frac{\dot{s}}{\zeta}\frac{\partial \hat{f}}{\partial \mathbf{T}} \quad , \tag{17}$$

$$\frac{\partial \hat{f}}{\partial \mathbf{T}} = \frac{1}{f}(\boldsymbol{\mathcal{H}}_0 + \varphi\,\boldsymbol{\mathcal{A}})[(\mathbf{T} - \boldsymbol{\xi})^D] \quad , \tag{18}$$

$$\zeta := \sqrt{\frac{2}{3}}\left\|\frac{\partial \hat{f}}{\partial \mathbf{T}}\right\| \quad , \tag{19}$$

$$\dot{s} := \sqrt{\frac{2}{3}\dot{\mathbf{E}}_p \cdot \dot{\mathbf{E}}_p} \quad . \tag{20}$$

It is common use in the framework of unified viscoplasticity to assume $\dot{s}$ as a function of the overstress F. In particular, we set

$$\dot{s} = \frac{\langle F \rangle^m}{\eta} \tag{21}$$

where m, η are positive material parameters, and the function $\langle x \rangle$ is defined by

$$\langle x \rangle := \begin{cases} x & \text{if} \quad x \geq 0 \\ 0 & \text{if} \quad x < 0 \end{cases} \quad , \tag{22}$$

for all real x.

3.3 Hardening Rules

Consider the case that kinematic hardening and distortional hardening are not coupled. This may be taken into account by an additive decomposition of ψ_p into two parts, ψ_p^{kin} and ψ_p^{dist}, which are related to the kinematic hardening and the distortional hardening effects, respectively,

$$\psi_p = \psi_p^{kin} + \psi_p^{dist} \quad . \tag{23}$$

During inelastic flow, a part of the plastic power $\mathbf{T} \cdot \dot{\mathbf{E}}_p$ will be dissipated into heat, while the remainder will be stored in the material in form of internal structure rearrangements. The parts $\rho\ \psi_p^{kin}$ and $\rho\ \psi_p^{dist}$ represent just the energy stored in the material due to kinematic hardening and distortional hardening, respectively. Following CHABOCHE ET AL. [5], we assume the existence of internal, second-order strain tensors $\mathbf{Y}_i$, $i = 1, .., k$, so that ψ_p^{kin} is a function of the strains $\mathbf{Y}_i$. For simplicity, we assume ψ_p^{kin} to be an isotropic function of the form

$$\psi_p^{kin} = \hat{\psi}_p^{kin}(\mathbf{Y}_i) = \sum_{i=1}^{k} \frac{c_i}{2\rho} \mathbf{Y}_i \cdot \mathbf{Y}_i \quad . \tag{24}$$

The internal stress tensors thermodynamically conjugate to the strains $\mathbf{Y}_i$ are

$$\boldsymbol{\xi}_i := \rho \frac{\partial \hat{\psi}_p^{kin}}{\partial \mathbf{Y}_i} = c_i \mathbf{Y}_i \quad \text{(no sum on i)} \quad . \tag{25}$$

According to this approach, the back stress tensor $\boldsymbol{\xi}$ is given by the formula

$$\boldsymbol{\xi} = \sum_{i=1}^{k} \boldsymbol{\xi}_i \quad . \tag{26}$$

In a similar fashion, we assume the existence of internal, symmetric fourth-order tensors $\boldsymbol{\mathcal{D}}_j$, $j = 1, .., d$, and set

$$\psi_p^{dist} = \hat{\psi}_p^{dist}(\boldsymbol{\mathcal{D}}_j) = \sum_{j=1}^{d} \frac{\alpha_j}{2\rho} \boldsymbol{\mathcal{D}}_j \cdot \boldsymbol{\mathcal{D}}_j \quad . \tag{27}$$

Henceforth, the inner product between two fourth-order tensors $\boldsymbol{\mathcal{P}}$ and $\boldsymbol{\mathcal{M}}$ is given by $\boldsymbol{\mathcal{P}} \cdot \boldsymbol{\mathcal{M}} = \mathcal{P}_{ijkl}\mathcal{M}_{ijkl}$, where $\mathcal{P}_{ijkl}$, $\mathcal{M}_{ijkl}$ are the components of $\boldsymbol{\mathcal{P}}, \boldsymbol{\mathcal{M}}$, respectively, relative to the orthonormal basis $\{\boldsymbol{e}_i\}$. We denote by $\boldsymbol{\mathcal{A}}_j$ the fourth-order tensors, which are thermodynamically conjugate to $\boldsymbol{\mathcal{D}}_j$,

$$\boldsymbol{\mathcal{A}}_j := \rho \frac{\partial \hat{\psi}_p^{dist}}{\partial \boldsymbol{\mathcal{D}}_j} = \alpha_j \boldsymbol{\mathcal{D}}_j \quad \text{(no sum on j)} \quad , \tag{28}$$

and define $\boldsymbol{\mathcal{A}}$ by

$$\boldsymbol{\mathcal{A}} = \sum_{j=1}^{d} \boldsymbol{\mathcal{A}}_j \quad . \tag{29}$$

In relations (24), (27), c_i, α_j are non-negative material parameters. To accomplish the hardening laws, evolution equations for the internal variables $\mathbf{Y}_i$ and $\boldsymbol{\mathcal{D}}_j$ must be formulated, which must be compatible with the dissipation inequality.
After using (23)-(29) in (10),

$$\begin{aligned}\mathcal{D}_d &= \mathbf{T} \cdot \dot{\mathbf{E}}_p - \sum_{i=1}^{k} \boldsymbol{\xi}_i \cdot \dot{\mathbf{Y}}_i - \sum_{j=1}^{d} \boldsymbol{\mathcal{A}}_j \cdot \dot{\boldsymbol{\mathcal{D}}}_j \\ &= (\mathbf{T} - \boldsymbol{\xi}) \cdot \dot{\mathbf{E}}_p + \sum_{i=1}^{k} \boldsymbol{\xi}_i \cdot (\dot{\mathbf{E}}_p - \dot{\mathbf{Y}}_i) - \sum_{j=1}^{d} \boldsymbol{\mathcal{A}}_j \cdot \dot{\boldsymbol{\mathcal{D}}}_j \geq 0 \quad ,\end{aligned} \tag{30}$$

or, by virtue of (17)-(20),

$$\begin{aligned}\mathcal{D}_d = & \frac{\dot{s}}{f\,\zeta} (\mathbf{T} - \boldsymbol{\xi})^D \cdot \boldsymbol{\mathcal{H}}_0[(\mathbf{T} - \boldsymbol{\xi})^D] \\ & + \sum_{j=1}^{d} \boldsymbol{\mathcal{A}}_j \cdot \left\{ \frac{\dot{s}\,\varphi}{f\,\zeta} (\mathbf{T} - \boldsymbol{\xi})^D \otimes (\mathbf{T} - \boldsymbol{\xi})^D \right\} \\ & + \sum_{i=1}^{k} \boldsymbol{\xi}_i \cdot (\dot{\mathbf{E}}_p - \dot{\mathbf{Y}}_i) - \sum_{j=1}^{d} \boldsymbol{\mathcal{A}}_j \cdot \dot{\boldsymbol{\mathcal{D}}}_j \\ = & \mathcal{D}_d^{(0)} + \mathcal{D}_d^{(kin)} + \mathcal{D}_d^{(dist)} \geq 0 \quad ,\end{aligned} \tag{31}$$

where

$$\mathcal{D}_d^{(0)} = \frac{\dot{s}}{f\,\zeta} (\mathbf{T} - \boldsymbol{\xi})^D \cdot \boldsymbol{\mathcal{H}}_0[(\mathbf{T} - \boldsymbol{\xi})^D] \quad , \tag{32}$$

$$\mathcal{D}_d^{(kin)} = \sum_{i=1}^{k} \boldsymbol{\xi}_i \cdot (\dot{\mathbf{E}}_p - \dot{\mathbf{Y}}_i) \quad , \tag{33}$$

$$\mathcal{D}_d^{(dist)} = \sum_{j=1}^{d} \boldsymbol{\mathcal{A}}_j \cdot \left\{ \frac{\dot{s}\,\varphi}{f\,\zeta} (\mathbf{T} - \boldsymbol{\xi})^D \otimes (\mathbf{T} - \boldsymbol{\xi})^D - \dot{\boldsymbol{\mathcal{D}}}_j \right\} \quad . \tag{34}$$

Since $\boldsymbol{\mathcal{H}}_0$ is positive definit for all deviatoric second-order tensors, we have $\mathcal{D}_d^{(0)} \geq 0$. Therefore, (31) will always be satisfied if

$$\mathcal{D}_d^{(kin)} \geq 0 \quad , \quad \mathcal{D}_d^{(dist)} \geq 0 \quad . \tag{35}$$

Clearly the relations (no sum on i, j)

$$\dot{\mathbf{E}}_p - \dot{\mathbf{Y}}_i = \dot{s}\, b_i\, \boldsymbol{\xi}_i \quad ,$$

$$\frac{\dot{s}\,\varphi}{f\,\zeta} (\mathbf{T} - \boldsymbol{\xi})^D \otimes (\mathbf{T} - \boldsymbol{\xi})^D - \dot{\boldsymbol{\mathcal{D}}}_j = \dot{s}\, B_j \boldsymbol{\mathcal{A}}_j \quad ,$$

or

$$\dot{\mathbf{Y}}_i = \dot{\mathbf{E}}_p - \dot{s}\, b_i\, \boldsymbol{\xi}_i \quad , \tag{36}$$

$$\dot{\boldsymbol{\mathcal{D}}}_j = \dot{s} \left\{ \frac{\varphi}{f\,\zeta} (\mathbf{T} - \boldsymbol{\xi})^D \otimes (\mathbf{T} - \boldsymbol{\xi})^D - B_j \boldsymbol{\mathcal{A}}_j \right\} \tag{37}$$

are sufficient conditions for (35) to hold, with b_i, B_j being non-negative material parameters.

Eqs. (36), together with (25), (26), represent the kinematic hardening law introduced by CHABOCHE ET AL. [5]. For $k = 1$, one obtains the so-called ARMSTRONG-FREDERICK kinematic hardening model (see [1]). MARQUIS [25] showed that the ARMSTRONG-FREDERICK rule may be derived in a purely mechanical context by using a two-surface model (see also [7]). Later, TSAKMAKIS [38] derived the model of CHABOCHE ET AL. from a so-called multisurface model. Generally, the concept of multisurface plasticity has been introduced by MROZ [28]. DAFALIAS and POPOV [7], [8] were the first to introduce the two-surface (yield and bounding) model in order to describe cyclic loading processes. Today, a large number of similar approaches can be found in the literature.

The equations governing the response of distortional hardening are given by (28), (29) and (37). Alternatively, (37) may be rewritten as (no sum on j)

$$\dot{\boldsymbol{\mathcal{A}}}_j = \dot{s} \left\{ \Theta_j (\mathbf{T} - \boldsymbol{\xi})^D \otimes (\mathbf{T} - \boldsymbol{\xi})^D - \alpha_j\, B_j \boldsymbol{\mathcal{A}}_j \right\} \quad , \tag{38}$$

$$\Theta_j := \frac{\varphi\, \alpha_j}{f\, \zeta} \quad , \tag{39}$$

in view of (28). On assuming homogeneous initial conditions, $\boldsymbol{\mathcal{A}}_j$ may be integrated to get (no sum on j)

$$\boldsymbol{\mathcal{A}}_j(s) = \int_0^s e^{-\alpha_j\, B_j (s - \bar{s})} \Theta_j(\bar{s}) (\mathbf{T}(\bar{s}) - \boldsymbol{\xi}(\bar{s}))^D \otimes (\mathbf{T}(\bar{s}) - \boldsymbol{\xi}(\bar{s}))^D d\bar{s} \quad . \tag{40}$$

From this, it is not difficult to see that every $\boldsymbol{\mathcal{A}}_j$ satisfies the required properties (7), (15), and therefore $\boldsymbol{\mathcal{A}}$ too.

4 Comparison with Experiments - Concluding Remarks

Figs. 8-14 illustrate yield loci predicted by the proposed model for the loading paths given in Figs. 1-7, respectively. ISHIKAWA's experimental data are also displayed in Figs. 8-14. The predicted responses are calculated by assuming $\boldsymbol{\xi}$ and $\boldsymbol{\mathcal{A}}$ to consist of two parts, respectively, i.e., $\boldsymbol{\xi} = \boldsymbol{\xi}_1 + \boldsymbol{\xi}_2$, $\boldsymbol{\mathcal{A}} = \boldsymbol{\mathcal{A}}_1 + \boldsymbol{\mathcal{A}}_2$. The material parameters are chosen as shown in Table 1.

Table 1. Material parameters

$\mu = 7.88 \times 10^4$ [MPa]	$\lambda = 1.18 \times 10^5$ [MPa]
$m = 2.25$ [-]	$\eta = 1.50 \times 10^7$ [MPa^{m}s]
$k_0 = 1.94 \times 10^2$ [MPa]	
$\varphi_0 = 1.00$ [-]	$\varphi_1 = 1.80 \times 10^{-2}$ [MPa^{-1}]
$\alpha_1 = 4.50$ [MPa^{-1}]	B_1=1.60 [MPa s^{-1}]
$\alpha_2 = 0.80$ [MPa^{-1}]	B_2=0.50 [MPa s^{-1}]
$c_1 = 4.50 \times 10^4$ [MPa^{-1}]	$b_1 = 3.30 \times 10^{-2}$ [-]
$c_2 = 9.00 \times 10^3$ [MPa^{-1}]	$b_2 = 0$ [-]

It is emphasized that these values are chosen based on trial and error. A systematic identification of material parameters by using established optimization algorithms is beyond the scope of the paper. Therefore, comparison of the predicted responses with the experimental data has qualitative meaning only.

For monotonous tension, Fig. 8b indicates that the translation of the yield locus, controlled by the back stress $\boldsymbol{\xi}$, is well predicted. The subsequent yield locus, placed in the origin of stress space in Fig. 8c shows a good agreement between experimental and predicted results for the distortion of the yield surface. The same is also true for monotonous torsional loading (see Figs. 9b, 9c). Figs. 10b, 10c reveal that translation and distortion of the yield locus are in essence well predicted for the case of monotonous radial loading conditions. However, from Figs. 10b and 10c, we recognize some differences between the predicted and experimental results, which arise from a missing explicit rotation in the theoretical model. In the case of a combined tension-torsion loading history, the numerical results in Figs. 11b and 11c show a very good agreement with the experiment for the tensile part A, but less good for the following torsional part C due to the aforementioned missing explicit rotation in the model. During uniaxial cyclic loading, the model predicts the behaviour of the yield surface very well for the tension and compression phase (see Figs. 12b, 12c, 12d), but less good for the reloading tensile part (see Fig. 12e). This may be due to the chosen values of the material parameters φ_0 and φ_1, or due to the chosen constitutive function φ itself or due to the fact that the present model is not amplified by further internal variables describing cyclic loading effects. More complex loading histories, including tensile loading followed by torsional loading, are displayed in Figs. 13 and 14. It can be seen that the initial tensile loading behaviour is described well, whereas the subsequent yield surfaces after the torsional loading part are not very well predicted, which may be interpreted to be caused by the missing explicit rotation in the constitutive model.

From this discussion, we conclude that the model is generally able to predict the experimentally measured yield loci. For the observed deviations from the experimental results in essence two reasons are likely to account for. On the one hand, since the model is highly nonlinear, it is very difficult to choose the material parameters appropriately. This means, one may assume that other values for the material parameters, which will be identified by using established optimization procedures, may furnish better agreement with the experimental data. On the other hand, only translation and distortion of the yield surface can be described by the proposed constitutive theory. The measured yield loci, however, translate, distort and rotate, dependent on the imposed loading history. Therefore, one may expect improved predicted results, if the constitutive theory will be amplified to model explicitly rotations of the yield surface as well. To clarify this point will be the subject of future work.

Acknowledgement

The authors want to thank Professor ISHIKAWA for further information concerning the experimental results.

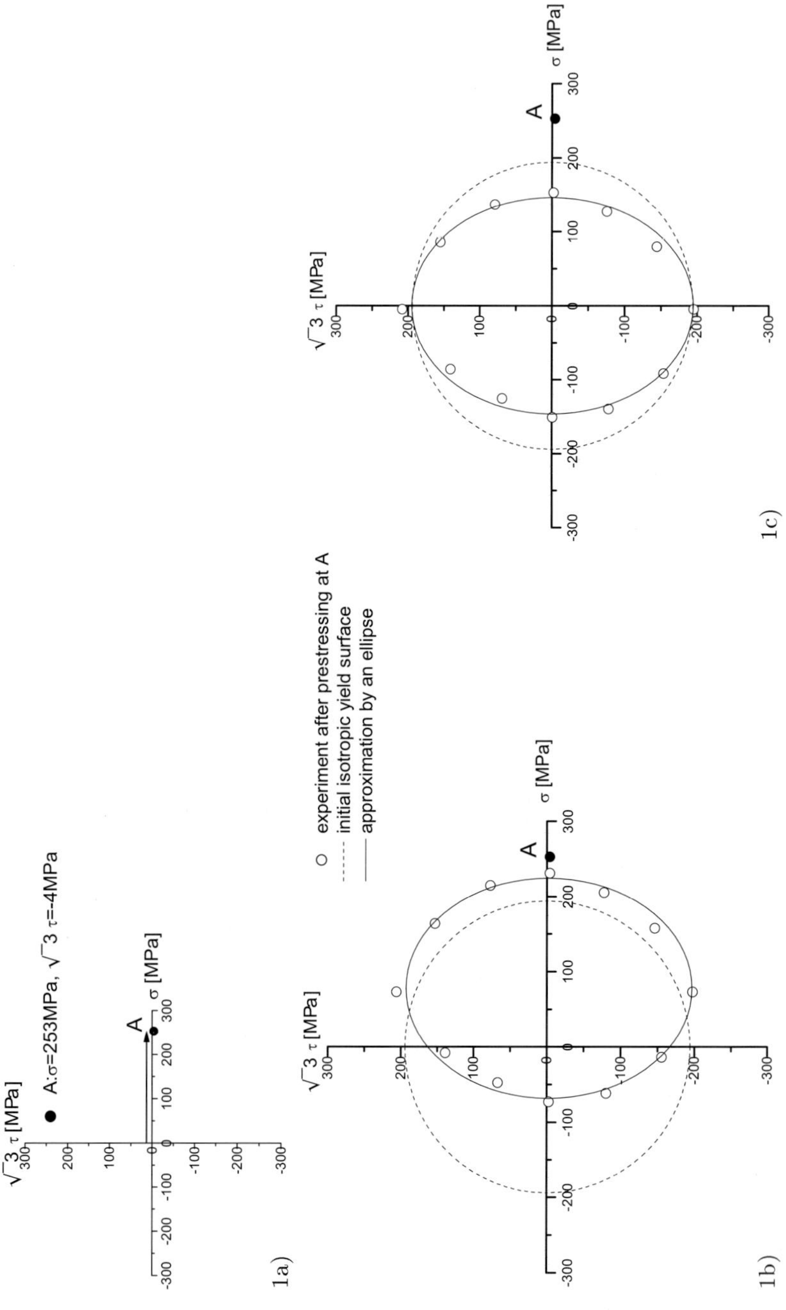

Fig. 1. Monotonous tensile loading. 1a) Loading path imposed. 1b) Yield locus after prestressing at A. 1c) Subsequent yield locus from 1b) placed in the origin of the stress space

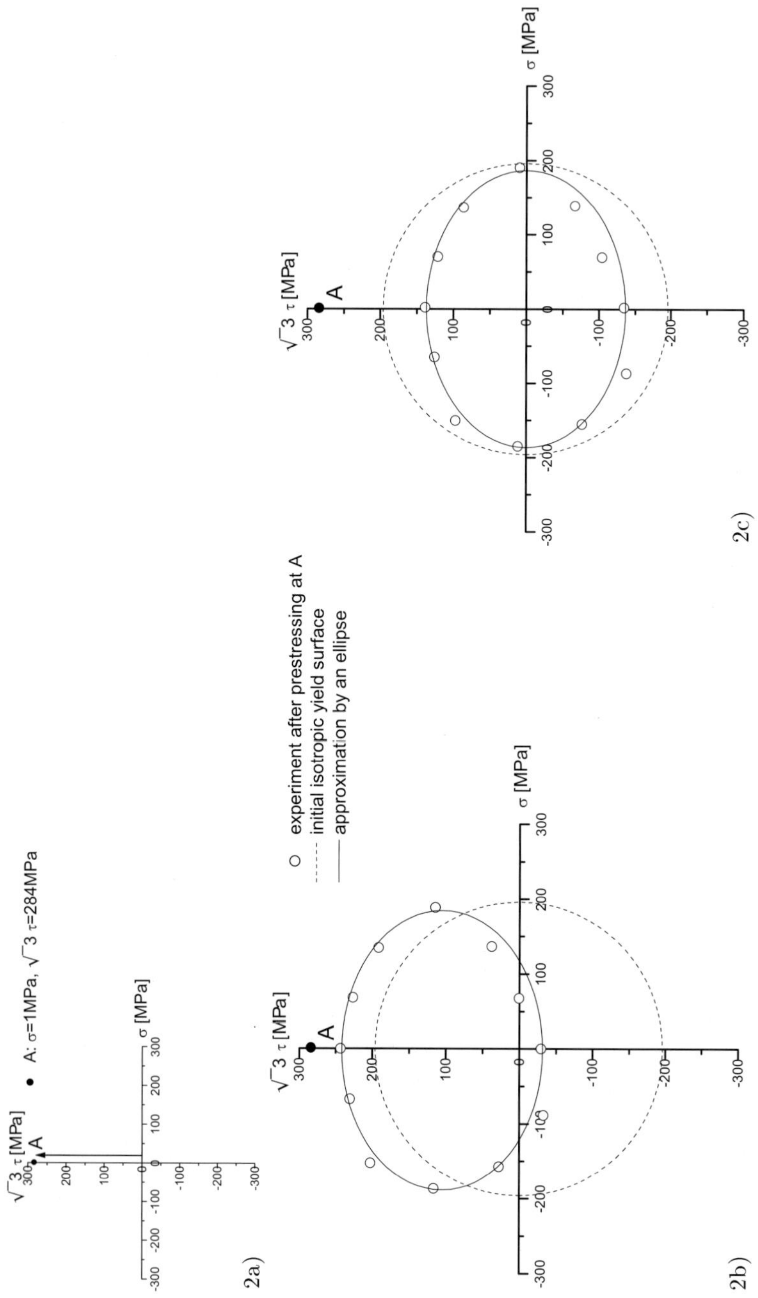

Fig. 2. Monotonous torsional loading. 2a) Loading path imposed. 2b) Yield locus after prestressing at A. 2c) Subsequent yield locus from 2b) placed in the origin of the stress space

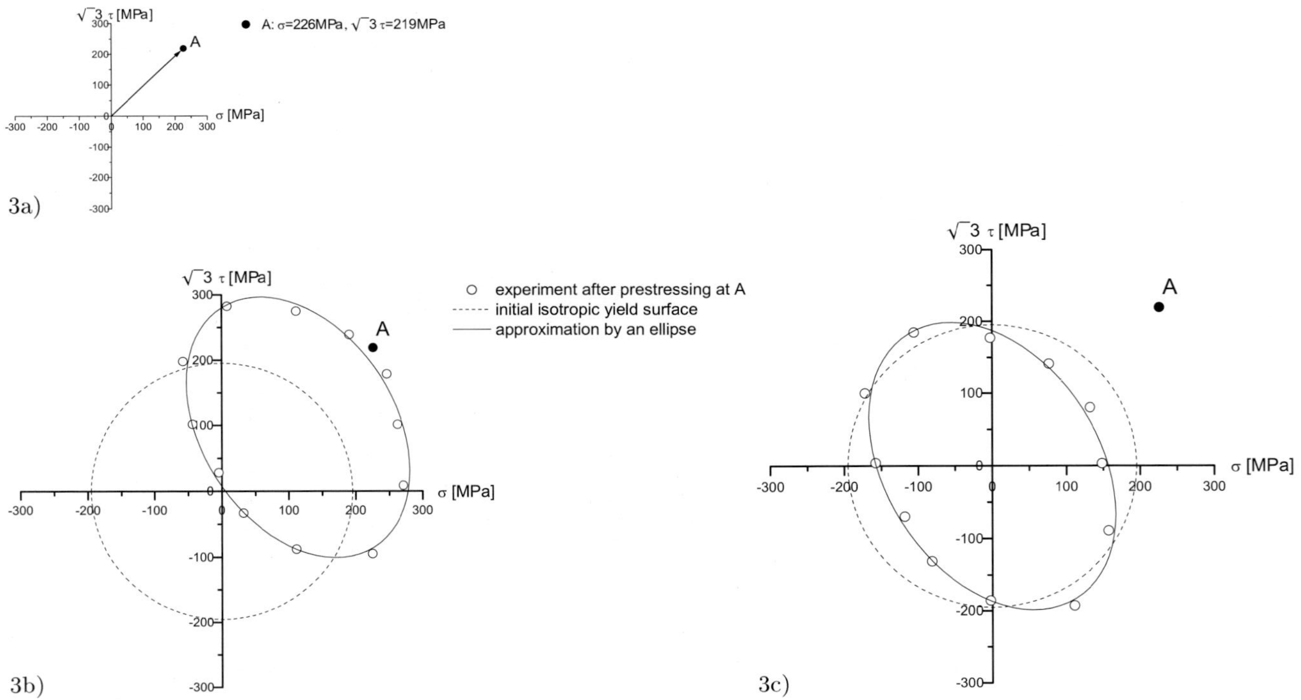

Fig. 3. Monotonous radial loading. 3a) Loading path imposed. 3b) Yield locus after prestressing at A. 3c) Subsequent yield locus from 3b) placed in the origin of the stress space

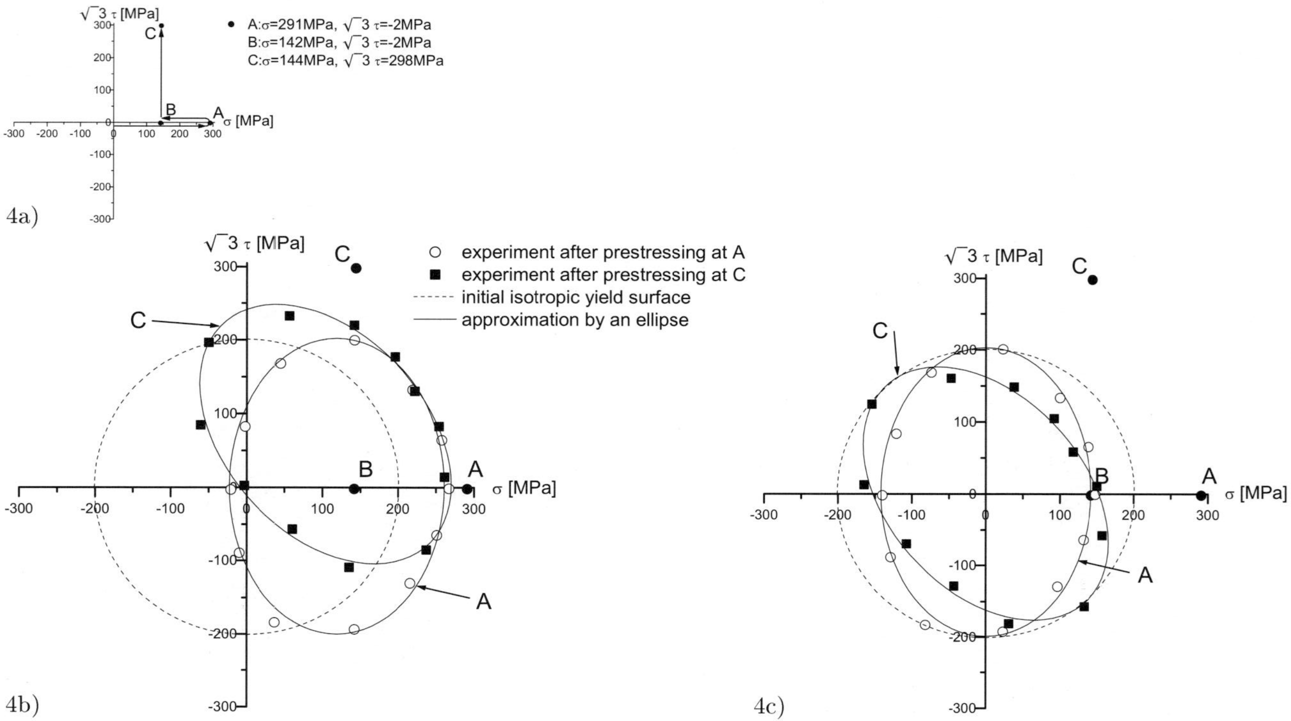

Fig. 4. Combined tension-torsion loading history. 4a) Loading path imposed. 4b) Yield loci after prestressing at A and C. 4c) Subsequent yield loci from 4b) placed in the origin of the stress space

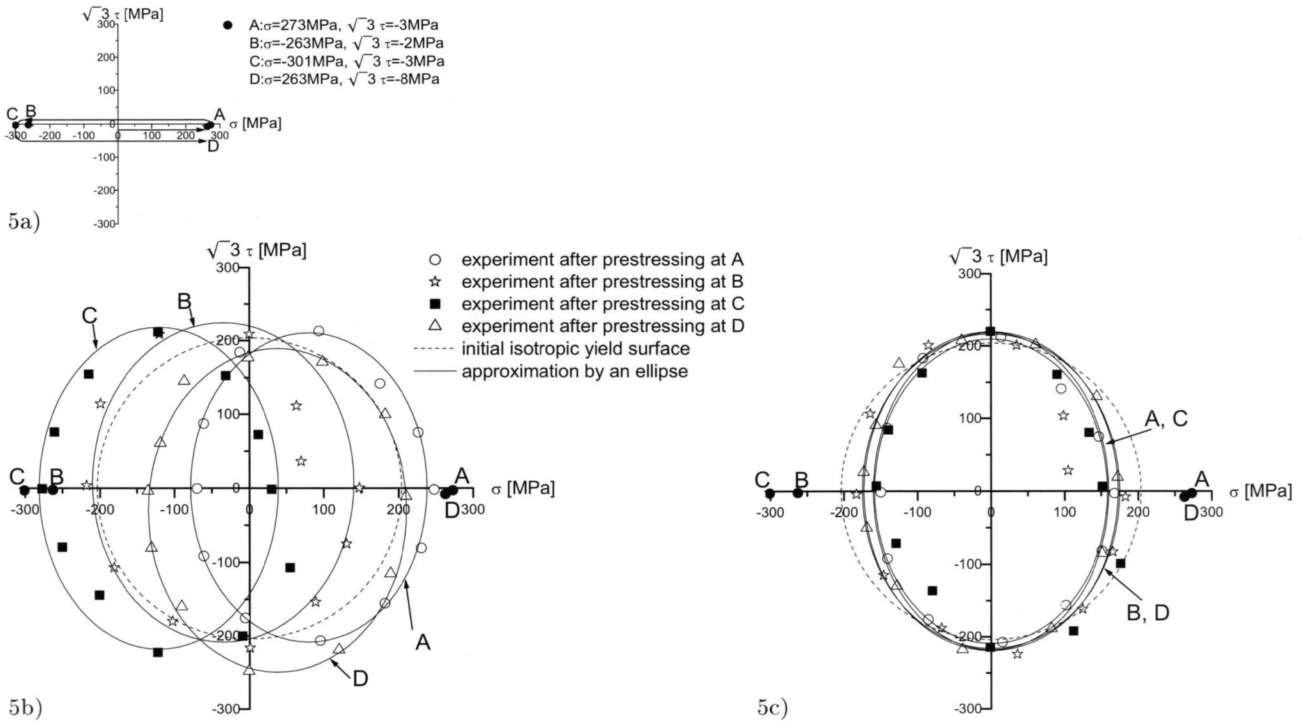

Fig. 5. Uniaxial, cyclic loading. 5a) Loading path imposed. 5b) Yield loci after prestressing at A, B, C and D. 5c) Subsequent yield loci from 5b) placed in the origin of the stress space

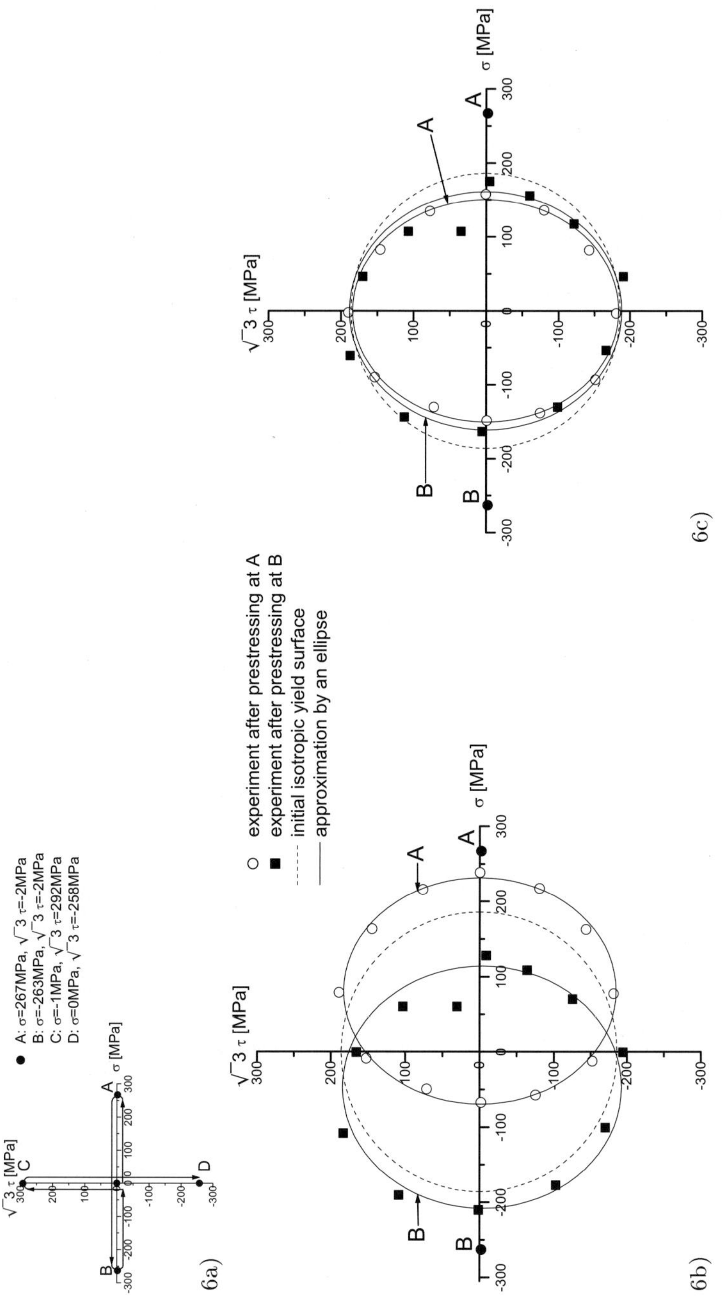
A: σ=267MPa, √3 τ=-2MPa
B: σ=-263MPa, √3 τ=-2MPa
C: σ=-1MPa, √3 τ=292MPa
D: σ=0MPa, √3 τ=-258MPa
√3 τ [MPa]
σ [MPa]
experiment after prestressing at A
experiment after prestressing at B
initial isotropic yield surface
approximation by an ellipse
A
B
C
D

6a) 6b) 6c)

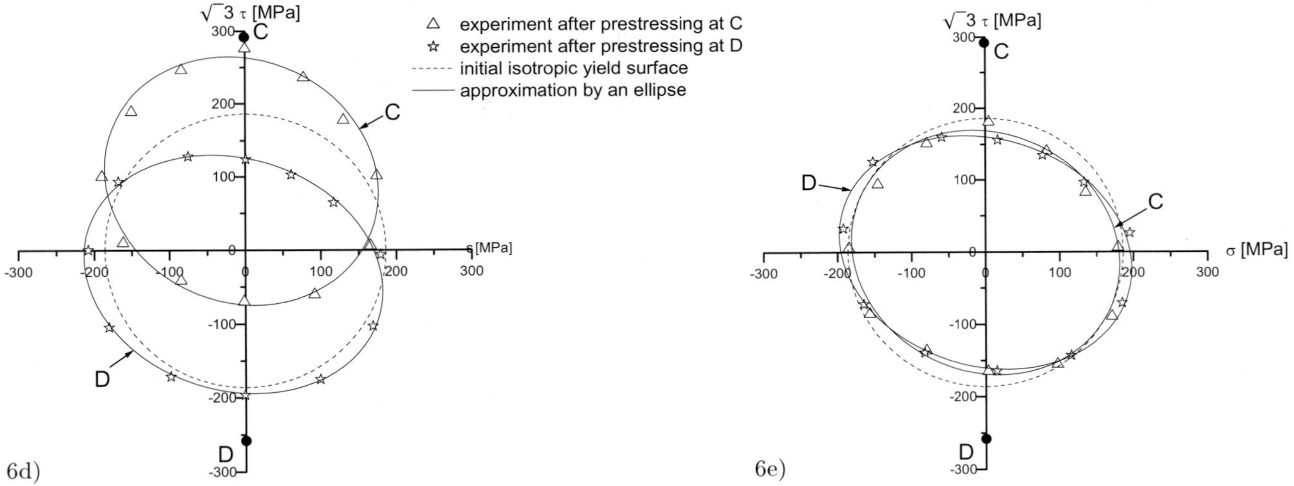

Fig. 6. Combined tension-torsion loading history. 6a) Loading path imposed. 6b) Yield loci after prestressing at A and B. 6c) Subsequent yield loci from 6b) placed in the origin of the stress space. 6d) Yield loci after prestressing at C and D. 6e) Subsequent yield loci from 6d) placed in the origin of the stress space

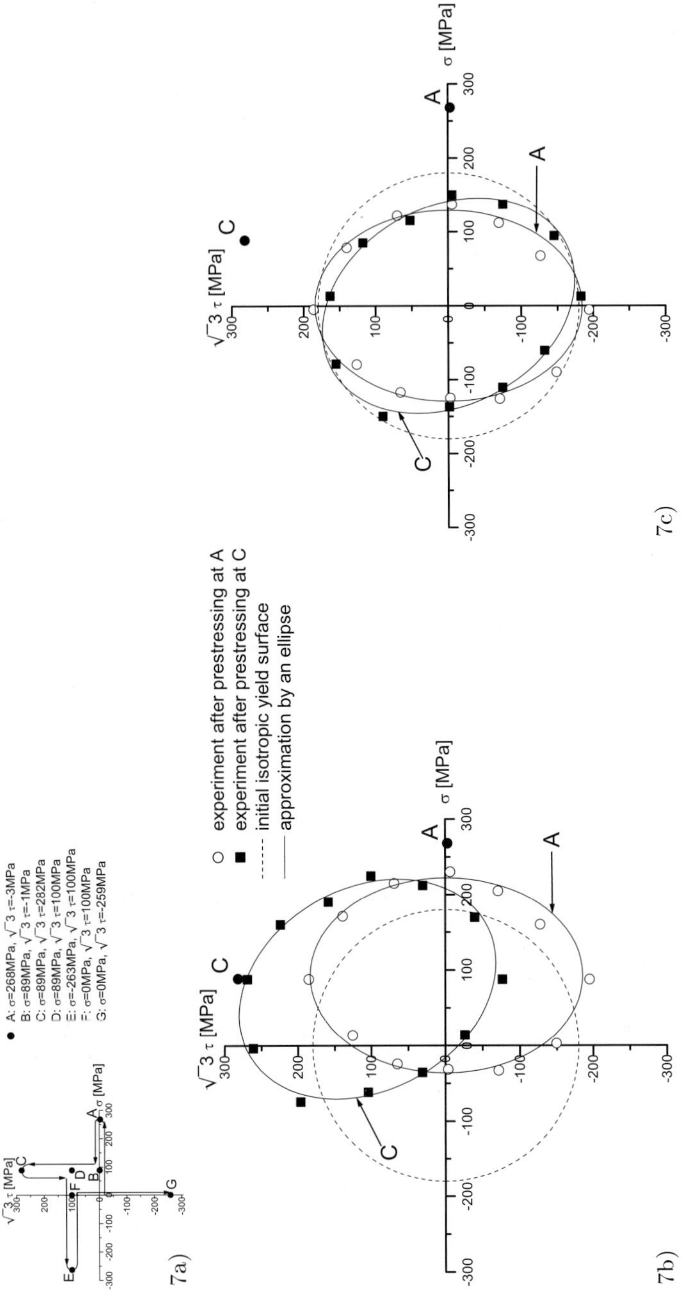

A: σ=268MPa, √3 τ=-3MPa
B: σ=89MPa, √3 τ=-1MPa
C: σ=89MPa, √3 τ=282MPa
D: σ=89MPa, √3 τ=100MPa
E: σ=-263MPa, √3 τ=100MPa
F: σ=0MPa, √3 τ=100MPa
G: σ=0MPa, √3 τ=-259MPa
√3 τ [MPa]
σ [MPa]
experiment after prestressing at A
experiment after prestressing at C
initial isotropic yield surface
approximation by an ellipse

7a) 7b) 7c)

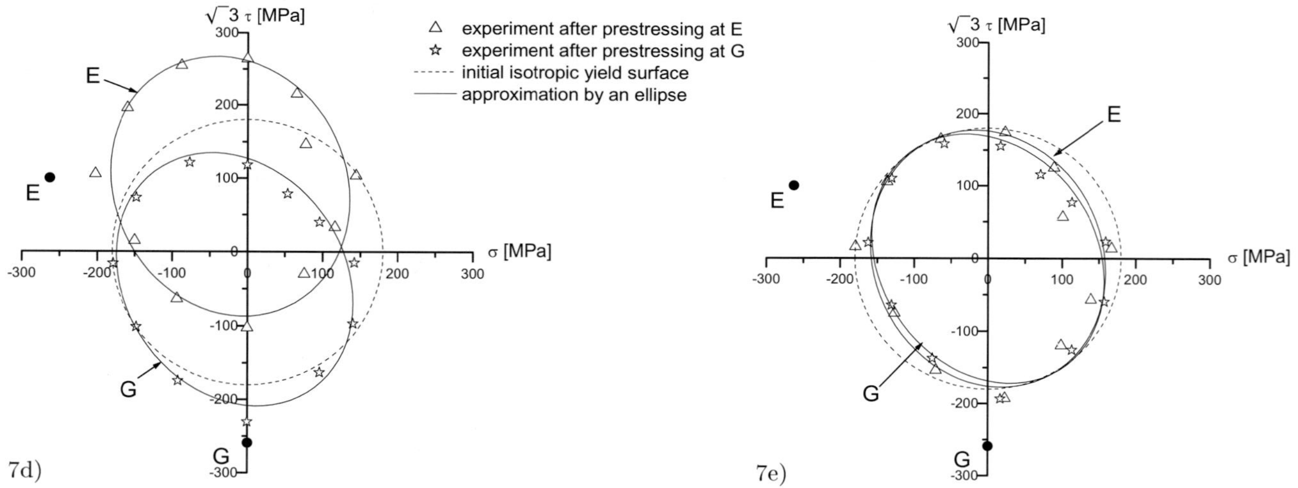

Fig. 7. Combined tension-torsion loading history. 7a) Loading path imposed. 7b) Yield loci after prestressing at A and C. 7c) Subsequent yield loci from 7b) placed in the origin of the stress space. 7d) Yield loci after prestressing at E and G. 7e) Subsequent yield loci from 7d) placed in the origin of the stress space

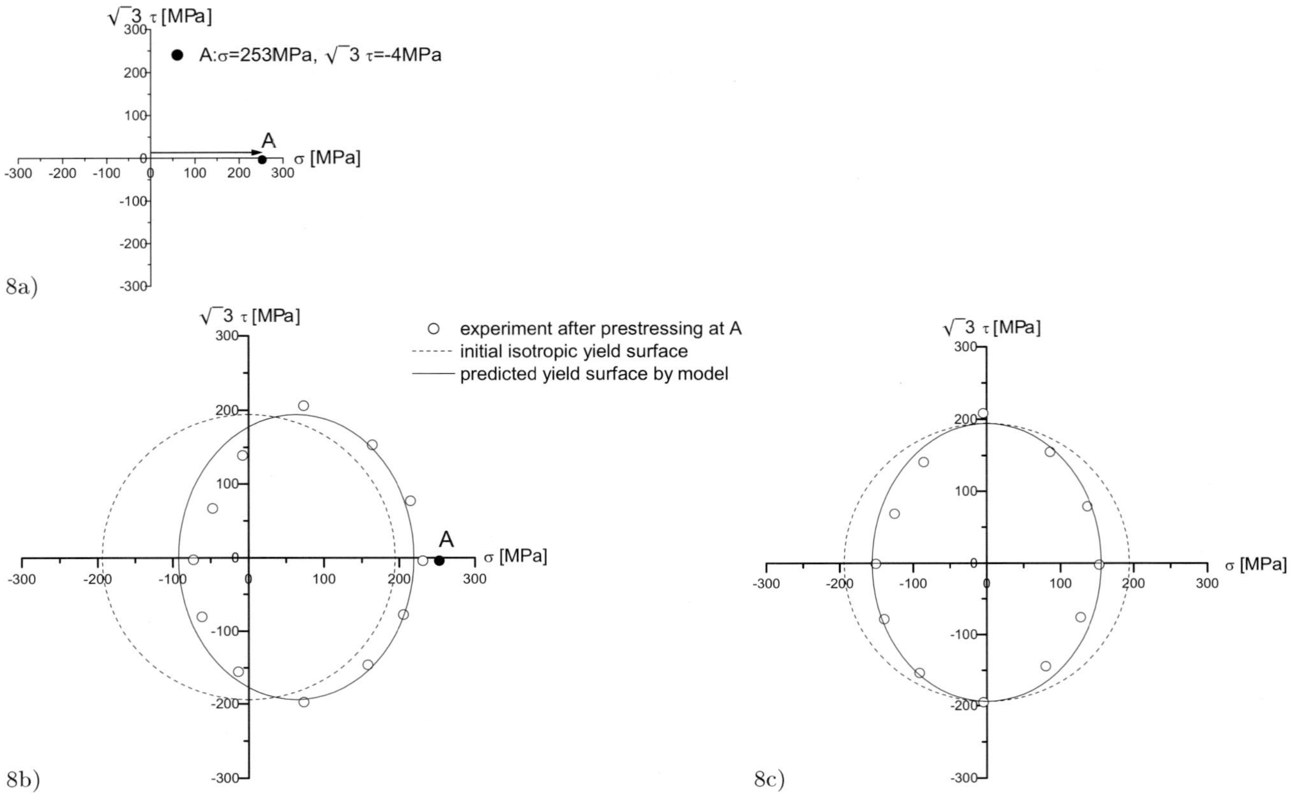

Fig. 8. Comparison of predicted responses with experimental data, displayed in Fig. 1. 8a) Loading path imposed. 8b) Yield locus after prestressing at A. 8c) Subsequent yield locus from b) placed in the origin of the stress space

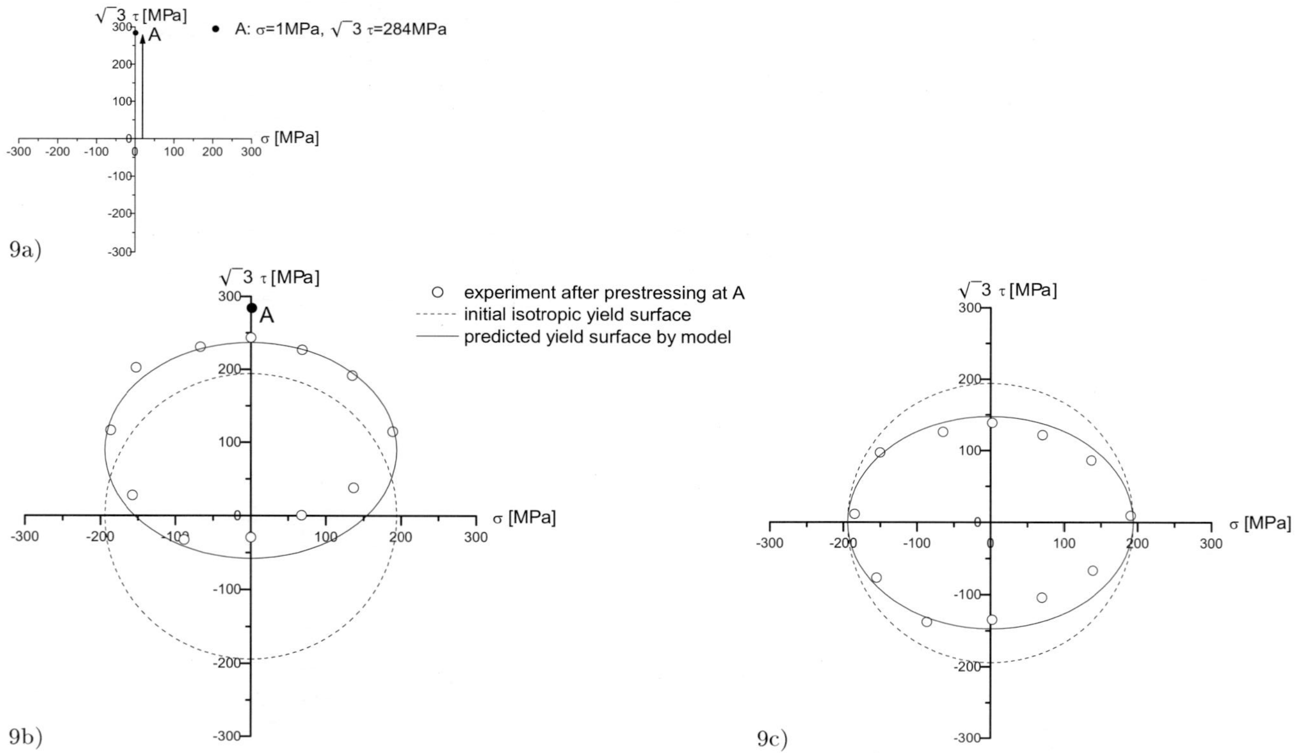

Fig. 9. Comparison of predicted responses with experimental data, displayed in Fig. 2. 9a) Loading path imposed. 9b) Yield locus after prestressing at A. 9c) Subsequent yield locus from 9b) placed in the origin of the stress space

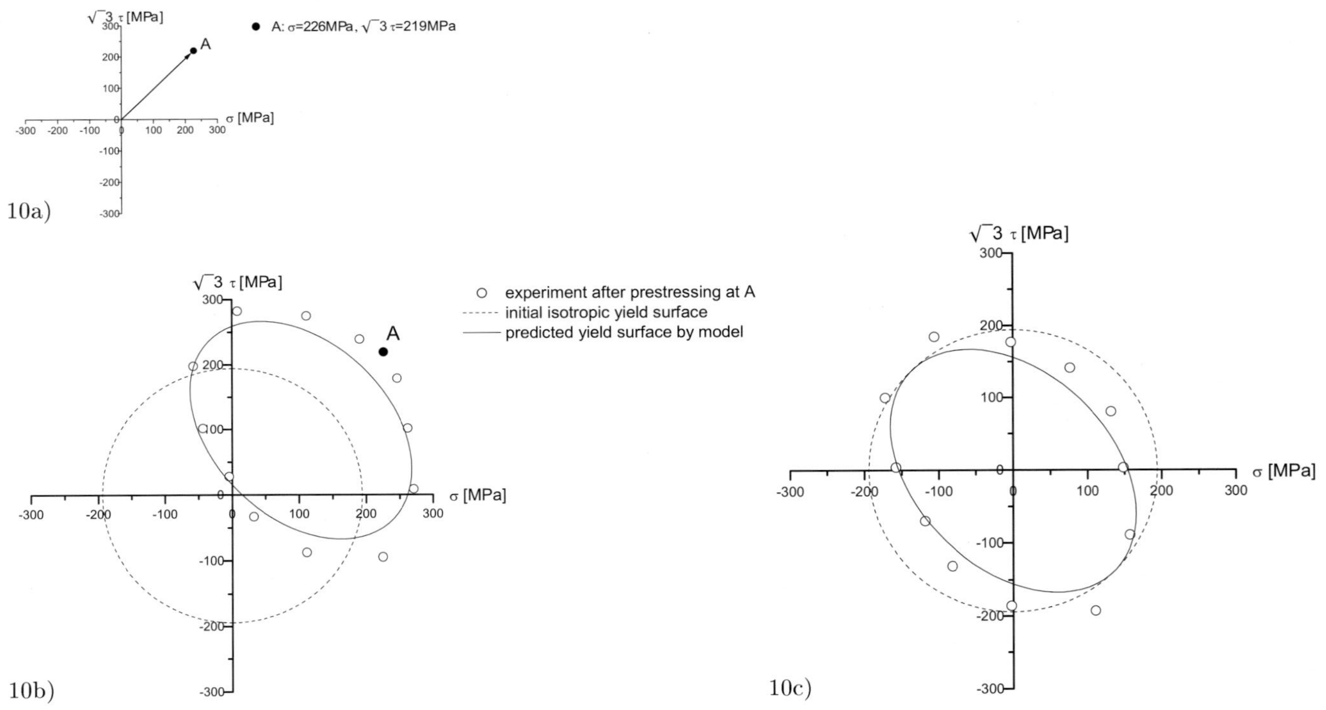

Fig. 10. Comparison of predicted responses with experimental data, displayed in Fig. 3. 10a) Loading path imposed. 10b) Yield locus after prestressing at A. 10c) Subsequent yield locus from 10b) placed in the origin of the stress space

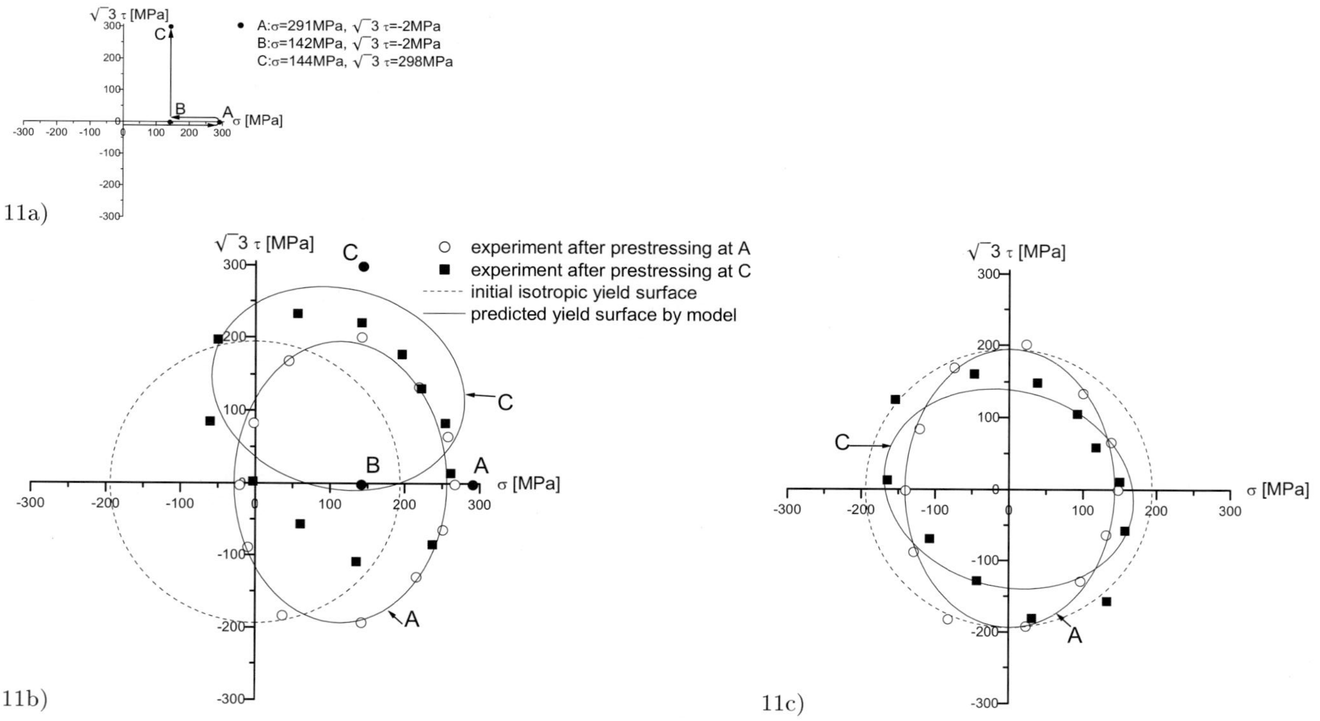

Fig. 11. Comparison of predicted responses with experimental data, displayed in Fig. 4. 11a) Loading path imposed. 11b) Yield loci after prestressing at A and C. 11c) Subsequent yield loci from 11b) placed in the origin of the stress space

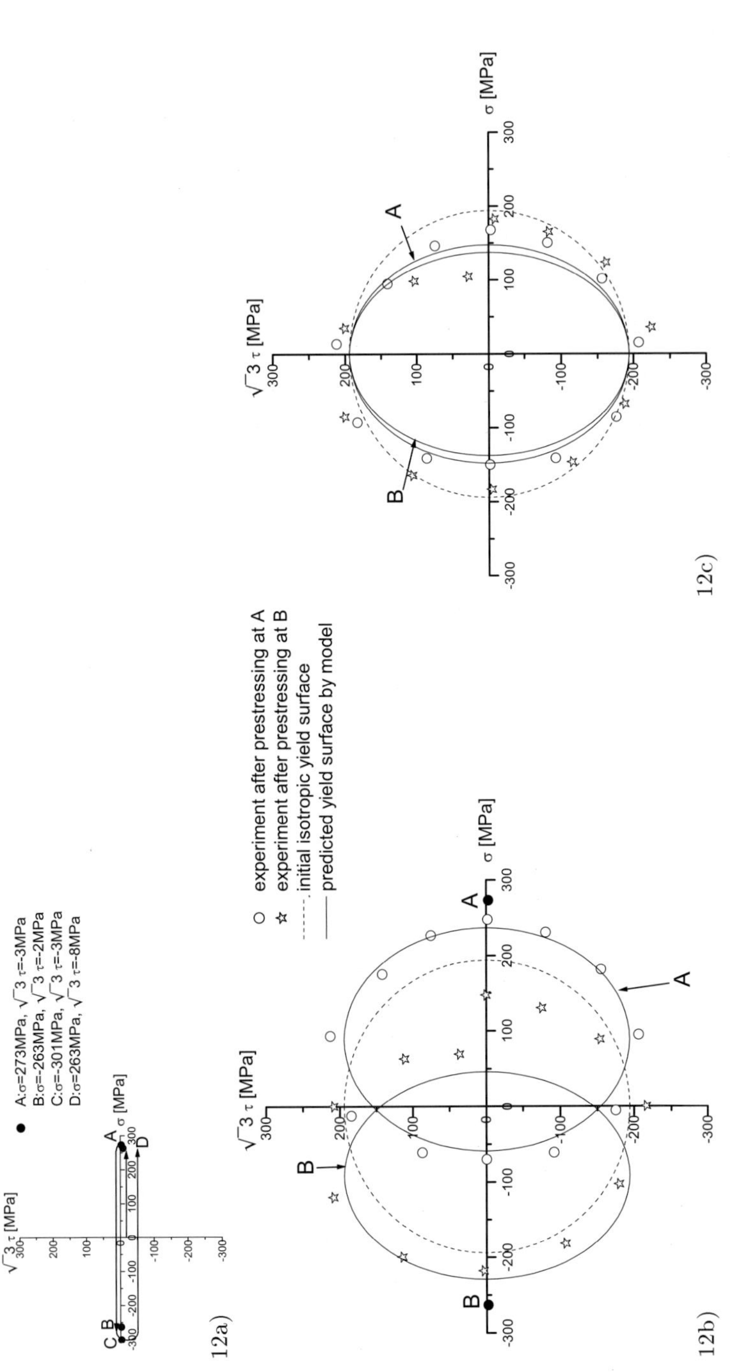
A:σ=273MPa, √3 τ=-3MPa
B:σ=-263MPa, √3 τ=-2MPa
C:σ=-301MPa, √3 τ=-3MPa
D:σ=263MPa, √3 τ=-8MPa
√3 τ [MPa]
σ [MPa]
12a)
experiment after prestressing at A
experiment after prestressing at B
initial isotropic yield surface
predicted yield surface by model
12b)
12c)

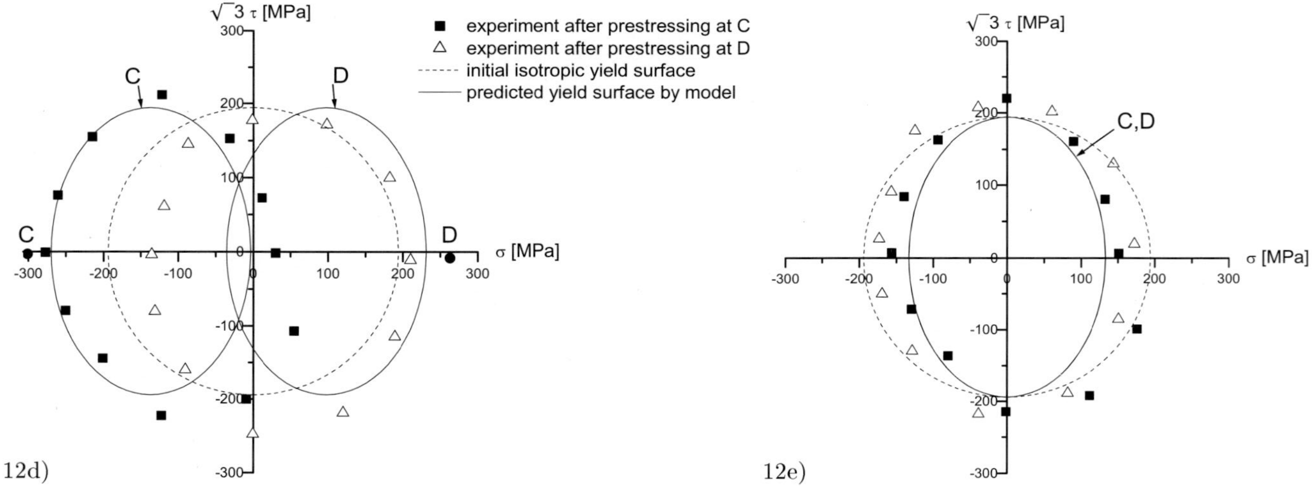

Fig. 12. Comparison of predicted responses with experimental data, displayed in Fig. 4. 12a) Loading path imposed. 12b) Yield loci after prestressing at A and B. 12c) Subsequent yield loci from 12b) placed in the origin of the stress space. 12d) Yield loci after prestressing at C and D. 12e) Subsequent yield loci from 12d) placed in the origin of the stress space

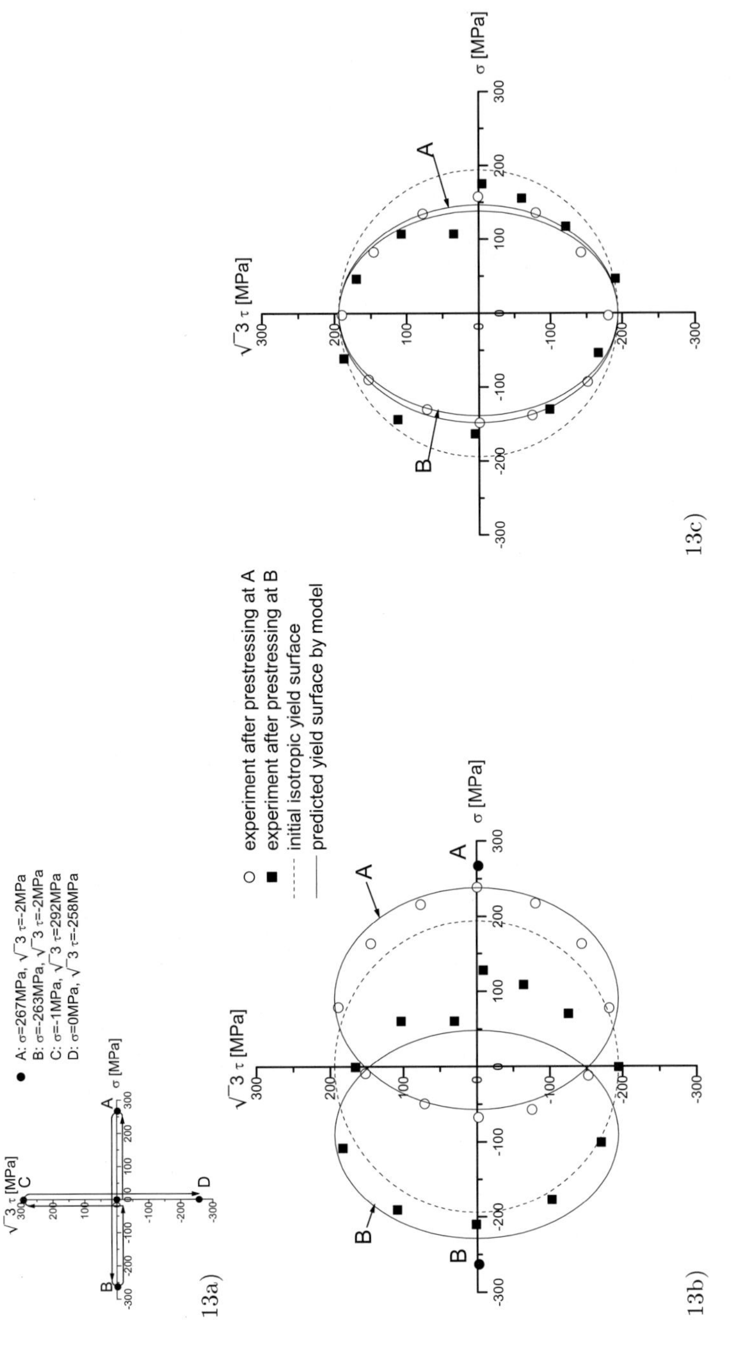
A: σ=267MPa, √3 τ=-2MPa
B: σ=-263MPa, √3 τ=-2MPa
C: σ=-1MPa, √3 τ=292MPa
D: σ=0MPa, √3 τ=-258MPa
√3 τ [MPa]
σ [MPa]
13a)
experiment after prestressing at A
experiment after prestressing at B
initial isotropic yield surface
predicted yield surface by model
13b)
13c)

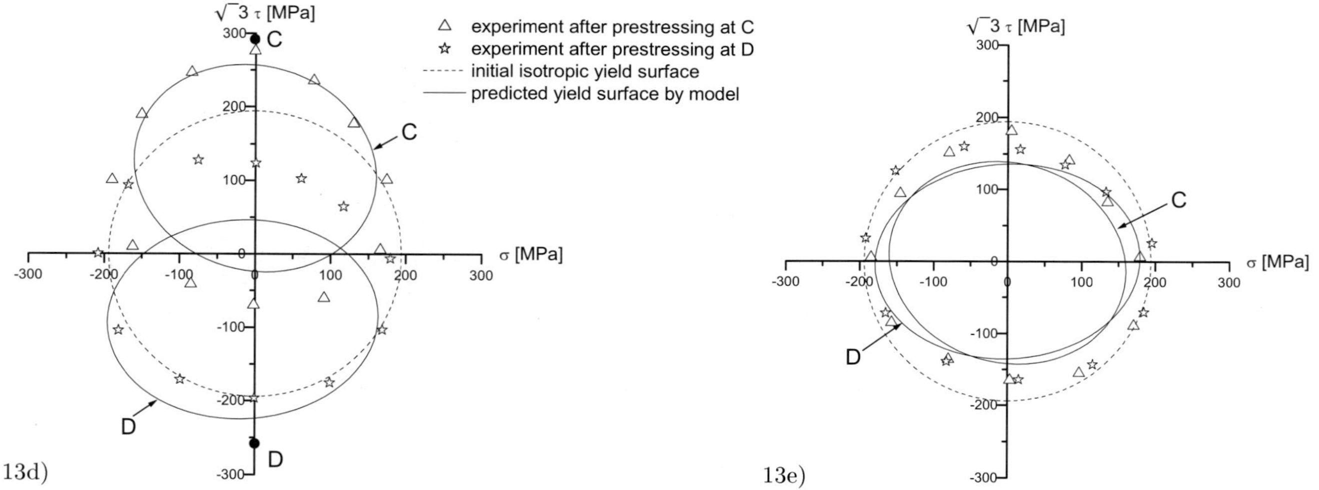

Fig. 13. Comparison of predicted responses with experimental data, displayed in Fig. 6. 13a) Loading path imposed. 13b) Yield loci after prestressing at A and B. 13c) Subsequent yield loci from 13b) placed in the origin of the stress space. 13d) Yield loci after prestressing at C and D. 13e) Subsequent yield loci from 13d) placed in the origin of the stress space

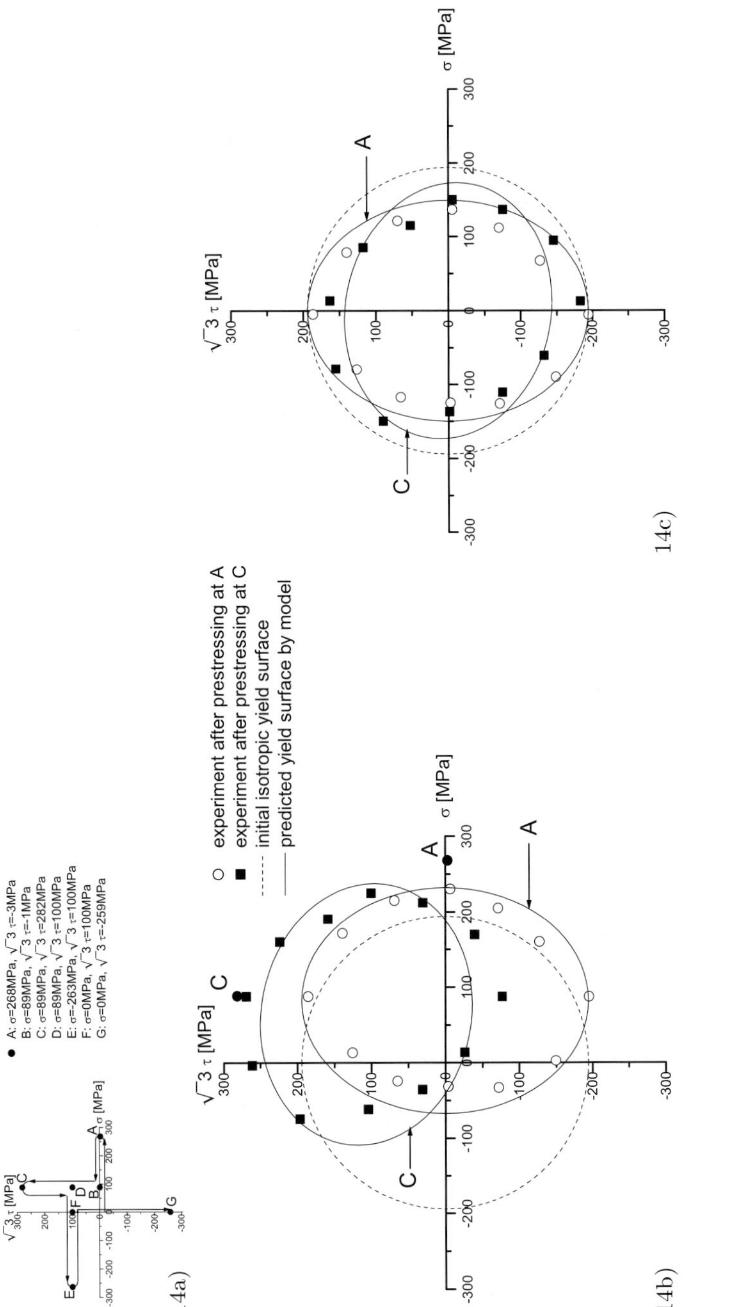
A: σ=268MPa, √3 τ=-3MPa
B: σ=89MPa, √3 τ=-1MPa
C: σ=89MPa, √3 τ=282MPa
D: σ=89MPa, √3 τ=100MPa
E: σ=-263MPa, √3 τ=100MPa
F: σ=0MPa, √3 τ=100MPa
G: σ=0MPa, √3 τ=-259MPa
experiment after prestressing at A
experiment after prestressing at C
initial isotropic yield surface
predicted yield surface by model
√3 τ [MPa]
σ [MPa]

14a) 14b) 14c)

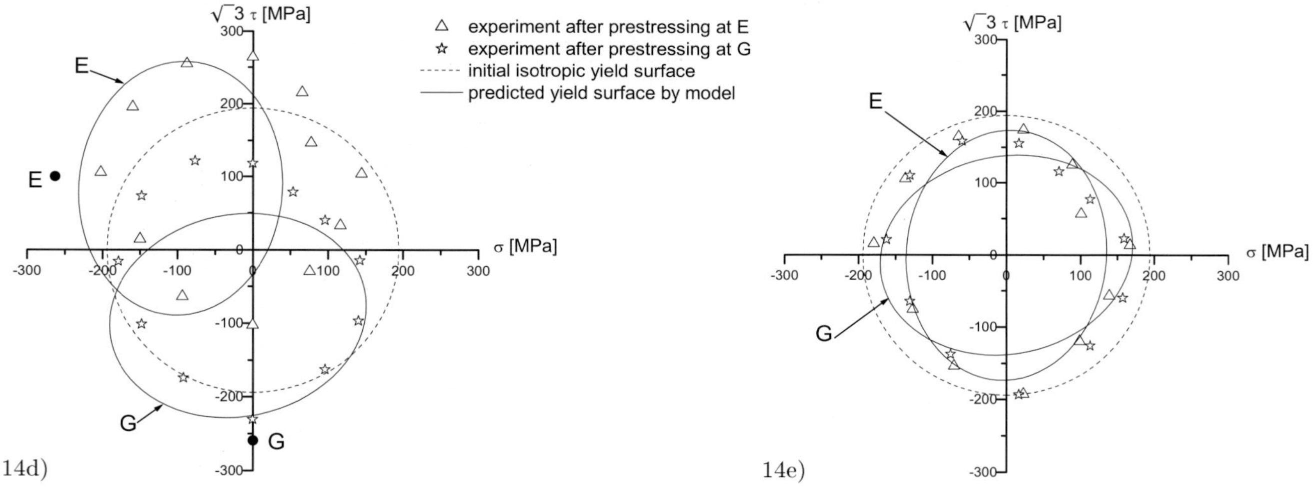

Fig. 14. Comparison of predicted responses with experimental data, displayed in Fig. 7. 14a) Loading path imposed. 14b) Yield loci after prestressing at A and C. 14c) Subsequent yield loci from 14b) placed in the origin of the stress space. 14d) Yield loci after prestressing at E and G. 14e) space

References

1. Armstrong, P. J., Frederick, C. O. (1966) A mathematical representation of the multiaxial Bauschinger effect. Gen. Electric Gen. Board, report RD/B/N **731**
2. Backhaus, G. (1968) Zur Fließgrenze bei allgemeiner Verfestigung. ZAMM **48**, 99–108
3. Baltov, A., Sawczuk, A. (1964) A rule of anisotropic hardening. Act. Mech.**1**, 81–92
4. Casey, J., Naghdi, P. M. (1981) On the characterization of strain-hardening in plasticity. Journal of Applied Mechanics **48**, 285–295
5. Chaboche, J. L., Dang-Van, K., Cordier, G. (1979) Modelization of the strain memory effect on the cyclic hardening of 316 stainless steel. SMIRT-5, Division L, Berlin
6. Dafalias, Y. F. (1979) Anisotropic hardening of initially orthotropic materials. ZAMM **59**, 437–446
7. Dafalias, Y. F., Popov, E. P. (1975) A model of nonlinearly hardening materials for complex loading. Acta Mechanica **21**, 173–192
8. Dafalias, Y. F., Popov, E. P. (1976) Plastic internal variables formalism of cyclic plasticity. Journal of Applied Mechanics **98**, 645–651
9. Dafalias, Y. F. (1987) Issues on the constitutive formulation at large elasto-plastic deformations, Part I: Kinematics. Archives of Mechanics **69**, 119–138
10. Dafalias, Y. F. (1993) On multiple spins and texture development. Case study: Kinematic and orthotropic hardening. Archives of Mechanics **100**, 171–194
11. Dafalias, Y. F. (1998) Plastic spin: Necessity and redundancy. International Journal of Plasticity **14**, 909–931
12. Häusler, O., Schick, D., Tsakmakis, Ch. (2002) Description of plastic anisotropy effects at large deformations. Part II: The case of transverse isotropy. In press in International Journal of Plasticity
13. Hecker, S. S. (1971) Yield surfaces in prestrained aluminum and copper. Metallurgical Transactions **2**, 2077–2086
14. Hecker, S. S. (1973) Influence of deformation history on yield locus and stress-strain behaviour of aluminum and copper. Metallurgical Transactions **4**, 985–989
15. Helling, D. E., Miller, A. K., Stout, M. G. (1986) An experimental investigation of the yield loci of 1100-0 aluminum, 70:30 brass, and an overaged 2024 aluminum alloy after various prestrains. J. Eng. Mat. Tech., ASME **108**, 313–320
16. Henshall, G. A., Helling, D. E., Miller, A. K. (1996) Improvements in the MATMOD equations for modeling solute effects and yield-surface distortion. In "Unified Constitutive laws of plastic deformation", (Eds. Krausz, A. S., Krausz, K.) Academic Press, New York, 153–227
17. Ikegami, K. (1982) Experimental plasticity on the anisotropy of metals. In "Proceedings of the Euromech Colloquium 115" (Ed. Boehler, J. P.) Éditions du Centre National de la Recherche Scientifique, Paris, 201–227
18. Ishikawa, H. (1997) Subsequent yield surface probed from its current center. International Journal of Plasticity **13**, 533–549
19. Ishikawa, H., Sasaki, K. (1988) Yield surface of SUS304 under cyclic loading. Journal of Engineering Materials and Technology, ASME **110**, 364–371
20. Ishikawa, H., Sasaki, K. (1998) Deformation induced anisotropy and memorized back stress in constitutive model. International Journal of Plasticity **14**, 627–646

21. Khan, A. S., Huang, S. (1995) Continuum theory of plasticity. Wiley, New York
22. Khan, A. S., Wang, X. (1993) An experimental study on subsequent yield surface after finite shear prestraining. Int. J. of Plasticity **9**, 889–905
23. Kowalewski, Z. L., Śliwowski, M. (1997) Effect of cyclic loading on yield surface evolution of 18G2A low-alloy steel. Int. J. of Mech. Sci. **39**, 51–68
24. Krempl, E. (1987) Models of viscoplasticity - some comments on equilibrium (back)stress and drag stress. Acta Mechanica **69**, 25–42
25. Marquis, D. (1979) Modélisation et identification de l'écrouissage anisotrope des métaux. General Electric Generating Board, report RD/B/N **731**
26. Miastkowski, J. (1968) Analysis of the memory effect of plastically prestrained material. Archiwum Mechaniki Stosowanej **20**, 257–276
27. Miastkowski, J., Szczepiński, W. (1965) An experimental study of yield surfaces of prestrained brass. International Journal of Solids and Structures **1**, 189–194
28. Mroz, Z. (1967) On the description of anisotropic workhardening. Journal of the Mechanics and Physics of Solids **15**, 163ff
29. Naghdi, P. M. (1960) Stress-strain relations in plasticity and thermoplasticity. In "Proc. Second Symp. Naval Structural Mechanics" (Eds. Lee, E. H. and Symonds, P. S.), Pergamon, Oxford, 121–169
30. Perzyna, P. (1963) The constitutive equations for rate sensitive plastic materials. Quarterly of Applied Mathematics **20**, 321–332
31. Phillips, A. (1979) The foundations of plasticity. Experiments. Theory and selected applications. CISM, Udine, 189–271
32. Phillips, A., Das, P. K. (1985) Yield surfaces and loading surfaces of aluminum and brass: an experimental investigation at room and elevated temperatures. International Journal of Plasticity **1**, 89–109
33. Phillips, A., Moon, H. (1977) An experimental investigation concerning yield surfaces and loading surfaces. Acta Mechanica **27**, 91–102
34. Phillips, A., Tang, J. L. (1972) The effect of loading path on the yield surface at elevated temperatures. International Journal of Solids and Structures **8**, 463–474
35. Rees, D. W. A. (1982) Yield functions that account for the effects of initial and subsequent plastic anisotropy. Acta Mechanica **43**, 223–241
36. Stout, M. G., Martin, P. L., Helling, D. E., Canova, G. R. (1985) Multiaxial yield behaviour of 1100 aluminum following various magnitudes of prestrain. International Journal of Plasticity **1**, 163–174
37. Trampczynski, W. (1992) The experimental verification of the unloading technique for the yield surface determination. Archives of Mechanics **44**, 171–190
38. Tsakmakis, Ch. (1987) Über inkrementelle Materialgleichungen zur Beschreibung großer inelastischer Deformationen. Fortschrittsberichte VDI Reihe **18** Nr. **36**, Düsseldorf: VDI-Verlag
39. Tsakmakis, Ch. (1995) Kinematic hardening rules in finite plasticity Part I: A constitutive approach. Continuum Mech. Thermodyn. **8**, 215–231
40. Tsakmakis, Ch. (1996) Formulation of viscoplasticity laws using overstress. Acta Mechanica **115**, 179–202
41. Tsakmakis, Ch. (2002) Description of plastic anisotropy effects at large deformations. Part I: Restrictions imposed by the second law and the postulate of Il'iushin. In press in International Journal of Plasticity
42. Wegener, K., Schlegel, M. (1996) Suitability of yield functions for the approximation of subsequent yield surfaces. Int. J. of Plasticity **12**, 1151–1177

43. Williams, J. F., Svensson, N. L. (1970) Locus of 1100-F aluminum. Journal of Strain Analysis **5**, 128–139
44. Williams, J. F., Svensson, N. L. (1971) Effect of torsional prestrain on the yield locus of 1100-f aluminum. Journal of Strain Analysis **6**, 263–272
45. Wu, H. C., Hong, H. K., Lu, J. K. (1995) An endochronic theory accounted for deformation induced anisotropy. International Journal of Plasticity **11**, 145–162
46. Yoshimura, Y. (1959) Hypothetical theory of anisotropy and the Bauschinger effect due to plastic strain history. Aeronautical Research Institute, University of Tokyo, report No. 349, 221-247

Damage and Failure of Ceramic Metal Composites: Experimental and Numerical Investigations

Thomas Emmel[3], Ulrich Stiefel[2], Dietmar Gross[1], and Jürgen Rödel[2]

[1] Institute of Mechanics, Darmstadt University of Technology
Hochschulstr. 1, D–64289 Darmstadt, Germany
gross@mechanik.tu-darmstadt.de

[2] Institute of Material Science, Darmstadt University of Technology
D–64287 Darmstadt, Germany
roedel@ceramics.tu-darmstadt.de

[3] Institute of Mechanics, now: ABACOM Software GmbH, Aachen, Germany

Abstract. Failure in ceramic metal composites due to temperature induced stresses during production and external loading is investigated. Two different structures, a metal layer between two ceramic supports and an inter-penetrating network, correlated to two different failure modes are studied in detail. Cavitation is found to be one significant mode of (internal) damage, which decreases the strength of the composite and subsequently leads to failure. Interface debonding, in conjunction with crack branching at the interface is another failure mode. Experimental observations are compared with numerical calculations, leading to a theoretical description to predict the failure mode and the critical stresses in the composite.

1 Introduction

Ceramic-metal composites are widely considered to overcome some of the disadvantages of pure ceramics, mainly the low toughness of these materials. However, the complex mechanical behaviour and the properties of the composite can not be explained by mixture theory or other simple models. There is a huge amount of literature concerning the behaviour of composites which shed some light on the process of failure in this class of materials, see e.g. Hill [8,9], Ashby et. al. [1], Hutchinson et. al. [13] or Varias et. al. [19].

In a previous work [20] the authors have shown, that, in the case of a strong interface between the metal and ceramic, the nucleation and the subsequent growth of cavities in the metal strongly affects the overall strength and the further damage process of the composite. It was illustrated how local failure mechanisms influence the macroscopic behaviour of the composite. Some additional aspects will be discussed in this work. First, a model to predict the stresses at the interface in a perfectly bonded composite will be described. Two different composite structures are investigated: inter-penetrating networks manufactured by pressure infiltration of a liquid metal phase into a porous ceramic pre-form, and – on the other side – a metal layer

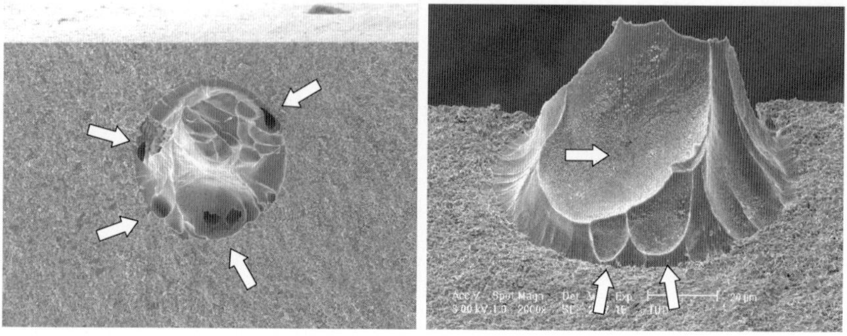

Fig. 1. Experimental observations of failure in ceramic-metal-composites containing inclusions with a given diameter of approx. 100 μm [20]. The arrows point to cavities mainly formed near the interface between the composite and the pure metal

between two ceramic supports. The first shows a significant enhancement of the toughness compared to the pure ceramic. The latter could be used, for example, to bond ceramical parts together. In the second part of the paper, experimental results for two different failure modes are presented and discussed: cavity formation in the pressure infiltrated metal phase from model experiments (Fig. 1) and interface debonding between the metal layer and the ceramic (Fig. 2).

The theory of cavity nucleation and growth is applied to the layered composite. According to this theory, cavitation should be the cause of damage and failure in the case of a strong interface. Neglecting size effects, it predicts cavity appearance for sufficiently high hydrostatic tension. In contrast, in relatively fine structures (typical length scales $< 10\,\mu m$) no cavities can be detected in experiments. This will be discussed in some detail.

In case of a relatively weak interface, precipitates at the interface between the ceramic and the metal can lead to an enhanced fracture toughness of the composite. Temporal and local crack branching within the precipitates (Fig. 2) seems to cause this enhancement. It is therefore studied by a numerical simulation of the stress-field at the interface between the materials. It is intended to improve the understanding of the different, complex failure modes observed in these composites and to propose some rules to enhance the properties of future designed materials.

2 Cavitation in Ductile Materials

The theory of cavity nucleation and growth was mainly founded by Ball [2]. It is used as a theoretical basis for micro-void appearance and growth in hyper-elastic as well as in elastic-plastic materials. Both descriptions are closely related and they differ only in the constitutive assumptions. Within this work, deformation theory is used to describe the elastic-plastic material

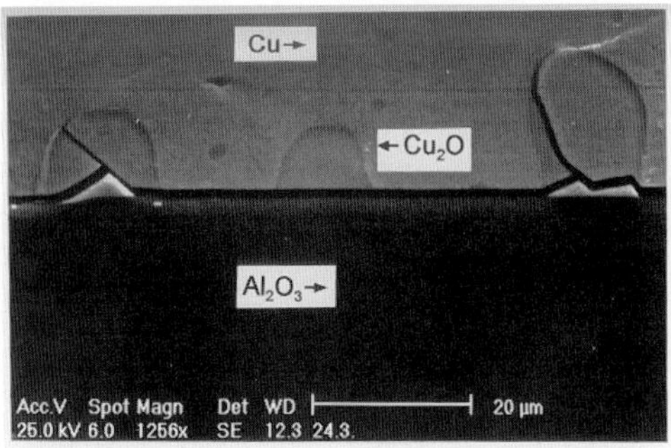

Fig. 2. Failure in a layered composite at the interface (Cu/Al_2O_3) with Cu_2O-precipitates [3]

behaviour in conjunction with the theory of cavitation as it was done in [5,12]. A detailed description of that theory is not repeated here, but can be found in the referenced papers and in Horgan and Poligone [11].

Following [5,11,12] it is assumed that under hydrostatic tension, instantaneous cavity nucleation may appear in a material without any pre-existing defects, or that unstable growth of a micro-cavity can happen. This hydrostatic load can be originated by an external load or e.g. by thermal stresses, caused, for example, by the initial cooling process during fabrication due to the different thermal expansion coefficients of metal and ceramic ($\alpha_{\mathrm{T_{cer}}} < \alpha_{\mathrm{T_{met}}}$).

The relevance of cavity formation is pertinent. It can be observed whenever a sufficiently large amount of metal is exposed to a high hydrostatic stress state. Metal enclosed by ceramic will be exposed to those high stresses generated by the temperature difference between fabrication and use. A model experiment with metal inclusions of a predefined size and shape illustrates the relevance of cavitation (see Fig. 1).

2.1 Critical Stress-State

At a critical stress state a cavity nucleates in a hyper-elastic material or a pre-assumed small cavity grows unstable in an elastic-plastic material. It should be mentioned that there is no difference between a pre-existing cavity for elastic-plastic materials or a prescribed defect with another shape in the material [5], since both can be regarded as a seed for a larger cavity.

Starting from deformation theory, the general formula

$$p_{\mathrm{cr}} = \int\limits_{1}^{v_0} \frac{W^{\mathrm{el.}}_{,v}(v)}{v^3 - 1}\,\mathrm{d}v + \int\limits_{v_0}^{\infty} \frac{W^{\mathrm{el.pl.}}_{,v}(v)}{v^3 - 1}\,\mathrm{d}v \tag{1}$$

can be derived. It relates the external hydrostatic critical tensile load p_{cr} to the overall weighted derivative of the strain-energy density W with respect to v, a radial strain measure[1] integrated over the elastic and the plastic portion of the body.

An alternative, more practical approach was derived by Huang [12] using

$$\frac{p_{\mathrm{cr}}}{\sigma_0} = -\int_0^\infty \left(e^{\frac{3\xi}{2}} - 1\right)^{-1} f(-\xi)\, \mathrm{d}\xi \ , \tag{2}$$

and relates the critical load p_{cr} directly to a given stress-strain relation $\sigma = f(\varepsilon)$. Here, ε is the logarithmic strain, σ is the Cauchy stress and σ_0 is the initial yield stress.

Furthermore, the relation between the cavity radius c and the applied load p_0 can be calculated[2] as

$$\frac{p_0}{\sigma_0} = -2\int_1^\infty f\left(\frac{2}{3}\ln\left[1 - \frac{1-(c_0/c)^3}{\eta^3}\right]\right)\frac{\mathrm{d}\eta}{\eta} \ . \tag{3}$$

Figure 3 illustrates the increasing growth of a pre-existing cavity of initial size c_0 as a function of the applied load and for different hardening exponents n. Here the function $f(\varepsilon)$ was chosen as

$$\sigma = f(\varepsilon) = \begin{cases} E\varepsilon, & \varepsilon \le \varepsilon_0 \\ b\varepsilon^{1/n}, & \varepsilon > \varepsilon_0 \end{cases} \ , \quad b = \sigma_0 \varepsilon_0^{-1/n}, \ \varepsilon_0 = \frac{\sigma_0}{E} \ , \tag{4}$$

combining Hooke's law with an exponential relationship between stress and strain in the plastic range.

2.2 Thermally Induced Stresses and Cavitation

Ceramics infiltrated under pressure with a liquid metal phase exhibit significant eigenstresses as a result of the different thermal expansion coefficients of the two constituents. They can be estimated by using the Eshelby solution for an elliptic inclusion in an infinite elastic matrix [16]. From them the stress in the inclusion is determined by the following relation [14,15]

$$\boldsymbol{\sigma} = \left(\mathbb{C}_{\mathrm{i}}\mathbb{G}^{-1}\mathbb{P}\mathbb{C}_{\mathrm{m}}^{-1} - \mathbb{1}\right)\boldsymbol{\sigma}^0 \ , \tag{5}$$

where

$$\mathbb{G} = \mathbb{1} + \mathbb{P}\left(\mathbb{C}_{\mathrm{m}}^{-1}\mathbb{C}_{\mathrm{i}} - \mathbb{1}\right) \ , \quad \boldsymbol{\sigma}^0 = \mathbb{C}_{\mathrm{i}}\boldsymbol{\varepsilon}^0 \ .$$

---

[1] Radial displacement versus resulting cavity radius.

[2] Indeed, this relation is used to calculate the critical stress state.

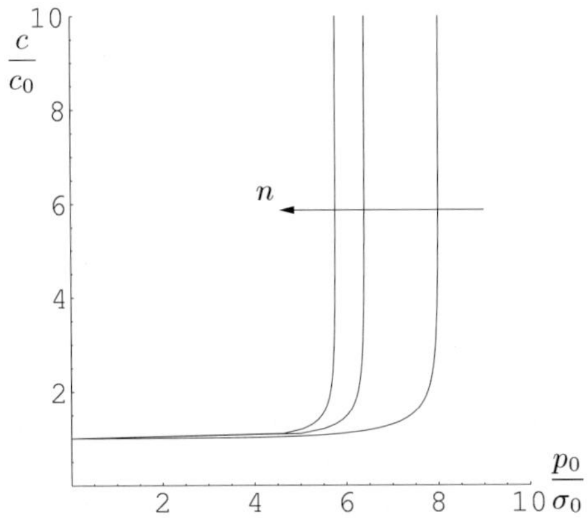

Fig. 3. Cavity growth in elastic-plastic materials for different hardening exponents n [5,12] ($n = 8,\ 16,\ 32,\ E = 70\,\text{GPa},\ \nu = 0.5$ (Incompressibility), $\sigma_0 = 50\,\text{MPa}$)

Here, $\mathbb{P}$ identifies the Eshelby-tensor, $\mathbb{C}_\mathrm{m}$ and $\mathbb{C}_\mathrm{i}$ are the elasticity tensors for the matrix and the inclusion, respectively. The eigenstrain-tensor, denoted by ε^0, can easily be calculated from the difference between the thermally induced strains

$$\varepsilon^0 = \varepsilon^{\mathrm{th}}_{\mathrm{diff.}}(T_\mathrm{e}, T_0) = \varepsilon^{\mathrm{th}}_{\mathrm{i}}(T_\mathrm{e}, T_0) - \varepsilon^{\mathrm{th}}_{\mathrm{m}}(T_\mathrm{e}, T_0) \quad , \tag{6}$$

for an initial temperature T_0 and the known end-temperature T_e.

Equation (5) can be simplified in the special case of a spherical symmetric inclusion, leading to the hydrostatic pressure in the inclusion

$$p_\mathrm{i}(T_\mathrm{e}, T_0) = \frac{-2\,E_\mathrm{m}\,E_\mathrm{i}\,\varepsilon^{\mathrm{th}}_{\mathrm{diff.}}(T_\mathrm{e}, T_0)}{2\,E_\mathrm{m} + E_\mathrm{i} + E_\mathrm{i}\,\nu_\mathrm{m} - 4\,E_\mathrm{m}\,\nu_\mathrm{i}} \quad . \tag{7}$$

Here, all material parameters (Young's moduli E_i, E_m and the corresponding Poisson's ratio ν) may depend on temperature.

Equating the hydrostatic pressure (7) with the predicted values from (2) and (3), again with temperature dependent parameters, results in the condition

$$p_\mathrm{i}(T_\mathrm{e}, T_0) = p_\mathrm{cr}(T_\mathrm{e}) \tag{8}$$

for a critical state. It implicitly describes the dependence of the cavity size on the temperature and can be evaluated pointwise in the material. In this sense it can be regarded as a criterion for an increased growth and nucleation of flaws in the material. Note that $p_\mathrm{cr}(T_\mathrm{e})$ was calculated for an infinite domain with a single cavity, i.e. no interaction with other cavities was taken into account. As a result, (8) is only an upper bound. The real, radial load near the cavity will be somewhat lower.

The stresses calculated using deformation theory are

$$\sigma_{rr}(R) = \int_{(1+c^3/R^3)^{1/3}}^{v_0} \frac{W^{\text{el.}}_{,v}(v)}{v^3-1}\mathrm{d}v + \int_{v_0}^{(1+c^3/c_0^3)^{1/3}} \frac{W^{\text{el.pl.}}_{,v}(v)}{v^3-1}\mathrm{d}v \tag{9}$$

and

$$p(R) = -\sigma_{rr}(R) - \frac{1}{2}\left(1+\frac{c^3}{R^3}\right)^{1/3} W_{,v} \quad , \tag{10}$$

where c_0 is the initial radius of a pre-existing cavity, c is the actual cavity radius and R is the distance from the origin. Once the radial stress-component $\sigma_{rr}(R)$ and the pressure $p(R)$ is known, the circumferential stress-components can be easily calculated.

The boundary conditions for thermally induced cavitation differ from those for the classic theory of cavitation. There is no unstable growth of a cavity in a metallic inclusion surrounded by a ceramic matrix. To calculate the stresses at the interface of such an inclusion and the ceramic we need to know the size of the cavity which is a-priori unknown. This difficulty can be overcome by estimating the resulting cavity size by the volumetric difference of two cooled spheres, one a metallic, the other a ceramic sphere,

$$V_{\text{cer}} - V_{\text{met}} = V_{\text{diff.}} = \frac{4}{3}\pi A^3\left((1+\varepsilon^{\text{th}}_{\text{cer}})^3 - (1+\varepsilon^{\text{th}}_{\text{met}})^3\right) \quad . \tag{11}$$

Here A is the initial radius which also can be regarded as a quantity which describes the overall size of many cavities. In case of one cavity we get

$$c_{\max}(T) = A\left((1+\varepsilon^{\text{th}}_{\text{cer}})^3 - (1+\varepsilon^{\text{th}}_{\text{met}})^3\right)^{\frac{1}{3}} \quad . \tag{12}$$

It was shown in [5], that this value is a good approximation of the actual size, but if necessary, the size also can be adjusted by experiments and/or numerical calculations.

2.3 Simulation of Cavity Growth

The introduced model for the critical temperature is now used to calculate the stresses in an Al/Al_2O_3-composite using the parameters given in Table 1. In addition, the same numerical model (finite elements), introduced in the previous work [20], is used to calculate the stresses at the interface between metal and ceramic. A model experiment with defined spherical inclusions is used to compare all results with experiments.

The (numerical) integration of (9) has been carried out using MATHEMATICA. For this purpose, the material data from Table 1 were represented by curve-fitted functions. On the other hand, the numerical simulation was done with piecewise linear material constants.

Table 1. Material parameters used for the calculations

	Aluminum (Al)					Alumina (Al_2O_3)		
T (°C)	E (GPa)	ν	α_T ($\times 10^{-5}$)	σ_0 (MPa)	n	E (GPa)	ν	α_T ($\times 10^{-5}$)
27	70.0	0.347	2.50	50.0	5	400.0	0.24	0.655
127	.	0.35	.	50.0	5	.	.	0.732
227	.	0.355	.	47.6	6	.	.	.
327	.	0.361	.	42.7	7	.	.	.
427	.	0.367	.	40.8	8	.	.	.
527	.	0.373	.	21.9	9	.	.	.
627	70.0	0.38	3.40	10.8	10	400.0	0.24	0.836

2.4 Results

For a perfect configuration without any cavity, an inclusion in the matrix is strongly constrained in its deformation and therefore follows Eshelby's solution for an elastic inclusion. Inelastic processes will not occur under the assumed thermal loading. This will result in an almost linear stress-temperature relation which can be seen in Fig. 4 for the initial slope up to 580°C. Here, aluminum in an alumina matrix was cooled down from an initial stress-free state at 600°C to room temperature.

At approx. 580°C the stress reaches the upper limit $p_{\mathrm{cr}} \approx 9.2\sigma_0$ (dashed line, Fig. 4). For a real composite the stress at the interface will be somewhat lower on account of interaction effects which are neglected in this analysis. In fact the pressure at the interface is lower: $p \approx 8\sigma_0$. In addition, the radial stress is computed from (9). Both values agree sufficiently well with those obtained from the numerical simulation. However, the wavy pattern only arises in the numerical simulation, where the material data is only linear interpolated in stages, as mentioned above. The calculated pressure at room temperature $p \approx 9\sigma_0$ corresponds to the experimental results in the range between 60-75% of the relative density of the ceramic preform (Fig. 5) independent of the structure and for the fine structure in the complete range. For finer structures other effects may be relevant to be treated below.

Suppose that the stress in the metal is bounded by the calculated critical stress. As a consequence, more than one cavity will nucleate and grow in a sufficiently large inclusion, resulting in a pattern-like cavity-formation. Therefore, the stress in the composite will be bounded everywhere. The size of these distributed cavities is bounded by the size of one single cavity with the radius c_{max} from (12). For the considered composite (Al/Al_2O_3) this is $c_{\mathrm{max}} \approx 0.35A$, where A is the characteristic size parameter of the metal phase.

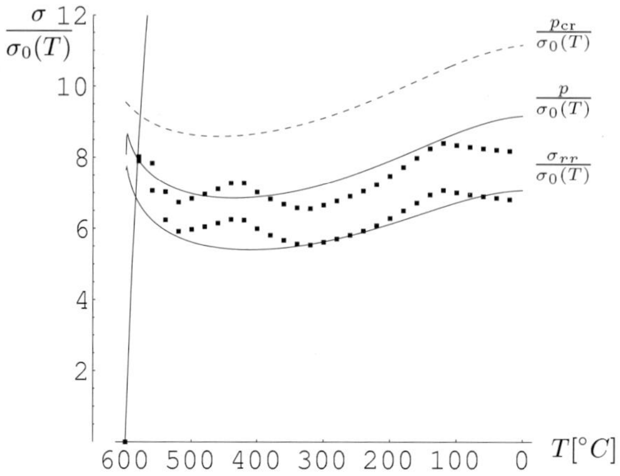

Fig. 4. Calculated stresses (deformation-theory) compared with numerical results obtained by finite element simulations (dotted curves) for a single cavity [5].

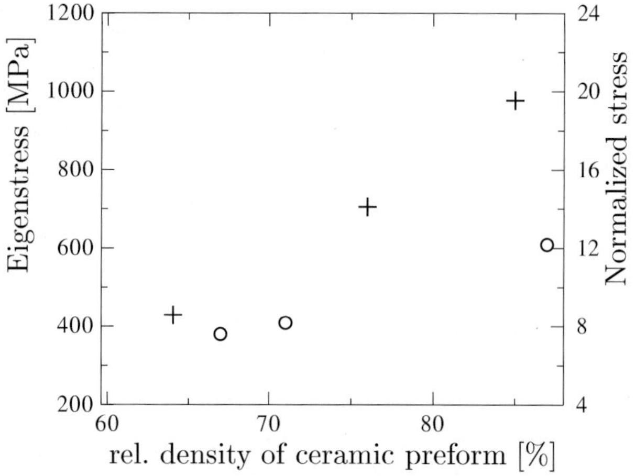

Fig. 5. Eigenstress in Al/Al_2O_3-composite [20] as a function of the preform-density. Circles: coarse structure, crosses: fine structure. Right scale: Results normalized with an assumed flow stress of $\sigma_0 = 50\,\text{MPa}$

3 Failure in Layered Composites

In addition to the prescribed inter-penetrating composites, layered composites were examined. This class of composites is known to introduce very high hydrostatic stresses into the metallic layer in front of an interfacial crack-tip. Even though cavitation was not observed in the experimentally tested composites, it is an imaginable process of failure in these structures. Therefore,

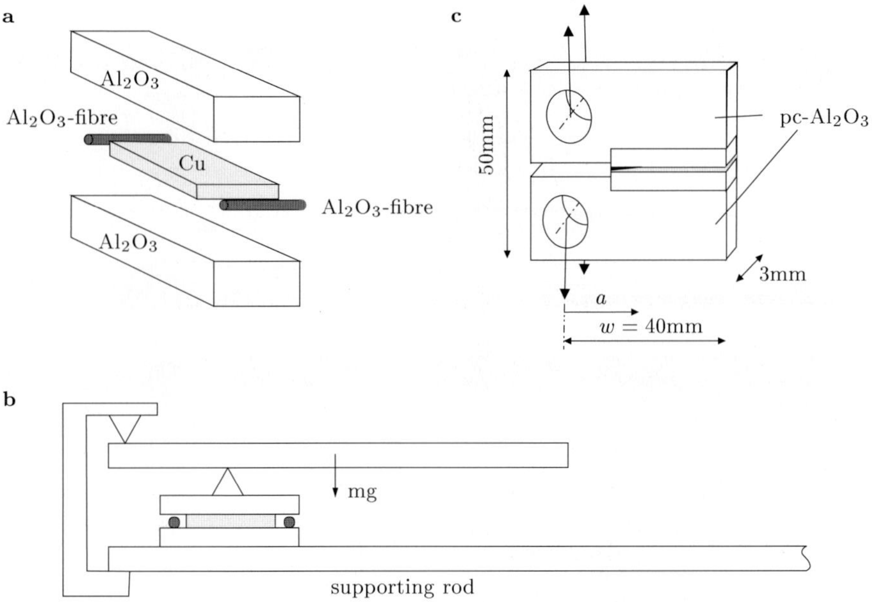

Fig. 6. Sketch of the samples' setup (**a**), the lever construction in the furnace (**b**), and the glued CT-specimen (**c**).

a simulation of cavitation growth was carried out, to determine conditions where this process may occur.

On the other hand, another failure mechanism was observed experimentally, which influences the fracture toughness of the composite as described below.

3.1 Processing and Experimental Procedure

Layered composites of $Al_2O_3/Cu/Al_2O_3$ were produced by liquid-phase-bonding. For this purpose two plates ($30 \times 8 \times 4\,mm^3$) of Al_2O_3 single-crystals[3] were stacked on each other with a slice of 99.95% pure copper[4] in between. Sapphire fibres[4] with a thickness of 110 µm were used to achieve a constant distance between the two sapphire plates. Figure 6a shows a sketch of this construction. Sapphire was used instead of polycrystalline alumina because of its transparency, which gives the possibility to control the quality of the metal layer produced, in particular with respect to interfacial porosity.

Because of the low wetting behaviour of Cu on sapphire, it was necessary to apply a load onto the sapphire plates (Fig. 6b). The whole sample setup was then heated to temperatures between 1150°C and 1400°C under

[3] sapphire, manufactured by CRYSTAL GmbH Berlin.
[4] Goodfellow, Bad Nauheim, Germany.

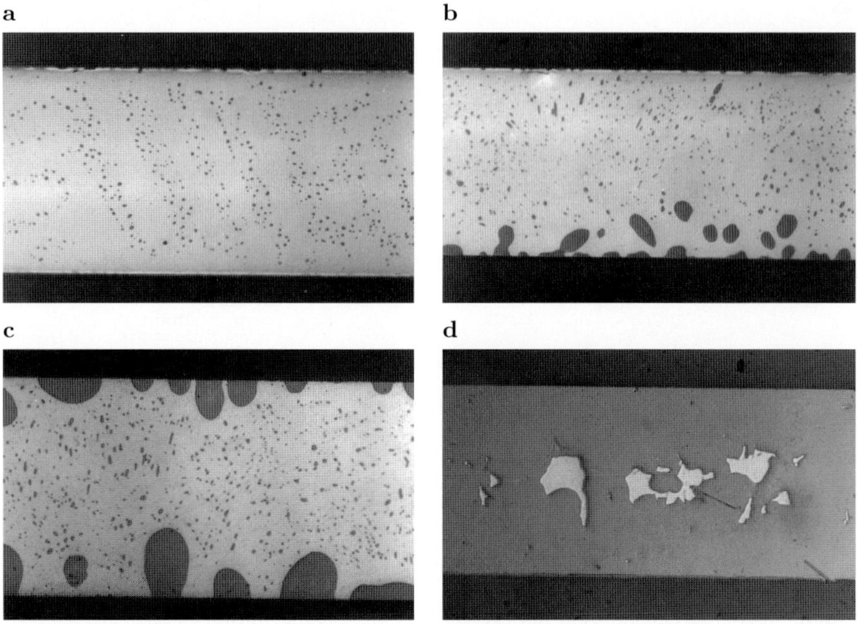

Fig. 7. Microstructures of the copper layer developing during cooling at different oxygen partial pressures. The dark regions on top and bottom are sapphire. The bright and dark phases in between are copper and Cu_2O, respectively

a controlled oxygen atmosphere [3]. The typical holding time at the joining temperature was 2 hours. Depending on the amount of oxygen dissolved in the liquid copper in the joining process different microstructures develop during cooling (Fig. 7a-d). The partial pressure increases from picture a) to d). Sample a) was produced at near-eutectic conditions ($p_{O_2} \approx 10^{-4.5}$bar), b) and c) at over-eutectic conditions ($p_{O_2} \approx 10^{-4} - 10^{-3.5}$bar), and d) at a very high partial pressure of $p_{O_2} \approx 10^{-3}$bar at 1300°C [18]. At extremely low partial pressures ($p_{O_2} < 10^{-8}$bar), which can be achieved by adding H_2 to the atmosphere, no precipitation can be observed.

The alumina/copper/alumina sandwiches fabricated this way were cut into 3mm thick plates and polished afterwards. Polycrystalline alumina frames were glued to these plates to get the necessary CT-specimens for R-curve measurement (Fig. 6c).

Prior to the R-curve measurement a pre-crack was produced using a half chevron notch as described in [17]. Polishing materials of very different hardness forms a step at the interface of these materials. Hence, and due to the high contrast between the metal layer and the surrounding sapphire, it is not possible to determine the crack position with the necessary accuracy under

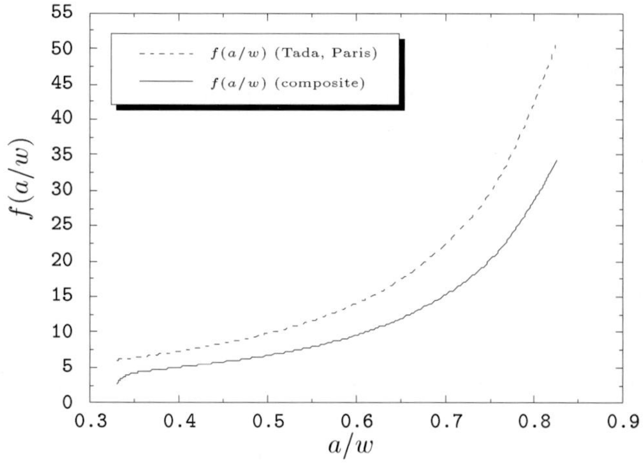

Fig. 8. $f(a/w)$ calculated for a homogeneous CT-sample (dashed line) and the geometry used in this study (solid line)

an optical microscope. Therefore penetration dye and ultraviolet light were used to render the crack visible.

The standard formula for calculating the K-values for a given crack-length in a CT-specimen was determined for homogeneous materials (ASTM E 399). Because of the wide difference in elastic constants for this material combination (see Table 2) a new geometry function $f(a/w)$ had to be calculated for these specimens. Therefore, a detailed model of the new geometry was built using ANSYS. Stress intensities at the crack-tip were calculated for the copper- and the sapphire-side of the crack using the ANSYS-specific `kcalc`-command. It extrapolates the stress intensity at the crack tip from the crack opening [17] for an elastic material

$$u(x) = \sqrt{\frac{8x}{\pi}\frac{K_{\mathrm{I}}}{E'}}, \quad v(x) = \sqrt{\frac{8x}{\pi}\frac{K_{\mathrm{II}}}{E'}} \quad . \tag{13}$$

Here, $u(x)$ describes the displacement of a node at position x in y-direction under an applied K_{I}, while $v(x)$ describes the displacement of this node in x-direction under an applied K_{II}, with $E' = E/(1-\nu^2)$ (plane strain). All calculations were performed for purely elastic deformations. In addition, the phase angle is calculated from the values of K_{I} and K_{II} at the interface by the formula [6]

$$\tan \Psi = \frac{K_{\mathrm{II}}}{K_{\mathrm{I}}} \quad . \tag{14}$$

Figure 8 shows the new calculated function f (solid line) compared with the standard formula (dashed line). The values for the composite material are lower by an amount of about 33%±2% compared to the standard values.

A bigger difference only occurs for $a/w < 0.35$ due to glueing, but this is irrelevant in practice because the material in this range is removed after precracking. The following formula is the result of a least square fit to the values of the composite material:

$$f(\xi) = \frac{(2+\xi)\left(-9.29 + 62.63\xi - 142.59\xi^2 + 140.71\xi^3 - 49.62\xi^4\right)}{(1-\xi)^{3/2}} \tag{15}$$

with $\xi = a/w$.

Figure 9 shows the phase angle calculated for the material combination investigated. It can be seen, that the phase angle remains constant at about 10° for a/w between 0.4 and 0.9. This means, that a constant failure mechanism can be assumed for this range.

3.2 Experimental Observations

Figure 10 shows examples of the R-curves measured in this study. It can be seen that, depending on the oxygen content of the metal layer, the curves show a strong toughening. Plotting the average values of each curve as a function of the oxygen content of the metal layer yields Fig. 11. It is obvious, that there is a maximum toughness for this material at around 5at% oxygen content. This is a region, where the microstructure of the metal layer shows some large precipitates of Cu_2O in the copper layer and at the interface. For precipitates at the interface, crack-branching could be observed (Fig. 2). Once the precipitates get too large in number and size and the interface is completely covered with oxide, the toughening effect vanishes.

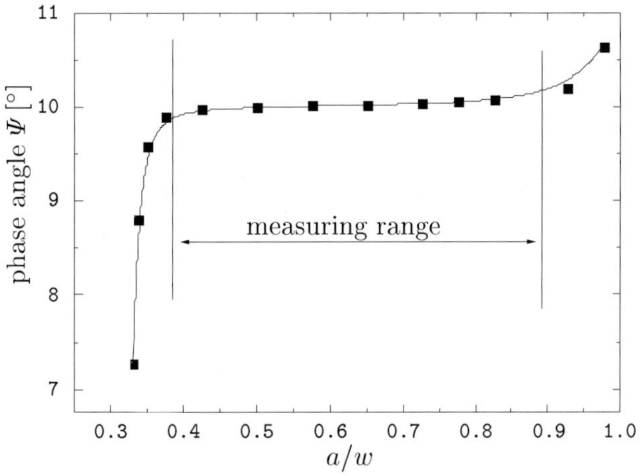

Fig. 9. Phase angle at the interface

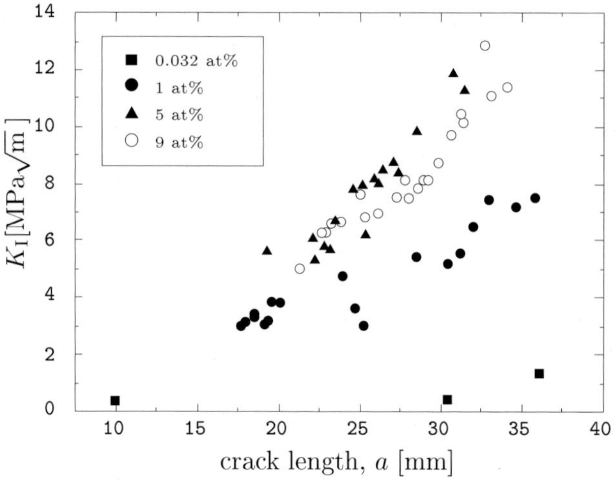

Fig. 10. Examples of R-curves with different content of oxygen in the metal layer

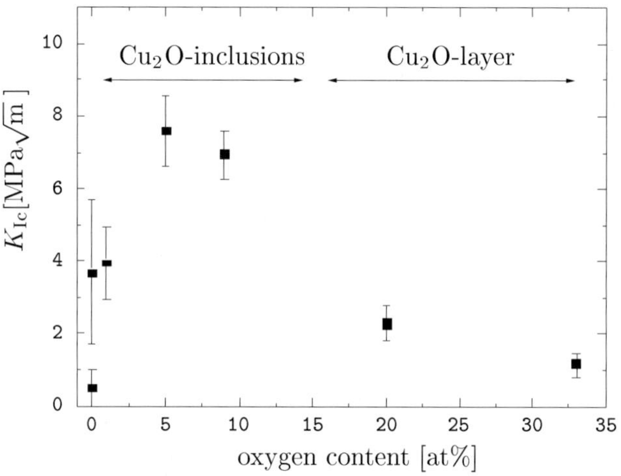

Fig. 11. Average K-values observed in the measured R-curves with increasing oxygen content in the metal layer

3.3 Simulated Stress Fields

As already mentioned, cavity nucleation and subsequent growth can introduce local failures, which will reduce the stresses in the surrounding region and determine the damage process of the whole composite.

The experimental observations suggest that the stress field at the interface due to a precipitate is changed such that it protects the interface from debonding, or that it turns the crack aside, preferably to the metal phase. A first step to get insight into the real behaviour is to calculate the stress field

Table 2. Material parameters for the ceramic (Al_2O_3), the layer (Cu), the precipitates (Cu_2O) and a fictive, stiff precipitate (X)

	Al_2O_3	Cu	Cu_2O	X
E(GPa)	400.0	130.0	28.0	300.0
ν	0.235	0.35	0.45	0.23
σ_0(MPa)	–.–	100.0	–.–	–.–
n	–.–	8.0	–.–	–.–
p_{cr}(MPa)	–.–	790.0	–.–	–.–

for several typical cases. Therefore, a similar procedure as used by Varias [19] has been applied: a circular region, surrounding the crack tip, with a radius of approximately 25 times the width of the layer has been loaded by displacements reflecting a remote mode I crack-tip-field. The material data in Table 2 are used, where a fictitious material (X) for the precipitates is added in order to compare a very stiff precipitate with the real precipitates. Furthermore, a number of cavities were introduced in a series of simulations to compare these results with those without cavities. As an example, Figs. 12 and 13 show the hydrostatic stress for different configurations and loads.

For a cavity and precipitate-free metallic layer (Fig. 12 top), a typical stress maximum develops at approximately twice the width of the layer ahead of the crack tip. This may be a potential region for cavity nucleation, provided that the interface is sufficiently strong.

The second case shows a layer containing one single, stiff precipitate (material X) which does not change the stresses far away, but partly shields the interface between the crack tip and the precipitate. A shear band will be formed between the crack tip and the top of the precipitate (not shown in the figure), which deflects the crack into the metal.

The third panel in Fig. 12 tries to model the characteristic crack pattern in the precipitate observed experimentally (Fig. 2). Again, the maximum stress develops at twice the layer width. Only in a small region near the precipitate the stress field is modified. It can be expected that a cracked precipitate may lead to crack bifurcation or to the formation of a secondary crack, increasing the work needed to drive the crack.

The last panel in Fig. 12 shows a series of cavities in front of the crack – nucleated for example by a sufficiently large hydrostatic pressure as seen in the first panel. It is obvious that these cavities reduce the stresses in their surrounding. In addition, they may attract or shield the crack against external load.

Figure 13 shows the growth process of the cavities with increasing load. The critical stress limit for the copper is calculated as $p_{cr} = 790$MPa. This value will be reached for a loading between the third and the fourth contour

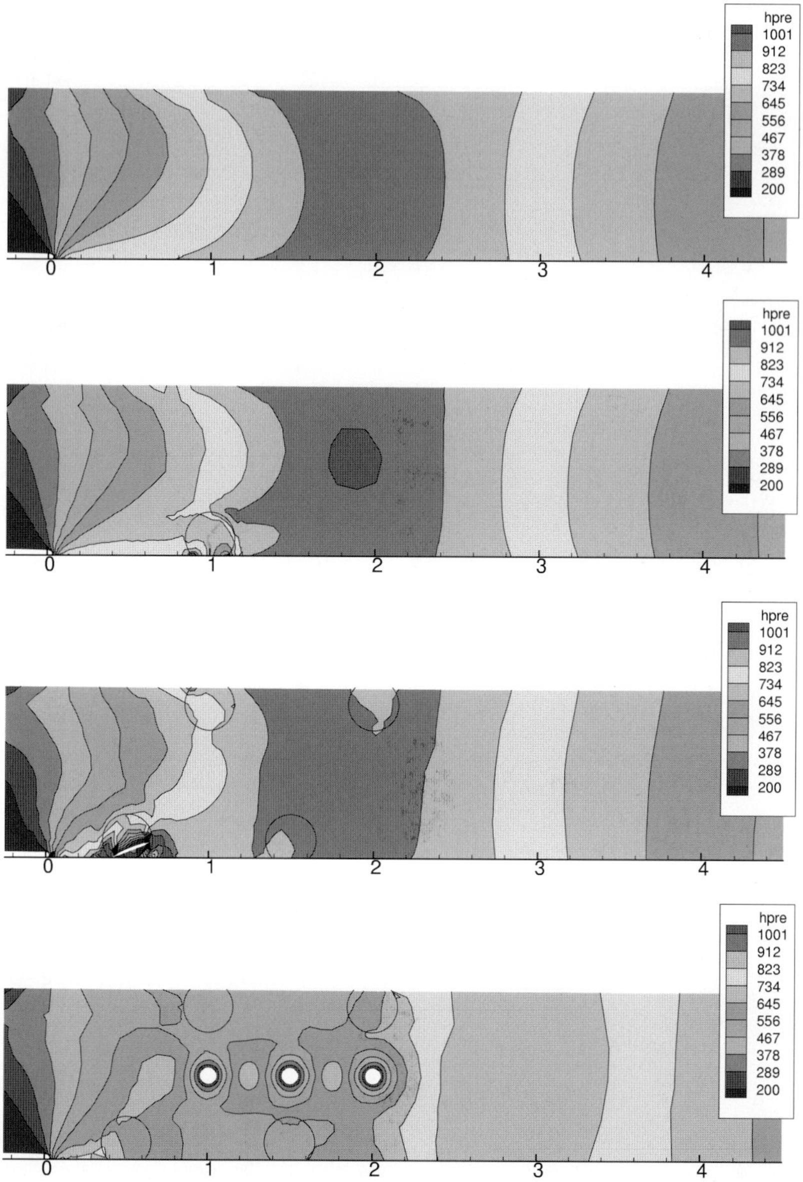

Fig. 12. Hydrostatic stress ($-p$) in the metallic layer and the precipitates. From top to bottom: pure layer; single stiff precipitate; some weak precipitates, one cracked; formation of cavities [5]

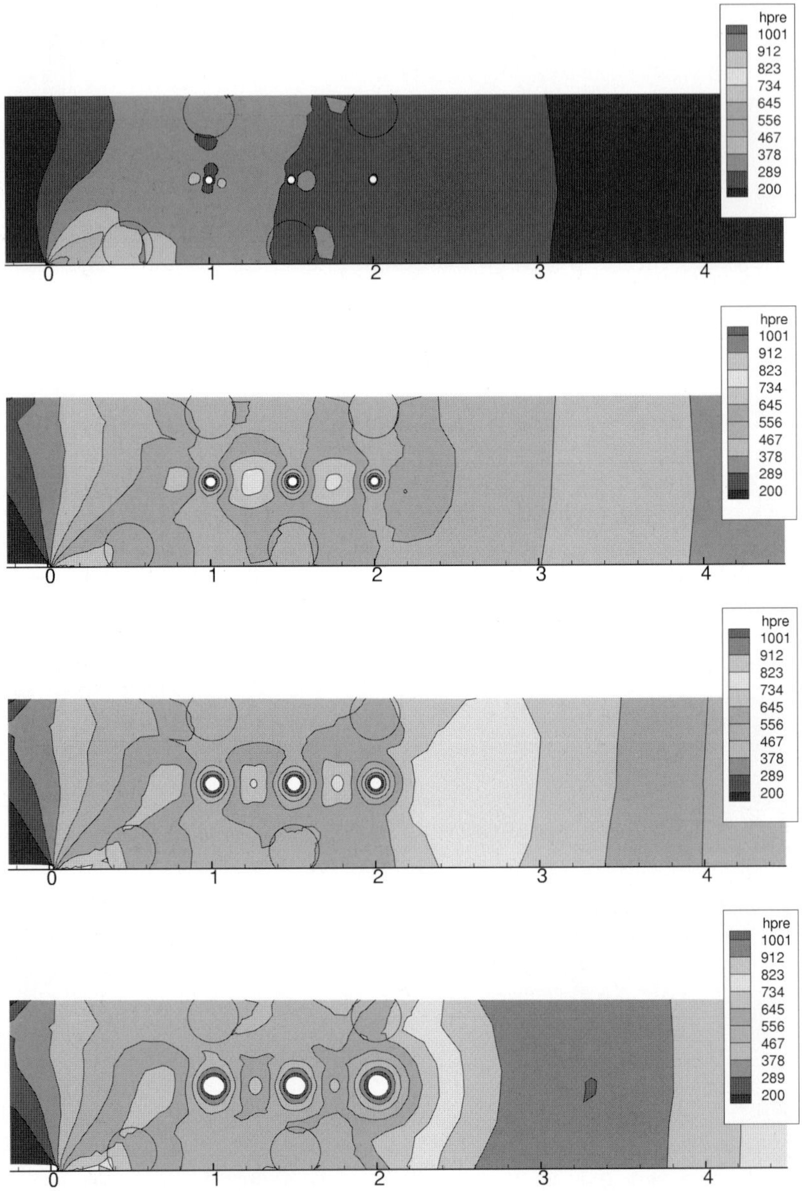

Fig. 13. Hydrostatic stress $(-p)$ in the metallic layer and precipitates for increased load and size of growing cavities from top to bottom [5]

plot in Fig. 13 and a new cavity nucleates in front of the existing cavities. This will result in a recursive process, where cavities nucleate at a given distance. The resulting pattern may be observable in experiments.

4 Conclusion

The current work presents a method to predict the stresses in a pressure infiltrated metal-ceramic composite. Specifically, it can be used to determine the stress components at an interface which might be compared with a known interface toughness and strength.

It has been shown that cavitation is one main failure mechanism in metal ceramic composites exposed to a large thermally induced stress or an external loading. Cavitation excludes debonding and vice versa. This explains the fact that cavities will not be observed for a weak interface. In contrast, a strong interface will not debond and therefore the nucleation of cavities is preferred.

In summary, these observations point out some goals in designing an optimized composite:

- Increase the interface strength, but not necessarily above the critical stress limit.
- Use a material with high hardening and high initial yield stress and by this, increase the limit stress without preventing the material to cavitate.
- Reduce the difference of the thermal expansion coefficients to lower the eigenstrains and -stresses.
- Avoid large inclusions. They will reduce the stresses in a large volume.

The simulation of a metal layer under external load confirms the results. In detail several possible global failure modes can be obtained:

- A strong interface provided,
  - relatively weak precipitates will attract the crack and deflect it from the interface, resulting in an increased fracture toughness. The crack may "jump" from one precipitate to another. Moreover,
  - cavitation will attract the crack, resulting in an increased fracture toughness. The crack path results in a characteristic pattern of joined cavities. Finally,
  - stiff precipitates only deflect the crack temporally, resulting in an increased fracture toughness.
- A weak interface provided,
  - weak precipitates tend to crack and deflect the main crack or just branch the main crack, resulting in an increased fracture toughness. Furthermore,
  - cavities will not nucleate and
  - stiff precipitates will not have any significant effect.

An estimation of the critical, hydrostatic stress p_{cr} is easily achieved by integrating Equation (2) numerically. Therefore, a stress-strain relation $\sigma = f(\varepsilon)$ fitting the given material data should be used (e.g. (4)).

4.1 Comparison with Widely Used Damage Models

The study of cavities as single and observable defects that influence the local surrounding stress field will shed some light on the use and the misuse of continuum damage models (for example Gurson's damage model [7]). Most of them are based on the stress field near a small defect, which is homogenized and averaged, resulting in a "continuous" representation of the defect. The resulting model can be used to calculate a damage parameter, for example the void volume of the material. This is a legitimate procedure, provided all characteristic length parameters of a particular problem are much larger than the characteristic micro length scale appearing in the model.

Consequently, continuum damage models have to be used very carefully in cases with other typical small length scales, e.g. micro cracks. Direct modeling of real defects (cavities, small cracks, inclusions, precipitates ...) will provide a deeper understanding of mechanisms on the microlevel, where the interest is focused to a scale below or of the same size of the micro defects. For example, the pattern of equally spaced cavities in the metal layer or the structure in Fig. 1 cannot easily be simulated by a continuum damage model.

4.2 Some Aspects of Micro-Cavitation

For relatively fine structures we can conclude from (12) that the formation of cavities is difficult to verify, since most techniques to experimentally observe those details will fail. The resulting cavities exert only a very local influence to the surrounding stress field, provided micro-cavitation occurs. In fact, it is not possible to separate clearly between micro-cavitation and interface debonding.

Anyway, the assumption of a perfect, defect-free material is not useful. A predicted increase of 50 times the initial yield stress has not been detected in experiments; compare Fig. 5. For a fine structure a stress of max. 20 times the yield stress was observed.

Micro-cavitation can explain the reduced stresses, but the theory underestimates the stresses by a factor of about 2 to 3. A possible source of this difference might be the difference in the material properties at different scales.

References

1. Ashby M. F., Blunt F. J., Bannister M. (1989) Flow Characteristics of Highly Constrained Metal Wires. *Acta metallurgica* **37**, 1847–1857
2. Ball J. M. (1982) Discontinuous Equilibrium Solutions and Cavitation in Nonlinear Elasticity. *Philosophical Transactions of the Royal Society of London* Series A **306**, 557–611
3. Diemer M. (1998) *Benetzungsverhalten und mechanische Eigenschaften der Cu/Al_2O_3-Grenzfläche*, TU Darmstadt, Dissertation.
4. Emmel T. (1995) *Untersuchung des Bruchverhaltens von Metall-Keramik-Verbundwerkstoffen*, TH Darmstadt, Master Thesis.

5. Emmel T. (2002) *Theoretische und numerische Untersuchung von Versagensmechanismen in Metall-Keramik-Verbundwerkstoffen*, TU Darmstadt, Inst. für Mechanik, Dissertation.
6. Gross D., Seelig Th. (2001) Bruchmechanik mit einer Einführung in die Mikromechanik Springer Verlag, Berlin Heidelberg.
7. Gurson A. L. (1977) Continuum Theory of Ductile Rupture by Void Nucleation and Growth: Part I - Yield Criteria and Flow Rules of Porous Ductile Media. *Journal for Engineering Materials in Technology* **99**, 2–15
8. Hill R. (1964) Theory of Mechanical Properties of Fibre-Strengthened Materials: I. Elastic Behaviour. *Journal of the Mechanics and Physics of Solids* **12**, 199–212
9. Hill R. (1964) Theory of Mechanical Properties of Fibre-Strengthened Materials: II. Inelastic Behaviour. *Journal of the Mechanics and Physics of Solids* **12**, 213–218
10. Hoffman M., Fiedler B., Emmel T., Prielipp H., Claussen N., Gross D., Rödel J. (1997) Fracture Behaviour in Metal Fibre Reinforced Ceramics. *Acta materialia* **45**, 3609–3623
11. Horgan C. O., Poligone D. A. (1995) Cavitation in Nonlinearly Elastic Solids: A Review. *Applied Mechanics Reviews* **48**, 471–485
12. Huang Y., Hutchinson, J. W., Tvergaard V. (1991) Cavitation Instabilities in Elastic-Plastic Solids. *Journal of the Mechanics and Physics of Solids* **39**, 223–241
13. Hutchinson J. W., Mear M. E., Rice J. R. (1987) Crack Paralleling an Interface Between Dissimilar Materials. *Journal of Applied Mechanics* **55**, 828–832
14. Kolling S. (2001) *Zur numerischen Simulation von Morphologieänderungen in mikroheterogenen Materialien*, TU Darmstadt, Inst. für Mechanik, Dissertation.
15. Müller R. (2001) *3D-Simulationen der Mikrostrukturentwicklung in Zwei-Phasen-Materialien*, TU Darmstadt, Inst. für Mechanik, Dissertation.
16. Mura T. (1982) *Micromechanics of Defects in Solids.* The Hague Boston London : Martinus Nijhoff, (Mechanics of elastic and inelastic solids 3).
17. Rödel J., Kelly J. F., Lawn B. R. (1993) In-Situ Measurement of Bridged crack Interfaces in the Scanning Electron Microscope. *Journal of the American Ceramic Society* **76**, 369–375
18. Schmid R. (1983) A Thermodynamic Analysis of the Cu-O System with an associated Solution Model. *Metall. Trans.* **14B**, 473–481
19. Varias A. G., Suo Z., Shih C. F. (1991) Ductile Failure of a Constrained Metal Foil. *Journal of the Mechanics and Physics of Solids* **39**, 963–986
20. Zimmermann A., Hoffman, M., Emmel T., Gross D., Rödel J. (2001) Failure of metal-ceramic composites with spherical inclusions. *Acta materialia* **49**, 3177–3187

Part III

Constitutive Behaviour — Mathematical Foundation

Monotone Constitutive Equations in the Theory of the Inelastic Behaviour of Metals - A Summary of Results

Krzysztof Chełmiński

[1] Department of Mathematics, Darmstadt University of Technology
Schlossgartenstr. 7, 64289 Darmstadt, Germany
chelminski@mathematik.tu-darmstadt.de
[2] Department of Mathematics, Cardinal Stefan Wyszyński University
Dewajtis 5, 01-815 Warsaw, Poland

received 24 Jul 2002 — accepted 31 Oct 2002

Abstract. This article presents an existence theory of global in time solutions to systems describing the inelastic behaviour of metals. It is assumed that the material under consideration is geometrically linear and the inelastic constitutive equation is of monotone type ([3]). At the end of this summary some classes containing nonmonotone models are presented.

This contribution is mainly concerned with the dynamical problem. In the last section some remarks on the quasistatic case are given.

1 Introduction and Formulation of the Initial-Boundary Value Problem

Systems of equations describing the inelastic response of metals, under the assumption of infinitesimal deformations, consist of linear partial differential equations coupled with nonlinear differential inclusions (or ordinary differential equations) for the vector of internal variables. The partial differential equations result from general mechanical laws. The differential inclusions are experimental, and depend on the kind of material considered. Therefore in engineering sciences there are many inelastic constitutive equations, always specially adapted to the material under consideration. The first part of the system is in all models the same and represents balance of linear momentum

$$\rho u_{tt}(x,t) = \mathrm{div}_x T(x,t) + F(x,t) \qquad (x,t) \in \Omega \times (0,T). \tag{1}$$

Here $T > 0$ is a fixed length of a time interval, $\Omega \subset \mathbb{R}^3$ is a bounded domain with a smooth boundary $\partial\Omega$, $u : \Omega \times (0,T) \to \mathbb{R}^3$ denotes the displacement field, $T : \Omega \times (0,T) \to \mathcal{S}^3 = \mathbb{R}^{3\times 3}_{\mathrm{sym}}$ is the Cauchy stress tensor, $\rho > 0$ is the mass density and $F : \Omega \times (0,T) \to \mathbb{R}^3$ describes external forces acting on the body. The other parts of the system consist of constitutive relations

$$T(x,t) = \mathcal{F}_{0\le s\le t}\Big(\nabla u(x,s)\Big)\,, \tag{2}$$

where the right-hand side denotes a functional depending on the history of the displacement gradient. The theory, which we are going to present, assumes that the functional $\mathcal{F}_{0\le s\le t}$ consists of the elastic constitutive equation

$$T(x,t) = \mathcal{D}\Big(\varepsilon(x,t) - \varepsilon^p(x,t)\Big)\,, \tag{3}$$

based on the additive split of the deformation into elastic and plastic parts where $\varepsilon = \frac{1}{2}(\nabla u + \nabla^T u)$ is the symmetrized displacement gradient, $\varepsilon^p : \Omega\times(0,T) \to \mathcal{S}^3$ describes the plastic part of the deformation and $\mathcal{D} : \mathcal{S}^3 \to \mathcal{S}^3$ is the elasticity tensor, which is assumed to be constant with respect to $x \in \Omega$ and $t \in (0,T)$, symmetric and positive definite. This equation is coupled with the inelastic constitutive relation formulated in general terms as a differential inclusion

$$z_t(x,t) \in f\Big(\varepsilon(x,t), z(x,t)\Big) \tag{4}$$

where $z : \Omega \times (0,T) \to \mathbb{R}^N$ is the vector of internal variables. z consists of ε^p and other components $\tilde{z}$ which are introduced to describe more appropriately the deformation process. $f : D(f) \subset \mathcal{S}^3 \times \mathbb{R}^N \to \mathcal{P}(\mathbb{R}^N)$ is the constitutive multifunction and causes the system of equations (1)+(3)+(4) to become nonlinear. Thermodynamical considerations show (for a deeper discussion we refer to [3, Appendix A, p.143]) that there exists a free energy function $\psi : D(f) \subset \mathcal{S}^3 \times \mathbb{R}^N \to \mathbb{R}_+$ such that for all $(\varepsilon, z) \in D(f)$

$$T = \frac{\partial \rho\psi(\varepsilon,z)}{\partial \varepsilon} \qquad \text{(hyperelasticity)}\,, \tag{5}$$

$$\frac{\partial \rho\psi(\varepsilon,z)}{\partial z}\cdot w^* \le 0 \quad \text{for all} \quad w^* \in f(\varepsilon,z)\,. \tag{6}$$

The existence of the free energy function ψ implies that the problem (1)+(3)+(4) considered with $F = 0$ and with homogeneous boundary condition of the Dirichlet or of the Neumann type possesses a natural semi-invariant, namely the total energy associated with the problem does not increase in time

$$\mathcal{E}(u,z)(t) \stackrel{\mathrm{df}}{=} \frac{1}{2}\int_\Omega \rho|u_t|^2\,dx + \int_\Omega \rho\psi(\varepsilon,z)\,dx \le \mathcal{E}(u,z)(0) \tag{7}$$

(for details see the monograph [3, Chapter 3.1 p.23]). Therefore inequality (6) is called the reduced dissipation inequality. Using the properties of the elasticity tensor $\mathcal{D}$, the assumed hyperelasticity (5) implies that the free energy function has to be of the form

$$\rho\psi(\varepsilon,z) = \frac{1}{2}\mathcal{D}(\varepsilon - \varepsilon^p)\cdot(\varepsilon - \varepsilon^p) + \psi_1(z)\,, \tag{8}$$

where the function ψ_1 depends on the vector z only. There is no precise relationship between free energy functions and constitutive multifunctions such that the reduced dissipation inequality would hold. Therefore we restrict our considerations to a subclass of problems, for which (6) will be satisfied automatically. We say that our problem (1)+(3)+(4) is of **pre-monotone type** (shortly: belongs to $\mathcal{PM}$) if the constitutive multifunction is of the form

$$f(\varepsilon, z) = g\Big(- \rho \nabla_z \psi(\varepsilon, z)\Big) \tag{9}$$

with a multifunction $g : D(g) \subset \mathbb{R}^N \to \mathcal{P}(\mathbb{R}^N)$ satisfying

$$\forall\, z \in D(g)\,, \quad \forall\, z^* \in g(z) \qquad z^* \cdot z \geq 0 \tag{10}$$

and with the free energy function ψ given as a positive semi-definite quadratic form

$$\rho\psi(\varepsilon, z) = \tfrac{1}{2}\mathcal{D}(\varepsilon - Bz) \cdot (\varepsilon - Bz) + \tfrac{1}{2} Lz \cdot z. \tag{11}$$

Here $L \in \mathbb{R}^{N\times N}_{\mathrm{sym}}$, $L \geq 0$ and $Bz = B(\varepsilon^p, \tilde{z}) \overset{\mathrm{df}}{=} \varepsilon^p$ is the orthogonal projection of the vector z on the direction ε^p. Moreover we assume that the symmetric operator $L + B^T\mathcal{D}B$ is **positive definite**. Note that the operators L and $B^T\mathcal{D}B$ are both positive semi-definite and the last requirement implies that $\mathrm{span}\{\mathrm{Range}(B^TB), \mathrm{Range}(L)\} = \mathbb{R}^N$. It is easy to see that for all pre-monotone models the requirements (5) and (6) are satisfied and that such models are thermodynamically admissible. This class of models was introduced by H.-D. ALBER in the monograph [3, Definition 3.1.1]. Assuming, moreover, that $0 \in g(0)$ we call (10) the monotonicity of g in the point 0. This observation motivates the used name for this class of models. The class $\mathcal{PM}$ is very large and contains all models known to the author and used in the engineering sciences. Unfortunately, an existence theory for the whole class $\mathcal{PM}$ has not been developed until today and it is our believe that the monotonicity of the multifunction g in the point zero only is not sufficient to obtain an existence result for global in-time-solutions of the problem (1)+(3)+(9).

2 Models of Monotone Type

The main idea used in many proofs of existence theorems for nonlinear problems consists of deriving a priori estimates for differences of two solutions or of two approximation steps related to the considered problem. H.-D. ALBER has observed that the monotonicity of the multifunction g is sufficient to obtain the energy estimate for differences of two smooth solutions (u_1, z_1) and (u_2, z_2) of the problem (1)+(3)+(9). Thus for monotone g we obtain

$$\begin{aligned}\mathcal{E}(u_1 - u_2, z_1 - z_2)(t) &= \tfrac{1}{2} \int_\Omega \rho |\partial_t u_1 - \partial_t u_2|^2 \, dx \\ &+ \int_\Omega \rho\psi(\varepsilon_1 - \varepsilon_2, z_1 - z_2)\, dx \leq \mathcal{E}(u_1 - u_2, z_1 - z_2)(0).\end{aligned}$$

This observation led to the definition of a very important subclass of the class $\mathcal{PM}$ (compare with [3, Definition 3.1.1]).

Definition 1 (models of monotone type). We say that the problem (1)+(3)+(9) with free energy in the form (11) is of monotone type (shortly: belongs to the class $\mathcal{M}$) if the constitutive multifunction $g : D(g) \subset \mathbb{R}^N \to \mathcal{P}(\mathbb{R}^N)$ is monotone, i.e.,

$$\forall\, z_1, z_2 \in D(g)\,, \quad \forall\, z_1^* \in g(z_1), z_2^* \in g(z_2) \quad (z_1^* - z_2^*, z_1 - z_2) \geq 0\,,$$

and additionally satisfies $0 \in D(g)$ and $0 \in g(0)$.

Let us denote by v the time derivative of the displacement vector, then the problem (1)+(3)+(9) of monotone type can be written in the form

$$\begin{aligned} \rho v_t(x,t) &= \operatorname{div}_x \mathcal{D}\Big(\varepsilon(x,t) - Bz(x,t)\Big) + F(x,t)\,, \\ \varepsilon_t(x,t) &= \tfrac{1}{2}\Big(\nabla_x v(x,t) + \nabla_x^T v(x,t)\Big)\,, \\ z_t(x,t) &\in g\Big(-\rho\nabla_z\psi(\varepsilon(x,t), z(x,t))\Big) \end{aligned} \tag{12}$$

with the quadratic free energy function ψ given by (11) and with the monotone multifunction g satisfying the requirements of Def. 1. This system (12) will be studied with Dirichlet boundary conditions

$$v(x,t) = g_D(x,t) \quad \text{for } x \in \partial\Omega\,,\ t \geq 0\,, \tag{13}$$

or with Neumann boundary conditions

$$\mathcal{D}\Big(\varepsilon(x,t) - Bz(x,t)\Big) \cdot n(x) = g_N(x,t) \quad \text{for } x \in \partial\Omega,\ t \geq 0\,, \tag{14}$$

where $n(x)$ denotes the outer normal vector to the boundary $\partial\Omega$ at the point $x \in \partial\Omega$, and the functions $g_D, g_N : \Omega \times \mathbb{R}_+ \to \mathbb{R}^3$ are given boundary data. Finally, the initial conditions are in the form

$$v(x,0) = v^0(x)\,, \quad \varepsilon(x,0) = \varepsilon^0(x)\,, \quad z(x,0) = z^0(x) \tag{15}$$

where $v^0 : \Omega \to \mathbb{R}^3$, $\varepsilon^0 : \Omega \to \mathcal{S}^3$, $z^0 : \Omega \to \mathbb{R}^N$ are given initial data.

2.1 Examples of Monotone Models

Next we present three examples of monotone models that are well known in the engineering sciences.

The Norton-Hoff model. In this model the inelastic flow for the vector z of internal variables contains only the plastic strain tensor $z = \varepsilon^p$ and the evolution of the vector ε^p is given by

$$\varepsilon_t^p(x,t) = \frac{1}{\eta}\Big[|\operatorname{dev} T(x,t)| - \sigma_y\Big]_+^n \frac{\operatorname{dev} T(x,t)}{|\operatorname{dev} T(x,t)|}$$

where $\sigma_y > 0$ is the yield limit and η, n are material constants. Moreover $\operatorname{dev} T = T - \frac{1}{3}\mathrm{tr}(T) \cdot I$ is the stress deviator and the symbol $h_+ = \max\{0, h\}$ denotes the nonnegative part of the scalar function h. If we define the free energy function ψ by

$$\rho\psi(\varepsilon, \varepsilon^p) = \tfrac{1}{2}\mathcal{D}(\varepsilon - \varepsilon^p) \cdot (\varepsilon - \varepsilon^p)\,,$$

then the Norton-Hoff inelastic flow can be written in the form

$$\varepsilon_t^p(x,t) = g\Big(- \rho\nabla_z\psi(\varepsilon(x,t), \varepsilon^p(x,t))\Big)$$

with the constitutive function $g(T) = \hat{g}(\operatorname{dev} T)$ where $\hat{g}(S) = \frac{1}{\eta}[|S| - \sigma_y]_+^n \frac{S}{|S|}$. From the definition of g we immediately see that g is monotone and $g(0) = 0$. The free energy ψ associated with this model is positive semi-definite only. Moreover, the model is viscoplastic of gradient type (see the definition in Subsection 4.3) and the constitutive function grows polynomially. An existence result of global in time solutions for this model will be obtained in Subsections 4.2 and 4.3. This model in the quasistatic setting was studied in [41] and in the dynamical setting in [12].

The Prandtl-Reuss model. This model is also called elasto-perfect plasticity. The vector z of internal variables contains only the plastic strain tensor $z = \varepsilon^p$, and the inelastic flow is given in the form of a differential inclusion

$$\varepsilon_t^p(x,t) \in \partial I_{\mathcal{E}}\Big(T(x,t)\Big)\,,$$

where the elastic domain is a convex, bounded and closed set $\mathcal{E} \subset \operatorname{dev}\mathcal{S}^3$ with $0 \in \operatorname{int}\mathcal{E}$. Moreover, $\partial I_{\mathcal{E}}$ is the subgradient of the indicator function $I_{\mathcal{E}}$. In the von Mises flow rule this set is defined by $\mathcal{E} = \{T \in \mathcal{S}^3 : |\operatorname{dev} T| \le \sigma_y\}$. The free energy function associated with the model is given by

$$\rho\psi(\varepsilon, \varepsilon^p) = \tfrac{1}{2}\mathcal{D}(\varepsilon - \varepsilon^p) \cdot (\varepsilon - \varepsilon^p)$$

and the flow rule can be written in the form

$$\varepsilon_t^p(x,t) = g\Big(- \rho\nabla_z\psi(\varepsilon(x,t), \varepsilon^p(x,t))\Big)$$

where the constitutive multifunction g is defined by $g(T) = \partial I_{\mathcal{E}}(T)$. This multifunction is monotone and $0 \in \partial I_{\mathcal{E}}(0)$. Similarly to the Norton-Hoff inelastic flow the free energy is positive semi-definite only. The Prandtl-Reuss model possesses a very important property: the inelastic flow is rate independent. Thus an existence result of global in time, so called weak-type solutions, will follow from Subsection 4.4. The model of Prandtl-Reuss in several settings was studied in [1,16,17,26,29,30,37,41].

The Melan-Prager model. In this model the vector z contains the plastic strain ε^p and the backstress $b \in \mathcal{S}^3$. The inelastic constitutive equation is given by

$$\varepsilon_t^p(x,t) \in \partial I_{\mathcal{E}}\Big(T(x,t) - b(x,t)\Big)\,,$$
$$b_t(x,t) = \gamma \varepsilon_t^p(x,t) \quad \gamma \geq 0\,,$$

where the multifunction $\partial I_{\mathcal{E}}$ is the same as in the Prandtl-Reuss model and γ is a positive material constant. If we integrate the equation for b, then $b = \gamma\varepsilon^p + \hat{T}$ and the variable b can be eliminated from the system. Inserting the result into the equation for ε^p we have

$$\varepsilon_t^p(x,t) \in \partial I_{\mathcal{E}}\Big(T(x,t) - \gamma\varepsilon^p(x,t) - \hat{T}(x)\Big)\,.$$

Thus the free energy associated with the model is of the form

$$\rho\psi(\varepsilon,\varepsilon^p) = \tfrac{1}{2}\mathcal{D}(\varepsilon - \varepsilon^p)\cdot(\varepsilon - \varepsilon^p) + \tfrac{1}{2}\gamma|\varepsilon^p|^2\,.$$

We conclude that this monotone model possesses a positive definite free energy. We shall see that this property is very important from the mathematical point of view. An existence result of strong $\mathbb{L}^2$ solutions for this model will be given in the next section. Moreover, a mathematical analysis of the model can be found in [16,28,33].

3 Existence Theory for Coercive Models of Monotone Type

In this section we start to present the existence theory for the class containing monotone models. First we are going to formulate precisely in what sense the system (12) will be satisfied. For the given boundary data g_D or g_N and for the external force F we require the following regularity for all $T > 0$:

$$g_D \in \mathbb{W}^{2,\infty}((0,T);\mathbb{H}^{\frac{1}{2}}(\partial\Omega;\mathbb{R}^3))\,,\ g_N \in \mathbb{W}^{2,\infty}((0,T);\mathbb{H}^{-\frac{1}{2}}(\partial\Omega;\mathbb{R}^3)) \\ \text{and} \ \ F \in \mathbb{W}^{1,\infty}((0,T);\mathbb{L}^2(\Omega;\mathbb{R}^3))\,, \tag{16}$$

where we have used the classical notation: $\mathbb{W}^{k,p}$ is the Sobolev space constructed over $\mathbb{L}^p$ and $\mathbb{H}^s = \mathbb{W}^{s,2}$.

Definition 2 (strong $\mathbb{L}^2$ solutions). We say that a triple (v,ε,z) from the space $\mathbb{L}^2((0,T);\mathbb{L}^2(\Omega;\mathbb{R}^3\times\mathcal{S}^3\times\mathbb{R}^N))$ satisfies the system (12) in the strong $\mathbb{L}^2$ sense if

a) $(v_t,\varepsilon_t,z_t) \in \mathbb{L}^\infty((0,T);\mathbb{L}^2(\Omega;\mathbb{R}^3\times\mathcal{S}^3\times\mathbb{R}^N))$,
b) $(\mathrm{div}_x\,\mathcal{D}(\varepsilon - Bz), \frac{1}{2}(\nabla v + \nabla^T v)) \in \mathbb{L}^\infty((0,T);\mathbb{L}^2(\Omega;\mathbb{R}^3\times\mathcal{S}^3))$,
c) there exists $z^* \in \mathbb{L}^\infty((0,T);\mathbb{L}^2(\Omega;\mathbb{R}^N))$ with

$$z^*(x,t) \in g\Big(-\rho\nabla_z\psi(\varepsilon(x,t),z(x,t))\Big) \ \text{for almost all}\ (x,t)\in\Omega\times(0,T)$$

and the equalities and the inclusion in (12) hold pointwise almost everywhere. Moreover, the boundary condition (13) or (14) and the initial conditions (15) are satisfied in the sense of traces.

We are not able to prove an existence result of global strong $\mathbb{L}^2$ solutions for all models of monotone type. Moreover, we believe that such results do not hold. Nevertheless, in this section we present a class of models for which such an existence result can be given. If the free energy function ψ associated with the model under consideration is positive definite (or equivalently $L > 0$) then we say that our model is **coercive**. In this case the bilinear form

$$\langle (v, \varepsilon, z), (\hat{v}, \hat{\varepsilon}, \hat{z}) \rangle_{\mathcal{E}} = \int_{\Omega} \Big(\rho v \cdot \hat{v} + \mathcal{D}(\varepsilon - Bz) \cdot (\hat{\varepsilon} - B\hat{z}) + Lz \cdot \hat{z} \Big)\, dx$$

defines a scalar product in the space $\mathbb{L}^2(\Omega; \mathbb{R}^3 \times \mathcal{S}^3 \times \mathbb{R}^N)$. Before we formulate the result for coercive models we define the class of admissible initial data for the problem (12).

Definition 3 (admissibility of initial data).
1. We say that the initial data $(v^0, \varepsilon^0, z^0) \in \mathbb{L}^2(\Omega; \mathbb{R}^3 \times \mathcal{S}^3 \times \mathbb{R}^N)$ are admissible for problem (12) if

$$\Big(\operatorname{div}_x \mathcal{D}(\varepsilon^0 - Bz^0),\ \tfrac{1}{2}(\nabla v^0 + \nabla^T v^0) \Big) \in \mathbb{L}^2(\Omega; \mathbb{R}^3 \times \mathcal{S}^3),$$

and there exists $z^* \in \mathbb{L}^2(\Omega; \mathbb{R}^N)$ such that

$$z^*(x) \in g\Big(-\rho \nabla_z \psi(\varepsilon^0(x), z^0(x)) \Big) \quad \text{for a.e. } x \in \Omega .$$

2. We say that the initial data $(v^0, \varepsilon^0, z^0) \in \mathbb{L}^2(\Omega; \mathbb{R}^3 \times \mathcal{S}^3 \times \mathbb{R}^N)$ satisfy the compatibility condition if

$$v^0(x) = g_D(x, 0) \quad \text{for a.e. } x \in \partial\Omega$$

for the Dirichlet boundary value problem, or

$$\mathcal{D}\Big(\varepsilon^0(x) - Bz^0(x) \Big) \cdot n(x) = g_N(x, 0) \quad \text{for a.e. } x \in \partial\Omega$$

for the Neumann boundary value problem.

In the coercive case the following general result was obtained which completes the existence theory for the entire class of coercive and monotone models.

Theorem 1 (existence of strong $\mathbb{L}^2$ solutions in the coercive case). *Let us assume that the considered model is coercive and the given boundary data and external force have the regularity (16). Moreover, suppose that the initial data $(v^0, \varepsilon^0, z^0) \in \mathbb{L}^2(\Omega; \mathbb{R}^3 \times \mathcal{S}^3 \times \mathbb{R}^N)$ are admissible and satisfy the compatibility condition. Finally, suppose that the multifunction g is maximal monotone. Then problem (12) with boundary condition (13) or (14) and with initial condition (15) possesses a global in time, unique, strong $\mathbb{L}^2$ solution.*

This theorem is proved in [23, Theorem1, p. 3]. The idea of the proof is based on the so called partial Yosida approximation. In this approximation procedure the maximal monotone multifunction g is approximated by its pointwise Yosida approximation G_λ. This new problem possesses global Lipschitz nonlinearities only and therefore possesses a unique, global in time strong $\mathbb{L}^2$ solution. The coerciveness of the problem allows to pass to the limit $\lambda \to 0^+$. In the case $g_D \equiv 0$ or $g_N \equiv 0$, Theorem 1 is proved in the book [3, Theorem 4.2.1 p. 48]. The proof in this case is based on the theory of evolution equations to maximal monotone operators (for details we refer to [5,7]). Some special coercive models were studied for example in [6,16,28,33,39].

4 Noncoercive Case — Coercive Approximation Procedure

To prove an existence result in the noncoercive case we approximate the system (12) by the following sequence of problems. Let $k > 0$ be a positive integer. We consider now the system

$$\begin{aligned} \rho v_t^k(x,t) &= \operatorname{div}_x \mathcal{D}\Big(\varepsilon^k(x,t) - Bz^k(x,t) + k^{-1}\varepsilon^k(x,t)\Big) + F(x,t)\,, \\ \varepsilon_t^k(x,t) &= \tfrac{1}{2}\Big(\nabla_x v^k(x,t) + \nabla_x^T v^k(x,t)\Big)\,, \\ z_t^k(x,t) &= g\Big(- \rho\nabla_z \psi^k(\varepsilon^k(x,t), z^k(x,t))\Big) \end{aligned} \tag{17}$$

with the free energy function ψ^k given by the formula

$$\rho\psi^k(\varepsilon^k, z^k) = \tfrac{1}{2}\mathcal{D}(\varepsilon^k - Bz^k)\cdot(\varepsilon^k - Bz^k) + \tfrac{1}{2}Lz^k\cdot z^k + \tfrac{1}{2}k^{-1}\mathcal{D}\varepsilon^k\cdot\varepsilon^k\,.$$

Moreover, the boundary condition for system (17) coincides with the boundary condition for system (12) with the Dirichlet boundary value problem

$$v^k(x,t) = g_D(x,t) \quad \text{for } x \in \partial\Omega\,,\ t \geq 0.$$

In the case of the Neumann boundary value problem we change the boundary condition slightly and require that

$$\mathcal{D}\Big(\varepsilon^k(x,t) - Bz^k(x,t) + k^{-1}\varepsilon^k(x,t)\Big)\cdot n(x) = g_N(x,t) \quad \text{for } x \in \partial\Omega,\ t \geq 0.$$

Finally, the initial conditions are prescribed in the form

$$v^k(x,0) = v_k^0(x)\,, \quad \varepsilon^k(x,0) = \varepsilon_k^0(x)\,, \quad z^k(x,0) = z_k^0(x)$$

with suitably chosen functions $v_k^0 : \Omega \to \mathbb{R}^3$, $\varepsilon_k^0 : \Omega \to \mathcal{S}^3$, $z_k^0 : \Omega \to \mathbb{R}^N$. Let us note that in problem (17) the free energy function ψ^k is positive definite inspite of the fact that the operator L does not control the whole $\mathbb{R}^N$. The lack of coerciveness of the function ψ is avoided by the additional "elastic

term" $(2k)^{-1}\mathcal{D}\varepsilon^k \cdot \varepsilon^k$.
The sequence of initial data $\{(v_k^0, \varepsilon_k^0, z_k^0)\}$ for problem (17) have to approximate in some sense the initial data $(v^0, \varepsilon^0, z^0)$ for problem (12). Hence we restrict the class of admissible initial data and set the following requirements.

Definition 4 (coercive limit property of initial data). We say that the initial data $(v^0, \varepsilon^0, z^0)$ have the coercive limit property if there exists a sequence $\{(v_k^0, \varepsilon_k^0, z_k^0)\}$ such that

1. $(v_k^0, \varepsilon_k^0, z_k^0)$ are admissible for problem (17) and satisfy the compatibility condition,
2. $(v_k^0, \varepsilon_k^0, z_k^0) \longrightarrow (v^0, \varepsilon^0, z^0)$ weakly in $\mathbb{L}^2(\Omega; \mathbb{R}^3 \times \mathcal{S}^3 \times \mathbb{R}^N)$,
3. the sequence $\|\mathcal{A}^k(v_k^0, \varepsilon_k^0, z_k^0)\|_{\mathcal{E}^k}$ is bounded i.e.,
$$\sup_{(v^*, \varepsilon^*, z^*) \in \mathcal{A}^k(v_k^0, \varepsilon_k^0, z_k^0)} \|(v^*, \varepsilon^*, z^*)\|_{\mathcal{E}^k} < \infty ,$$
where $\mathcal{A}^k$ denotes the operator generated by the right-hand side of problem (17) and the norm $\|\cdot\|_{\mathcal{E}^k}$ is induced by the total energy of this problem
$$\begin{aligned}\mathcal{E}^k(v^k, \varepsilon^k, z^k)(t) &= \tfrac{1}{2}\int_\Omega \rho |v^k(t)|^2 \, dx \\ &+ \tfrac{1}{2}\int_\Omega \mathcal{D}(\varepsilon^k(t) - Bz^k(t)) \cdot (\varepsilon^k(t) - Bz^k(t)) \, dx \\ &+ \tfrac{1}{2}\int_\Omega Lz^k(t) \cdot z^k(t) dx + \tfrac{1}{2}k^{-1}\int_\Omega \mathcal{D}\varepsilon^k(t) \cdot \varepsilon^k(t) \, dx .\end{aligned}$$

The solvability of problem (17) for all $k > 0$ follows from Theorem 1. Next we have to prove a priori estimates for the sequence of approximate solutions. The main difficulty to obtain such estimates is the noncoerciveness of the problem. In general by energy estimates in the k-th step we can control the sequence of stresses $\{T^k\}$ and the sequence of the directions $\{Lz^k\}$ of the vector z^k in the $\mathbb{L}^2$ norm. Consequently, we control the difference $\varepsilon^k - \varepsilon^{p,k}$, but in general we cannot control the strains ε^k and the inelastic strains $\varepsilon^{p,k}$ separately. The total energy function $\mathcal{E}^k$ associated with system (17) admits directions which cannot be controlled independently of k. Therefore using for example a special form of the inelastic constitutive relation we have to obtain some a priori estimates for the uncontrolled directions. Assuming that $\mathbb{L}^p$ estimates for some $p \geq 1$ for all functions $(v^k, \varepsilon^k, z^k)$ are done we have to pass to the limit in the nonlinear terms of the system. In the next four subsections we present subclasses of the class $\mathcal{M}$ in which both steps can be performed.

4.1 Self-controlling Inelastic Constitutive Equations

The first subclass of noncoercive models for which the general existence theory of global in time solutions can be obtained contains the so called self-controlling models. This notion was introduced in [12]. Moreover, some results for models from this class can be found in [13] or in [22]. We say that the model (12) is **self-controlling** or belongs to the class $\mathcal{SC}$ if the constitutive function g possesses a very special structure. Namely, we require that the uncontrolled directions Bg should be a priori estimatable by the controlled directions Lg. This means mathematically the existence of a continuous function $\mathcal{F} : \mathbb{R}_+ \times \mathbb{R}_+ \to \mathbb{R}_+$ with

$$||Bg(y)|| \leq \mathcal{F}(||Lg(y)||, ||y||) \text{ for all } y \in \mathbb{L}^2(\Omega; \mathbb{R}^N) . \tag{18}$$

The condition (18) is a kind of a priori estimate for the nonlinear vector field g. It is easy to see that Lipschitz constitutive functions satisfy (18). Moreover, all coercive models with a single-valued multifunction g are self-controlling. If the function $\mathcal{F}$ grows linearly then we say that the model is **linear self-controlling**.

Theorem 2 (existence of strong $\mathbb{L}^2$ solutions for the class $\mathcal{SC}$). *Let us assume that the considered model is self-controlling and the given boundary data g_D or g_N and external force F have the regularity required in Theorem 1. Let us, moreover, suppose that the initial data $(v^0, \varepsilon^0, z^0)$ satisfy the compatibility condition and additionally have the coercive limit property. If the vector field $g : \mathbb{R}^N \to \mathbb{R}^N$ is monotone and continuous, then problem (12) with boundary condition (13) or (14) and with initial condition (15) possesses a global in time, unique, strong $\mathbb{L}^2$ solution, provided that in the case of the Neumann boundary value problem, g satisfies the linear self-controlling inequality.*

The proof of this theorem is given in [22] and is based on Theorem 1 and on the idea of coercive approximations.
The last section in [22] contains a generalization of Theorem 2. Namely, for Neumann boundary value problems a class of boundary data is found for which Theorem 2 holds without the assumption that the self-controlling property of g is linear. Such data have to satisfy the so called safe-load condition. We define this condition in the next subsections.

4.2 Models with Polynomial Growth

The next subclass of models for which the general existence theory works, contains models for which the constitutive function grows polynomially. In this case an a priori $\mathbb{L}^p$ estimate for the sequence of strains (equivalently for the sequence of inelastic strains) follows from the growth condition. In [15] the following noncoercive problem is studied

$$\begin{aligned} \rho v_t(x,t) &= \operatorname{div}_x \mathcal{D}\big(\varepsilon(x,t) - \varepsilon^p(x,t)\big) , \\ \varepsilon_t(x,t) &= \tfrac{1}{2}\big(\nabla_x v(x,t) + \nabla_x^T v(x,t)\big) , \\ \varepsilon_t^p(x,t) &= \mathcal{G}\big(\operatorname{dev} T(x,t)\big) \end{aligned}$$

where $\mathcal{G} : \mathcal{S}^3 \to \operatorname{dev}\mathcal{S}^3$ is a continuous monotone vector field satisfying $\mathcal{G}(0) = 0$. If $|\mathcal{G}(\operatorname{dev} T)| \to \infty$ faster than linear for $|\operatorname{dev} T| \to \infty$ then the problem above does not possesses the self-controlling property. Using the theory of Orlicz spaces ([32]) and the Minty-Browder method ([27]) and assuming that the vector field $\mathcal{G}(\operatorname{dev} T)$ satisfies a polynomial growth condition for $|\operatorname{dev} T| \to \infty$, existence and uniqueness of global in time solutions for this system was proved. Thus, [15] enlarges the class of monotone plastic constitutive equations for which the coercive approximation process converges and the limit solves the original noncoercive problem. For example the Norton-Hoff model belongs to the class of problems studied in [15]. A general situation, where the vector z of internal variables contains more components than just ε^p only, needs a theory of anisotropic vectorial Orlicz spaces; this means spaces containing vectorial functions whose components possess different growths. Such spaces are studied in [35].

4.3 Viscoplasticity of Gradient Type

In this subsection we summarise the results obtained for viscoplastic models. Before we start to present the results, we give a mathematical definition of what we mean by viscoplasticity. We say that the pre-monotone model (12) is viscoplastic if the model is noncoercive and

$$\forall_{R>0} \quad \sup_{|z|\le R} |g(z)| < \infty \,, \tag{19}$$

where $|g(z)| = \sup_{z^* \in g(z)} |z^*|$.

Hence, for viscoplastic models the constitutive multifunction g does not blow up on finite sets. Note that for maximal monotone g condition (19) is equivalent to $D(g) = \mathbb{R}^N$. To prove an existence result for such models we use the coercive approximation procedure presented above. The fundamental result used to prove convergence of the coercive approximation is an energy estimate uniform with respect to k. For homogeneous boundary data the energy method works without additional assumptions. In the general situation we have to restrict the class of given boundary data and recall from [22] the definition of the "weak safe-load condition".

Definition 5 (weak safe-load property of boundary data). We say that the boundary data g_D or g_N satisfy the weak safe-load condition if there exist functions

$$(\tilde{v}^0, \tilde{\varepsilon}^0, \tilde{z}^0) \in \mathbb{H}^1(\Omega; \mathbb{R}^3) \times \mathbb{L}^2(\Omega; \mathcal{S}^3 \times \mathbb{R}^N) \text{ with } \operatorname{div} \mathcal{D}\tilde{\varepsilon}^0 \in \mathbb{L}^2(\Omega; \mathbb{R}^3)$$

such that the following elastic problem

$$\begin{aligned} \rho\tilde{v}_t(x,t) &= \operatorname{div}_x \mathcal{D}\Big(\tilde{\varepsilon}(x,t) - B\tilde{z}(x,t)\Big)\,, \\ \tilde{\varepsilon}_t(x,t) &= \tfrac{1}{2}\Big(\nabla_x \tilde{v}(x,t) + \nabla_x^T \tilde{v}(x,t)\Big)\,, \\ \tilde{z}_t(x,t) &= 0 \end{aligned} \tag{20}$$

with Dirichlet boundary conditions

$$v^k(t,x) = g_D(t,x) \quad \text{for a.e. } x \in \Omega \text{ and } t \geq 0,$$

or with Neumann boundary conditions

$$\mathcal{D}\Big(\tilde{\varepsilon}(x,t) - B\tilde{z}(x,t)\Big) \cdot n(x) = g_N(x,t) \quad \text{for a.e. } x \in \Omega \text{ and } t \geq 0,$$

and with the initial conditions

$$(\tilde{v}(0), \tilde{\varepsilon}(0), \tilde{z}(0)) = (\tilde{v}^0, \tilde{\varepsilon}^0, \tilde{z}^0)\,,$$

compatible with the boundary data, possesses a solution $(\tilde{v}, \tilde{\varepsilon}, \tilde{z})$ which satisfies

$$\begin{aligned} \forall_{T>0} \qquad (\tilde{v}, \tilde{\varepsilon}, \tilde{z}) &\in \mathbb{L}^\infty((0,T); \mathbb{L}^2(\Omega; \mathbb{R}^3 \times \mathcal{S}^3 \times \mathbb{R}^N))\,, \\ (\tilde{v}_t, \tilde{\varepsilon}_t) &\in \mathbb{L}^\infty((0,T); \mathbb{L}^2(\Omega; \mathbb{R}^3 \times \mathcal{S}^3))\,, \\ -\rho\nabla_z\psi(\tilde{\varepsilon}, \tilde{z}) &= B^T\mathcal{D}\tilde{\varepsilon} - (B^T\mathcal{D}B + L)\tilde{z} \in D(g)\,, \end{aligned}$$

and there exists $z^* \in \mathbb{L}^\infty((0,T); \mathbb{L}^2(\Omega; \mathbb{R}^N))$ such that

$$z^*(t,x) \in g(-\rho\nabla_z\psi(\tilde{\varepsilon}(t,x), \tilde{z}(t,x)))$$

for a.e. $(t,x) \in (0,T) \times \Omega$.

The notion of the weak safe-load condition allows to prove energy estimates for the differences between the approximate solution $(v^k, \varepsilon^k, z^k)$ and the triple $(\tilde{v}, \tilde{\varepsilon}, \tilde{z})$ from Def. 5.

Theorem 3 (energy estimate). *Suppose that the given functions g_D, g_N, F have the regularity (16) and the boundary data g_D or g_N satisfy the weak safe-load condition. If the sequence $\mathcal{E}^k(v_k^0, \varepsilon_k^0, z_k^0)$ is bounded, then for each $T > 0$ there exists a positive constant $C(T)$ independent of k such that*

$$\sup_{t\in(0,T)} \mathcal{E}^k(v^k - \tilde{v}, \varepsilon^k - \tilde{\varepsilon}, z^k - \tilde{z})(t)$$

$$+ \sup_{t\in(0,T)} \int_0^t \int_\Omega (z_t^k - z^*) \cdot \Big(- \rho\nabla_z\psi^k(\varepsilon^k, z^k) + \rho\nabla_z\psi(\tilde{\varepsilon}, \tilde{z})\Big)\, dx\, d\tau \leq C(T)\,.$$

The proof of this theorem is given in [16] and in [22]. The idea of the proof is based on energy estimates for the differences between the sequence of coercive approximation $(v^k, \varepsilon^k, z^k)$ and the triple $(\tilde{v}, \tilde{\varepsilon}, \tilde{z})$. This works well because the boundary data for this difference are almost homogeneous. To pass to the limit $k \to \infty$ an estimate of the energy $\mathcal{E}^k(v_t^k, \varepsilon_t^k, z_t^k)$ independent of k should be proved. This is done in the next theorem.

Theorem 4 (energy estimate for time derivatives). *Let us assume that the given data g_D, g_N, F have the regularity (16) and, additionally, that the boundary data g_N "are bounded", i.e.,*

$$\forall_{T>0} \qquad g_{N,tt},\, g_{N,t} \in \mathbb{L}^\infty((0,T)\times\partial\Omega)\,. \tag{21}$$

Moreover, assume that the initial data $(v^0, \varepsilon^0, z^0) \in \mathbb{L}^2(\Omega; \mathbb{R}^3 \times \mathcal{S}^3 \times \mathbb{R}^N)$ have the coercive limit property. Then the energy for the time derivatives can be estimated as follows

$$\mathcal{E}^k(v_t^k, \varepsilon_t^k, z_t^k)(t) = \tfrac{1}{2}\int_\Omega \mathcal{D}(\varepsilon_t^k(t) - Bz_t^k(t))\cdot(\varepsilon_t^k(t) - Bz_t^k(t))\,dx$$

$$+\tfrac{1}{2}\int_\Omega \rho|v_t^k(t)|^2\,dx + \tfrac{1}{2}\int_\Omega Lz_t^k(t)\cdot z_t^k(t)\,dx + \tfrac{1}{2}k^{-1}\int_\Omega \mathcal{D}\varepsilon_t^k(t)\cdot\varepsilon_t^k(t)\,dx$$

$$\leq C(T)(1 + \sup_{t\in(0,T)} \|Bz_t^k(t)\|_{\mathbb{L}^1(\Omega)})$$

where the constant $C(T)$ does not depend on k.

The proof of this theorem is based on the energy estimates for the differences between $(v^k(t+h), \varepsilon^k(t+h), z^k(t+h))$ and $(v^k(t), \varepsilon^k(t), z^k(t))$. For details we refer to [16] or to [22].

According to Theorems 3 and 4 for the initial data $(v^0, \varepsilon^0, z^0)$ possessing the coercive limit property and for the boundary data g_D or g_N satisfying the weak safe-load condition and the assumption (21) we conclude that there exists a positive constant $C(T)$ independent of k such that

$$\sup_{t\in(0,T)} \Big(\|v_t^k(t)\|_{\mathbb{L}^2(\Omega)}\,, \|\varepsilon_t^k(t) - Bz_t^k(t)\|_{\mathbb{L}^2(\Omega)}\,, \|Lz_t^k(t)\|_{\mathbb{L}^2(\Omega)}\,,$$

$$k^{-\frac{1}{2}}\|\varepsilon_t^k(t)\|_{\mathbb{L}^2(\Omega)}\Big) \leq C(T)(1 + \sup_{t\in(0,T)} \|Bz_t^k(t)\|_{\mathbb{L}^1(\Omega)}) \text{ and} \tag{22}$$

$$\sup_{t\in(0,T)} \Big(\|v^k(t)\|_{\mathbb{L}^2(\Omega)}\,, \|\varepsilon^k(t) - Bz^k(t)\|_{\mathbb{L}^2(\Omega)}\,, \|Lz^k(t)\|_{\mathbb{L}^2(\Omega)}\,,$$

$$k^{-\frac{1}{2}}\|\varepsilon^k(t)\|_{\mathbb{L}^2(\Omega)}\Big) \leq C(T)\,.$$

The estimate (22) is not sufficient to prove convergence of the coercive approximation process because the sequence $\|Bz_t^k\|_{\mathbb{L}^1(\Omega)}$ is not apriori bounded and the direction ε^k (and consequently the direction Bz^k) is not controlled in $\mathbb{L}^\infty(\mathbb{L}^2)$. For self-controlling models the very special structure of the constitutive function g gave the possibility to cancel the factor $k^{-\frac{1}{2}}$ in the last inequalities and to prove the convergence result for the coercive approximation in the space $\mathbb{L}^2$. In the general situation we cannot expect that the sequence $\{\varepsilon^k\}$ will be bounded in the space $\mathbb{L}^2(\Omega; \mathcal{S}^3)$ and therefore we have to find a larger space than $\mathbb{L}^2$ in which the sequence $\{\varepsilon^k\}$ will be sequentially weakly precompact. In the next theorem we prove that for **viscoplastic materials** the coercive approximation converges weakly in the space $\mathbb{L}^1$.

Theorem 5 (weak precompactness in $\mathbb{L}^1$). *Let us assume that the initial data $(v^0, \varepsilon^0, z^0)$ have the coercive limit property, the given functions g_D, g_N, F have the regularity (16) and the boundary data g_D or g_N satisfy the weak safe-load condition. If in addition*

$$\nabla_z \psi(\tilde{\varepsilon}, \tilde{z}) \in \mathbb{L}^\infty \Big(\Omega \times (0,T) \Big) , \tag{23}$$

where $(\tilde{v}, \tilde{\varepsilon}, \tilde{z})$ is a solution of problem (20) satisfying all requirements of Def. 5, then for viscoplastic models the following statements hold: for all $T > 0$

- *the sequence $\{Bz_t^k\}$ is bounded in the space $\mathbb{L}^\infty((0,T); \mathbb{L}^1(\Omega; \mathcal{S}^3))$*
- *the sequences $\{\varepsilon^k\}$ and $\{\varepsilon^{p,k}\}$ are sequentially weakly precompact in the space $\mathbb{L}^1(\Omega \times (0,T))$.*

The proof of this theorem is found in [16]. The main idea of this proof is similar to that used in Lemma 3.3 from [24]. Using the last result we have

$$\begin{aligned} z_t^k &\rightharpoonup z_t && \text{in } \mathbb{L}^1(\Omega \times (0,T), \mathbb{R}^N) \quad \text{and} \\ -\rho \nabla_z \psi^k(\varepsilon^k, z^k) &\overset{*}{\rightharpoonup} \rho \nabla_z \psi(\varepsilon, z) && \text{in } \mathbb{L}^\infty((0,T); \mathbb{L}^2(\Omega; \mathbb{R}^N)) \,. \end{aligned} \tag{24}$$

The sequence $(v^k, \varepsilon^k, z^k)$ solves problem (17) in the strong sense, and therefore for almost all $(t,x) \in (0,T) \times \Omega$ the following inequality is satisfied:

$$(z_t^k(x,t) - w^*, -\rho \nabla_z \psi(\varepsilon^k(x,t), z^k(x,t)) - w) \geq 0 \,, \tag{25}$$

for all $w \in \mathbb{R}^N$ (we are in the viscoplastic case) and for all $w^* \in g(w)$. It is easy to see that the convergence results (24) are not sufficient to prove an inequality similar to (25) for the limit functions. Therefore we need to improve the convergence of the sequence $\{\rho \nabla_z \psi^k(\varepsilon^k, z^k)\}$. To do that, we have to restrict the class of our materials. Subsequently we assume that the monotone model (12) is viscoplastic and of gradient type. This means that there exists a convex flow potential $\mathcal{X} : \mathbb{R}^N \to \mathbb{R}_+$ such that

$$g(z) = \partial \mathcal{X}(z) \quad \text{for all } z \in \mathbb{R}^N$$

where $\partial \mathcal{X}$ denotes the subgradient in the sense of convex analysis

$$w \in \partial \mathcal{X}(z) \Leftrightarrow \mathcal{X}(z+y) - \mathcal{X}(z) \geq (y, w) \quad \text{for all } y \in \mathbb{R}^N \,.$$

According to the requirement $0 \in g(0)$, we additionally suppose that $0 \in \partial \mathcal{X}(0)$. Finally, we can assume that $\mathcal{X}(0) = 0$.

Theorem 6 (strong convergence of $\{\rho \nabla_z \psi^k(\varepsilon^k, z^k)\}$). *Let us assume that the initial data $(v^0, \varepsilon^0, z^0)$ have the coercive limit property and, additionally, that the initial data $(v_k^0, \varepsilon_k^0, z_k^0)$ for the problem (17) are so chosen that*

$$(v_k^0, \varepsilon_k^0, z_k^0) \to (v^0, \varepsilon^0, z^0) \quad \textit{strongly in} \quad \mathbb{L}^2(\Omega; \mathbb{R}^3 \times \mathcal{S}^3 \times \mathbb{R}^N) \,.$$

Moreover, suppose that the given functions g_D, g_N, F have the regularity (16) and the boundary data g_D or g_N fulfill the weak safe-load condition, (21) and (23). Then, for viscoplastic models of monotone-gradient type for all $T > 0$ the sequence $\{-\rho\nabla_z\psi^k(\varepsilon^k, z^k)\}$ defined by the solutions of the problem (17) converges to $-\rho\nabla_z\psi(\varepsilon, z)$ in the space $\mathbb{L}^\infty((0,T); \mathbb{L}^2(\Omega; \mathbb{R}^N))$. Moreover, for all $T > 0$ and for almost all $(x,t) \in \Omega \times (0,T)$

$$z_t(x,t) \in \partial\chi\Big(- \rho\nabla_z\psi(\varepsilon(x,t), z(x,t))\Big) .$$

This theorem is a union of Theorems 5 and 6 from [16]. The proof is based on a special energy estimate for models with gradient structure and on the notion of the generalized directional derivatives defined by F. H. Clarke in [25]. The assumption (23) of $\mathbb{L}^\infty$ boundedness of the trajectory $\rho\nabla_z\psi(\tilde{\varepsilon}, \tilde{z})$ seems to be very restrictive. However, if we compare this property with the safe-load condition given in the next subsection then we can easyly see that (23) is similar to the first of the two last requirements from Def. 7.

4.4 General Models of Monotone Type and the Rate Independent Case

The lack of coerciveness of the free energy function ψ induces a loss of regularity of the inelastic strain tensor ε^p and equivalently a loss of regularity of the strain tensor ε. Both tensor fields may not belong to the space $\mathbb{L}^2(\Omega; \mathcal{S}^3)$ and, even worse, they can become bounded vector-valued Radon measures only. Therefore, we should introduce a new notion of solutions of the problem (12). First of all, we see that the second equality in (12) has to be considered in the sense of measures, only. But this is always well defined. It is more difficult in what sense the differential inclusion occuring in (12) has to be satisfied. The differential inclusion is a pointwise relation, but the measure $\varepsilon^p = Bz$ and its time derivative ε^p_t do not have a pointwise meaning. Therefore, we must first drop the pointwise meaning of this inclusion. Let us assume that a triple (v, ε, z) is a strong $\mathbb{L}^2$ solution of system (12) and the constitutive multifunction g is maximal monotone. Then we have that the inelastic constitutive relation is equivalent to the following pointwise statement

$$\forall\, w \in D(g)\,,\ \forall\, w^* \in g(w) \quad \Big(z_t(x,t) - w^*, -\rho\nabla_z\psi(\varepsilon(x,t), z(x,t)) - w\Big) \geq 0$$

for almost all $(x,t) \in \Omega \times (0,T)$. (Note that here $D(g) = \{w \in \mathbb{R}^N : g(w) \neq \emptyset\}$ is the domain of the pointwise maximal monotone multifunction $g : D(g) \subset \mathbb{R}^N \to \mathcal{P}(\mathbb{R}^N)$.) If we denote by

$$\begin{aligned}\mathbb{D}(g) = \{&w \in \mathbb{L}^2(\Omega; R^N) : \text{there exists } w^* \in \mathbb{L}^2(\Omega; R^N) \text{ with}\\ &w^*(x) \in g(w(x)) \text{ for almost all } x \in \Omega\}\end{aligned}$$

the domain of the maximal monotone operator g in the space $\mathbb{L}^2(\Omega; R^N)$, then for strong $\mathbb{L}^2$ solutions (v, ε, z) we have that the differential inclusion is equivalent to

$$\forall\, w \in \mathbb{D}(g)\,,\ \forall\, w^* \in g(w) \quad \Big(z_t - w^*, -\rho\nabla_z\psi(\varepsilon, z) - w \Big)_{\mathbb{L}^2} \geq 0 \tag{26}$$

for almost all $t \in (0, T)$. Inequality (26) does not have a pointwise character, but, unfortunately, if Bz_t is only a bounded measure then the product of this measure with an $\mathbb{L}^2$ function is not well defined. In this case we have to change the form of (26) slightly. For the gradient of ψ with respect to z we obtain

$$-\rho\nabla_z\psi(\varepsilon, z) = B^T\mathcal{D}(\varepsilon - Bz) - Lz = B^T T - Lz\,,$$

and the assumption $B^T\mathcal{D}B + L > 0$ yields that there exist $\tilde{w}, \hat{w} \in \mathbb{L}^2(\Omega; R^N)$ with $w = B^T B\tilde{w} - L\hat{w}$. Then, inequality (26) can be written in the form

$$\begin{aligned} &\forall\, w \in \mathbb{D}(g),\ w = B^T B\tilde{w} - L\hat{w}\,, \quad \forall\, w^* \in g(w) \\ &\qquad (\varepsilon^p_t - Bw^*, T - B\tilde{w}) - (z_t - w^*, Lz - L\hat{w}) \geq 0. \end{aligned} \tag{27}$$

Finally, using the constitutive relation $T = \mathcal{D}(\varepsilon - Bz)$ and integrating by parts we conclude that

$$\begin{aligned} &(\varepsilon^p_t, T - B\tilde{w}) = (\varepsilon_t - \mathcal{D}^{-1}T_t, T - B\tilde{w}) \\ &= -(v, \operatorname{div} T - \operatorname{div} B\tilde{w}) - (\mathcal{D}^{-1}T_t, T - B\tilde{w}) + \langle\, v,\, (T - B\tilde{w}) \cdot n\,\rangle_{\partial\Omega}\,, \end{aligned} \tag{28}$$

where the bracket $\langle\,\cdot\,,\,\cdot\,\rangle_{\partial\Omega}$ denotes the duality mapping between the spaces $\mathbb{H}^{\frac{1}{2}}(\partial\Omega)$ and $\mathbb{H}^{-\frac{1}{2}}(\partial\Omega)$. Then inserting (28) into (27) we see that for all strong $\mathbb{L}^2$ solutions of system (12) the condition

$$\begin{aligned} &\forall\, w \in \mathbb{D}(g),\ w = B^T B\tilde{w} - L\hat{w}\,, \quad \forall\, w^* \in g(w) \\ &\qquad (v, \operatorname{div} T - \operatorname{div} B\tilde{w}) + (\mathcal{D}^{-1}T_t + Bw^*, T - B\tilde{w}) \\ &\qquad + (z_t - w^*, Lz - L\hat{w}) \leq \langle\, v,\, (T - B\tilde{w}) \cdot n\,\rangle_{\partial\Omega} \end{aligned} \tag{29}$$

must hold for almost all $t \in (0, T)$, provided that the test stress $B\tilde{w} \in \mathbb{L}^2(\Omega; \mathbb{R}^3)$ possesses the weak divergence $\operatorname{div} B\tilde{w}$. Note that this yields that the trace of the test stress $B\tilde{w}$ in the normal direction to the boundary $\partial\Omega$ is well defined in the space $\mathbb{H}^{-\frac{1}{2}}(\partial\Omega)$ (see [40, p.14]). Hence, using the boundary condition (13) or (14) we rewrite the boundary term in (29) in the form

$\langle\, g_D,\, (T - B\tilde{w}) \cdot n\,\rangle_{\partial\Omega}$ for the Dirichlet boundary value problem or

$\langle\, v,\, (g_N - B\tilde{w}) \cdot n\,\rangle_{\partial\Omega}$ for the Neumann boundary value problem.

Unfortunately, in the case of the Neumann boundary value problem we encounter at this point the following difficulty. In the space $\mathbb{BD}(\Omega)$, containing functions with bounded deformations, the trace operator is not weakly sequentially continuous ([40, Theorem 3.1, p.160]). Hence, for a sequence $\{v^n\}$ in this space, which converges only weakly to v, the boundary values $v^n_{|\partial\Omega}$ may not converge weakly to the trace $v_{|\partial\Omega}$. In order to avoid this difficulty we will require additionally that in the case of the Neumann boundary value problem (29) has to be satisfied for all test stresses $B\tilde{w}$ with $B\tilde{w} \cdot n = g_N$. Let us denote by $\mathcal{M}^{3\times 3}_{\text{sym}}(\Omega)$ the Banach space containing all bounded Radon measures in Ω with values in $\mathcal{S}^3$, equipped with the norm

$$\|\mu\|_{\mathcal{M}^{3\times 3}_{\text{sym}}(\Omega)} = \sup_{\substack{f \in C_0(\Omega; \mathcal{S}^3) \\ |f| \le 1}} \int_\Omega f \cdot \sigma \, d|\mu| \,,$$

where the measure $|\mu|$ is the total variation measure of the measure μ and σ is the Radon-Nikodym derivative $d\mu/d|\mu|$ The space $\mathbb{L}^1(\Omega; \mathcal{S}^3)$ is isometrically imbedded into the space $\mathcal{M}^{3\times 3}_{\text{sym}}(\Omega)$. This space is isomorphic to the dual of $C_0(\Omega; \mathcal{S}^3)$ which is separable, hence the weak compactness and weak sequential compactness coincide. Moreover, let us denote by $\mathbb{L}^\infty_w((0,T); \mathcal{M}^{3\times 3}_{\text{sym}}(\Omega))$ the space of essentially bounded, weakly measurable functions $\mu : (0,T) \to \mathcal{M}^{3\times 3}_{\text{sym}}(\Omega)$. This means that for all $\phi \in C_0(\Omega; \mathcal{S}^3)$ the function $\int_\Omega \phi(x)\, d\mu(x)$ is measurable and essentially bounded. The norm in this space is defined by

$$\sup_{t \in (0,T)} \sup_{\substack{\phi \in C_0(\Omega; \mathcal{S}^3) \\ |\phi| \le 1}} \int_\Omega \phi \cdot \sigma \, d|\mu| \,.$$

Definition 6 (weak-type solutions). We say that a triple (v, ε, z) satisfies the system (12) in the weak-type sense, if

a) $(v, T = \mathcal{D}(\varepsilon - Bz), Lz), (v_t, T_t, Lz_t) \in \mathbb{L}^\infty\Big((0,T); \mathbb{L}^2(\Omega; \mathbb{R}^3 \times \mathcal{S}^3 \times \mathbb{R}^N)\Big)$,

b) $(\varepsilon, \varepsilon^p = Bz), (\varepsilon_t, \varepsilon^p_t) \in \mathbb{L}^\infty_w\Big((0,T); \mathcal{M}^{3\times 3}_{\text{sym}}(\Omega)\Big)$,

c) $\operatorname{div} T \in \mathbb{L}^\infty\Big((0,T); \mathbb{L}^2(\Omega; \mathbb{R}^3)\Big)$, $\frac{1}{2}(\nabla v + \nabla^T v) \in \mathbb{L}^\infty_w\Big((0,T); \mathcal{M}^{3\times 3}_{\text{sym}}(\Omega)\Big)$;

the first equality in (12) holds pointwise for almost all $(x,t) \in \Omega \times (0,T)$ and the second equality in (12) holds in the space $\mathbb{L}^\infty_w((0,T); \mathcal{M}^{3\times 3}_{\text{sym}}(\Omega))$. Finally, in the case of the Dirichlet boundary value problem we require that

$$\begin{aligned} &\forall\, w \in \mathbb{D}(g) \;\; w = B^T B\tilde{w} - L\hat{w}\,, \quad \forall\, w^* \in g(w) \\ &(v, \operatorname{div} T - \operatorname{div} B\tilde{w}) + (\mathcal{D}^{-1} T_t + Bw^*, T - B\tilde{w}) \\ &\qquad + (z_t - w^*, Lz - L\hat{w}) \le \langle\, g_D, (T - B\tilde{w}) \cdot n \,\rangle_{\partial\Omega} \end{aligned} \tag{30}$$

holds for almost all $t \in (0,T)$, provided that w and w^* have $\mathbb{L}^\infty(\mathbb{L}^2)$ regularity and $\operatorname{div} B\tilde{w} \in \mathbb{L}^\infty((0,T); \mathbb{L}^2(\Omega; \mathbb{R}^3))$.
In the case of the Neumann boundary value problem we require

$$\begin{aligned} &\forall\, w \in \mathbb{D}(g) \;\; w = B^T B\tilde{w} - L\hat{w}\,, \quad \forall\, w^* \in g(w) \\ &(v, \operatorname{div} T - \operatorname{div} B\tilde{w}) + (\mathcal{D}^{-1} T_t + Bw^*, T - B\tilde{w}) \\ &+ (z_t - w^*, Lz - L\hat{w}) \le 0 \end{aligned} \tag{31}$$

for almost all $t \in (0,T)$, assuming, as in the previous case, that w and w^* have $\mathbb{L}^\infty(\mathbb{L}^2)$ regularity, $\operatorname{div} B\tilde{w} \in \mathbb{L}^\infty((0,T);\mathbb{L}^2(\Omega;\mathbb{R}^3))$ and, additionally, $B\tilde{w}\cdot n_{|\partial\Omega} = g_N$.
Moreover, the initial conditions (15) are satisfied in the sense of traces and
$v\cdot n_{|\partial\Omega} = g_D\cdot n$ for the Dirichlet boundary condition or
$T\cdot n_{|\partial\Omega} = g_N$ for the Neumann boundary condition.

Note that in the case of the Dirichlet boundary-value problem we are not able to satisfy the boundary condition in the tangent direction to the boundary $\partial\Omega$. This difficulty is connected with the loss of the weak sequential continuity of the trace operator in the space $\mathbb{BD}(\Omega)$. The validity of the boundary condition in the normal direction to the boundary $\partial\Omega$ will be obtained by an additional assumption on the constitutive multifunction g. Namely, we will further assume that the values $Bg(z)\subset\mathcal{S}^3$ are trace free, this means that for all $\sigma\in Bg(z)$ $\operatorname{dev}\sigma = \sigma - \frac{1}{3}(\operatorname{tr}\sigma)\cdot I = \sigma$. The notion of the weak-type solutions introduced in Def. 6 was based on the idea of the weak solution defined for the Prandtl-Reuss model of elasto-perfect plasticity. For details we refer the reader to [26,29,30,37,38]. We extend this idea to general models of monotone type.
Next we restrict the class of the given boundary data. Following the idea of C. Johnson ([30] or [31]) we formulate the so called safe-load condition. Mathematically, this property is the main piece of information in the proof of a priori estimates for the coercive approximate sequence. Physically, the condition says which boundary data are well-tolerated by the considered material.

Definition 7 (safe-load condition). We say that the boundary data g_D or g_N satisfy the safe-load condition if there exist functions

$$(\tilde{v}^0,\tilde{\varepsilon}^0,\tilde{z}^0)\in\mathbb{H}^1(\Omega;\mathbb{R}^3)\times\mathbb{L}^2(\Omega;\mathcal{S}^3\times\mathbb{R}^N) \text{ with } \operatorname{div}\mathcal{D}\tilde{\varepsilon}^0\in\mathbb{L}^2(\Omega;\mathbb{R}^3)$$

such that the following elastic problem

$$\begin{aligned}
\rho\tilde{v}_t(x,t) &= \operatorname{div}_x \mathcal{D}\Big(\tilde{\varepsilon}(x,t) - B\tilde{z}(x,t)\Big),\\
\tilde{\varepsilon}_t(x,t) &= \tfrac{1}{2}\Big(\nabla_x\tilde{v}(x,t) + \nabla_x^T\tilde{v}(x,t)\Big),\\
\tilde{z}_t(x,t) &= 0
\end{aligned}$$

with the Dirichlet boundary condition

$$\tilde{v}(t,x) = g_D(t,x) \quad \text{for a.e. } x\in\Omega \text{ and } t\geq 0,$$

or with the Neumann boundary condition

$$\mathcal{D}\Big(\tilde{\varepsilon}(x,t) - B\tilde{z}(x,t)\Big)\cdot n(x) = g_N(x,t) \quad \text{for a.e. } x\in\Omega \text{ and } t\geq 0,$$

and with the initial conditions

$$(\tilde{v}(0),\tilde{\varepsilon}(0),\tilde{z}(0)) = (\tilde{v}^0,\tilde{\varepsilon}^0,\tilde{z}^0)$$

compatible with the boundary data, possesses a strong $\mathbb{L}^2$ solution $(\tilde{v}, \tilde{\varepsilon}, \tilde{z})$ satisfying

$$\exists_{\delta>0}\ \forall_{|\sigma|<\delta}\ \ (-\rho\nabla_z\psi(\varepsilon^*, z^*) + \sigma) \in D(g) \ \text{ and}$$

$$\forall_{|\sigma|<\delta}\ \exists_{z^*_\sigma \in g(-\rho\nabla_z\psi(\varepsilon^*,z^*)+\sigma)}\ \|\sup_{|\sigma|<\delta} |z^*_\sigma|\,\|_{\mathbb{L}^2} < C(T)\,,$$

where the constant $C(T)$ does not depend on σ.

Note that in Def. 5 we do not require that the distance of the trajectory $-\rho\nabla_z\psi(\varepsilon^*, z^*)$ from the boundary of the of the domain $D(g)$ must be positive.

Theorem 7 (existence of global weak-type solutions).
1. Let us fix $T > 0$ and assume that problem (12) is of monotone type with a maximal monotone constitutive multifunction g satisfying: $0 \in g(0)$ and $Bg \subset \operatorname{dev} \mathcal{S}^3 = \{\sigma \in \mathcal{S}^3 : \operatorname{tr}\sigma = 0\}$. Further more, assume that the given boundary data g_D or g_N and the external force F have the property (16). Moreover, suppose that the boundary data g_D or g_N satisfy the safe-load condition and in the case of the Neumann boundary value problem the boundary data g_N "are bounded", i.e.,

$$g_{N,tt}, g_{N,t} \in \mathbb{L}^\infty(\partial\Omega \times (0,T))\,.$$

Then for all initial data $(v^0, \varepsilon^0, z^0) \in \mathbb{L}^2(\Omega; \mathbb{R}^3 \times \mathcal{S}^3 \times \mathbb{R}^N)$, which are admissible for system (12) and have the coercive limit property, there exists a triple (v, ε, z) satisfying all requirements of Def. 6 except one, namely inequality (30) is satisfied in the following weaker form:

$$\forall\, w \in \mathbb{D}(g)\,,\ w = B^T B\tilde{w} - L\hat{w}\,, \quad \forall\, w^* \in g(w)$$

$$\int_0^t \{(v, \operatorname{div} T - \operatorname{div} B\tilde{w}) + (\mathcal{D}^{-1}T_t + Bw^*, T - B\tilde{w})\}d\tau \tag{32}$$

$$+ \int_0^t (z_t - w^*, Lz - L\hat{w})d\tau \le \int_0^t \langle\, g_D, (T - B\tilde{w})\cdot n\,\rangle_{\partial\Omega}\, d\tau$$

for all $t \in (0,T)$, provided that w and w^* possess the properties from Def. 6. Similarly, in the case of the Neumann boundary value problem inequality (31) is satisfied in the following weaker form:

$$\forall\, w \in \mathbb{D}(g)\,,\ w = B^T B\tilde{w} - L\hat{w}\,, \quad \forall\, w^* \in g(w)$$

$$\int_0^t \{(v, \operatorname{div} T - \operatorname{div} B\tilde{w}) + (\mathcal{D}^{-1}T_t + Bw^*, T - B\tilde{w})\}d\tau \tag{33}$$

$$+ \int_0^t (z_t - w^*, Lz - L\hat{w})\, d\tau \le 0$$

for all $t \in (0,T)$, assuming that, as in the previous case, w and w^* possess the properties from Def. 6.

2. If additionally the constitutive multifunction g satisfies

$$\forall\, w \in D(g) \qquad g(w) \text{ is a conical set with vertex } 0 \tag{34}$$

then system (12) possesses global, unique weak-type solutions.

Models possessing property (34) are rate-independent. For the precise definition we refer to [3, Definition 2.1.1]. Theorem 7 is the main Theorem in [18]. This theorem yields, for a general model of monotone type, existence of functions which satisfy the weak-type inequality integrated in time. Moreover, Theorem 7 yields the existence of global in time weak-type solutions for the Prandtl-Reuss model.

5 General Theory for some Classes Containing Nonmonotone Models

Next we are interested to enlarge the class of models for which an existence result of global in time solutions can be proved. Therefore we try to study models which are not of monotone type. The first class containing such models was defined in the monograph [3]. This class contains models which can be transformed into the class $\mathcal{M}$.

5.1 Transformation of Internal Variables

We study transformations of internal variables in the system

$$\begin{aligned} \rho u_{tt}(x,t) &= \operatorname{div}_x T(x,t)\,, \\ T(x,t) &= \mathcal{D}(\varepsilon(\nabla_x u(x,t)) - Bz(x,t))\,, \\ z_t(x,t) &= f(\varepsilon(\nabla_x u(x,t)), z(x,t))\,. \end{aligned}$$

Let $H : \mathbb{R}^N \to \mathbb{R}^N$ be a continuously differentiable vector field and let

$$h(x,t) = H(z(x,t))$$

be transformed interior variables. If (u, z) is a solution of the system above, then the function (u, h) satisfies the transformed system

$$\begin{aligned} \rho u_{tt}(x,t) &= \operatorname{div}_x T(x,t)\,, \\ T(x,t) &= \mathcal{D}(\varepsilon(\nabla_x u(x,t)) - BH^{-1}(h(x,t)))\,, \\ h_t(x,t) &= f_H(\varepsilon(\nabla_x u(x,t)), h(x,t)) \end{aligned}$$

with the transformed function $f_H : D(f_H) \subset \mathcal{S}^3 \times \mathbb{R}^N \to \mathbb{R}^N$ defined by

$$f_H(\varepsilon, h) = H'(H^{-1}(h))\, f(\varepsilon, H^{-1}(h))\,.$$

We say that a model is transformable into the class $\mathcal{M}$ or belongs to the class $\mathcal{TM}$ if the transformed function f_H is of monotone type. Results for models from $\mathcal{TM}$ can be found in the monograph [3].

5.2 Lipschitz Perturbations of Monotone Models

The next generalisation of the monotonicity requirement is very simple. We say that an inelastic constitutive equation belongs to the class $\mathcal{LM}$ if and only if it is of pre-monotone type and there exists a global Lipschitz operator $\mathcal{L} : \mathbb{R}^N \to \mathbb{R}^N$ such that the inelastic constitutive equation with the constitutive function $g + \mathcal{L}$ is of monotone type. The assumption $g(0) = 0$ implies that the operator $\mathcal{L}$ satisfies the same condition. It is easy to see that an inelastic constitutive equation belongs to the class $\mathcal{LM}$ if and only if

$$\exists_{l \geq 0} \forall_{z,z' \in D(g)} \qquad (g(z) - g(z')) \cdot (z - z') \geq -l|z - z'|^2 .$$

In [21] it is proved that the nonmonotone model of Bodner-Partom with the inelastic constitutive relation in the form

$$\begin{aligned} \varepsilon_t^p &= F\Big(\frac{|\mathrm{dev}\, T|}{y}\Big) \frac{\mathrm{dev}\, T}{|\mathrm{dev}\, T|} , \\ y_t &= \gamma(y)\, F\Big(\frac{|\mathrm{dev}\, T|}{y}\Big)\, |\mathrm{dev}\, T| - \delta(y) , \end{aligned}$$

where y is the isotropic hardening of the material

$$F(s) = d\, e^{-\alpha(\frac{1}{s})^n} , \; \gamma(y) = m(y_1 - y) , \; \delta(y) = A y_1 \Big(\frac{y - y_2}{y_1}\Big)^r ,$$

and the material constants occuring in the model satisfy $\alpha, n, r > 1$, $d, m > 0$, $A \geq 0$, $y_1 > y_2 > 0$, belongs to the class $\mathcal{LM}$. Moreover, in the monograph [3] we find that this model cannot be transformed into the class $\mathcal{M}$ and this shows that the set $\mathcal{LM} \setminus \mathcal{TM}$ is not empty. Mathematical analysis of the Bodner-Partom model in several settings can be found in [8–11,21,23].

6 Remarks on the Quasistatic Case

In the quasistatic case the model equations for materials of monotone type can be written in the form

$$\begin{aligned} -\mathrm{div}_x\, T(x,t) &= F(x,t), \\ T(x,t) &= \mathcal{D}(\varepsilon(\nabla_x u(x,t)) - Bz(x,t)), \\ z_t(x,t) &\in g(B^T T(x,t) - Lz(x,t)), \\ z(x,0) &= z^{(0)}(x) \end{aligned}$$

and g is a monotone constitutive multifunction. In [4] an existence result for a subclass of problems for which g grows polynomially is proved. Thus this article yields a similar result to that of [15]. The methods used in [4] are different from the methods in [15]. The authors work with the $X \to X^*$ monotonicity where X is a reflexive Banach space. This allows to enlarge the class of initial and boundary data for which existence of global solutions can be shown. Nevertheless, in the last section of this article it is proved that methods used in the dynamical setting of the problem work also in the quasistatic case.

6.1 Convex Composite Inelastic Constitutive Equations

The special structure of the quasistatic equations is used to find some models for which this formulation allows to prove an existence result which in a dynamical setting fails in general. One of such sublasses contains models called of **convex composite type**. This class is studied in [19]. We say that a model is of convex composite type if the constitutive multifunction has the form

$$\forall\, z \in D(g) \quad g(z) = \partial_c(\mathcal{X} \circ \Phi)(z)\,,$$

where $\mathcal{X}$ is a convex function, Φ is a global Lipschitz $\mathbb{C}^{1,1}$ diffeomorphism and ∂_c denotes the Clarke generalised gradient. By the properties of Φ we see that the right-hand side of the last equality is equivalent to

$$\forall\, z \in D(g) \quad g(z) = D^T\Phi(z) \cdot \partial\mathcal{X}(\Phi(z))\,.$$

For convex composite models the following generalisation of the monotonicity condition holds (see [19, Lemma 2.2])

$$\forall\, z, z' \in D(g) \quad (g(z) - g(z'), z - z') \geq -C(\|g(z)\| + \|g(z')\|)\|z - z'\|^2\,.$$

A disadvantage of this class, from a practical point of view, is the fact that all convex composite models possess an associated flow rule. From the mathematical point of view this property is extremely useful and yields existence of an additional semi-invariant of the problem. Assume that we have to solve in a Hilbert space the equation

$$\dot{x} = -\nabla\mathcal{X}(x)\,,$$

where $\mathcal{X}$ is a nonnegative function. Then this problem possesses a natural semi-invariant: the function $\mathcal{X}(x)$

$$\frac{d}{dt}\mathcal{X}(x) = -\nabla\mathcal{X}(x) \cdot \dot{x} = -|\dot{x}|^2 \leq 0\,.$$

For convex composite models this observation yields

$$\int_0^t \int_\Omega \rho\psi(\varepsilon_t, z_t)\, dx\, d\tau \leq C\,,$$

and this shows the boundedness in $\mathbb{L}^2(\mathbb{L}^2)$ of the time derivatives. Based on this information in the article [19] it is proved that coercive and linear self-controlling models of convex composite type possess global in time strong $\mathbb{L}^2$ solutions.

7 Remarks on the Armstrong-Frederick Model

At the end of this summary we present a model which is not of monotone type, does not belong to the class $\mathcal{LM}$, cannot be transformed into the class $\mathcal{M}$ and is not of convex composite type. Thus an existence result of global in time solutions does not follow from the general theory presented in this article. The main difficulty, from the mathematical point of view, is the missing monotonicity of the model. The flow rule for the vector of internal variables (here this vector consists of the plastic strain and the backstress) is not in the direction normal to the yield surface (in the literature such inelastic constitutive relations are called nonassociated flow rules).

$$\varepsilon_t^p(x,t) \in \partial I_{\mathcal{E}(b(x,t))}\Big(T(x,t)\Big),$$

$$b_t(x,t) = c\varepsilon_t^p(x,t) - d|\varepsilon_t^p(x,t)|b(x,t).$$

Here the set of admissible stresses $\mathcal{E}(b(x,t))$ is defined in the same way as in the Melan-Prager model (for the mathematical analysis of this model we refer for example to [16]) $\mathcal{E}(b) = \{T \in \mathcal{S}^3 : |\operatorname{dev} T - b| \le \sigma_y\}$ and σ_y is the yield stress. The function $I_{\mathcal{E}(b)}$ is the indicator function of the set $\mathcal{E}(b)$ and $\partial I_{\mathcal{E}(b)}$ is the subgradient of the convex function $I_{\mathcal{E}(b)}$. In the case $d = 0$ the Armstrong-Frederick model coincides with the form of the Melan-Prager model. Hence, the last term in the evolution equation for the backstress can be called a perturbation term. In comparison with the Melan-Prager model this term yields the $\mathbb{L}^\infty$ boundedness of the backstress and from the mathematical point of view this property is an important advantage of the model. Unfortunately, the perturbation term violates the monotonicity of the flow rule and this is an important disadvantage of the model. In the article [20] it is proved that this nonmonotone flow rule belongs to the class of pre-monotone models. Moreover, an approximation procedure for this model is proposed. This procedure uses the maximal monotonicity of the multifunction $\partial I_{\mathcal{E}(b)}$ and is given by

$$\operatorname{div}_x T^\lambda(x,t) = -F(x,t),$$

$$T^\lambda(x,t) = \mathcal{D}(\varepsilon^\lambda(x,t) - \varepsilon^{p,\lambda}(x,t)),$$

$$\varepsilon^\lambda(x,t) = \tfrac{1}{2}(\nabla_x u^\lambda(x,t) + \nabla_x^T u^\lambda(x,t)),$$

$$\varepsilon_t^{p,\lambda}(x,t) = \lambda^{-1}[|\operatorname{dev} T^\lambda(x,t) - b^\lambda(x,t)| - \sigma_y]_+ \frac{\operatorname{dev} T^\lambda(x,t) - b^\lambda(x,t)}{|\operatorname{dev} T^\lambda(x,t) - b^\lambda(x,t)|},$$

$$b_t^\lambda(x,t) = c\varepsilon_t^{p,\lambda}(x,t) - d|\varepsilon_t^{p,\lambda}(x,t)|b^\lambda(x,t),$$

where $\lambda > 0$, and the symbol $[r]_+$ denotes the positive part of the scalar function r. This approximation is called of Yosida-type because we use here the Yosida approximation of the multifunction $\partial I_{\mathcal{E}(b)}$. An existence result for each $\lambda > 0$ follows from the fact that for admissible initial backstress the approximate model possesses the linear self-controlling property and belongs to the class $\mathcal{LM}$. In [20] a different method yielding existence of global in time strong $\mathbb{L}^2$ solutions for the approximate problem is presented. This method is based on the cutting off of large values of the inelastic strain rates in the last equation of the approximated problem. Using the special geometrical properties of the Armstrong-Frederick flow rule in [20] it is proved that the sequence of stresses $\{T^{\lambda}\}$ is bounded in the space $\mathbb{L}^{\infty}(\mathbb{L}^2)$, the sequence of the inelastic strain rates $\{\varepsilon_t^{p,\lambda}\}$ is bounded in the space $\mathbb{L}^1(\mathbb{L}^1)$ and the sequence of the stress rates $\{T_t^{\lambda}\}$ is bounded in the space $\mathbb{L}^2(\mathbb{L}^2)$. Unfortunately, for the sequence $\{b_t^{\lambda}\}$ only the very weak boundedness in the space $\mathbb{L}^1(\mathbb{L}^1)$ was proved. Nevertheless, all these information together are enough to show that the approximation sequence converges to functions which satisfy the so called reduced energy inequality for this model. This inequality does not contain the nonlinear term $\int_{\Omega} |\varepsilon_t^p||b|^2\, dx$ and, therefore, such limit functions cannot be called weak-type solutions of the original Armstrong-Frederick problem. For details we refer to [20].

8 Acknowledgements

This article summarizes the results obtained by the author in the theory of inelastic deformations. The stimulating environment of the SFB 298 "Deformation and failure in metallic and granular materials" is gratefully acknowledged. The presented results are obtained while the author held a SFB position at Darmstadt University of Technology in the group of H.-D. Alber. The author would like to express his gratitude to H.-D. Alber, all members of the group AG6, K. Hutter and to other SFB 298 members for their kind hospitality.

References

1. Anzellotti G., Luckhaus S., (1987) Dynamical evolution of elasto-perfectly plastic bodies, Appl. Math. Optim. **15**, 121–140.
2. Alber H.-D., (1995) Mathematische Theorie des inelastischen Materialverhaltens von Metallen. Mitt. Ges. Angew. Math. Mech. **18**, 9–38
3. Alber H.-D., (1998) Materials with memory, Lecture Notes in Math., **vol. 1682**, Springer, Berlin Heidelberg New York, 1998.
4. Alber H.-D., Chełmiński K. (2002) Quasistatic problems in viscoplasticity theory, Fachbereich Math. TU Darmstadt, Preprint **2190**, submitted to Mathematische Zeitschrift.
5. Aubin J. P., Cellina A., (1984) Differential Inclusions. Springer Berlin, Heidelberg.
6. Blanchard D., at al. (1989) Numerical analysis of evolution problems in nonlinear small strains elastoviscoplasticity, Numer. Math. **55**, 177–195.
7. Brézis H. (1973) Operateurs maximaux monotones, North Holland, Amsterdam.
8. Chełmiński K., (1995) Global in time existence of solutions to the constitutive model of Bodner – Partom with isotropic hardening, Dem. Math. **28**, 667-688.

9. Chełmiński K., (1996) On large solutions for the quasistatic problem in nonlinear viscoelasticity with the constitutive equations of Bodner – Partom, Math. Meth. in App. Sci. **19**, 933-942.
10. Chełmiński K., (1996) Energy estimates and global in time results for a problem from nonlinear viscoelasticity, Bull. of Pol. Acad. of Sci.: Math. **44**, 465-477.
11. Chełmiński K., (1997) Stress L^∞– estimates and the uniqueness problem for the quasistatic equations to the model of Bodner–Partom, Math. Meth. in App. Sci. **20**, 1127-1134.
12. Chełmiński K., (1997) Coercive limits for a subclass of monotone constitutive equations in the theory of inelastic material behaviour of metals, Roczniki PTM: App. Math. **40**, 41–81.
13. Chełmiński K., (1998) On self-controlling models in the theory of inelastic material behaviour of metals, Contin. Mech. Thermodyn. **10**, 121–133.
14. Chełmiński K., (1998) Stress L^∞-estimates and the uniqueness problem for the equations to the model of Bodner-Partom in the two dimensional case, Math. Meth. in App. Sci. **21**, 43-58.
15. Chełmiński K., (1999) On monotone plastic constitutive equations with polynomial growth condition, Math. Methods Appl. Sci. **22**, 547–562.
16. Chełmiński K. (2001) Coercive approximation of viscoplasticity and plasticity, Asymptotic Analysis **26**, 115-135.
17. Chełmiński K. (2001) Perfect plasticity as a zero relaxation limit of plasticity with isotropic hardening, Math. Meth. in App. Sci. **24**, 117-136.
18. Chełmiński K. (2001) Global existence of weak-type solutions for models of monotone type in the theory of inelastic deformations, Fachbereich Math. TU Darmstadt, Preprint **2133**, accepted for publication in Math. Methods Appl. Sci.
19. Chełmiński K. (2002) Coercive and self-controlling quasistatic models of the gradient type with convex composite inelastic constitutive equations, Fachbereich Math. TU Darmstadt, Preprint **2164**, submitted to J. Nonlinear Science.
20. Chełmiński K. (2002) Mathematical analysis of the Armstrong - Frederick model from the theory of inelastic deformations of metals. First results and open problems, Fachbereich Math. TU Darmstadt, Preprint **2191**, submitted to Contin. Mech. Thermodyn.
21. Chełmiński K., Gwiazda P., (1999) Monotonicity of operators of viscoplastic response: application to the model of Bodner-Partom, Bull. Polish Acad. Sci.: Tech. Sci. **47**, 191–208.
22. Chełmiński K., Gwiazda P., (2000) Nonhomogeneous initial–boundary value problems for coercive and self–controlling models of monotone type, Contin. Mech. Thermodyn. **12**, 217-234.
23. Chełmiński K., Gwiazda P., (2000) On the model of Bodner-Partom with nonhomogeneous boundary data, Math. Nachrichten **214**, 5-23.
24. Chełmiński K., Naniewicz Z., (2002) Coercive limits for constitutive equations of monotone-gradient type, Nonlinear Analysis TM&A **48**, 1197-1214.
25. Clarke F.H., (1989) Optimization and nonsmooth analysis, CMR Université de Montréal, Montréal.
26. Duvaut G., Lions J.L., (1972) Les inéquations en méchanique et en physique, Dunod, Paris.
27. Evans L. C., (1990) Weak convergence methods for partial differential equations, Conference Board of Math. Sci. **74** AMS, Providence

28. Han W., Reddy B. D., (1999) Plasticity: mathematical theory and numerical analysis, Interdisciplinary Applied Mathematics, vol. 9, Springer, Berlin Heidelberg New York.
29. Ionescu I. R., Sofonea M., (1993) Functional and numerical methods in viscoplasticity, Oxford Science Publications.
30. Johnson C., (1976) Existence theorems for plasticity problems, J. Math. Pures Appl. **55**, 431–444.
31. Johnson C., (1978) On plasticity with hardening, J. Math. Anal. Appl. **62**, 325–336.
32. Krasnosielskii M, A., Rutickii Y. B., (1961) Convex functions and Orlicz spaces, P. Noordhoff Ltd.-Groningen
33. Nečas J., Hlaváček I., Mathematical theory of elastic and elastico–plastic bodies: An introduction. Elsevier Amsterdam, Oxford, New York.
34. Nouri A., Rascle M. (1995) A global existence and uniqueness theorem for a model problem in dynamic elasto–plasticity with isotropic strain–hardening. SIAM J. Math. Anal. **26**, 850–868.
35. Pompe W. (2002) Thesis, Fachbereich Mathematik TU Darmstadt, in preparation.
36. M. Rascle M. (1996) Global existence of L^2–solutions in dynamical elasto–plasticity. Mat. Contemp. **11**, 121–134.
37. Suquet P.-M., (1980) Evolution problems for a class of dissipative materials, Quart. Appl. Math. **38**, 391–414.
38. Suquet P.-M., (1988) Discontinuities and plasticity, Nonsmooth mechanics and applications (J.J. Moreau and P.D. Panagiotopoulos, eds.), CISM Courses Lect., vol. 302, Springer, Wien, New York, 279–340.
39. Le Tallec P., (1990) Numerical analysis of viscoelastic problems, Masson Paris, Springer Berlin, Heidelberg, New York.
40. Temam R., (1983) Problèmes mathématiques en plasticité, Gauthier–Villars, Paris.
41. Temam R., (1986) A generalized Norton-Hoff model and the Prandtl-Reuss law of plasticity, Arch. Rational Mech. Anal. **95**, 137–183.

Some Results Concerning the Mathematical Treatment of Finite Plasticity.

Patrizio Neff

Department of Mathematics, Darmstadt University of Technology
Schlossgartenstr. 7, D–64289 Darmstadt, Germany
neff@mathematik.tu-darmstadt.de

received 24 Jun 2002 — accepted 21 Oct 2002

Abstract. The initial-boundary value problems arising in the context of finite elasto-plasticity models relying on the multiplicative split $F = F_e\, F_p$ are investigated. First, we present such a model based on the elastic Eshelby tensor. We highlight the behaviour of the system at frozen plastic flow. It is shown how the direct methods of variations can be applied to the resulting boundary value problem. Next the coupling with a viscoplastic flow rule is discussed. With stringent elastic stability assumptions and with a nonlocal extension in space local existence in time can be proved.

Subsequently, a new model is introduced suitable for small elastic strains. A key feature of the model is the introduction of an independent field of elastic rotations R_e. An evolution equation for R_e is presented which relates R_e to F_e. The equilibrium equations at frozen plastic flow are now linear elliptic leading to a local existence and uniqueness result without further stability assumptions or other modifications. An extended Korn's first inequality is used taking the plastic incompatibility of F_p into account.

1 Notation

Let $\Omega \subset \mathbb{R}^3$ be a bounded domain with smooth Lipschitz boundary $\partial\Omega$ and let Γ be a smooth subset of $\partial\Omega$ with non-vanishing 2-dimensional Hausdorff measure. For $a, b \in \mathbb{R}^3$ we let $\langle a, b\rangle_{\mathbb{R}^3}$ denote the scalar product on $\mathbb{R}^3$ with associated vector norm $||a||^2_{\mathbb{R}^3} = \langle a, a\rangle_{\mathbb{R}^3}$. We denote by $\mathbb{M}^{3\times 3}$ the set of real 3×3 tensors. The standard Euclidean scalar product on $\mathbb{M}^{3\times 3}$ is given by $\langle A, B\rangle_{\mathbb{M}^{3\times 3}} = \mathrm{tr}\left[AB^T\right]$ and thus the Frobenius tensor norm is $||A||^2 = \langle A, A\rangle_{\mathbb{M}^{3\times 3}}$. In the following we omit the index $\mathbb{R}^3, \mathbb{M}^{3\times 3}$. The identity tensor on $\mathbb{M}^{3\times 3}$ will be denoted by $1\!\!1$, so that $\mathrm{tr}\,[A] = \langle A, 1\!\!1\rangle$. We let $\mathbb{S}ym$ and $P\mathbb{S}ym$ denote the symmetric and positive definite symmetric tensors, respectively. We adopt the usual abbreviations of Lie-Group theory, i.e., $\mathrm{GL}(3,\mathbb{R}) := \{X \in \mathbb{M}^{3\times 3}\ |\det[X] \neq 0\}$ the general linear group, $\mathrm{SL}(3,\mathbb{R}) := \{X \in \mathrm{GL}(3,\mathbb{R})\ |\det[X] = 1\}$, $\mathrm{O}(3) := \{X \in \mathrm{GL}(3,\mathbb{R})\ |\ X^TX = 1\!\!1\}$, $\mathrm{SO}(3,\mathbb{R}) := \{X \in \mathrm{GL}(3,\mathbb{R})\ |X^TX = 1\!\!1,\ \det[X] = 1\}$ with corresponding Lie-Algebras $\mathfrak{so}(3) := \{X \in \mathbb{M}^{3\times 3}\ |X^T = -X\}$ of skew symmetric tensors and $\mathfrak{sl}(3) := \{X \in \mathbb{M}^{3\times 3}\ |\mathrm{tr}\,[X] = 0\}$ of traceless tensors.

With Adj A we denote the tensor of transposed cofactors Cof(A) such that Adj $A = \det[A]\, A^{-1} = \mathrm{Cof}(A)^T$ if $A \in \mathrm{GL}(3,\mathbb{R})$. We set $\mathrm{sym}(A) = \frac{1}{2}(A^T + A)$ and $\mathrm{skew}(A) = \frac{1}{2}(A - A^T)$ such that $A = \mathrm{sym}(A) + \mathrm{skew}(A)$. For $X \in \mathbb{M}^{3\times 3}$ we set $\operatorname{dev} X = X - \frac{1}{3}\operatorname{tr}[X]\, 1\!\!1 \in \mathfrak{sl}(3)$ and for vectors $\xi, \eta \in \mathbb{R}^n$ we have $(\xi \otimes \eta)_{ij} = \xi_i\, \eta_j$. We write the polar decomposition in the form $F = R\, U = \mathrm{polar}(F)\, U$. In general we work in the context of nonlinear, finite elasticity. For the total deformation $\varphi \in C^1(\overline{\Omega}, \mathbb{R}^3)$ we have the deformation gradient $F = \nabla\varphi \in C(\overline{\Omega}, \mathbb{M}^{3\times 3})$. Furthermore $S_1(F)$ and $S_2(F)$ denote the first and second Piola Kirchhoff stress tensors, respectively. Total time derivatives are written $\frac{d}{dt}A(t) = \dot{A}$. The first and second differential of a scalar valued function $W(F)$ are written $D_F W(F).H$ and $D_F^2 W(F).(H,H)$ respectively. $\partial\chi$ is the (possibly set valued) subdifferential of the scalar valued function χ. We set $C = F^T F$, $C_p = F_p^T F_p$, $C_e = F_e^T F_e$, $E = \frac{1}{2}(C - 1\!\!1)$, $E_p = \frac{1}{2}(C_p - 1\!\!1)$, $E_e = \frac{1}{2}(C_e - 1\!\!1)$. We employ the standard notation of Sobolev spaces, i.e. $L^2(\Omega), H^{1,2}(\Omega), H_\circ^{1,2}(\Omega)$ which we use indifferently for scalar-valued functions as well as for vector-valued and tensor-valued functions. Moreover we set $\|A\|_\infty = \sup_{x\in\Omega} \|A(x)\|$. For $A \in C^1(\overline{\Omega}, \mathbb{M}^{3\times 3})$ we define $\operatorname{Rot} A(x) = \operatorname{Curl} A(x)$ as the operation curl applied row wise. We define $H_\circ^{1,2}(\Omega,\Gamma) := \{\phi \in H^{1,2}(\Omega) \mid \phi_{|_\Gamma} = 0\}$, where $\phi_{|_\Gamma} = 0$ is to be understood in the sense of traces and by $C_0^\infty(\Omega)$ we denote infinitely differentiable functions with compact support in Ω. We use capital letters to denote possibly large positive constants, e.g.,C^+, K and lower case letters to denote possibly small positive constants, e.g., c^+, d^+. The smallest eigenvalue of a positive definite symmetric tensor P is abbreviated by $\lambda_{\min}(P)$. Finally, w.r.t. abbreviates "with respect to".

2 Introduction

In the nonlinear theory of elasto-visco-plasticity at large deformation gradients it is often assumed that the deformation gradient $F = \nabla\varphi$ splits multiplicatively into an elastic and plastic part Lee [26], Mandel [30]

$$\nabla\varphi(x) = F(x) = F_e(x)\, F_p(x), \quad F_e, F_p \in \mathrm{GL}^+(3,\mathbb{R})\,, \tag{1}$$

where F_e, F_p are explicitly understood to be incompatible configurations, i.e. $F_e, F_p \neq \nabla\Psi$ for any $\Psi : \Omega \subset \mathbb{R}^3 \mapsto \mathbb{R}^3$. Thus F_p introduces in a natural way a **non-Riemannian manifold** structure. In our context we assume that this decomposition is uniquely defined only up to a **global** rigid rotation, since for arbitrary $\overline{Q} \in \mathrm{SO}(3)$ we have

$$\nabla\varphi(x) = F(x) = F_e(x)\, F_p(x) = F_e(x)\, \overline{Q}\, \overline{Q}^T\, F_p(x) = \tilde{F}_e(x)\, \tilde{F}_p(x)\,, \tag{2}$$

implying invariance under $F_p \mapsto \overline{Q}\, F_p$, $\forall\, \overline{Q} \in \mathrm{SO}(3)$. This multiplicative split, which has gained more or less permanent status in the literature, is micromechanically motivated by the kinematics of single crystals where dislocations move along fixed slip systems through the crystal lattice. The source for the incompatibility are those dislocations which did not completely traverse the crystal and consequently give rise to an inhomogeneous plastic deformation. Therefore, in the case of single crystal plasticity it is reasonable to introduce the deviation of the plastic intermediate configuration F_p from compatibility as a kind of plastic **dislocation density**. This deviation should be related somehow to the quantity $\operatorname{Curl} F_p$ and indeed in NEFF [43] we see the important role played by $\operatorname{Curl} F_p$ in the existence theory related to models in this area.

The constitutive assumption (1) is incorporated into balance equations governing the elastic response of the material and supplemented by flow rules in the form of ordinary differential equations determining the evolution of the plastic part.

We refer the reader to BLOOM [6], KONDO [22], KRÖNER [23,24] and MAUGIN [31], STEINMANN [54], CERMELLI and GURTIN [9] for more details on the subject of dislocations and incompatibilities and to ORTIZ,REPETTO and STAINIER [47] and ORTIZ and STAINIER [48] for an account of the occurrence of the microstructure related to dislocations. A summary presentation of the theory for single crystals can be found in GURTIN [18]. For applications of the general theory of polycrystalline materials in the engineering field we refer to the non exhaustive list DAFALIAS [12,13], MIEHE [35] and SIMO and HUGHES [51], SIMO [52], SIMO and ORTIZ [53]. An introduction to the theory of materials and inelastic deformations can be found in HAUPT [20], BESSELING and GIESSEN [5] and LEMAITRE and CHABOCHE [27]. Abstract mathematical treatments concerning the modelling of elasto-plasticity may be found in SILHAVY [50] and LUCCHESI and PODIO–GUIDUGLI [29].

3 The General Finite Elasto-Plastic Model

To begin with let us first introduce the considered finite compressible 3D-model. In most applications inertia effects can be safely neglected; one confines attention to the so called quasistatic case. Moreover, we restrict our considerations to the adiabatic problem without hardening. In general, hardening laws can be incorporated and will not affect the subsequent mathematical results. For simplicity the exposition is based on the phenomenological approach for isotropic polycrystals with associated flow rule, but the single crystal case as well as non-associated flow rules and general anisotropies can also be treated in the same spirit. We have opted to present a theory with elastic domain and yield function, but unified constitutive models cf. BODNER and PARTOM [7], SANSOUR and KOLLMANN [49] can also be considered. The inclusion of dead load body forces is standard and for brevity omitted.

In the quasi-static setting without body forces we are therefore led to study the following system of coupled partial differential and evolution equations for the deformation $\varphi : [0,T] \times \overline{\Omega} \mapsto \mathbb{R}^3$ and the plastic variable $F_p : [0,T] \times \overline{\Omega} \mapsto \mathrm{GL}^+(3,\mathbb{R})$:

$$\begin{aligned}
&\int_\Omega W(F_e) \det[F_p]\, dx \mapsto \min . \quad \text{w.r.t. } \varphi \text{ at given } F_p\,,\\
&0 = \mathrm{Div}\ D_F\left[W(F_e)\det[F_p]\right] = \mathrm{Div}\left[S_1(F_e)\det[F_p]\right],\\
&W(F_e) = \frac{\mu}{2}\,\|F_e\|^2 + \frac{\lambda}{4}\det[F_e]^2 - \frac{2\mu+\lambda}{2}\ln\det[F_e]\,,\\
&F_e = \nabla\varphi\, F_p^{-1},\ \Sigma_E = F_e^T D_{F_e}W(F_e)\det[F_p] - W(F_e)\det[F_p]\,\mathbb{1}\,,\\
&\frac{d}{dt}\left[F_p^{-1}\right](t) \in -F_p^{-1}(t)\, f(\Sigma_E)\,,\\
&\varphi_{|\Gamma}(t,x) = g(t,x) \quad x\in\Gamma, \quad F_p^{-1}(0) = F_{p_0}^{-1}, \quad F_{p_0} \in \mathrm{GL}^+(3,\mathbb{R})\,,
\end{aligned} \tag{3}$$

with the constitutive monotone multifunction $f : \mathbb{M}^{3\times3} \mapsto \mathbb{M}^{3\times3}$, that governs the plastic evolution and which is motivated by the principle of maximal dissipation relevant for the thermodynamical consistency of the model. Subsequently, f will be obtained as $f = \partial\chi$, with a nonlinear flow potential $\chi : \mathbb{M}^{3\times3} \mapsto \mathbb{R}$ (associated plasticity). $W(F_e)$ is the underlying elastic free energy which is already specified to be of Neo-Hooke type and $\mu, \lambda > 0$ are the Lamé constants of the material. Here Σ_E denotes the elastic Eshelby energy momentum tensor which may be reduced to $\Sigma_M = F_e^T D_{F_e} W(F_e)$, the elastic Mandel stress tensor in case of a deviatoric flow rule according to isochoric plasticity. F_{p_0} is the initial condition for the plastic variable. The inclusion sign $\in$ indicates that rate-independent, ideal plasticity is covered in this formulation.

The peculiar form of the elastic ansatz

$$\int_\Omega W(\nabla\varphi\, F_p^{-1}) \det[F_p]\, dx \mapsto \min . \quad \text{w.r.t. } \varphi \text{ at given } F_p\,, \tag{4}$$

is motivated by the following observation. If $F_p(x) = \nabla\Psi_p(x)$ is compatible and Ψ_p is a diffeomorphism, then the multiplicative decomposition (1) turns into

$$\begin{aligned}
\nabla\varphi(x) &= \nabla_x\left[\Psi_e(\Psi_p(x))\right] = \nabla_\xi\Psi_e(\Psi_p(x))\,\nabla_x\Psi_p(x) = F_e(x)\,F_p(x)\,,\\
F_e &= \nabla\varphi(x)\,F_p^{-1} = \nabla_\xi\Psi_e(\Psi_p(x))
\end{aligned} \tag{5}$$

and (4) is nothing but the change of variables formula

$$\begin{aligned}
\int_\Omega W(\nabla\varphi\, F_p^{-1})\det[F_p]\, dx &= \int_\Omega W(F_e)\det[\nabla\Psi_p]\, dx\\
&= \int_\Omega W(\nabla_\xi\Psi_e(\Psi_p(x)))\det[\nabla\Psi_p]\, dx = \int_{\xi\in\Psi_p(\Omega)} W(\nabla_\xi\Psi_e(\xi))\, d\xi .
\end{aligned} \tag{6}$$

If $DW(1\!\!1) = 0$, as is usually the case, we see that $\Psi_e(\xi) = \xi$ induces a globally stress free compatible new (intermediate) reference configuration $\Psi_p(\Omega)$ and the invariance requirement (2) preserves the compatibility of Ψ_p. Hence, locally, F_p induces a change of coordinates to a stress free reference configuration.

4 Infinitesimal Model - Linearized Kinematics

If we identify $F = 1\!\!1 + \nabla u$ and $F_p = 1\!\!1 + p$, with both displacement gradient ∇u and plastic variable p infinitesimally small, then the finite model (3) may be approximated by the reduced, partially linearized system

$$\begin{aligned}
&\int_\Omega \tfrac{1}{2}\langle \mathcal{D}.\varepsilon_e, \varepsilon_e\rangle \left[1 + \operatorname{tr}[\varepsilon_p]\,\right] dx \mapsto \min . \quad \text{w.r.t. } u \text{ at given } \varepsilon_p\,, \\
&0 = \operatorname{Div}\, D_\varepsilon\left[\tfrac{1}{2}\langle \mathcal{D}.\varepsilon_e, \varepsilon_e\rangle \left[1 + \operatorname{tr}[\varepsilon_p]\,\right]\right] = \operatorname{Div}\left[T\left[1 + \operatorname{tr}[\varepsilon_p]\,\right]\right], \\
&\Psi(\varepsilon_e) = \tfrac{1}{2}\langle \mathcal{D}.\varepsilon_e, \varepsilon_e\rangle = \mu\, \|\varepsilon_e\|^2 + \frac{\lambda}{2}\operatorname{tr}[\varepsilon_e]^2, \quad T = \mathcal{D}.\varepsilon_e = \frac{\partial \Psi(\varepsilon_e)}{\partial \varepsilon}\,, \\
&\varepsilon_e = \varepsilon - \varepsilon_p, \quad \varepsilon(\nabla u(x)) = \tfrac{1}{2}(\nabla u^T + \nabla u), \quad \varepsilon_p = \tfrac{1}{2}(p^T + p)\,, \qquad (7)\\
&\Psi^{thermo}(\varepsilon, \varepsilon_p) = \Psi(\varepsilon_e)\left[1 + \operatorname{tr}[\varepsilon_p]\,\right], \\
&\dot{\varepsilon}_p(t) \in f(T_E)\,, \\
&T_E = -\partial_{\varepsilon_p}\left[\Psi^{thermo}(\varepsilon, \varepsilon_p)\right] = T\left[1 + \operatorname{tr}[\varepsilon_p]\,\right] - \tfrac{1}{2}\langle \mathcal{D}.\varepsilon_e, \varepsilon_e\rangle\, 1\!\!1\,, \\
&u_{|_\Gamma}(t, x) = \tilde{g}(t, x) \quad x \in \Gamma, \quad \varepsilon_p(0) \in \mathbb{S}ym(3)\,,
\end{aligned}$$

where $\mathcal{D}$ is the 4th. order elasticity tensor, T is the Cauchy stress tensor and the multiplicative decomposition (1) has been replaced by the additive decomposition of the infinitesimal strains into elastic and plastic parts

$$\varepsilon = \varepsilon_e + \varepsilon_p\,. \qquad (8)$$

Here, Ψ^{thermo} acts as thermodynamic potential for the plastic flow. The new reduced system (7) remains intrinsically thermodynamically correct. There is a rich mathematical literature successfully treating models based on (8) with $\operatorname{tr}[\varepsilon_p] = 0$, i.e., $\operatorname{tr}[f(T_E)] = 0$ or $T_E = T$, in which case $\Psi^{thermo} = \Psi(\varepsilon_e)$ and the model is of pre-monotone type in the sense of Alber. See, e.g., ALBER [1], HAN and REDDY [19], IONESCU and SOFONEA [21], CHELMINSKI [10,11] and references therein.

A general mathematical treatment either of the reduced system (7) or of finite plasticity is, however, largely wanting. In the following we want to contribute some partial results in respect of finite plasticity.

Remark 4.1 (Linearization) *The reduced system* (7) *is not the exact formal linearization of* (3). *However, the performed reduction yields a system* *of equations which is, where different from the formal linearization, correct of higher order and remains intrinsically thermodynamically admissible. In addition it retains the Eshelbian like structure.*

5 Thermodynamically Consistent Plastic Flow Rules

In this part of the paper we would like to indicate how to obtain the **canonical** flow rules of finite multiplicative elasto-plasticity. In our context we use the term canonical in the sense of MIEHE [34] meaning that fundamental dissipation principles together with tools from convex analysis are invoked to get an overall framework for multiplicative elasto-plasticity. In a more abstract setting this fits into the framework of a Thermodynamics with Internal Variables (TIV) as in MAUGIN [32]. The development is in principle well known but the use of the Eshelby tensor has only surfaced recently. We will see how the ansatz (4) naturally leads to the use of the Eshelby tensor. We include therefore the following for the presentation to be sufficiently self-contained.

A word of caution may be in order. Contrary to some other papers concerned with multiplicative elasto-plasticity we use as independent set of variables F, F_p^{-1} leading to left-rate flow rules of the form $F_p \frac{d}{dt}\left[F_p^{-1}\right] =: L_{F_p^{-1}}$. The traditional approach would use as independent variables F, F_p (C, C_p) leading to flow rules of the form $\frac{d}{dt}F_p\, F_p^{-1} =: L_{F_p}$. However, there is no mathematical reason to prefer one representation over the other. The main point is that both types lead to similar mathematical structures. We employ throughout a material description, any quantity being defined with respect to the reference configuration; thus avoiding any discussion on consistent stress rates.

Let $W = W(F_e) = W(F\,F_p^{-1})$ be the given hyperelastic energy. The first Piola-Kirchhoff stress tensor S_1 is then $S_1(F, F_p^{-1}) = D_F W(FF_p^{-1}) = DW(FF_p^{-1})F_p^{-T}$. Using the objective Lie derivative and the principle of maximal dissipation one arrives at the canonical flow rule

$$-F_p \frac{d}{dt}\left[F_p^{-1}\right] \in \partial\chi(\Sigma_E)\,, \tag{9}$$

where $\partial\chi$ is the set valued subdifferential of the indicator function χ of a convex set $\mathcal{E}$ in the stress space related to Σ_E. Thus for rate-independent ideal plasticity

$$\chi(\Sigma_E) = \begin{cases} 0 & \Sigma_E \in \mathcal{E} \\ \infty & \Sigma_E \notin \mathcal{E}\,. \end{cases} \tag{10}$$

This flow rule can accommodate the assumption of isochoric plasticity, i.e., $\det[F_p] = 1$ (which replaces $\operatorname{tr}[\varepsilon_p] = 0$ in (8)) by defining the convex set $\mathcal{E}$ to be

$$\mathcal{E} := \{\Sigma_E \in \mathbb{M}^{3\times 3} \mid \ \|\operatorname{dev}\operatorname{sym}(\Sigma_E)\| \le \sigma_{\mathrm{y}}\}\,, \tag{11}$$

where σ_y is the yield limit. Moreover with $0 = \frac{d}{dt}\left[F_p F_p^{-1}\right] = \dot{F}_p F_p^{-1} + F_p \frac{d}{dt}\left[F_p^{-1}\right]$ and

$$\begin{aligned}\dot{F}F^{-1} &= \frac{d}{dt}\left[F_e F_p\right](F_e F_p)^{-1} = \dot{F}_e F_e^{-1} + F_e \dot{F}_p F_p^{-1} F_e^{-1} \\ &= \dot{F}_e F_e^{-1} + F_e \left[-F_p \frac{d}{dt}\left[F_p^{-1}\right]\right] F_e^{-1}\end{aligned} \tag{12}$$

the spatial velocity gradient $\dot{F}F^{-1}$ may be additively decomposed into elastic and plastic parts as usual.

This is a straightforward generalization of the classical von Mises type J_2-plasticity for infinitesimal strains to finite strains. The macroscopic yield limit σ_y corresponds conceptually to the microscopic activation level of dislocation glide. Observe that the choice $\mathrm{sym}(\Sigma_E)$ instead of Σ_E sets the so called plastic material spin DAFALIAS [14] to zero. But for isotropic W the elastic Eshelby tensor is already symmetric, which has been noted previously MAUGIN and EPSTEIN [33]. A shortcut way to see that this flow rule is thermodynamically admissible proceeds as follows: Let the deformation gradient F be constant in time and consider

$$\begin{aligned}&\frac{d}{dt}\left[W(FF_p^{-1}(t))\,\det[F_p]\right] = \\ &= \langle DW(FF_p^{-1}(t)), F\frac{d}{dt}\left[F_p^{-1}\right]\rangle \det[F_p] + W(F_e)\,\langle \mathrm{Adj}\, F_p{}^T, \frac{d}{dt}\left[F_p\right]\rangle \\ &= \det[F_p]\left[\langle DW(FF_p^{-1}(t)), FF_p^{-1}F_p\frac{d}{dt}\left[F_p^{-1}\right]\rangle + W(F_e)\,\langle F_p^{-T}, \frac{d}{dt}\left[F_p\right]\rangle\right] \\ &= \det[F_p]\left[\langle F_p^{-T}F^T DW(FF_p^{-1}(t)), F_p\frac{d}{dt}\left[F_p^{-1}\right]\rangle + W(F_e)\,\langle \mathbb{1}, \frac{d}{dt}[F_p]\,F_p^{-1}\rangle\right] \\ &= \det[F_p]\left[\langle \underbrace{F_e^T DW(F_e(t))}_{\Sigma_M}, F_p\frac{d}{dt}\left[F_p^{-1}\right]\rangle - W(F_e)\,\langle \mathbb{1}, F_p\frac{d}{dt}\left[F_p^{-1}\right]\rangle\right] \\ &= \langle \underbrace{\det[F_p]\left(F_e^T DW(F_e(t)) - W(F_e)\,\mathbb{1}\right)}_{\Sigma_E}, F_p\frac{d}{dt}\left[F_p^{-1}\right]\rangle\,.\end{aligned} \tag{13}$$

In the absence of thermal effects, classical continuum mechanics may be based on a second law in the form

$$\forall V \subset \mathbb{R}^3: \quad \frac{d}{dt}\int_V W\,dx \le \int_{\partial V} \langle S_1.n, \dot{\varphi}\rangle\,dS + \int_V \langle f, \dot{\varphi}\rangle\,dx\,, \tag{14}$$

where W is the free energy, n is the unit outward normal to the control volume V and f are the body forces GURTIN [17, p.41]. A sufficient condition for (14) to hold is the reduced dissipation inequality which is fulfilled whenever

$$\frac{d}{dt}\left[W(F\,F_p^{-1}(t))\,\det[F_p]\right] \le 0\,, \tag{15}$$

for arbitrary F fixed in time. Thus, when choosing

$$-F_p \frac{d}{dt}\left[F_p^{-1}\right] = \lambda^+ \, \partial\chi(\Sigma_E)\,, \tag{16}$$

with χ convex, the reduced dissipation inequality (15) is guaranteed, since from convex analysis $\langle \Sigma_E, \partial\chi(\Sigma_E)\rangle \geq 0$. Moreover, if

$$\partial\chi(\Sigma_E) = \lambda^+ \ \mathrm{dev}(\mathrm{sym}(\Sigma_E))\,, \tag{17}$$

the right-hand side is traceless implying that $\det[F_p^{-1}] = 1 = \det[F_p]$.

Observe that the choice of the elastic Eshelby tensor Σ_E as the relevant stress measure can be conveniently related to the local configurational driving forces on the inherent inhomogeneities introduced by the local change of reference through the plastic variable F_p. Moreover, this flow rule is in a natural way invariant under the change of plastic variable $F_p^{-1} \mapsto F_p$, as indeed it should be inconsequential which form of independent variable we take.

In order to formulate a viscoplastic regularization of the above evolution equation (9) the traditional approach proceeds as follows: instead of χ take the following function χ_η:

$$\chi_\eta(\Sigma_E) = \begin{cases} 0 & \Sigma_E \in \mathcal{E} \\ \frac{1}{2\eta}\|\Sigma_E - P_{\mathcal{E}}.(\Sigma_E)\|^2 & \Sigma_E \notin \mathcal{E}. \end{cases} \tag{18}$$

Here $P_{\mathcal{E}}$ denotes the orthogonal projection onto the convex set $\mathcal{E}$ which is uniquely defined. Obviously, in the limit $\eta \to 0$ we recover the rate-independent evolution equation, at least formally. As it stands, $\partial\chi_\eta$ is just the well known Yosida approximation (linear viscosity) of the subdifferential $\partial\chi$ and it holds true HAN and REDDY[19, p.184] that

$$\forall\, H \in \mathbb{M}^{3\times 3} : \ \langle \partial\chi_\eta(\Sigma_E), H\rangle = \frac{1}{\eta}\langle \Sigma_E - P_{\mathcal{E}}(\Sigma_E), H\rangle \tag{19}$$

and, moreover, $\partial\chi_\eta$ is a monotone function due to the convexity of χ_η. Unfortunately, the ordinary differential evolution equation $-F_p \frac{d}{dt}\left[F_p^{-1}\right] = \partial\chi_\eta(\Sigma_E)$ does not possess the advantageous monotonicity properties with respect to the plastic variable F_p, in marked contrast to the properties of the infinitesimal flow rule according to (7) and $\mathrm{tr}\,[\varepsilon_p] = 0$. Reformulating the last equation we have

$$\frac{d}{dt}\left[F_p^{-1}\right] = -F_p^{-1}\,\partial\chi_\eta(\Sigma_E) = -\frac{1}{\eta}F_p^{-1}\,\left(\Sigma_E - P_{\mathcal{E}}.(\Sigma_E)\right)\,. \tag{20}$$

In particular, we see that $\partial\chi_\eta$ being Lipschitz continuous in Σ_E does not entail that the right-hand side is Lipschitz continuous altogether with respect to F_p^{-1} since F_p^{-1} enters again multiplicatively. However, this is one of the main features which made the Yosida approximation so valuable in infinitesimal plasticity. For multiplicative plasticity it is typically the case that either the right-hand side is not Lipschitz continuous (rate independent case), or the right-hand side is neither monotone nor possesses potential structure.

6 Polyconvexity Conditions in Finite Plasticity

In order to investigate the boundary value problem which arises in the formulation of (3) if one freezes the plastic variable F_p it is convenient to place this in the context of the direct methods of variations.

In the purely elastic case it is usually a convexity condition, the polyconvexity condition in the sense of BALL [3], that is used together with some coerciveness condition to ensure that a minimization problem has at least one solution. Let us recall this notion. We say that

Definition 6.1 (Polyconvexity) *The free energy density $W(x, F)$ is polyconvex whenever there exists a (possibly non-unique) function $P(x, X, Y, Z)$: $\mathbb{R}^3 \times \mathbb{M}^{3\times3} \times \mathbb{M}^{3\times3} \times \mathbb{R}^+ \mapsto \mathbb{R}$ such that $P(x, , ,)$ is convex for each $x \in \mathbb{R}^3$ and*

$$W(x, F) = P(x, F, \operatorname{Adj} F, \det[F]) \, .$$

Example 6.2 *The Neo-Hooke energy density*

$$W(F) = \frac{\mu}{p\,(\sqrt{3})^{p-2}}\,||F||^p + \frac{\lambda}{4}\,\det[F]^2 - \frac{2\mu + \lambda}{2}\,\ln\det[F]$$

is polyconvex for $p \geq 1$.

Corollary 6.3 (Polyconvexity and Ellipticity) *It is well known that every smooth strictly polyconvex free energy density $W(x, F)$ is automatically ensuring overall Legendre-Hadamard ellipticity of the corresponding boundary value problem in the sense that*

$$\forall F \in \mathrm{GL}^+(3, \mathbb{R}) : \forall\, \xi, \eta \in \mathbb{R}^3 : \; D_F^2 W(x, F).(\xi \otimes \eta, \xi \otimes \eta) \geq c^+ ||\xi||^2\, ||\eta||^2 .$$

Example 6.4 *Free energies defined on the Hencky strain tensor $\ln C$ such as*

$$W(F) = \frac{\mu}{4}||\operatorname{dev}\ln C||^2 + \frac{3\lambda + 2\mu}{3}\,\operatorname{tr}\,[\ln C]^2$$

are in general not elliptic and therefore not polyconvex, see NEFF [38]. However, these energies are very popular among engineers due to certain advantages in a numerical implementation. Quadratic expressions in E such as the St. Venant-Kirchhoff density

$$W(F) = \mu||E||^2 + \frac{\lambda}{2}\,\operatorname{tr}\,[E]^2$$

are neither elliptic nor polyconvex, they loose ellipticity at finite elastic compression. Energy densities which are convex functions of generalized strain measures LUBARDA*[28, p. 33] are equally non elliptic.*

Lemma 6.5 (Polyconvexity and Multiplicative Decomposition) *Let $W(F)$ be polyconvex and assume that $F_p \in L^\infty(\Omega, \mathrm{GL}^+(3,\mathbb{R}))$ is given. Then the function*

$$\tilde{W}(x,F) := W(F\,F_p^{-1}(x))\,\det[F_p(x)]$$

is itself polyconvex.

Proof. This is accomplished by a direct check of the polyconvexity condition. Since W is polyconvex, we know that there is some function P such that $W(F) = P(F, \mathrm{Adj}\,F, \det[F])$ with P convex. This yields

$$\begin{aligned}\tilde{W}(x,F) &= W(F\,F_p^{-1}(x))\,\det[F_p(x)]\\ &= P(F\,F_p^{-1}(x), \mathrm{Adj}\,F\,F_p^{-1}(x), \det[F\,F_p^{-1}(x)])\,\det[F_p(x)]\\ &= P(F\,F_p^{-1}(x), \mathrm{Adj}\,F_p^{-1}(x)\,\mathrm{Adj}\,F, \det[F]\det[F_p^{-1}(x)])\,\det[F_p(x)].\end{aligned}$$

Now define

$$\tilde{P}(x,X,Y,Z) := P(X\,F_p^{-1}(x), \mathrm{Adj}\,F_p^{-1}(x)\,Y, Z\det[F_p^{-1}(x)])\,\det[F_p(x)]\,.$$

It is easy to see that $\tilde{P}(x,\cdot,\cdot,\cdot)$ is a convex function since P is. The essence is that F_p introduces merely local inhomogeneity into the formulation. See also NEFF[38].

Definition 6.6 (Coercivity) *We say that W leads to a p-coercive problem whenever*

$$\int_\Omega W(\nabla\varphi)\,dx \le K_1 \Rightarrow \|\varphi\|_{1,p,\Omega} \le K_2\,.$$

Thus p-coercivity implies that a finite elastic energy level necessitates a finite value of the $W^{1,p}(\Omega)$-norm of the deformation φ.

Example 6.7 *It is easily seen that the Neo-Hooke energy density is p-coercive for $p \ge 1$ if $\lambda > 0$. In this case the term $\frac{\lambda}{4}\det[F]^2 - \frac{2\mu+\lambda}{2}\ln\det[F]$ is pointwise bounded from above. An application of Poincaré's inequality completes the argument if Dirichlet boundary conditions are prescribed. However, energies defined on the Hencky strain tensor $\ln C$ such as*

$$W(F) = \frac{\mu}{4}\|\mathrm{dev}\ln C\|^2 + \frac{3\lambda+2\mu}{3}\,\mathrm{tr}\,[\ln C]^2$$

are, typically, not coercive for any p, perhaps indicating a more serious deficiency. See NEFF *[38, p.185].*

7 On the Choice of the Elastic Free Energy

Freezing F_p is typically involved in computations of elasto-plasticity where this is called the elastic trial step. It seems to be a reasonable requirement in finite plasticity that the elastic trial step at frozen plastic variable F_p should lead to a well posed elastic minimization problem as long as F_p is invertible and sufficiently smooth. Whether this is indeed the case depends entirely on the chosen elastic free energy.

The only general method in finite elasticity to ascertain well posedness is based on the direct methods of the calculus of variations. The successful application relies on polyconvexity and coercivity. It is expedient to impose these conditions a priori to guarantee that the elastic trial step can be treated adequately.

In view of the invariance property expressed in Lemma 6.5 it suffices to specify some polyconvex free energy $W(F)$ and then to substitute the elastic part F_e instead of F to get a polyconvex minimization problem at frozen F_p. Thus, the multiplicative decomposition of the deformation gradient and polyconvexity are mutually compatible. It can easily be seen that p-coercivity is preserved as well under the multiplicative decomposition. Following this approach, elastic energies based on the Hencky tensor $\ln C$ or based on E cannot be used. Formulations based on an additive decomposition $E = E_e + E_p$ which have been advocated by GREEN and NAGHDI [16] and NAGHDI [36], are also excluded since ellipticity of the elastic trial step may again be lost.

8 Existence Results in the General Finite Case - the Flow Based Approach

We therefore freeze the plastic variable F_p and analyse the elastic trial step. The corresponding boundary value problem has a variational structure in the sense that the equilibrium part of (3) is formally equivalent to the minimization problem

$$\begin{aligned}
&\forall\, t \in [0,T]: \quad I(\varphi(t), F_p^{-1}(t)) \mapsto \min, \quad \varphi(t) \in g(t) + W_\circ^{1,p}\,, \\
&I(\varphi, F_p^{-1}) = \int_\Omega W(\nabla\varphi\, F_p^{-1})\, \det[F_p(x)]\, dx\,, \qquad (21)\\
&W(F_e) = \frac{\mu}{p\,(\sqrt{3})^{p-2}}\, ||F||^p + \frac{\lambda}{4}\det[F_e]^2 - \frac{2\mu+\lambda}{2}\ln\det[F_e]\,.
\end{aligned}$$

We have the following preliminary result for fixed time:

Theorem 8.1 (Existence for the static elastic trial step) *Assume that $F_p \in L^\infty(\Omega, \mathrm{GL}^+(3,\mathbb{R}))$ and $g \in W^{1,p}(\Omega,\mathbb{R}^3)$ with $p \geq 2$ is given. Then the elastic minimization problem (21) admits at least one minimizer.*

Proof. We sketch the proof and apply the direct methods of variations. The elastic free energy is of polyconvex Neo-Hooke type. By the invariance of polyconvexity under the multiplicative decomposition the elastic minimization problem is still polyconvex. The energy is also coercive over $W^{1,p}(\Omega, \mathbb{R}^3)$, cf. Example(6.7). Thus infimizing sequences $\varphi_k \in W^{1,p}(\Omega, \mathbb{R}^3)$ exist and admit weakly converging subsequences. The functional I is lower semicontinuous due to its polyconvexity. Hence the weak limit $\varphi \in W^{1,p}(\Omega, \mathbb{R}^3)$ minimizes I. For the details, see NEFF [38].

However, no statement is made as to how this solution varies if F_p is varied or how it changes if the boundary data g are varied. Due to the nonlinear nature of the problem at hand general theories of this kind cannot be expected to hold.

What would be most convenient is to assume that the solution of the minimization problem depends continuously on F_p and the boundary data, i.e., elastic stability with respect to the data at least locally. This can be achieved by assuming that the minimizer lies in a uniform potential well for all plastic variables in a certain given set. By a uniform potential well we mean

Definition 8.2 (Uniform potential well) *Assume that φ is a global minimizer of (21). Whenever there exists a nondecreasing function $\gamma^+(s) > 0$, (e.g. $\gamma^+(s) = c^+ |s|^2$), such that*

$$\forall\, 0 < s \leq s_0 : \inf_{\|h\|_{W_\circ^{1,p}}=1} \int_\Omega W(\nabla(\varphi + s\,h)\, F_p^{-1}) \det[F_p]\, dx \geq \int_\Omega W(\nabla\varphi\, F_p^{-1}) \det[F_p]\, dx + \gamma^+(s)\,,$$

we say that φ lies in a uniform potential well.

A sufficient condition for φ to lie in a uniform potential well is that the global minimizer of (21) is locally unique. This definitely does not imply that W needs to be convex since there might be other local or global minimizers or stationary points, see NEFF [38, p.173] and BALL and MARSDEN [4]. Under these circumstances it can be shown, that it is possible to define a local solution operator $\varphi = T(F_p, g)$, such that

$$\inf_{\varphi \in g + W_\circ^{1,p}} \int_\Omega W(\nabla\varphi F_p^{-1}) \det[F_p(x)]dx = \int_\Omega W(\nabla T(F_p, g) F_p^{-1}) \det[F_p(x)]dx\,, \tag{22}$$

which is Hoelder continuous if $p \geq 6$, but may in general not be Lipschitz continuous. Be that as it may, by introducing T we can dispose of the boundary value problem and concentrate on the flow rule (**the flow based approach**). In order to approach the flow problem we introduce a further modification. Instead of considering Σ_E we replace Σ_E with a space averaged $\widehat{\Sigma}_E^{\varepsilon}(x)$ where ε indicates the average over some small ε-ball $B_\varepsilon(x) := \{y \in \Omega \mid \|y - x\| \leq \varepsilon\}$ centred at $x \in \Omega$. In this fashion we introduce a nonlocal dependence into the model. This ensures at the same time that the averaged Eshelby stresses are smoothly distributed which would not be necessarily true for the non averaged quantities due to a general lack of regularity for the nonlinear elliptic problem. This type of averaging preserves frame indifference and could even be argued for on physical grounds. It is also necessary to remove the possible singularity inherent in the elastic free energy through $-\ln\det[F_e]$ if $\det[F]$ approaches zero. This can be done in a consistent manner by replacing $-\ln$ with a smooth convex function $h : \mathbb{R} \mapsto \mathbb{R}$ such that $h \geq 0$, $h'(1) = -1$.

With these assumptions and modifications it is possible to prove a local existence result for a fully viscoplastic formulation of the model. The modified nonlocal model reads then

$$
\begin{aligned}
&\int_\Omega W(F_e)\,\det[F_p]\,dx \mapsto \min. \quad \text{w.r.t. } \varphi \text{ at given } F_p\,,\\
&W(F_e) = \frac{\mu}{p\,(\sqrt{3})^{p-2}}\,\|F\|^p + \frac{\lambda}{4}\det[F_e]^2 - \frac{2\mu+\lambda}{2}\,h(\det[F_e])\,, \qquad (23)\\
&\Sigma_E = F_e^T D_{F_e} W(F_e)\det[F_p] - W(F_e)\det[F_p]1\!\!1\,,\\
&\frac{d}{dt}\left[F_p^{-1}\right](t) \in -F_p^{-1}(t)\,\partial\chi(\widehat{\Sigma}_E)\,,\\
&\varphi_{|\partial\Omega}(t,x) = g(t,x) \quad x \in \partial\Omega, \quad F_p^{-1}(0) = F_{p_0}^{-1}\,, \quad F_{p_0} \in \mathrm{GL}^+(3,\mathbb{R})\,,
\end{aligned}
$$

where the nonlinear flow potential $\chi : \mathbb{M}^{3\times 3} \mapsto \mathbb{R}$ is assumed to have a local Lipschitz subdifferential $\partial\chi$, e.g. of the form (25). It is possible to prove the following result:

Theorem 8.3 (Local existence for nonlocal model) *Let W be as above with $p = 6$ and let $g \in C^1(\mathbb{R}^+; W^{1,\infty}(\Omega,\mathbb{R}^3))$; moreover, assume that solutions of the elastic minimization problem at fixed plastic variable F_p lie in a uniform potential well. Then, there exists a time $T > 0$ such that (23) admits a (possibly non unique) local solution $F_p^{-1} \in C^1([0,T]; C(\overline{\Omega}, \mathrm{GL}^+(3,\mathbb{R}))$ and $\varphi \in C([0,T], W^{1,p}(\Omega,\mathbb{R}^3))$.*

Proof. Consider the following iterative scheme:

$$
\begin{aligned}
&\int_\Omega W(\nabla\varphi^{n+1}(x,t)\,F_p^{-n}(x,t))\,\det[F_p^n]\,dx \mapsto \min.\text{w.r.t. } \varphi^{n+1} \text{ at given } F_p^n\,,\\
&\frac{d}{dt}\left[F_p^{-1,n+1}\right](t) \in -F_p^{-1,n+1}(t)\,\partial\chi(\widehat{\det[F_p^n]\Sigma^n})\,, \qquad (24)
\end{aligned}
$$

$$\Sigma^n = F_p^{-T,n}\, \nabla\varphi^{T,n+1}\, D_{F_e} W(\nabla\varphi^{n+1} F_p^{-n}) - W(\nabla\varphi^{n+1}\, F_p^{-n})\, 1\!\!1\,,$$
$$\widehat{\Sigma}(x) := \frac{1}{|B_\varepsilon(x)|}\int_{y\in B_\varepsilon(x)} \Sigma(y)\, dy\,,$$
$$\varphi^{n+1}_{|\partial\Omega}(t,x) = g(t,x) \quad x\in\partial\Omega, \quad F_p^{-1,n+1}(0) = F_{p_0}^{-1}, \quad F_{p_0}\in \mathrm{GL}^+(3,\mathbb{R})\,.$$

The direct methods of variations (Theorem 8.1) show the existence of a minimizer $\nabla\varphi^{n+1}$. By the elastic stability assumptions we have that $\nabla\varphi^{n+1}(x,t)$ is well defined. The evolution equation at given $\nabla\varphi^{n+1}(x,t)$ has a unique local solution $F_p^{n+1}(x,t)$ due to Banach's fixed point principle (linear ordinary differential equation). This defines an operator

$$P:\ C([0,T];C(\overline{\Omega},\mathrm{GL}^+(3,\mathbb{R}))) \mapsto C([0,T];C(\overline{\Omega},\mathrm{GL}^+(3,\mathbb{R})))\,,$$
$$\frac{d}{dt}\left[P.F_p^n\right]^{-1}(t) \in -\left[P.F_p^n\right]^{-1}(t)\,\partial\chi(\widehat{\Sigma}_E^n)\,, \qquad \text{(ODE)}$$
$$P:\ F_p^n \mapsto F_p^{n+1}\,.$$

It is then possible to show that this operator is indeed compact. First, since the solutions of the boundary value problem $\nabla\varphi^{n+1}$ can be independently bounded in $L^2(\Omega)$ and the averaging procedure for fixed $\varepsilon > 0$ delivers smooth solutions F_p^{n+1}, the operator P is continuous. Gronwall's inequality applied to the ODE together with an application of the Arcela-Ascoli theorem shows that P maps bounded sets into equicontinuous sets and individual arguments are transformed to Hoelder continuous functions. Schauder's fixpoint principle for continuous, compact operators yields then the existence of at least one fixed point of P. This proves the claim, for details compare NEFF [38]. Observe in passing that F_p^{n+1} is not to be confused with a time incremental update but is a new solution over the whole time interval.

Remark 8.4 (Flow Based Vs. Time Incremental Formulation) *In what we like to call the* **flow based** *approach priority is given to the plastic flow rule, and the elastic balance equation is rather treated as a (static) side condition. This is the approach followed in this contribution. In the extreme case of rigid plastic behaviour the elastic problem is completely discarded, which is a well known strategy in the literature. Similar ideas are used on the numerical side, where the resulting discretized initial boundary value problem is interpreted as a differential algebraic equation (DAE) and the algebraic constraint corresponds to the fulfilment of the elastic balance equation.*

In the opposite case, the **time incremental formulation**, *priority is given to the elastic balance equation. The flow rule is implicitly discretized and the updated plastic variable is inserted into the balance equation. The resulting field problem in general looses ellipticity in the finite problem but may retain a variational structure in the case of an associated flow rule. Here, the flow rule acts merely as a side condition,* CARSTENSEN *and* HACKL *[8]. In the* *infinitesimal case the time incremental formulation leads to the deformation theory of Hencky plasticity retaining ellipticity.*

Remark 8.5 *The above result (Theorem 8.3) is only a very weak statement in view of the following: we had to assume local elastic stability (uniform potential well) of the minimization problem related to the elastic trial step. This can in general not be proved. In addition we needed to modify the formulation to a nonlocal one through the space averaged elastic Eshelby stresses. And due to the use of Schauder's principle we get only a local solution which might not even be locally unique. In summary, we see that the solution of the elastic trial step was easily found but that the properties of this solution are not sufficient to couple it as such with the flow problem. No attempt has been made to consider the limit behaviour* $\varepsilon \to 0$.

In order to overcome these serious technical difficulties which were entirely due to the nature of the finite elastic free energy we shall now introduce a new model which reflects more closely one aspect of the physics of the problem at least for metals. The chosen Neo-Hooke elastic free energy is in principle appropriate for arbitrarily large elastic strains. However, the prevailing deformation mode for metals is known to comply to small elastic strains in most practical cases.

9 A Model for Small Elastic Strains

In the three-dimensional case it is easily seen that small elastic strains, i.e., $||F_e^T F_e - 1\!\!1||$ pointwise small, imply that F_e is approximately a rotation $R_e \in \mathrm{SO}(3)$. If we assume that R_e is known, all quantities can be linearized with respect to $R_e = \mathrm{polar}(F_e)$. This is a nonlinear constraint. It is possible to relax this static constraint into a dynamic evolution equation such that a rotation R_e is determined which coincides approximately with $\mathrm{polar}(F_e)$ whenever F_e is approximately a rotation. The static constraint $R_e = \mathrm{polar}(F_e)$ turns out to be a global attractor of the evolution equation. These modifications significantly simplify the mathematical structure without loosing the main ingredients of finite multiplicative visco-plasticity notably frame indifference and invariance with respect to superposed rotations of the so called intermediate configuration are preserved. In addition, the model allows for finite elastic rotations, finite plastic deformations and overall finite deformations. Let us introduce the considered new 3D-model which modifies the exposition in Neff [41,37] to include in a consistent manner non-isochoric plasticity, i.e., $\det[F_p] \neq 1$.

In the quasi-static setting without body forces we are led to study the following system of coupled partial differential and evolution equations for the deformation $\varphi : [0,T] \times \overline{\Omega} \mapsto \mathbb{R}^3$, the plastic variable $F_p : [0,T] \times \overline{\Omega} \mapsto$

$\mathrm{GL}^+(3,\mathbb{R})$ and the independent elastic rotation $R_e : [0,T]\times\overline{\Omega}\mapsto \mathrm{SO}(3)$:

$$\begin{aligned}
&\int_\Omega W(F_e,R_e)\det[F_p]\,dx \mapsto \min. \quad \text{w.r.t. } \varphi \text{ at given } R_e, F_p\,,\\
&0 = \mathrm{Div}\; D_F\left[W(F_e,R_e)\det[F_p]\right] = \mathrm{Div}\left[S_1(F_e,R_e)\det[F_p]\right],\\
&W(F_e,R_e) = \frac{\mu}{4}\,\|F_e^TR_e + R_e^TF_e - 2 1\!\!1\|^2 + \frac{\lambda}{8}\,\mathrm{tr}\left[F_e^TR_e + R_e^TF_e - 2 1\!\!1\right]^2,\\
&F_e = \nabla\varphi\,F_p^{-1}\,,\\
&\Sigma_E = F_e^T D_{F_e}W(F_e,R_e)\det[F_p] - W(F_e,R_e)\det[F_p]1\!\!1\,,\\
&S_1(F_e,R_e) = R_e\left[\mu(F_e^TR_e + R_e^TF_e - 2 1\!\!1) + \lambda\,\mathrm{tr}\left[F_e^TR_e - 1\!\!1\right]1\!\!1\right]F_p^{-T}\,,\\
&\varphi_{|\Gamma}(t,x) = g(t,x) \quad x\in\Gamma\,, \qquad\qquad \text{(P3)}\\
&\frac{d}{dt}\left[F_p^{-1}\right](t) \in -F_p^{-1}(t)\,\partial\chi(\Sigma_E)\,,\\
&\frac{d}{dt}R_e(t) = \nu^+\;\mathrm{skew}\left(F_eR_e^T\right)R_e(t)\,, \quad \nu^+ = \nu^+(F_e,R_e)\in\mathbb{R}^+\,,\\
&F_p^{-1}(0) = F_{p_0}^{-1}\,, \quad F_{p_0}\in\mathrm{GL}^+(3,\mathbb{R}), \quad R_e(0) = R_e^0, \quad R_e^0\in\mathrm{SO}(3)\,.
\end{aligned}$$

The term $\nu^+ := \nu^+(F_e,R_e)$ represents a scalar valued penalty function introducing elastic viscosity. $F_{p_0}^{-1}$ and R_e^0 are the initial conditions for the plastic variable and elastic rotation part, respectively. If we identify $F_p = 1\!\!1 + p$ and set $R_e \equiv 1\!\!1$, the model settles also down to (7).

In the (vanishing elastic viscosity) limit $\nu^+ \to \infty$, the model (P3) approaches formally the problem

$$\begin{aligned}
&\int_\Omega W_\infty(U_e)\det[F_p]\,dx \mapsto \text{stationary w.r.t. } \varphi \text{ at given } F_p\,,\\
&0 = \mathrm{Div}\; D_F\left[W_\infty(F_e)\det[F_p]\right],\\
&W_\infty(U_e) = \mu\,\|U_e - 1\!\!1\|^2 + \frac{\lambda}{2}\,\mathrm{tr}\left[U_e - 1\!\!1\right]^2, \qquad\qquad \text{(Biot)}\\
&\frac{d}{dt}\left[F_p^{-1}\right](t) \in -F_p^{-1}(t)\,\partial\chi(\Sigma_{E,\infty})\,,\\
&\Sigma_{E,\infty} = U_e^T D_{U_e}W_\infty(U_e)\det[F_p] - W_\infty(U_e)\det[F_p]1\!\!1\,,
\end{aligned}$$

with $U_e = (F_e^TF_e)^{\frac{1}{2}}$ the elastic stretch and $U_e - 1\!\!1$ the elastic Biot strain tensor. The system (Biot) is an exact model for small elastic strains and finite plastic deformations. The transition from (P3) to (Biot) is not entirely trivial since it is not just the replacement of R_e by $R_e = \mathrm{polar}(F_e)$. Moreover, note the tacit change from minimization in (P3) to stationarity in (Biot). For a detailed account of the derivation and the properties of the new model (P3) we refer to NEFF [37,41]. Now in NEFF [39,42] it has been shown that the model (P3) in a viscous form is locally wellposed. Moreover, first numerical computations NEFF and WIENERS [44] confirm the general applicability of the model (P3) for structural applications compared with standard models. To grasp the main idea in the above approximation we look at the corresponding evolution equation for R_e. The following theorem can be proved.

Theorem 9.1 (Exact dynamic polar decomposition) *Let $F_e \in \mathrm{GL}^+(3,\mathbb{R})$ and assume that $R_0 \in \mathrm{SO}(3)$ is given with $\|R_0 - \mathrm{polar}(F_e)\|^2 < 8$. Then the evolution equation*

$$\frac{d}{dt} R_e(t) = \nu^+ \ \mathrm{skew}\left(F_e R_e^T\right) \ R_e(t), \quad R_e(0) = R_0 \,,$$

has a unique global in time solution $R_e(t) \in \mathrm{SO}(3)$ which converges to

$$R_\infty = \mathrm{polar}(F_e) \,.$$

Proof. See NEFF [37,41].

The guiding idea is then to relax the algebraic constraint $R_e = \mathrm{polar}(F_e)$ inherent in the formulation with U_e in (Biot) into an associated evolution equation which locally approximates this constraint.

In practice the flow rule for the (independent) elastic rotations R_e introduces merely reversible nonlinear viscoelastic behaviour restricted to multiaxial deformations well below an assumed elastic yield limit. In contrast to micropolar theories an additional field equation for R_e is avoided.

10 Existence and Uniqueness for Small Elastic Strains

Now we specify the flow potential χ and the function ν^+. For von Mises type J_2-viscoplasticity with elastic domain $\mathcal{E}$ (see (11)) and yield stress σ_{y}, we take as visco-plastic potential $\chi : \mathbb{M}^{3\times 3} \mapsto \mathbb{R}$ of generalized Norton-Hoff overstress type the following function:

$$\chi(\Sigma_E) = \begin{cases} 0 & \Sigma_E \in \mathcal{E} \\ \frac{1}{(r+1)(m+1)\,\eta} \left(1 + (\|\mathrm{dev}(\mathrm{sym}\,\Sigma_E)\| - \sigma_{\mathrm{y}})^{r+1}\right)^{m+1} & \Sigma_E \notin \mathcal{E} \end{cases},$$

where $\eta > 0$ is a viscosity parameter and $r, m \geq 0$. An easy calculation shows that this leads to the single valued locally Lipschitz continuous subdifferential

$$\partial\chi(\Sigma_E) = \frac{1}{\eta}\left(1 + [\|\mathrm{dev}(\mathrm{sym}\,\Sigma_E)\| - \sigma_{\mathrm{y}}]_+^{r+1}\right)^m \times \\ [\|\mathrm{dev}(\mathrm{sym}\,\Sigma_E)\| - \sigma_{\mathrm{y}}]_+^r \ \frac{\mathrm{dev}(\mathrm{sym}\,\Sigma_E)}{\|\mathrm{dev}(\mathrm{sym}\,\Sigma_E)\|} \,. \tag{25}$$

The parameter r allows to adjust the smoothness of the flow rule when passing the boundary of the elastic domain $\mathcal{E}$. The special choice $m = 0, r = 1$ corresponds precisely to the Yosida approximation (19). With $r > 5$ and $m \geq$

1 it is clear that $\partial \mathcal{X} \in C^5(\mathbb{M}^{3\times 3}, \mathbb{M}^{3\times 3})$. A typical range for m in engineering applications is $m \in \{0, \ldots, 80\}$. For $m \to \infty$ we recover formally ideal rate independent plasticity. For simplicity, we choose the positive parameter ν^+ in the elastic flow part according to the same level of viscosity as is used in the plastic flow part and set formally

$$\nu^+ = \frac{1}{\eta}\left(1+[\|\operatorname{skew}(F_e R_e^T)\| - 0]_+^{r+1}\right)^m \times \\ [\|\operatorname{skew}(B)\| - 0]_+^r \frac{1}{\|\operatorname{skew}(B)\|}, \tag{26}$$

similar to (25). This choice makes the flow rule altogether a C^5 function. In this setting we can prove the following result:

Theorem 10.1 (Local existence and uniqueness) *Suppose for the displacement boundary data* $g \in C^1(\mathbb{R}^+, H^{5,2}(\Omega, \mathbb{R}^3))$. *Then there exists a time* $T > 0$ *such that the initial boundary value problem* (P3) *with* (25) *and* (26) *admits a unique local solution*

$$\varphi \in C([0,T], H^{5,2}(\Omega, \mathbb{R}^3)),$$
$$(F_p, R_e) \in C^1([0,T], H^{4,2}(\Omega, SL(3,\mathbb{R})), H^{4,2}(\Omega, SO(3)))\,.$$

Proof. See NEFF [39]. We basically repeat the ideas of Theorem 8.3. At frozen variables (F_p, R_e) the (elastic) equilibrium system in (P3) is a linear, second order, strictly Legendre-Hadamard elliptic boundary value problem with nonconstant coefficients. The nonlinearity has been shifted into the appended evolution equation. This system has variational structure in the sense that the equilibrium part of (P3) is formally equivalent to the minimization problem

$$\begin{aligned} &\forall\, t \in [0,T]: \quad I(\varphi(t), F_p^{-1}(t), R_e(t)) \mapsto \min, \quad \varphi(t) \in g(t) + H_\circ^{1,2}, \\ &I(\varphi, F_p^{-1}, R_e) = \int_\Omega W(\nabla\varphi\, F_p^{-1}, R_e)\, \det[F_p]\, dx, \\ &W(F_e, R_e) = \frac{\mu}{4}\,\|F_e^T R_e + R_e^T F_e - 2\mathbb{1}\|^2 + \frac{\lambda}{8}\operatorname{tr}\left[F_e^T R_e + R_e^T F_e - 2\mathbb{1}\right]^2. \end{aligned} \tag{27}$$

The main task in proving that (P3) is well posed consists of showing uniform estimates for solutions of linear, elliptic systems whose coefficients are time dependent and do not induce a pointwise positive bilinear form. This problem does not arise in infinitesimal elasto-viscoplasticity (7) with $\operatorname{tr}[\varepsilon_p] = 0$, since there, the elasticity tensor $\mathcal{D}$ is assumed to be a constant positive definite 4th. order tensor.

We are first concerned with the static situation where (F_p, R_e) are assumed to be known. We prove existence, uniqueness and regularity of solutions to the related (elastic) boundary value problem. In addition we elucidate in which manner these solutions depend on (F_p, R_e). These investigations rely heavily on a theorem recently proved by the author extending Korn´s first inequality to nonconstant coefficients.

Theorem 10.2 (Extended Korn's first inequality) *Let $\Omega \subset \mathbb{R}^3$ be a bounded Lipschitz domain and let $\Gamma \subset \partial\Omega$ be a smooth part of the boundary with nonvanishing 2-dimensional Lebesgue measure. Define $H_\circ^{1,2}(\Omega,\Gamma) := \{\phi \in H^{1,2}(\Omega) \mid \phi_{|\Gamma} = 0\}$ and let $F_p, F_p^{-1} \in C^1(\overline{\Omega}, GL^+(3,\mathbb{R}))$ be given with $\det[F_p(x)] \geq \mu^+ > 0$. Moreover suppose that for the* **dislocation density** $\operatorname{Curl} F_p \in C^1(\overline{\Omega}, \mathbb{M}^{3\times 3})$. *Then*

$$\exists\, c^+ > 0 \quad \forall\, \phi \in H_\circ^{1,2}(\Omega,\Gamma): \\ \|\nabla\phi\, F_p^{-1}(x) + F_p^{-T}(x)\, \nabla\phi^T\|^2_{L^2(\Omega)} \geq c^+\, \|\phi\|^2_{H^{1,2}(\Omega)} \,.$$

Clearly this result generalizes the classical Korn's first inequality

$$\exists\, c^+ > 0 \quad \forall\, \phi \in H_\circ^{1,2}(\Omega,\Gamma): \quad \|\nabla\phi + \nabla\phi^T\|^2_{L^2(\Omega)} \geq c^+\, \|\phi\|^2_{H^{1,2}(\Omega)} \,,$$

which is just our result with $F_p = 1\!\!1$.

Proof. See NEFF [43]. Recently it was possible to significantly relax the continuity assumptions necessary for this theorem to hold. Precisely in the case $\Gamma = \partial\Omega$ it can be shown that

$$\exists\, c^+ > 0 \quad \forall\, \phi \in H_\circ^{1,2}(\Omega) \\ \|\nabla\phi\, F_p^{-1}(x) + F_p^{-T}(x)\, \nabla\phi^T\|^2_{L^2(\Omega)} \geq c^+\, \|\phi\|^2_{H^{1,2}(\Omega)} \,,$$

if only $F_p \in L^\infty(\Omega, \mathrm{GL}^+(3,\mathbb{R}))$ with $\det[F_p(x)] \geq \mu^+ > 0$ and $\operatorname{Curl} F_p \in L^2(\Omega, \mathbb{M}^{3\times 3})$. In addition one has to assume that with $F_p = R_p\, U_p$, the polar factorization of F_p, the orthogonal part R_p has locally jumps of maximal height $C^+ \frac{\lambda_{min,\Omega}(U_p)}{\lambda_{max,\Omega}(U_p)}$, where C^+ is a given non small constant. This allows F_p to be quite discontinuous. In the general case of mixed boundary data one needs, moreover, that F_p is smooth in an arbitrarily small boundary layer NEFF [40].

Now, the minimization problem (27) can be easily solved by applying the direct methods of variations. We show that I is strictly convex over the affine space $\{g + H_\circ^{1,2}(\Omega)\}$. This is done by computing the second derivative. We have

$$\begin{aligned} &D^2_\varphi I(\varphi, F_p^{-1}, R_e).(\phi,\phi) = \\ &\int_\Omega \left(\frac{\mu}{2}\, \|F_p^{-T}\nabla\phi^T R_e + R_e^T\nabla\phi F_p^{-1}\|^2 + \lambda \operatorname{tr}\left[R_e^T\nabla\phi F_p^{-1}\right]^2\right) \det[F_p(x)]\, dx \\ &\geq \int_\Omega \frac{\mu}{2}\, \|(R_e F_p)^{-T}\nabla\phi^T + \nabla\phi(R_e F_p)^{-1}\|^2 \det[F_p(x)]\, dx \\ &\geq \mu\, c^+(F_p, R_e, \Omega)\, \|\phi\|^2_{1,2,\Omega} \,, \end{aligned} \tag{28}$$

by applying Theorem 10.2 with $F_p := R_e F_p$. Since the Lamé constant $\mu > 0$ we see that $D^2_\varphi I(\varphi, F_p^{-1}, R_e).(\phi, \phi)$ is uniformly positive. Hence $I(\varphi, F_p^{-1}, R_e)$ is strictly convex.

We can write the evolution part of (P3) in the following block diagonal form with $A = (F_p^{-T}, R_e)$:

$$\frac{d}{dt}\begin{pmatrix} F_p^{-T}(t) \\ R_e(t) \end{pmatrix} = \begin{pmatrix} -\partial\chi(\Sigma_E(t))^T & 0 \\ 0 & \nu^+ \operatorname{skew}(B(t)) \end{pmatrix} \begin{pmatrix} F_p^{-T}(t) \\ R_e(t) \end{pmatrix}. \tag{29}$$

Thus, the system (P3) is equivalent to

$$\frac{d}{dt}A(t) = \hat{h}\,(\nabla_x T(A(t), g(t)), A(t))\; A(t)\,, \tag{30}$$

with $\hat{h} : \mathbb{M}^{3\times 3} \times \mathbb{M}^{3\times 3} \mapsto \mathbb{M}^{6\times 6}$

$$\hat{h}\,(\nabla_x T(A(t), g(t)), A(t)) = \begin{pmatrix} -\partial\chi(\Sigma_E(t))^T & 0 \\ 0 & \nu^+ \operatorname{skew}(B(t)) \end{pmatrix}, \tag{31}$$

where Σ_E and B are expressions depending on $A = (F_p^{-T}, R_e)$ and on

$$F = \nabla_x \varphi = \nabla_x T(A, g)\,, \tag{32}$$

where $\varphi = T(A, g)$ is the unique solution of the (elastic) elliptic boundary value problem at given A whose existence has been established by making use of the extended Korn inequality. By use of refined elliptic regularity results Ebenfeld [15] it can be shown that T is indeed locally Lipschitz (note that in the general finite case of Theorem 8.3 we had instead to **assume** that T is well defined and locally Hoelder continuous). It remains to show that the right hand side as a function of A is locally Lipschitz in some properly defined Banach space allowing to apply the well known local existence and uniqueness theorem. This part then is standard since the subdifferential is sufficiently smooth.

In a recent numerical study Neff and Wieners [44] the approximation inherent in (P3) has been completely justified for structural applications.

11 Discussion

In the first part of the contribution we saw that finite elasto-plasticity based on the multiplicative decomposition allows the successful application of the direct methods of variations to the static elastic trial step. This feature is generally lost for additive ansatzes.

The only additional requirement necessary is polyconvexity of the elastic free energy, which can be easily met. However, elastic free energies defined on Hencky strains are not polyconvex, leading to a loss of ellipticity.

Polyconvexity in itself, however, is not sufficient to treat the coupled problem. Certain mathematical problems can be circumvented in the case of small elastic strains but finite deformations. The investigated model for small elastic strains combines mathematical simplifications with additional physical mechanisms. The introduced evolution equation for 'elastic' rotations R_e leads to deformation induced texture evolution and R_e can conceptually be interpreted as the elastic part of the total rotation of grains in a polycrystal, see NEFF [41].

All presented mathematical results are obtained for essentially fully rate dependent viscous models. Our investigations suggest that introducing viscous behaviour in the finite deformation regime is still enough to regularize the initial boundary value problem. Yet, the smoothness of the plastic variables may deteriorate in finite time leading to bifurcation or fracture. In the viscous case, however, this is entirely a problem of the smoothness of the elastic moduli set by the internal variables. It remains to investigate which type of mechanism could prevent this catastrophic loss of smoothness.

The flow based approach hinges on the assumption that the flow rule is locally Lipschitz thus excluding rate independent material behaviour. It may therefore turn out that the qualitative picture for rate independent behaviour changes dramatically, e.g. microstructure could immediately develop even for smooth data. Numerical calculations based on the time incremental formulation seem to indicate this possibility, cf. CARSTENSEN and HACKL [8] and ORTIZ and REPETTO [46].

First mathematical ideas suggest that a physically motivated backstress evolution, based on augmenting the thermodynamic potential with quantities measuring the local incompatibility, such as $||F_p^T \operatorname{Curl} F_p||$, could prevent the above mentioned failure process and at the same time introducing a length scale into the model.

The ensuing coupled system, however, is drastically changed: instead of the ordinary differential evolution equation, one has to solve a degenerate parabolic system for the evolution of the plastic variable. The standard numerical treatment of elasto-plasticity does not any longer apply.

12 Acknowledgements

This paper mostly summarizes the authors recent work on problems in finite plasticity. The stimulating environment of the SFB 298: 'Deformation and failure in metallic and granular materials' is gratefully acknowledged. The paper was compiled while the author held a visiting faculty position under the ASCI program in Michael Ortiz' group at the California Institute of Technology, Graduate Aeronautical Laboratories. Their kind hospitality has been a great help.

References

1. Alber HD. (1998) *Materials with Memory. Initial-Boundary Value Problems for Constitutive Equations with Internal Variables.*, volume 1682 of *Lecture Notes in Mathematics.* Springer, Berlin
2. Anthony KH. (1971) Die Theorie der nichtmetrischen Spannungen in Kristallen. *Arch. Rat. Mech. Anal.*, 40:50–78
3. Ball JM. (1977) Convexity conditions and existence theorems in nonlinear elasticity. *Arch. Rat. Mech. Anal.*, 63:337–403
4. Ball JM., Marsden JE. (1984) Quasiconvexity at the boundary, positivity of the second variation and elastic stability. *Arch. Rat. Mech. Anal.*, 86:251–277
5. Besseling JF., van der Giessen E. (1994) *Mathematical Modelling of Inelastic Deformation*, volume 5 of *Applied Mathematics and Mathematical Computation.* Chapman Hall, London
6. Bloom F. (1979) *Modern Differential Geometric Techniques in the Theory of Continuous Distributions of Dislocations*, volume 733 of *Lecture Notes in Mathematics.* Springer, Berlin
7. Bodner SR., Partom Y. (1975) Constitutive equations for elastic-viscoplastic strainhardening materials. *J. Appl. Mech.*, 42:385–389
8. Carstensen C., Hackl K. (2000) On microstructure occuring in a model of finite-strain elastoplasticity involving a single slip system. *ZAMM*
9. Cermelli C., Gurtin ME. (2001) On the characterization of geometrically necessary dislocations in finite plasticity. *J. Mech. Phys. Solids*, 49:1539–1568
10. Chelminski K. (1998) On self-controlling models in the theory of inelastic material behaviour of metals. *Cont. Mech. Thermodyn.*, 10:121–133
11. Chelminski K. (1999) On monotone plastic constitutive equations with polynomial growth condition. *Math. Meth. App. Sc.*, 22:547–562
12. Dafalias YF. (1984) The plastic spin concept and a simple illustration of its role in finite plastic transformations. *Mechanics of Materials*, 3:223–233
13. Dafalias YF. (1987) Issues on the constitutive formulation at large elastoplastic deformations, Part I: Kinematics. *Acta Mechanica*, 69:119–138
14. Dafalias YF. (1998) Plastic spin: necessity or redundancy? *Int. J. Plasticity*, 14(9):909–931
15. Ebenfeld S. (2002) L^2-regularity theory of linear strongly elliptic Dirichlet systems of order $2m$ with minimal regularity in the coefficients. *to appear in Quart. Appl. Math.*
16. Green AE., P.M. Naghdi PM. (1965) A general theory of an elastic-plastic continuum. *Arch. Rat. Mech. Anal.*, 18:251–281
17. Gurtin ME. (2000) *Configurational Forces as Basic Concepts of Continuum Physics*, volume 137 of *Applied Mathematical Science.* Springer, Berlin, first edition
18. Gurtin ME. (2000) On the plasticity of single crystals: free energy, microforces, plastic-strain gradients. *J. Mech. Phys. Solids*, 48:989–1036
19. Han W., Reddy BD. (1999) *Plasticity. Mathematical Theory and Numerical Analysis.* Springer, Berlin
20. Haupt P. (2000) *Continuum Mechanics and Theory of Materials.* Springer, Heidelberg
21. Ionescu IR., Sofonea M. (1993) *Functional and Numerical Methods in Viscoplasticity.* Oxford Mathematical Monographs. Oxford University Press, Oxford, first edition

22. Kondo K. (1955) Geometry of elastic deformation and incompatibility. In K. Kondo, editor, *Memoirs of the Unifying Study of the Basic Problems in Engineering Science by Means of Geometry*, volume 1, Division C, pages 5–17 (361–373). Gakujutsu Bunken Fukyo-Kai,
23. Kröner E. (1958) *Kontinuumstheorie der Versetzungen und Eigenspannungen.*, volume 5 of *Ergebnisse der Angewandten Mathematik.* Springer, Berlin
24. Kröner E. (1962) Dislocation: a new concept in the continuum theory of plasticity. *J. Math. Phys.*, 42:27–37
25. Kröner E., Seeger A. (1959) Nichtlineare Elastizitätstheorie der Versetzungen und Eigenspannungen. *Arch. Rat. Mech. Anal.*, 3:97–119
26. Lee EH. (1969) Elastic-plastic deformation at finite strain. *J. Appl. Mech.*, 36:1–6
27. Lemaitre J., Chaboche JL. (1985) *Mecanique des materiaux solides.* Dunod, Paris
28. Lubarda VD. (2002) *Elastoplasticity Theory.* CRC Press, London
29. Lucchesi M, Podio-Guidugli P. (1988) Materials with elastic range: a theory with a view toward applications. Part I. *Arch. Rat. Mech. Anal.*, 102:23–43
30. Mandel J. (1973) Equations constitutives et directeurs dans les milieux plastiques et viscoplastiques. *Int. J. Solids Structures*, 9:725–740
31. Maugin G. (1993) *Material Inhomogeneities in Elasticity.* Applied Mathematics and Mathematical Computations. Chapman-Hall, London
32. Maugin G. (1999) *The Thermomechanics of Nonlinear Irreversible Behaviors.* Number 27 in Nonlinear Science. World Scientific, Singapore
33. Maugin G, Epstein M. (1998) Geometrical material structure of elastoplasticity. *Int. J. Plasticity*, 14:109–115
34. Miehe C. (1992) *Kanonische Modelle multiplikativer Elasto-Plastizität. Thermodynamische Formulierung und numerische Implementation.* Habilitationsschrift, Universität Hannover
35. Miehe C. (1995) A theory of large-strain isotropic thermoplasticity based on metric transformation tensors. *Archive Appl. Mech.*, 66:45–64
36. Naghdi PM. (1990) A critical review of the state of finite plasticity. *J. Appl. Math. Phys.(ZAMP)*, 41:315–394
37. Neff P. (2000) Formulation of visco-plastic approximations in finite plasticity for a model of small elastic strains, Part I: Modelling. Preprint 2127, TU Darmstadt
38. Neff P. (2000) *Mathematische Analyse multiplikativer Viskoplastizität. Ph.D. Thesis, TU Darmstadt.* Shaker Verlag, ISBN:3-8265-7560-1, Aachen
39. Neff P. (2001) Formulation of visco-plastic approximations in finite plasticity for a model of small elastic strains, Part IIa: Local existence and uniqueness results. Preprint 2138, TU Darmstadt
40. Neff P. (2002) An extended Korn's first inequality with discontinuous coefficents. in preparation, Darmstadt University of Technology
41. Neff P. (2002) Finite multiplicative plasticity for small elastic strains with linear balance equations and grain boundary relaxation. *to appear in Cont. Mech. Thermodynamics*
42. Neff P. (2002) Local existence and uniqueness for quasistatic finite plasticity with grain boundary relaxation. *submitted to SIAM J. Math. Anal.*
43. Neff P. (2002) On Korn's first inequality with nonconstant coefficients, Preprint 2080 TU Darmstadt. *Proc. Roy. Soc. Edinb.*, 132A:221–243

44. Neff P., Wieners C. (2002) Comparison of models for finite plasticity. A numerical study. *to appear in Computing and Visualization in Science*
45. Ortiz M., Repetto EA. 1993 Nonconvex energy minimization and dislocation structures in ductile single crystals, manuscript.
46. Ortiz M., Repetto EA. (1999) Nonconvex energy minimization and dislocation structures in ductile single crystals. *J. Mech. Phys. Solids*, 47:397–462
47. Ortiz M., Repetto EA., Stainier L. (2000) A theory of subgrain dislocation structures. *J. Mech. Phys. Solids*, 48:2077–2114
48. Ortiz M., Stainier L. (1999) The variational formulation of viscoplastic constitutive updates. *Comp. Meth. Appl. Mech. Engrg.*, 171:419–444
49. Sansour C., Kollmann FG. (1996) On theory and numerics of large viscoplastic deformation. *Comp. Meth. Appl. Mech. Engrg.*, 146:351–369
50. Silhavy M. (1976) On transformation laws for plastic deformations of materials with elastic range. *Arch. Rat. Mech. Anal.*, 63:169–182
51. Simo JC. (1998) Numerical analysis and simulation of plasticity. In P.G. Ciarlet and J.L. Lions, editors, *Handbook of Numerical Analysis*, volume VI. Elsevier, Amsterdam
52. Simo JC., Hughes JR. (1998) *Computational Inelasticity*, volume 7 of *Interdisciplinary Applied Mathematics*. Springer, Berlin
53. Simo JC., Ortiz M. (1985) A unified approach to finite deformation elastoplastic analysis based on the use of hyperelastic constitutive equations. *Comp. Meth. Appl. Mech. Engrg.*, 49:221–245
54. Steinmann P. (1996) Views on multiplicative elastoplasticity and the continuum theory of dislocations. *Int. J. Engrg. Sci.*, 34:1717–1735

Initial Boundary Value Problems in Continuum Mechanics

Stefan Ebenfeld

Department of Mathematics, Darmstadt University of Technology
Schlossgartenstr. 7, D–64289 Darmstadt, Germany
now: Landgrafenring 27, D–63071 Offenbach
ebenfeld@mathematik.tu-darmstadt.de

received 27 Sep 2002 — accepted 16 Oct 2002

Abstract. We study coupled systems of nonlinear initial boundary value problems from continuum mechanics, where each system is of higher order, and of hyperbolic or parabolic type. Our goal is to derive sufficient conditions on the underlying constitutive equations, such that the coupled problem admits a unique smooth solution. These conditions constitute the mathematical counterpart to those conditions obtained by exploiting the second law of thermodynamics. For the proofs of our results we refer the reader to [3], [4] and [5].

1 Introduction

In continuum mechanics we often wish to mathematically describe the accelerated motion of matter on the one hand, and diffusion phenomena on the other hand. A standard example for the first phenomenon is the deformation of an elastic body, whereas a standard example for the second phenomenon is the diffusion of heat in a body. However, such problems generally occur as coupled problems. A standard example for the latter is the problem of thermoelasticity, where the deformation of an elastic body and the diffusion of heat in the same body mutually influence each other. In many cases the accelerated motion of a matter field $u(x,t)$ (the displacement of an elastic body, say) is governed by a physical law of abstract hyperbolic type (Newton's law of motion, say),

$$\partial_t^2 u(x,t) - \sum_{i=1}^{n} \partial_{x_i} \hat{F}_i(x,t) = \hat{f}(x,t),$$

or, more general,

$$\partial_t^2 u(x,t) + \sum_{|\alpha| \le m_u} (-1)^{|\alpha|} \partial_x^\alpha \hat{F}_\alpha(x,t) = \hat{f}(x,t), \tag{H}$$

whereas the diffusion-type behaviour of a physical field $v(x,t)$ (the temperature density in a body, say) is governed by a physical law of abstract parabolic type (the heat equation, say),

$$\partial_t v(x,t) - \sum_{i=1}^{n} \partial_{x_i} \hat{G}_i(x,t) = \hat{g}(x,t),$$

or, more general,

$$\partial_t v(x,t) + \sum_{|\alpha|\le m_v} (-1)^{|\alpha|} \partial_x^\alpha \hat{G}_\alpha(x,t) = \hat{g}(x,t). \tag{P}$$

In the above equations, (H) and (P), we have used the standard multiindex notation, $\alpha = (\alpha_1, \ldots, \alpha_n)$, $|\alpha| = |\alpha_1| + \ldots + |\alpha_n|$, and $\partial_x^\alpha = \partial_{x_1}^{\alpha_1} \ldots \partial_{x_n}^{\alpha_n}$. Of course, in the context of continuum mechanics the dimension of the underlying space will be $n = 1, 2$ or 3. However, from a mathematical point of view such a restriction is not necessary. Now, continuum mechanics provides us with a multitude of different constitutive laws for the quantities $\hat{F}(x,t)$, $\hat{f}(x,t)$, $\hat{G}(x,t)$ and $\hat{g}(x,t)$ (the stress, the body force, the heat flux and the heat production, say). In the simplest case, the constitutive equations will be such that the quantities $\hat{F}(x,t)$, $\hat{f}(x,t)$, $\hat{G}(x,t)$ and $\hat{g}(x,t)$ depend on respective collections of partial derivatives of the physical fields $u(x,t)$ and $v(x,t)$,

$$\begin{aligned} \hat{F}(x,t) &= F(U^F, V^F, x, t), & \hat{f}(x,t) &= f(U^f, V^f, x, t), \\ \hat{G}(x,t) &= G(U^G, V^G, x, t), & \hat{g}(x,t) &= g(U^g, V^g, x, t), \end{aligned}$$

where

$$\begin{aligned} U^F &= (u(x,t), \ldots, \mathrm{D}_x^{M_u^{F,0}} u(x,t), \partial_t u(x,t), \ldots, \mathrm{D}_x^{M_u^{F,1}} \partial_t u(x,t)), \\ V^F &= (v(x,t), \ldots, \mathrm{D}_x^{M_v^F} v(x,t)), \\ U^f &= (u(x,t), \ldots, \mathrm{D}_x^{M_u^{f,0}} u(x,t), \partial_t u(x,t), \ldots, \mathrm{D}_x^{M_u^{f,1}} \partial_t u(x,t)), \\ V^f &= (v(x,t), \ldots, \mathrm{D}_x^{M_v^f} v(x,t)), \\ U^G &= (u(x,t), \ldots, \mathrm{D}_x^{M_u^{G,0}} u(x,t), \partial_t u(x,t), \ldots, \mathrm{D}_x^{M_u^{G,1}} \partial_t u(x,t)), \\ V^G &= (v(x,t), \ldots, \mathrm{D}_x^{M_v^G} v(x,t)), \\ U^g &= (u(x,t), \ldots, \mathrm{D}_x^{M_u^{g,0}} u(x,t), \partial_t u(x,t), \ldots, \mathrm{D}_x^{M_u^{g,1}} \partial_t u(x,t)), \\ V^g &= (v(x,t), \ldots, \mathrm{D}_x^{M_v^g} v(x,t)), \end{aligned}$$

and $\mathrm{D}_x^M w(x,t)$ denotes the collection of all partial derivatives w.r.t. x of a function $w(x,t)$ that are of order M. However, more complicated constitutive equations occur. It turns out that without additional conditions on the constitutive equations the coupled system (H)–(P) does not make sense, neither from a physical nor form a mathematical point of view. In physics and engineering sciences these conditions arise in a natural way from the principle of equipresence, saying that a priori every constitutive quantity should depend on the same collection of partial derivatives of the physical fields, and from the exploitation of the second law of thermodynamics, saying that every solution (u, v) to the coupled system (H)–(P) should satisfy an entropy inequality. In summary, the principle of equipresence and the exploitation of the second law of thermodynamics give an answer to the following question:

> *What is the most general form of the constitutive equations, such that the coupled system* (H)–(P) *is reasonable from a physical point of view?*

In this paper we address the same problem from a mathematical point of view, i.e. we give an answer to the following question:

> *What is the most general form of the constitutive equations, such that the coupled system* (H)–(P) *is reasonable from a mathematical point of view?*

More precisely, we consider an arbitrary number of abstract hyperbolic problems of type (H) together with an arbitrary number of abstract parabolic problems of type (P). We give sufficient conditions on the constitutive equations such that the coupled system (H)–(P), when complemented with suitable initial and boundary conditions, admits a unique smooth solution.

In view of its physical interpretation, our result is closely related to the exploitation of the second law of thermodynamics, as explained above. This technique originates from the work of Liu, see [16], and it has actually become a standard tool for the derivation of conditions on constitutive equations in continuum mechanics, see also Müller–Ruggeri, [18]. In particular, joining the present SFB 298, Hauser, Hutter, Kirchner, Rickenmann, Svendsen and Wang, together with Laloui and Vuillet, have produced a number of papers where this technique is applied, see [6], [7], [8], [9], [10], [11], [12] and [13]. In view of its mathematical content, our result is an application of the so called energy method. The energy method can be considered as a mathematical tool designed to prove existence results for nonlinear evolution equations. Many authors, such as Dafermos–Hrusa, Kato and Majda, have applied the energy method to single hyperbolic problems before, see e.g. [1], [15] and [17]. Moreover, Jiang–Racke have applied the energy method to problems in thermoelasticity, see [14]. The paper at hand differs from the afore mentioned works in that it presents a systematical investigation of the couplings between arbitrary many hyperbolic and parabolic problems of arbitrary order. It turns out that the energy method is applicable to such a hyperbolic-parabolic system provided that the underlying constitutive equations satisfy respective conditions. These conditions constitute precisely the mathematical counterpart to the conditions obtained by exploiting the second law of thermodynamics.

The paper is organized as follows. In Sect. 2 we give a general formulation of the above problem (H)–(P). We start with an abstract version (the 'abstract problem') allowing the constitutive quantities to be general history functionals of the underlying physical fields, and then we turn to the above case (the 'hyperbolic-parabolic problem'), where the constitutive quantities are actually functions of respective collections of partial derivatives of the underlying physical fields. We close the section with a discussion of a number of problems from continuum mechanics, showing how to rewrite these problems such that they fit into the scheme of our general theory. In Sect. 3 we formulate the above mentioned conditions on the constitutive quantities for both the abstract and the hyperbolic-parabolic problem. Moreover, we state the corresponding existence and uniqueness theorems. In Sect. 4 we give a sketch of proof of our main theorem that involves the energy method. For a detailed proof we refer the reader to [3], [4] and [5]. Finally, in Sect. 5 we exploit the conditions on the constitutive equations for the case of two coupled problems of hyperbolic-hyperbolic, hyperbolic-parabolic or parabolic-parabolic type, respectively.

2 Statement of the Problem

In this section we formulate both the abstract and the hyperbolic-parabolic problem. Therefore, we have to introduce some index notation. We use a first index $j = 1, 2, 3$ in order to indicate the type of the of the problem. More precisely,

$j = 1$ indicates a hyperbolic problem satisfying a symmetry condition,
$j = 2$ indicates a parabolic problem satisfying no symmetry condition,
$j = 3$ indicates a parabolic problem satisfying a symmetry condition.

For a precise definition of the symmetry condition see Sect. 3. As mentioned above, we consider an arbitrary number of hyperbolic and parabolic problems. Therefore, we use a second index $i_j = 1, \ldots, I_j$ in order to enumerate the problems of type j. Now, let $T > 0$, and let $\Omega \subset \mathbb{R}^n$ be a bounded domain with a smooth boundary. $[0, T]$ denotes the time interval of observation. In case of Lagrange coordinates Ω denotes a reference configuration, whereas in case of Euler coordinates Ω denotes a subset of the configuration space. According to our index notation we consider the following physical fields,

$$u_{ji_j} : \overline{\Omega \times (0, T)} \longrightarrow \mathbb{R}^{N_{ji_j}} : (x, t) \longmapsto u_{ji_j}(x, t).$$

Throughout this paper we assume that the physical fields satisfy the following initial and boundary conditions,

$$u_{1i_1}\Big|_{t=0} = 0, \qquad \partial_t u_{1i_1}\Big|_{t=0} = 0, \qquad u_{2i_2}\Big|_{t=0} = 0, \qquad u_{3i_3}\Big|_{t=0} = 0,$$

$$\partial_x^\alpha u_{ji_j}\Big|_{x\in\partial\Omega} = 0 \qquad (|\alpha| = 0, \ldots, m_{ji_j} - 1).$$

Of course, since our theory is of general character the $u_{ji_j}(x,t)$ can have various physical interpretations. In Subsect. 2.3 we consider a number of examples from continuum mechanics, showing how they fit into the scheme of our general theory.

2.1 The Abstract Problem

In this Subsection we consider the case where the constitutive quantities are general history functionals of the physical fields. More precisely, we consider the following coupled problem,

$$\partial_t^2 u_{1i_1} + \sum_{|\alpha|,|\beta|=0}^{m_{1i_1}} (-1)^{|\alpha|}\partial_x^\alpha\Big(A_{1i_1,\alpha\beta}[u](x,t)\,\partial_x^\beta u_{1i_1}\Big) = f_{1i_1}[u](x,t), \tag{H}$$

$$\partial_t u_{2i_2} + \sum_{|\alpha|,|\beta|=0}^{m_{2i_2}} (-1)^{|\alpha|}\partial_x^\alpha\Big(A_{2i_2,\alpha\beta}[u](x,t)\,\partial_x^\beta u_{2i_2}\Big) = f_{2i_2}[u](x,t), \tag{P1}$$

$$\partial_t u_{3i_3} + \sum_{|\alpha|,|\beta|=0}^{m_{3i_3}} (-1)^{|\alpha|}\partial_x^\alpha\Big(A_{3i_3,\alpha\beta}[u](x,t)\,\partial_x^\beta u_{3i_3}\Big) = f_{3i_3}[u](x,t). \tag{P2}$$

Throughout this paper the square bracket notation $\Phi[u]$ means that Φ acts as a nonlinear operator on the physical fields $u = (u_{11}, \ldots, u_{3I_3})$. According to our notation (H) is a collection of hyperbolic PDEs for the physical fields $u_{11}, \ldots, u_{1I_1}$, whereas (P1) and (P2) are collections of parabolic PDEs for the physical fields $u_{21}, \ldots, u_{2I_2}$ and $u_{31}, \ldots, u_{3I_3}$, respectively.

2.2 The Hyperbolic–Parabolic Problem

In this Subsection we consider the case where the constitutive quantities are actually functions of respective collections of partial derivatives of the underlying physical fields. More precisely, we consider the following coupled problem,

$$\partial_t^2 u_{1i_1} + \sum_{|\alpha|=0}^{m_{1i_1}} (-1)^{|\alpha|}\partial_x^\alpha\Big(F_{1i_1,\alpha}(U_{1i_1}^F, x, t)\Big) = f_{1i_1}(U_{1i_1}^f, x, t), \tag{H}$$

$$\partial_t u_{2i_2} + \sum_{|\alpha|=0}^{m_{2i_2}} (-1)^{|\alpha|}\partial_x^\alpha\Big(F_{2i_2,\alpha}(U_{2i_2}^F, x, t)\Big) = f_{2i_2}(U_{2i_2}^f, x, t), \tag{P1}$$

$$\partial_t u_{3i_3} + \sum_{|\alpha|=0}^{m_{3i_3}} (-1)^{|\alpha|} \partial_x^\alpha \Big(F_{3i_3,\alpha}(U^F_{3i_3}, x, t) \Big) = f_{3i_3}(U^f_{3i_3}, x, t). \tag{P2}$$

Analogous with the example in the introduction, the $U^F_{ji_j}$ and $U^f_{ji_j}$ denote collections of partial derivatives of all physical fields $u_{11}, \ldots, u_{3I_3}$. Since we consider arbitrary many problems that are of hyperbolic or parabolic type, and of higher order, the definition of the $U^F_{ji_j}$ and $U^f_{ji_j}$ becomes somewhat lengthy. However, for the sake of precision as well as for later use in Sect. 3 we write the definition explicitly,

$$\begin{aligned}
U^F_{ji_j} = (&u_{11}(x,t), \ldots, \mathrm{D}_x^{M^{F,0}_{ji_j,11}} u_{11}(x,t), \partial_t u_{11}, \ldots, \mathrm{D}_x^{M^{F,1}_{ji_j,11}} \partial_t u_{11}(x,t), \ldots \\
&u_{1I_1}(x,t), \ldots, \mathrm{D}_x^{M^{F,0}_{ji_j,1I_1}} u_{1I_1}(x,t), \partial_t u_{1I_1}, \ldots, \mathrm{D}_x^{M^{F,1}_{ji_j,1I_1}} \partial_t u_{1I_1}(x,t), \\
&u_{21}(x,t), \ldots, \mathrm{D}_x^{M^{F,0}_{ji_j,21}} u_{21}(x,t), \ldots, u_{2I_2}(x,t), \ldots, \mathrm{D}_x^{M^{F,0}_{ji_j,2I_2}} u_{2I_2}(x,t), \\
&u_{31}(x,t), \ldots, \mathrm{D}_x^{M^{F,0}_{ji_j,31}} u_{31}(x,t), \ldots, u_{3I_3}(x,t), \ldots, \mathrm{D}_x^{M^{F,0}_{ji_j,3I_3}} u_{3I_3}(x,t)), \\
U^f_{ji_j} = (&u_{11}(x,t), \ldots, \mathrm{D}_x^{M^{f,0}_{ji_j,11}} u_{11}(x,t), \partial_t u_{11}, \ldots, \mathrm{D}_x^{M^{f,1}_{ji_j,11}} \partial_t u_{11}(x,t), \ldots \\
&u_{1I_1}(x,t), \ldots, \mathrm{D}_x^{M^{f,0}_{ji_j,1I_1}} u_{1I_1}(x,t), \partial_t u_{1I_1}, \ldots, \mathrm{D}_x^{M^{f,1}_{ji_j,1I_1}} \partial_t u_{1I_1}(x,t), \\
&u_{21}(x,t), \ldots, \mathrm{D}_x^{M^{f,0}_{ji_j,21}} u_{21}(x,t), \ldots, u_{2I_2}(x,t), \ldots, \mathrm{D}_x^{M^{f,0}_{ji_j,2I_2}} u_{2I_2}(x,t), \\
&u_{31}(x,t), \ldots, \mathrm{D}_x^{M^{f,0}_{ji_j,31}} u_{31}(x,t), \ldots, u_{3I_3}(x,t), \ldots, \mathrm{D}_x^{M^{f,0}_{ji_j,3I_3}} u_{3I_3}(x,t)).
\end{aligned}$$

In particular, we have introduced the following notation,

- $M^{F,0}_{ji_j,lk_l}$ denotes the highest order of spatial derivatives of the physical field u_{lk_l} that occurs in the constitutive function $F_{ji_j,\alpha}$,
- $M^{F,1}_{ji_j,1k_1}$ denotes the highest order of spatial derivatives of the physical field $\partial_t u_{1k_1}$ that occurs in the constitutive function $F_{ji_j,\alpha}$,
- $M^{f,0}_{ji_j,lk_l}$ denotes the highest order of spatial derivatives of the physical field u_{lk_l} that occurs in the constitutive function f_{ji_j},
- $M^{f,1}_{ji_j,1k_1}$ denotes the highest order of spatial derivatives of the physical field $\partial_t u_{1k_1}$ that occurs in the constitutive function f_{ji_j}.

As in the previous Subsection, (H) is a collection of hyperbolic PDEs for the physical fields $u_{11}, \ldots, u_{1I_1}$, whereas (P1) and (P2) are collections of parabolic PDEs for the physical fields $u_{21}, \ldots, u_{2I_2}$ and $u_{31}, \ldots, u_{3I_3}$, respectively.

2.3 Applications in Continuum Mechanics

As mentioned in the introduction, the functions $u_{ji_j}(x,t)$ can be interpreted as physical fields of various kind. In this Subsection we consider some single problems from continuum mechanics, and we show how to rewrite these problems such that they fit into the scheme of our general theory. More precisely, we transform the underlying physical field, such that the equation of motion for the transformed field $u(x,t)$ is of one of the above types, (H) or (P), respectively. We note that the mathematical structure of some of the transformed problems can be identical. However, their mutual coupling can be quite different.

Elasticity and higher gradient materials. Let $\chi : \overline{\Omega \times (0,T)} \longrightarrow \mathbb{R}^n$. χ describes the deformation of a material body. We consider an action functional of the following form,

$$J[\chi] = \int_0^T \int_\Omega \Big(\frac{1}{2} |\partial_t \chi(x,t)|^2 - \hat{\Psi}(x,t) + b(x,t) \cdot \chi(x,t) \Big) \, \mathrm{d}x \, \mathrm{d}t,$$

where

$$\hat{\Psi}(x,t) = \Psi(\mathrm{D}_x \chi(x,t), \ldots, \mathrm{D}_x^m \chi(x,t), x, t).$$

Ψ describes the free energy, and b describes the body force. Applying Hamilton's principle the corresponding Euler–Lagrange equations read,

$$\partial_t^2 \chi(x,t) + \sum_{1 \le |\alpha| \le m} (-1)^{|\alpha|} \partial_x^\alpha \Big(\frac{\partial \Psi}{\partial w^\alpha} (\mathrm{D}_x \chi(x,t), \ldots, \mathrm{D}_x^m \chi(x,t), x, t) \Big)$$
$$= b(x,t).$$

We make the following definitions,

$$u(x,t) := \chi(x,t), \qquad U^F := (\mathrm{D}_x \chi(x,t), \ldots, \mathrm{D}_x^m \chi(x,t)).$$

$$F_\alpha(w,x,t) := \frac{\partial \Psi}{\partial w^\alpha}(w,x,t), \qquad f(x,t) := b(x,t).$$

Now the above Euler–Lagrange equations can be rewritten as a hyperbolic problem for the physical field $u(x,t)$ in the sense of Sect. 2.2,

$$\partial_t^2 u(x,t) + \sum_{1 \le |\alpha| \le m} (-1)^{|\alpha|} \partial_x^\alpha \Big(F_\alpha(U^F, x, t) \Big) = f(x,t). \tag{H}$$

Viscoelasticity. Let $\varphi : \overline{\Omega \times (0,T)} \longrightarrow \mathbb{R}^N$. φ describes the internal variables of a material body. We consider the problem of elasticity for the case of a free energy of the following form,

$$\Psi(\mathrm{D}_x \chi(x,t), \varphi(x,t), x, t).$$

The corresponding Euler–Lagrange equations read,

$$\partial_t^2 \chi(x,t) - \sum_{i=1}^n \partial_{x_i}\Big(\frac{\partial \Psi}{\partial w_i}(\mathrm{D}_x\chi(x,t), \varphi(x,t), x, t)\Big) = b(x,t).$$

Additionally, we assume that φ satisfies an ODE of the following form,

$$\partial_t \varphi(x,t) = g(\mathrm{D}_x\chi(x,t), \varphi(x,t), x, t).$$

In order to rewrite the equations of motion as an abstract problem we introduce the following new physical field,

$$u(x,t) := \partial_t \chi(x,t).$$

Solving the above ODE we can write,

$$\varphi(x,t) = \Phi[u](x,t).$$

Differentiating the Euler–Lagrange equations w.r.t. t and inserting the above solution operator we obtain,

$$\begin{aligned}
&\partial_t^2 u(x,t) - \sum_{i,j=1}^n \partial_{x_i}\Big(\frac{\partial^2 \Psi}{\partial w_i \partial w_j}(\mathrm{D}_x\chi(x,t), \Phi[u](x,t), x, t)\, \partial_{x_j} u(x,t)\Big) \\
&= \sum_{i=1}^n \partial_{x_i}\Big(\frac{\partial^2 \Psi}{\partial w_i \partial \varphi}(\mathrm{D}_x\chi(x,t), \Phi[u](x,t), x, t)\, g(\mathrm{D}_x\chi(x,t), \Phi[u](x,t), x, t)\Big) \\
&\quad + \sum_{i=1}^n \partial_{x_i}\Big(\frac{\partial^2 \Psi}{\partial w_i \partial t}(\mathrm{D}_x\chi(x,t), \Phi[u](x,t), x, t)\Big) + \partial_t b(x,t).
\end{aligned}$$

We make the following definitions,

$$A_{ij}[u](x,t) := \frac{\partial^2 \Psi}{\partial w_i \partial w_j}(\mathrm{D}_x\chi(x,t), \Phi[u](x,t), x, t),$$

$$\begin{aligned}
&f[u](x,t) \\
&:= \sum_{i=1}^n \partial_{x_i}\Big(\frac{\partial^2 \Psi}{\partial w_i \partial \varphi}(\mathrm{D}_x\chi(x,t), \Phi[u](x,t), x, t)\, g(\mathrm{D}_x\chi(x,t), \Phi[u](x,t), x, t)\Big) \\
&\quad + \sum_{i=1}^n \partial_{x_i}\Big(\frac{\partial^2 \Psi}{\partial w_i \partial t}(\mathrm{D}_x\chi(x,t), \Phi[u](x,t), x, t)\Big) + \partial_t b(x,t).
\end{aligned}$$

Now the above Euler–Lagrange equations can be rewritten as an abstract hyperbolic problem for the physical field $u(x,t)$ in the sense of Sect. 2.1,

$$\partial_t^2 u(x,t) - \sum_{i,j=1}^n \partial_{x_i}\Big(A_{ij}[u](x,t)\, \partial_{x_j} u(x,t)\Big) = f[u](x,t). \tag{H}$$

Heat flow. Let $\theta : \overline{\Omega \times (0,T)} \longrightarrow \mathbb{R}$. θ describes the temperature of a material body. We consider the following balance law for the energy,

$$\partial_t \theta(x,t) + \sum_{i=1}^{n} \partial_{x_i}\Big(q_i(\mathrm{D}_x\theta(x,t),x,t)\Big) = \pi(x,t).$$

q describes the heat flow, and π describes the heat production. We make the following definitions,

$$u(x,t) := \theta(x,t), \qquad U^F := \mathrm{D}_x\theta(x,t),$$

$$F_i(w,x,t) := q_i(w,x,t), \qquad f(x,t) := b(x,t).$$

Now the above equations of motion can be rewritten as a parabolic problem for the physical field $u(x,t)$ in the sense of Sect. 2.2,

$$\partial_t u(x,t) - \sum_{i=1}^{n} \partial_{x_i}\Big(F_i(U^F,x,t)\Big) = f(x,t). \tag{P}$$

Compressible fluid flow. Let $(\rho, v) : \overline{\Omega \times (0,T)} \longrightarrow \mathbb{R} \times \mathbb{R}^n$. ρ describes the mass density of a fluid, and v describes the corresponding velocity. We consider the following balance laws for mass and linear momentum,

$$\partial_t \rho(x,t) + \sum_{i=1}^{n} \partial_{x_i}\Big(\rho(x,t)\, v_i(x,t)\Big) = 0,$$

$$\begin{aligned}\rho(x,t)\Big(\partial_t v(x,t) + \sum_{i=1}^{n} v_i(x,t)\, \partial_{x_i} v(x,t)\Big)& \\ - \sum_{i=1}^{n} \partial_{x_i}\Big(\rho(x,t)\, \sigma_i(\mathrm{D}_x v(x,t),x,t)\Big)& \\ = -\mathrm{D}_x\Big(p(\rho(x,t),x,t)\Big) + \rho(x,t)\, b(x,t).&\end{aligned}$$

σ describes the frictional stress, p describes the pressure, and b describes the body force. In order to rewrite the above equations of motion as an abstract problem we introduce the following new physical field,

$$u(x,t) := \partial_t v(x,t).$$

Solving the balance of mass PDE for ρ we can write,

$$\rho(x,t) = \Phi[u](x,t).$$

Differentiating the balance of linear momentum PDE w.r.t. t and inserting the above solution operator we obtain,

$$\begin{aligned}
&\partial_t u(x,t) - \sum_{i=1}^{n} \partial_{x_i}\Big(\frac{\partial \sigma_i}{\partial w_j}(\mathrm{D}_x v(x,t),x,t)\,\partial_{x_j} u(x,t)\Big) \\
&= \sum_{i=1}^{n} \partial_{x_i}\Big(\frac{\partial \sigma_i}{\partial t}(\mathrm{D}_x v(x,t),x,t)\Big) \\
&\quad - \sum_{i=1}^{n} u_i(x,t)\,\partial_{x_i} v(x,t) - \sum_{i=1}^{n} v_i(x,t)\,\partial_{x_i} u(x,t) \\
&\quad + \sum_{i=1}^{n} \partial_t\Big(\frac{1}{\Phi[u](x,t)}\Big(\partial_{x_i}\Phi[u](x,t)\Big)\sigma_i(\mathrm{D}_x v(x,t),x,t)\Big) \\
&\quad - \partial_t\Big(\frac{1}{\Phi[u](x,t)}\mathrm{D}_x\Big(p(\Phi[u](x,t),x,t)\Big)\Big) + \partial_t b(x,t).
\end{aligned}$$

We make the following definitions,

$$A_{ij}[u](x,t) := \frac{\partial \sigma_i}{\partial w_j}(\mathrm{D}_x v(x,t),x,t),$$

$$\begin{aligned}
f[u](x,t) &:= \sum_{i=1}^{n} \partial_{x_i}\Big(\frac{\partial \sigma_i}{\partial t}(\mathrm{D}_x v(x,t),x,t)\Big) \\
&\quad - \sum_{i=1}^{n} u_i(x,t)\,\partial_{x_i} v(x,t) - \sum_{i=1}^{n} v_i(x,t)\,\partial_{x_i} u(x,t) \\
&\quad + \sum_{i=1}^{n} \partial_t\Big(\frac{1}{\Phi[u](x,t)}\Big(\partial_{x_i}\Phi[u](x,t)\Big)\sigma_i(\mathrm{D}_x v(x,t),x,t)\Big) \\
&\quad - \partial_t\Big(\frac{1}{\Phi[u](x,t)}\mathrm{D}_x\Big(p(\Phi[u](x,t),x,t)\Big)\Big) + \partial_t b(x,t).
\end{aligned}$$

Now the above balance of linear momentum PDE can be rewritten as an abstract parabolic problem for the physical field $u(x,t)$ in the sense of Sect. 2.1,

$$\partial_t u(x,t) - \sum_{i,j=1}^{n} \partial_{x_i}\Big(A_{ij}[u](x,t)\,\partial_{x_j} u(x,t)\Big) = f[u](x,t). \tag{P}$$

Remarks. The above examples demonstrate that some single PDE problems from continuum mechanics immediately fit into the scheme of our general theory, whereas others have to undergo some transformations first before our general theory applies. In particular, we see that in order to define such a transformation it is often necessary to solve a 'well behaved' PDE problem first, and that the resulting transformed problem rather fits into the scheme of our abstract theory than our hyperbolic–parabolic theory. Of course, there are many more single PDE problems from continuum mechanics to which our general theory is applicable. Some prominent examples are the plate theories of Reissner–Mindlin type and of Kirchhoff–Love type, the heat flow of Müller type, the multipolar compressible fluid flow, the phase transitions of Cahn–Allen type and of Cahn–Hilliard type, and magnetodynamics.

3 Statement of the Theorem

Now we come to the core of our work. In this Section we formulate sufficient conditions on the constitutive equations, such that the corresponding coupled system (H)–(P) admits a unique smooth solution. As mentioned in the introduction, these conditions constitute the mathematical counterpart to those conditions obtained by exploiting the second law of thermodynamics. For both, the abstract problem and the hyperbolic-parabolic problem, we assume a symmetry condition, an ellipticity condition and a compatibility condition. They can be considered as structural conditions that guarantee a reasonable mathematical structure for an isolated single problem. The other conditions concern the mutual couplings and consequently are of particular interest in view of their physical interpretation. For the case of the abstract problem we recall that the constitutive quantities are general history functionals in the physical fields. We assume a boundedness condition and a Lipschitz continuity condition saying that the constitutive quantities are bounded and Lipschitz continuous operators when acting on respective function spaces. Of course, the exploitation of these coupling conditions is technical and requires some mathematical background. However, for the case of the hyperbolic-parabolic problem the exploitation of the coupling conditions is very easy. Using the notation from Sect. 2, it turns out that the coupling conditions reduce to some linear algebraic inequalities for the orders $M^{F,0}_{ji_j,lk_l}, \ldots, M^{f,1}_{ji_j,1k_1}$ of the derivatives of the underlying physical fields that occur in the constitutive functions. Actually, these coupling conditions look most similar to the conditions obtained by exploiting the second law of thermodynamics.

3.1 The Abstract Problem

In this Subsection we formulate our conditions on the constitutive quantities for the case of the abstract problem (H)–(P) from Sect. 2.1. In view of the above classification, (A1), (A2) and (A3) denote the structural conditions, whereas (A4) and (A5) denote the proper coupling conditions.

(A1) Let the following symmetry condition hold for $j = 1, 3$,

$$A_{ji_j,\beta\alpha}[u](x,t) = \Big(A_{ji_j,\alpha\beta}[u](x,t)\Big)^T.$$

(A2) Let the following Legendre–Hadamard condition of strong ellipticity hold,

$$\sum_{|\alpha|,|\beta|=m_{ji_j}} \eta^T \Big(A_{ji_j,\alpha\beta}[u](x,t)\xi^\alpha \xi^\beta\Big)\eta \geq c|\xi|^{2m_{ji_j}} |\eta|^2.$$

(A3) Let the following implication hold,

$$\partial_t^\kappa u_{1i_1}\Big|_{t=0} = 0, \qquad \partial_t^{\kappa+1} u_{1i_1}\Big|_{t=0} = 0, \qquad \partial_t^\kappa u_{2i_2}\Big|_{t=0} = 0,$$

$$\partial_t^\kappa u_{3i_3}\Big|_{t=0} = 0 \qquad (\kappa = 0, \dots, k+1)$$

$$\Longrightarrow$$

$$\partial_t^k f_{ji_j}[u]\Big|_{t=0} = 0.$$

This is a compatibility condition.

(A4) Let an implication of the following form hold,

$$\|u_{lk_l}\|_{\mathcal{U}_{ji_j,lk_l}(T)} \leq R$$

$$\Longrightarrow$$

$$\big\|A_{ji_j,\alpha\beta}[u]\big\|_{\mathcal{A}_{ji_j}(T)} \leq \Phi(R,T),$$

$$\big\|f_{ji_j}[u]\big\|_{\mathcal{F}_{ji_j}(T)} \leq \Phi(R,T).$$

This is a boundedness condition.

(A5) Let an implication of the following form hold:

$$\|u^1_{lk_l}\|_{\overline{\mathcal{U}}_{ji_j,lk_l}(T)}, \|u^2_{lk_l}\|_{\overline{\mathcal{U}}_{ji_j,lk_l}(T)} \leq R,$$

$$\|u^2_{lk_l} - u^1_{lk_l}\|_{\underline{\mathcal{U}}_{ji_j,lk_l}(T)} \leq S$$

$$\Longrightarrow$$

$$\big\|A_{ji_j,\alpha\beta}[u^2] - A_{ji_j,\alpha\beta}[u^1]\big\|_{\mathcal{A}_{ji_j}(T)} \leq \Phi(R,T)\, S,$$

$$\big\|f_{ji_j}[u^2] - f_{ji_j}[u^1]\big\|_{\mathcal{F}_{ji_j}(T)} \leq \Phi(R,T)\, S.$$

This is a Lipschitz continuity condition.

In the above conditions the $\mathcal{U}_{ji_j,lk_l}(T)$, $\overline{\mathcal{U}}_{ji_j,lk_l}(T)$, $\underline{\mathcal{U}}_{ji_j,lk_l}(T)$, $\mathcal{A}_{ji_j}(T)$ and $\mathcal{F}_{ji_j}(T)$ denote suitable function spaces. For a precise definition of these function spaces we refer the reader to [3], [4], [5]. As mentioned above, the above conditions guarantee that the abstract problem (H)–(P) is reasonable from a mathematical point of view. More precisely, we have the following theorem.

Theorem 1 (Local Existence, Uniqueness, Regularity) *Let the assumptions* (A1),...,(A5), *hold. Then, the abstract problem* (H)–(P) *has a unique smooth local in time solution* $u = (u_{11}, \ldots, u_{3I_3})$.

For a precise formulation of the above theorem we refer the reader to [3], [4], [5].

3.2 The Hyperbolic–Parabolic Problem

In this Subsection we formulate our conditions on the constitutive quantities for the case of the hyperbolic-parabolic problem (H)–(P) from Sect. 2.2. Therefore, we make the following definition,

$$A_{ji_j,\alpha\beta}[u](x,t) := \frac{\partial F_{ji_j,\alpha}}{\partial(\partial_x^\beta u_{ji_j})}(U^F_{ji_j}, x, t).$$

In view of the above classification, (B1),...,(B4) denote the structural conditions, whereas (B5) denotes the proper coupling condition.

(B1) Let the constitutive functions $F_{ji_j,\alpha}$ and f_{ji_j} be smooth.
(B2) Let the following symmetry condition hold for $j = 1, 3$,

$$A_{ji_j,\beta\alpha}[u](x,t) = \Big(A_{ji_j,\alpha\beta}[u](x,t)\Big)^T.$$

We note that by Poincaré's lemma this symmetry condition is equivalent to the following integrability condition,

$$F_{ji_j,\alpha}(U^F_{ji_j}, x, t) = \frac{\partial \Psi_{ji_j}}{\partial(\partial_x^\alpha u_{ji_j})}(U^F_{ji_j}, x, t).$$

(B3) Let the following Legendre–Hadamard condition of strong ellipticity hold,

$$\sum_{|\alpha|,|\beta|=m_{ji_j}} \eta^T\Big(A_{ji_j,\alpha\beta}[u](x,t)\xi^\alpha\xi^\beta\Big)\eta \geq c|\xi|^{2m_{ji_j}}|\eta|^2.$$

(B4) Let the following compatibility condition hold,

$$\partial_t^k\Big(f_{ji_j}(0,x,t)\Big)\Big|_{t=0} = 0 \qquad (k = 0, \ldots, \bar{l} + \bar{k}).$$

(B5) Let the following linear algebraic inequalities hold,

$$\begin{array}{lll} M^{F,0}_{1i_1,1i_1} = m_{1i_1}, & M^{F,1}_{1i_1,1i_1} = -\infty, & M^{f,0}_{1i_1,1i_1} \le m_{1i_1}, \\ M^{f,1}_{1i_1,1i_1} \le 0. & M^{F,0}_{2i_2,2i_2} = m_{2i_2}, & M^{f,0}_{2i_2,2i_2} \le 2m_{2i_2} - 1, \\ M^{F,0}_{3i_3,3i_3} = m_{3i_3}, & M^{f,0}_{3i_3,3i_3} \le 2m_{3i_3} - 1. & \end{array}$$

Throughout this paper the notation $M^{\phi,p}_{ji_j,lk_l} = -\infty$ means that the constitutive function ϕ_{ji_j} is independent the physical field $\partial_t^p u_{lk_l}$. Moreover, let additional linear algebraic inequalities of the following form hold,

$$M^{\phi,\nu}_{ji_j,lk_l} \le \Phi^{\phi,\nu}_{ji_j,lk_l}(m_{ji_j} m_{lk_l}).$$

For a precise definition of the last inequalities in (B5) we refer the reader to [3], [4], [5]. We will exploit the inequalities in (B5) in detail for the case of two coupled problems in Sect. 5. As mentioned above, the above inequalities constitute the mathematical counterpart to the exploitation of the second law of thermodynamics. In particular, (B2) imposes an integrability condition. In applications to elasticity this integrability condition means precisely that we assume the a priori existence of a free energy. However, the existence of a free energy is often a result of the exploitation of the second law of thermodynamics. This is a nice example how physical conditions and mathematical conditions work hand in hand. Now, applying our abstract theory to the hyperbolic-parabolic problem, we obtain the following corollary.

Corollary 2 (Local Existence, Uniqueness, Regularity) *Let the assumptions* (B1),...,(B5), *hold. Then, the hyperbolic-parabolic problem* (H)–(P) *has a unique smooth local in time solution* $u = (u_{11}, \ldots, u_{3I_3})$.

For a precise formulation of the above corollary we refer the reader to [3], [4], [5].

4 Sketch of Proof (Energy Method)

In order to prove our main theorem we make use of the so called energy method. As mentioned above, Dafermos–Hrusa, Kato and Majda have applied the energy method to single hyperbolic problems before, see [1], [15] and [17], whereas Jiang–Racke have applied the energy method to problems in thermoelasticity, see [14]. The proof proceeds in several steps:

1. First, we define of a sequence $\{u^\nu\}_{\nu=0}^{\infty}$ of approximate solutions with the help of the following recursion scheme:
   (a) Let $u^0 := 0$.
   (b) Insert the previous solution u^ν into the coefficients $A_{ji_j,\alpha\beta}[\cdot]$ and right hand sides $f_{ji_j}[\cdot]$.

(c) Solve the linearized problem for $u^{\nu+1}$.
The crucial point in this step is that we cannot make use of the full regularity of u^ν as we insert it into the $A_{ji_j,\alpha\beta}[\,\cdot\,]$ and $f_{ji_j}[\,\cdot\,]$. Therefore, following the ideas of Jiang–Racke, see [14], we linearize the single systems successively. In order to simplify the notation, we rewrite our abstract problem (H)–(P) in the following way,

$$\partial_t v_k + \boldsymbol{B}_k[v](\mathrm{D}_x, x, t)v_k = g_k[v](x,t) \qquad (k = 1, \ldots, K),$$

where the $\boldsymbol{B}_k[v](\mathrm{D}_x, x, t)$ denote suitable differential operators. Now our linearization procedure is of the following form,

$$\partial_t v_k^{\nu+1} + \boldsymbol{B}_k[\ldots, v_{k-1}^{\nu+1}, v_k^\nu, \ldots](\mathrm{D}_x, x, t)v_k^{\nu+1}$$
$$= g_k[\ldots, v_{k-1}^{\nu+1}, v_k^\nu, \ldots](x,t).$$

2. Next, we derive a–priori estimates for the sequence $\{u^\nu\}_{\nu=0}^\infty$ of approximate solutions using Galerkin's method. In view of the regularity of the $u_{ji_j}^\nu$ we have the following correspondence,
   - $j = 1$ corresponds to testing $\partial_t^2 u + (-\Delta)^m u = f$ with $\partial_t u$,
   - $j = 2$ corresponds to testing $\partial_t u + (-\Delta)^m u = f$ with u,
   - $j = 3$ corresponds to testing $\partial_t u + (-\Delta)^m u = f$ with $\partial_t u$.

   One crucial point in this step is that we need an elliptic regularity theory for linear elliptic differential operators with minimal regularity in the coefficients. Such a theory was previously developed by the author, see [2]. Another crucial point in this step is that we have to find suitable scales of function spaces. Again following the ideas of Jiang–Racke, see [14], we observe that we can find some $\mu \geq 1$ and corresponding scales of function spaces such that for any double index (j, i_j) one temporal derivative of u_{ji_j} corresponds to μ spatial derivatives of u_{ji_j}.
3. Next, with the help of the above a–priori estimates we show that the sequence $\{u^\nu\}_{\nu=0}^\infty$ of approximate solutions is bounded with respect to some high norm and contractive with respect to some low norm.
4. Finally, we pass to the limit $\nu \longrightarrow \infty$. In particular, we show that the sequence $\{u^\nu\}_{\nu=0}^\infty$ of approximate solutions converges to some limit function u and that u is the unique solution to our abstract problem (H)–(P). The crucial point in this step is that we make use of interpolation inequalities in non–dual spaces. Consequently, we cannot make direct use of Banach's Fixed Point Theorem (Kato's direct method), but we rather pass to the limit 'by hand'. This procedure is similar to the procedure of Majda, see [17]. However, it is different from the procedures of Dafermos–Hrusa, Kato and also Jiang–Racke, see [1], [14] and [15].

5 Application to Two Coupled Systems

In this Section, we exploit the coupling conditions of Subsect. 3.2 for the case of two coupled problems. As we have pointed out, the conditions (B1),…,(B4) are mere structural conditions, whereas the proper coupling conditions are given by (B5). Therefore, we make the following definition.

Definition 1 *We say that the coupling in the hyperbolic–parabolic problem* (H)–(P) *is* admissible, *if the coupling conditions* (B5) *hold.*

Now, given the structural conditions (B1),...,(B4), the corollary in Subsect. 3.2 says that the hyperbolic–parabolic system (H)–(P) admits a unique smooth solution provided that the coupling is admissible. We recall that the coupling conditions (B5) are linear algebraic inequalities for the orders $M^{F,0}_{ji_j,lk_l},\dots,M^{f,1}_{ji_j,1k_1}$ of the derivatives of the underlying physical fields that occur in the constitutive functions. In summary, exploiting the coupling conditions of Subsect. 3.2 gives an answer to the following question:

> *How many derivatives of each physical field are allowed to occur in the respective constitutive functions, such that the coupled hyperbolic–parabolic problem* (H)–(P) *is reasonable from a mathematical point of view?*

We have already pointed out that our theory is of rather general character, and in Subsect. 2.3 we have given some examples from continuum mechanics that can be rewritten as single problems of type (H) or (P). Consequently, the following investigation is not dedicated to a particular problem from continuum mechanics, but rather serves as a guideline to characterize those mutual couplings between arbitrary physical fields that are reasonable from a mathematical point of view.

5.1 Hyperbolic–Hyperbolic Systems

In this subsection we assume that the equations of motion for the two physical fields, u and v, admit the following form,

$$\partial_t^2 u + \sum_{|\alpha|=0}^{m_u} (-1)^{|\alpha|}\partial_x^\alpha\Big(F_\alpha(U^F,V^F,x,t)\Big) = f(U^f,V^f,x,t), \tag{H}$$

$$\partial_t^2 v + \sum_{|\alpha|=0}^{m_v} (-1)^{|\alpha|}\partial_x^\alpha\Big(G_\alpha(U^G,V^G,x,t)\Big) = g(U^g,V^G,x,t), \tag{H}$$

where

$$
\begin{aligned}
U^F &= (u, \ldots, \mathrm{D}_x^{m_u} u), \\
V^F &= (v, \ldots, \mathrm{D}_x^{M_v^{F,0}} v, \partial_t v, \ldots, \mathrm{D}_x^{M_v^{F,1}} \partial_t v), \\
U^f &= (u, \ldots, \mathrm{D}_x^{m_u} u, \partial_t u), \\
V^f &= (v, \ldots, \mathrm{D}_x^{M_v^{f,0}} v, \partial_t v, \ldots, \mathrm{D}_x^{M_v^{f,1}} \partial_t v), \\
U^G &= (u, \ldots, \mathrm{D}_x^{M_u^{G,0}} u, \partial_t u, \ldots, \mathrm{D}_x^{M_u^{G,1}} \partial_t u), \\
V^G &= (v, \ldots, \mathrm{D}_x^{m_v} v), \\
U^g &= (u, \ldots, \mathrm{D}_x^{M_u^{g,0}} u, \partial_t u, \ldots, \mathrm{D}_x^{M_u^{g,1}} \partial_t u), \\
V^g &= (v, \ldots, \mathrm{D}_x^{m_v} v, \partial_t v).
\end{aligned}
$$

Exploiting the coupling conditions (B5) we obtain the following lemma.

Lemma 1 *The coupling in the above problem* (H)-(H) *is admissible for the following choice of the parameters* M*,*

$$
\begin{aligned}
&M_v^{F,0} \leq m_v - m_u, \qquad M_v^{F,1} = -\infty, \qquad M_v^{f,0} \leq m_v, \qquad M_v^{f,1} \leq 0, \\
&M_u^{G,0} \leq m_u - m_v, \qquad M_u^{G,1} = -\infty, \qquad M_u^{g,0} \leq m_u, \qquad M_u^{g,1} \leq 0.
\end{aligned}
$$

5.2 Hyperbolic–Parabolic Systems

In this subsection we assume that the equations of motion for the two physical fields, u and v, admit the following form,

$$
\partial_t^2 u + \sum_{|\alpha|=0}^{m_u} (-1)^{|\alpha|} \partial_x^\alpha \Big(F_\alpha(U^F, V^F, x, t) \Big) = f(U^f, V^f, x, t), \tag{H}
$$

$$
\partial_t v + \sum_{|\alpha|=0}^{m_v} (-1)^{|\alpha|} \partial_x^\alpha \Big(G_\alpha(U^G, V^G, x, t) \Big) = g(U^g, V^G, x, t), \tag{P}
$$

where

$$
\begin{aligned}
U^F &= (u, \ldots, \mathrm{D}_x^{m_u} u), \\
V^F &= (v, \ldots, \mathrm{D}_x^{M_v^F} v), \\
U^f &= (u, \ldots, \mathrm{D}_x^{m_u} u, \partial_t u), \\
V^f &= (v, \ldots, \mathrm{D}_x^{M_v^f} v), \\
U^G &= (u, \ldots, \mathrm{D}_x^{M_u^{G,0}} u, \partial_t u, \ldots, \mathrm{D}_x^{M_u^{G,1}} \partial_t u), \\
V^G &= (v, \ldots, \mathrm{D}_x^{m_v} v), \\
U^g &= (u, \ldots, \mathrm{D}_x^{M_u^{g,0}} u, \partial_t u, \ldots, \mathrm{D}_x^{M_u^{g,1}} \partial_t u), \\
V^g &= (v, \ldots, \mathrm{D}_x^{2m_v - 1} v).
\end{aligned}
$$

Exploiting the coupling conditions (B5) we obtain the following lemma.

Lemma 2 *The coupling in the above problem* (H)-(P) *is admissible for the following choice of the parameters M,*

$$M_v^F \le m_v - m_u, \qquad M_v^f \le m_v, \qquad M_u^{G,0} \le m_u, \qquad M_u^{G,1} \le 0,$$
$$M_u^{g,0} \le m_u + m_v, \qquad M_u^{g,1} \le m_v,$$

or

$$M_v^F \le 2m_v - m_u, \quad M_v^f \le 2m_v, \quad M_u^{G,0} \le m_u - m_v, \quad M_u^{G,1} = -\infty,$$
$$M_u^{g,0} \le m_u, \qquad M_u^{g,1} \le 0.$$

5.3 Parabolic–Parabolic Systems

In this subsection we assume that the equations of motion for the two physical fields, u and v, admit the following form,

$$\partial_t u + \sum_{|\alpha|=0}^{m_u} (-1)^{|\alpha|} \partial_x^\alpha \Big(F_\alpha(U^F, V^F, x, t) \Big) = f(U^f, V^f, x, t), \tag{P}$$

$$\partial_t v + \sum_{|\alpha|=0}^{m_v} (-1)^{|\alpha|} \partial_x^\alpha \Big(G_\alpha(U^G, V^G, x, t) \Big) = g(U^g, V^G, x, t), \tag{P}$$

where

$$\begin{aligned}
U^F &= (u, \ldots, \mathrm{D}_x^{m_u} u),\\
V^F &= (v, \ldots, \mathrm{D}_x^{M_v^F} v),\\
U^f &= (u, \ldots, \mathrm{D}_x^{2m_u - 1} u),\\
V^f &= (v, \ldots, \mathrm{D}_x^{M_v^f} v),\\
U^G &= (u, \ldots, \mathrm{D}_x^{M_u^G} u),\\
V^G &= (v, \ldots, \mathrm{D}_x^{m_v} v),\\
U^g &= (u, \ldots, \mathrm{D}_x^{M_u^g} u),\\
V^g &= (v, \ldots, \mathrm{D}_x^{2m_v - 1} v).
\end{aligned}$$

Exploiting the coupling conditions (B5) we obtain the following lemma.

Lemma 3 *The coupling in the above problem* (P)-(P) *is admissible for the following choice of the parameters M,*

$$M_v^F \le m_v - 1, \quad M_v^f \le m_u + m_v - 1, \quad M_u^G \le m_u, \quad M_u^g \le m_u + m_v,$$

or

$$M_v^F \le 2m_v - 1, \quad M_v^f \le m_u + 2m_v - 1, \quad M_u^G \le m_u - m_v, \quad M_u^g \le m_u,$$

or

$$M_v^F \le 2m_v, \quad M_v^f \le m_u + 2m_v, \quad M_u^G \le m_u - m_v - 1, \quad M_u^g \le m_u - 1,$$

or

$$M_v^F \le 2m_v - m_u - 1, \quad M_v^f \le 2m_v - 1, \quad M_u^G \le 2m_u - m_v, \quad M_u^g \le 2m_u.$$

References

1. Dafermos, C., and Hrusa, W. (1985) Energy methods for quasilinear initial boundary value problems. Arch. Rat. Mech. Analysis **87**, 267–292
2. Ebenfeld, S. (2002) L^2–regularity theory of linear strongly elliptic dirichlet systems of order $2m$ with minimal regularity in the coefficients. Quat. Appl. Math., to appear
3. Ebenfeld, S. (2002) Nonlinear initial boundary value problems of hyperbolic–parabolic type. A general investigation of admissible couplings between systems of higher order. Part 1: A–priori estimates for linear systems. Math. Methods Appl. Sci., **25** (3), 179–212
4. Ebenfeld, S. (2002) Nonlinear initial boundary value problems of hyperbolic–parabolic type. A general investigation of admissible couplings between systems of higher order. Part 2: Local existence, uniqueness and regularity of solutions Math. Methods Appl. Sci., **25** (3), 213–240
5. Ebenfeld, S. (2002) Nonlinear initial boundary value problems of hyperbolic–parabolic type. A general investigation of admissible couplings between systems of higher order. Part 3: applications to hyperbolic–parabolic systems Math. Methods Appl. Sci., **25** (3), 241–262
6. Hauser, R., and Kirchner, N. (2002) A historical note on the entropy principle of Müller and Liu. Cont. Mech. Thermodynamics, **14** (2), 223-226
7. Hutter, K., and Kirchner, N. (2002) Elasto-plastic behaviour of granular material with an additional scalar degree of freedom. Porous Media (Eds. Blum, J., and Ehlers, W.), Springer Verlag
8. Hutter, K., and Laloui, L., and Vulliet, L. (1999) Thermodynamically based mixture models of saturated and unsaturated soils. Mech. Cohesive-Frictional Mat. **4** (4), 295–338
9. Hutter, K., and Rickenmann, D., and Svendsen, B. (1996) Debris flow modeling: A review. Cont. Mech. Thermodynamics **8**, 1–35
10. Hutter, K., and Svendsen, B. (1995) On the Thermodynamics of a mixture of isotropic viscous materials with kinematic hardening. Int. J. Eng. Sci. **33**, 2021–2054
11. Hutter, K., and Wang, Y. (1999) Comparison of two entropy principles and their applications in granular flows with/without fluid. Arch. Mechanics **51** (5), 605–632

12. Hutter, K., and Wang, Y. (1999) A constitutive model of multiphase mixtures and its application in shearing flows of saturated soil-fluid mixtures. Granular Matter **1**, 163–181
13. Hutter, K., and Wang, Y. (1999) A constitutive theory of fluid saturated granular materials and its application in gravitational flows Rheologica Acty **38**, 214–223
14. Jiang, S., and Racke, R. (2000) Evolution equations in thermoelasticity. Chapman–Hall, Boca Raton
15. Kato, T. (1985) Abstract evolution equations and nonlinear mixed problems. Accademia Nazionale dei Lincei and Scuola Normale Superiore, Pisa
16. Liu, I. (1972) Methods of Lagrange multipliers for exploitation of the entropy principle. Arch. Rat. Mech. Analysis **46**, 131–148
17. Majda, A. (1984) Compressible fluid flow and systems of conservation laws in several space variables. Springer, New York
18. Müller, I., and Ruggeri, T. (1998) Rational extended thermodynamics. Springer, New York

Justification of Homogenized Models for Viscoplastic Bodies with Microstructure

Hans–Dieter Alber

Department of Mathematics, Darmstadt University of Technology
Schlossgartenstr. 7, D-64289 Darmstadt, Germany
alber@mathematik.tu-darmstadt.de

received 14 Aug 2002 — accepted 7 Nov 2002

Abstract. We justify the formal homogenization of the quasistatic initial boundary value problem with internal variables, called the microscopic problem, which models the deformation behavior of viscoplastic bodies. To this end it is first shown that the formally derived homogenized initial-boundary value problem has a solution. From this solution an asymptotic solution of the microscopic problem is constructed, and it is shown that the difference of the exact solution and the asymptotic solution tends to zero if the length scale of the microstructure converges to zero. Our results are proved for viscoplastic material behavior that can be modeled by constitutive equations of monotone type with linear hardening terms. For technical reasons we are only able to prove the convergence result locally in time and for smooth data.

1 Introduction and Statement of Results

The numerical simulation of viscoplastic material behavior is expensive, since the dependence of the stress field on the deformation history must be taken into account. The difficulties increase for viscoplastic bodies with a microstructure caused by phase changes or by other spatial variations of the material properties, because of the fine discretization required by the microstructure. If the length scale of the microstructure is small, effective numerical simulations can thus not be based on a mathematical model which faithfully describes this microstructure. For viscoplastic bodies it is therefore of particular interest to derive from this faithful model, which we call the microscopic model, a homogenized or macroscopic model, which describes a body without microstructure, but which shows the same overall behavior as the body with microstructure.

In this article we study the justification of the formally derived homogenized model for a viscoplastic body. To this end we show that from the solution of the homogenized model an asymptotic solution of the microscopic model can be derived. We prove that the difference of the exact solution and the asymptotic solution tends to zero if the length scale of the microstructure converges to zero. For technical reasons we are only able to prove this result locally in time and for smooth data.

The microscopic model consists of a quasistatic initial-boundary value problem. The formulation of this problem is based on the assumption that only small strains occur: Let $\Omega \subseteq \mathbb{R}^3$ denote the set of material points of the body, let $\mathcal{S}^3$ denote the set of symmetric 3×3–matrices, and let $u(x,t) \in \mathbb{R}^3$ be the unknown displacement of the material point x at time t. Furthermore, $T(x,t) \in \mathcal{S}^3$ is the unknown Cauchy stress tensor and $z(x,t) \in \mathbb{R}^N$ denotes the unknown vector of internal variables. The model equations of the microscopic problem are

$$-\operatorname{div}_x T(x,t) = b(x,t)\,, \tag{1}$$

$$T(x,t) = \mathcal{D}[\frac{x}{\eta}]\Big(\varepsilon\big(\nabla_x u(x,t)\big) - Bz(x,t)\Big)\,, \tag{2}$$

$$\frac{\partial}{\partial t}\, z(x,t) \in f\Big(\frac{x}{\eta}\,,\, \varepsilon\big(\nabla_x u(x,t)\big)\,,\, z(x,t)\Big)\,, \tag{3}$$

$$z(x,0) = z^{(0)}(x)\,, \tag{4}$$

which must hold for $x \in \Omega$ and $t \in [0,\infty)$. For simplicity we only consider the Dirichlet boundary condition

$$u(x,t) = \gamma_D(x,t)\,, \tag{5}$$

which must be satisfied for $(x,t) \in \partial\Omega \times [0,\infty)$. Here $\nabla_x u(x,t)$ denotes the 3×3–matrix of first order derivatives of u, the deformation gradient, $\big(\nabla_x u(x,t)\big)^T$ denotes the transposed matrix,

$$\varepsilon\big(\nabla_x u(x,t)\big) = \frac{1}{2}\Big(\nabla_x u(x,t) + \big(\nabla_x u(x,t)\big)^T\Big) \in \mathcal{S}^3$$

is the strain tensor, and $B : \mathbb{R}^N \to \mathcal{S}^3$ is a linear mapping, which assigns to the vector $z(x,t)$ the plastic strain tensor $\varepsilon_p(x,t) = Bz(x,t)$. For every $y \in \mathbb{R}^3$ we denote by $\mathcal{D}[y] : \mathcal{S}^3 \to \mathcal{S}^3$ a linear, symmetric, positive definite mapping, the elasticity tensor. It is assumed that the mapping $y \mapsto \mathcal{D}[y]$ is periodic with a rectangular periodicity cell $Y \subseteq \mathbb{R}^3$. The number $\eta > 0$ is the scaling parameter of the microstructure.

The given data are the volume force $b : \Omega \times [0,\infty) \to \mathbb{R}^3$, the boundary displacement $\gamma_D : \partial\Omega \times [0,\infty) \to \mathbb{R}^3$ and the initial values $z^{(0)} : \Omega \to \mathbb{R}^N$ of the vector of internal variables. $f : \mathbb{R}^3 \times \mathcal{S}^3 \times \mathbb{R}^N \to 2^{\mathbb{R}^N}$ in (3) is a given function. The equation (2) and the differential inclusion (3) together determine the dependence of the stress $T(x,t)$ on the strain history $s \mapsto \varepsilon\big(\nabla_x u(x,s)\big)$. They are the constitutive relations which model the inelastic behavior of the body. The choice of f is restricted by thermodynamical and mathematical requirements. In this article we assume that (3) belongs to the class of constitutive relations of monotone type with positive definite free energy. For this class the function f can be written in the form

$$f(y,\varepsilon,z) = g\big(y, -\rho\nabla_z\psi(y,\varepsilon,z)\big)\,, \quad (y,\varepsilon,z) \in \mathbb{R}^3 \times \mathcal{S}^3 \times \mathbb{R}^N\,, \tag{6}$$

with the constant mass density $\rho > 0$, with a suitable free energy ψ, which is a positive definite quadratic form

$$\rho\psi(y,\varepsilon,z) = \frac{1}{2}[\mathcal{D}[y](\varepsilon - Bz)] \cdot (\varepsilon - Bz) + \frac{1}{2}(Lz) \cdot z\,, \tag{7}$$

with respect to the variables (ε, z), and with a suitable function $g : \mathbb{R}^3 \times \mathbb{R}^N \to 2^{\mathbb{R}^N}$, which satisfies $0 \in g(y,0)$ and for which the function $z \mapsto g(y,z) : \mathbb{R}^N \to 2^{\mathbb{R}^N}$ is monotone for all $y \in \mathbb{R}^3$. In equation (7) we denote the scalar product of two matrices $\sigma, \tau \in \mathcal{S}^3$ by

$$\sigma \cdot \tau = \sum_{i,j=1}^{3} \sigma_{ij}\tau_{ij}\,,$$

and L denotes a symmetric $N \times N$–matrix. It is easily seen that this matrix is positive definite if and only if the quadratic form ψ is positive definite. We assume also that the function $y \mapsto g(y,z)$ is periodic with periodicity cell Y for all $z \in \mathbb{R}^N$.

We employ (2) and obtain by a simple computation $-\rho\nabla_z\psi(y,\varepsilon,z) = B^T\mathcal{D}[y](\varepsilon - Bz) - Lz = B^T T - Lz$, where $B^T : \mathcal{S}^3 \to \mathbb{R}^N$ is the mapping adjoint to B. Using this equation we can write the microscopic problem in the form

$$-\mathrm{div}_x\, T(x,t) = b(x,t)\,, \tag{8}$$

$$T(x,t) = \mathcal{D}[\frac{x}{\eta}]\Big(\varepsilon\big(\nabla_x u(x,t)\big) - Bz(x,t)\Big)\,, \tag{9}$$

$$z_t(x,t) \in g\big(\frac{x}{\eta},\, B^T T(x,t) - Lz(x,t)\big)\,, \tag{10}$$

$$z(x,0) = z^{(0)}(x)\,, \tag{11}$$

$$u(x,t) = \gamma_D(x,t), \quad (x,t) \in \partial\Omega \times [0,\infty)\,. \tag{12}$$

The class of constitutive equations of monotone type, which was introduced in the book [1], extends the class of generalized standard materials introduced by Halphen and Nguyen Quoc Son [23]. The class of generalized standard materials includes the classical constitutive equations such as the Prandtl-Reuss and Norton-Hoff laws, but it does not contain most constitutive equations developed in engineering in the last decades. To treat these constitutive equations it is therefore necessary to seek larger classes, for which existence theorems can be proved. One such class is the class of constitutive equations of monotone type. It has been shown in [1] that most constitutive equations lie outside even this larger class, and a further enlargement of this class by transformation methods has been discussed. Yet, a general mathematical existence theory for most of the constitutive equations used in practice is not available up to now. It is nevertheless an important mathematical goal to understand initial-boundary value problems to constitutive equations of monotone type as a basis for the investigation of still more general equations. For a discussion of these questions, for the existence theory and for an introduction to the mathematical literature in viscoplasticity we refer to [1,3,4,15–18,21] and to [19].

The class of constitutive equations of monotone type requires the free energy to be positive semi-definite. In this article we only consider the subclass of constitutive equations of monotone type with positive definite free energy because of the strong existence theorems available for this subclass, cf. [4,21], which allow to derive regularity and stability estimates. Constitutive equations with linear hardening are of this type.

The investigations in this article thus only form the beginning of the study of homogenization in viscoplasticity; in particular, homogenization of models to constitutive equations with positive semi-definite free energy remains to be considered in the future.

We are interested in the solution of (8) – (12) to quasiperiodic initial data of the form

$$z^{(0)}(x) = z_\eta^{(0)}(x) = z_0^{(0)}(x, \frac{x}{\eta}), \tag{13}$$

with a given function $z_0^{(0)} : \Omega \times \mathbb{R}^3 \to \mathbb{R}^N$, such that for all $x \in \Omega$ the function $y \mapsto z_0^{(0)}(x,y)$ is periodic with periodicity cell Y. We denote a solution of the microscopic problem to such initial data by (u_η, T_η, z_η). Since for small values of η the function $x \mapsto z_0^{(0)}(x, \frac{x}{\eta})$ is close to a periodic function with periodicity cell ηY, and since $x \mapsto \mathcal{D}[\frac{x}{\eta}]$ and $x \mapsto g(\frac{x}{\eta}, z)$ are periodic with this periodicity cell, one expects that also (u_η, T_η, z_η) will be close to a quasiperiodic function $(\hat{u}_\eta, \hat{T}_\eta, \hat{z}_\eta)$ of the form

$$\hat{u}_\eta(x,t) = u_0(x,t) + \eta u_1(x, \frac{x}{\eta}, t), \tag{14}$$

$$\hat{T}_\eta(x,t) = T_0(x, \frac{x}{\eta}, t), \tag{15}$$

$$\hat{z}_\eta(x,t) = z_0(x, \frac{x}{\eta}, t), \tag{16}$$

where the function $(x,y,t) \mapsto (u_1, T_0, z_0)(x,y,t) : \Omega \times \mathbb{R}^3 \times [0,\infty) \to \mathbb{R}^3 \times \mathcal{S}^3 \times \mathbb{R}^N$ is required to be periodic with respect to y and to have periodicity cell Y. In [2] it has been shown that if $(\hat{u}_\eta, \hat{T}_\eta, \hat{z}_\eta)$ is asymptotically equal to the solution (u_η, T_η, z_η) for $\eta \to 0$, then (u_0, u_1, T_0, z_0) and the overall stress T_∞ must satisfy the *homogenized initial-boundary value problem* formed by the equations

$$-\operatorname{div}_x T_\infty(x,t) = b(x,t)\,, \tag{17}$$

$$T_\infty(x,t) = \frac{1}{|Y|}\int_Y T_0(x,y,t)dy\,, \tag{18}$$

$$-\operatorname{div}_y T_0(x,y,t) = 0\,, \tag{19}$$

$$T_0(x,y,t) = \mathcal{D}[y]\Big(\varepsilon\big(\nabla_y u_1(x,y,t)\big) - Bz_0(x,y,t) + \varepsilon\big(\nabla_x u_0(x,t)\big)\Big), \tag{20}$$

$$\frac{\partial}{\partial t} z_0(x,y,t) \in g\big(y, B^T T_0(x,y,t) - Lz_0(x,y,t)\big)\,, \tag{21}$$

$$z_0(x,y,0) = z_0^{(0)}(x,y)\,, \tag{22}$$

which must hold for $(x,y,t) \in \Omega \times Y \times [0,\infty)$, and by the boundary condition

$$u_0(x,t) = \gamma_D(x,t)\,, \tag{23}$$

which must hold for $(x,t) \in \partial\Omega \times [0,\infty)$. The symbol $|Y|$ in (18) denotes the measure of Y.

Note that for x fixed the equations (19) – (22) together with the requirement that $y \mapsto (u_1, T_0)(x,y,t)$ must be periodic, which can be considered to be a boundary condition, define an initial-boundary value problem, the cell problem, in the domain $Y \times [0,\infty)$, the representative volume element. The cell problem is of the same form as the microscopic problem. u_1 is the microdisplacement, T_0 the microstress; the overall stress T_∞ is obtained via (18) by averaging of T_0 over the representative volume element, u_0 is the macrodisplacement. The term $\varepsilon(\nabla_x u_0(x,t))$ in (20) can be considered to be a homogeneous strain imposed on the representative volume element by the macrodisplacement. If the history $t \mapsto \varepsilon(\nabla_x u_0(x,t))$ of the macrostrain is known, then the function $(y,t) \mapsto (u_1, T_0, z_0)(x,y,t)$ and therefore also the function $t \mapsto T_\infty(x,t)$ can be determined from the cell problem. This dependence of T_∞ on $\varepsilon(\nabla_x u)$ defines a history functional

$$\big[t \mapsto \varepsilon\big(\nabla_x u_0(x,t)\big)\big] \mapsto \big[t \mapsto T_\infty(x,t) = \mathop{\mathcal{F}}_{s\le t}\big(\varepsilon\big(\nabla_x u_0(x,s)\big)\big)\big]\,,$$

the constitutive relation for the homogenized material modeled by the balance law (17), by this constitutive relation and by the boundary condition (23).

The goal of this article is to prove that indeed the solution (u_η, T_η, z_η) of (8) – (12) with initial data given by (13) is asymptotically equal to $(\hat{u}_\eta, \hat{T}_\eta, \hat{z}_\eta)$ for $\eta \to 0$, with u_0, u_1, T_0, z_0 determined from (17) – (23). However, since our estimates are not sharp enough to decide whether the term ηu_1 is present in (14), we actually prove that the solution is asymptotically equal to $(u_0, \hat{T}_\eta, \hat{z}_\eta)$. Moreover, because of technical reasons we are only able to prove that this result holds in a finite interval of time, and for smooth data and smooth functions $\mathcal{D}$ and g.

Rigorous mathematical investigation of homogenization has been carried out for many problems. Of particular importance for our problem are the investigations to the linear theory of elasticity. Other examples are the investigations to the nonlinear theory of elasticity, to the transport of neutrons, to problems of hydrodynamics and porous media, to linear viscoelasticity and electrodynamics, cf. for example [5–11,13,22,24–30,34,36]. However, the only rigorous mathematical investigations of homogenization in the theory of plastic or viscoplastic solids known to the author are [2,21]; this is in contrast to the importance of homogenization in solid mechanics, which is demonstrated by the many engineering publications devoted to the study of different aspects of homogenization in plasticity and viscoplasticity; we only mention [31–33,35,37–41].

In this article we study periodic microstructures. In [21] a homogenization result has been proved for a material with a random microstructure in one space dimension. This result is strictly one-dimensional and cannot be transferred to higher space dimensions, since in one space dimension the amplitude of the fast oscillations of the stress tends to zero when the length scale of the microstructure decreases to zero. This is not true in the higher dimensional case studied here.

Statement of the main results. To state the main results we need some notations and preparations.

If not stated otherwise we assume that $\Omega \subseteq \mathbb{R}^3$ is a bounded open set with C^1-boundary $\partial\Omega$. The periodicity cell $Y \subseteq \mathbb{R}^3$ is a cuboid. Let ∂Y_i and ∂Y_{-i} be parallel faces for $i = 1, 2, 3$, and let $y_i \in \mathbb{R}^3$ be the vector such that $\partial Y_i = y_i + \partial Y_{-i}$. We make Y into a manifold Y_{per} without boundary by identifying the points x and $x + y_i$ for all $x \in \partial Y_{-i}$, and by choosing the appropriate topology and parametrization. It is clear that every function on Y_{per} can be identified with a function on $\mathbb{R}^3$, which is periodic and has periodicity cell Y. If a function belongs to $C^m(Y_{\text{per}})$, then the corresponding periodic function belongs to $C^m(\mathbb{R}^3)$.

By T_e we denote a positive number (time of existence), and for $0 \le t \le T_e$ we set

$$\Omega_t = \Omega \times [0, t], \quad Y_{\text{per},t} = Y_{\text{per}} \times [0, t], \quad (\Omega \times Y_{\text{per}})_t = \Omega \times Y_{\text{per}} \times [0, t] .$$

If w is a function defined on $\Omega_t, Y_{\text{per},t}$ or $(\Omega \times Y_{\text{per}})_t$ and if $0 \le s \le t$, we denote the function $x \mapsto w(x, s)$ by $w(s)$. For a suitable subset Γ of $\mathbb{R}^n$ the scalar products on $L^2(\Gamma, \mathbb{R}^m)$ and on $L^2(\Gamma, \mathcal{S}^3)$ are denoted by

$$(\sigma, \tau)_\Gamma = \int_\Gamma \sigma(x) \cdot \tau(x)\, dx .$$

For $1 \le p \le \infty$ and for a Banach space V the Sobolev space of all functions which together with their weak derivatives up to order m belong to $L^p(\Gamma, V)$ is denoted by $H^p_m(\Gamma, V)$. The norm of $L^p(\Gamma, V)$ is $\|u\|_{\Gamma,p}$, and the norm of

$H^p_m(\Gamma, V)$ is $\|u\|_{\Gamma,p;m}$; for $p = 2$ we set $\|u\|_\Gamma = \|u\|_{\Gamma,2}$, $\|u\|_{\Gamma;m} = \|u\|_{\Gamma,2;m}$. Also, $\overset{\circ}{H}_m(\Gamma, V)$ denotes the closure of $C_0^\infty(\Gamma, V)$ in $H_m(\Gamma, V) = H^2_m(\Gamma, V)$.

We assume that the symmetric linear mapping $\mathcal{D}[y] : \mathcal{S}^3 \to \mathcal{S}^3$ is positive definite uniformly with respect to y, and that the mapping $y \mapsto \mathcal{D}[y]$ is bounded, periodic with periodicity cell Y and measurable. Measurability means that the coefficients in the tensorial representation of $\mathcal{D}[y]$ are measurable functions of y. These assumptions imply that a bounded, selfajoint, positive definite linear mapping $\sigma \mapsto \mathcal{D}\sigma : L^2(Y_{\text{per}}, \mathcal{S}^3) \to L^2(Y_{\text{per}}, \mathcal{S}^3)$ is defined by

$$(\mathcal{D}\sigma)(y) = D[y]\sigma(y) , \quad y \in Y_{\text{per}} .$$

With this mapping

$$[\sigma, \tau]_{\Omega\times Y} = (D\sigma, \tau)_{\Omega\times Y}$$

is a scalar product on $L^2(\Omega \times Y_{\text{per}}, \mathcal{S}^3)$. The norm associated to this scalar product is equivalent to the norm $\|\cdot\|_{\Omega\times Y}$.

If we fix t in the equations (17) – (20), (23) we obtain a linear boundary value problem, which slightly extends the classical *homogenized problem of linear elasticity theory.* Since we need solutions of this problem in the formulation of our results and in the proofs, we introduce and shortly discuss this problem here: To given functions $\hat{b} : \Omega \to \mathbb{R}^3$, $\hat{\gamma}_D : \partial\Omega \to \mathbb{R}^3$ and $\hat{\varepsilon}_p : \Omega \times Y \to \mathcal{S}^3$ we seek a solution $(u_0, u_1, T_\infty, T_0)$ with $(u_0, T_\infty) : \Omega \to \mathbb{R}^3 \times \mathcal{S}^3$ and $(u_1, T_0) : \Omega \times Y_{\text{per}} \to \mathbb{R}^3 \times \mathcal{S}^3$ of the equations

$$-\operatorname{div}_x T_\infty(x) = \hat{b}(x) , \tag{24}$$

$$T_\infty(x) = \frac{1}{|Y|} \int_Y T_0(x, y)\, dy , \tag{25}$$

$$-\operatorname{div}_y T_0(x, y) = 0 , \tag{26}$$

$$T_0(x, y) = \mathcal{D}[y]\Big(\varepsilon\big(\nabla_x u_0(x)\big) + \varepsilon\big(\nabla_y u_1(x, y)\big) - \hat{\varepsilon}_p(x, y)\Big) , \tag{27}$$

$$u_0(x) = \hat{\gamma}_D(x) , \quad x \in \partial\Omega . \tag{28}$$

To define a weak solution of this problem assume that $\hat{b} \in L^2(\Omega, \mathbb{R}^3)$, $\hat{\varepsilon}_p \in L^2(\Omega \times Y, \mathcal{S}^3)$ and $\hat{\gamma}_D \in H_1(\Omega, \mathbb{R}^3)$. We combine (24) and (25), multiply the resulting equation by $v_0 \in \overset{\circ}{H}_1(\Omega, \mathbb{R}^3)$ and integrate by parts. If we identify v_0 with the function $(x, y) \mapsto v_0(x)$, the resulting equation can be written as

$$\big(T_0, \varepsilon(\nabla_x v_0)\big)_{\Omega\times Y} = (\hat{b}, v_0)_{\Omega\times Y} , \tag{29}$$

where we also used that $T_0(x, y)$ is a symmetric matrix, hence $T_0 \cdot \nabla_x v_0 = T_0 \cdot \varepsilon(\nabla_x v_0)$. Furthermore, we multiply (26) by $v_1 \in L^2\big(\Omega, H_1(Y_{\text{per}}, \mathbb{R}^3)\big)$, integrate by parts and add the resulting equation to (29) to obtain

$$\big(T_0, \varepsilon(\nabla_x v_0) + \varepsilon(\nabla_y v_1)\big)_{\Omega\times Y} = (\hat{b}, v_0)_{\Omega\times Y} .$$

Insertion of (27) yields

$$[\varepsilon(\nabla_x u_0 + \nabla_y u_1) - \hat{\varepsilon}_p \,,\, \varepsilon(\nabla_x v_0 + \nabla_y v_1)]_{\Omega\times Y} = (\hat{b}, v_0)_{\Omega\times Y} \,. \tag{30}$$

A function

$$(u_0, u_1, T_\infty, T_0) \\ \in H_1(\Omega,\mathbb{R}^3) \times L^2(\Omega, H_1(Y_{\mathrm{per}},\mathbb{R}^3)) \times L^2(\Omega,\mathcal{S}^3) \times L^2(\Omega\times Y_{\mathrm{per}},\mathcal{S}^3)$$

is called weak solution of (24) – (28), if (25), (27) hold, if u_0 can be represented in the form $u_0 = \hat{\gamma}_D + w_0$ with $w_0 \in \overset{\circ}{H}_1(\Omega,\mathbb{R}^3)$, and if (30) is satisfied for all $(v_0, v_1) \in \overset{\circ}{H}_1(\Omega,\mathbb{R}^3) \times L^2(\Omega, H_1(Y_{\mathrm{per}},\mathbb{R}^3))$.

The following existence result is well known:

Lemma 1 *Let $\hat{b} \in L^2(\Omega,\mathbb{R}^3)$, $\hat{\gamma}_D \in H_1(\Omega,\mathbb{R}^3)$ and $\hat{\varepsilon}_p \in L^2(\Omega\times Y_{\mathrm{per}},\mathcal{S}^3)$. Then there is a unique weak solution $(u_0, u_1, T_\infty, T_0)$ of the Dirichlet problem* (24) – (28) *satisfying*

$$\int_Y u_1(x,y)\,dy = 0$$

for all $x \in \Omega$. Moreover, there is a constant C such that for $\hat{b} = \hat{\gamma}_D = 0$ the solution satisfies

$$\|u_0\|_{1,\Omega} + \Big(\int_\Omega \|u_1(x,\cdot)\|^2_{1,Y}\,dx\Big)^{1/2} \le C\|\hat{\varepsilon}_p\|_{\Omega\times Y} \,. \tag{31}$$

Proofs can be found for example in [7,30]. [6, pp. 1494] contains an existence proof for the corresponding scalar boundary value problem; the formulation of the homogenized boundary value problem given there is similar in spirit to (24) – (28), and the proof can be generalized to (24) – (28).

Before we can state our existence theorem for the homogenized problem of viscoplasticity, we finally need some assumptions and definitions for the function g: We assume that the mapping $y \mapsto g(y,z) : \mathbb{R}^3 \to 2^{\mathbb{R}^N}$ is periodic with periodicity cell Y for all $z \in \mathbb{R}^N$. As usual, $z \mapsto g(y,z)$ is said to be monotone if

$$(\zeta - \hat{\zeta}) \cdot (z - \hat{z}) \ge 0$$

for all $z, \hat{z} \in \mathbb{R}^N$ and all $\zeta \in g(y,z)$, $\hat{\zeta} \in g(y,\hat{z})$. This function is said to be maximal monotone if it does not have a proper monotone extension. It is well known that if $z \mapsto g(y,z)$ is maximal monotone, then the mapping $z \mapsto z + \lambda g(y,z)$ has a single valued inverse $j_\lambda[y] : \mathbb{R}^N \to \mathbb{R}^N$ for all $\lambda > 0$.

Theorem 2 (Existence and uniqueness of solutions for the homogenized problem of viscoplasticity) *Assume that the $N\times N$–matrix L in* (21) *is positive definite and that the mapping $g : \mathbb{R}^3 \times \mathbb{R}^N \to 2^{\mathbb{R}^N}$ satisfies the following three conditions:*

(i) $0 \in g(0)$.

(ii) $z \mapsto g(y,z)$ *is maximal monotone for all* $y \in \mathbb{R}^3$.

(iii) $y \mapsto j_\lambda[y](z) : \mathbb{R}^3 \to \mathbb{R}^N$ *is measurable for all* $\lambda > 0$ *and all* $z \in \mathbb{R}^N$.

Suppose that $b \in H_2^1(0,T_e;L^2(\Omega,\mathbb{R}^3))$ *and* $\gamma_D \in H_2^1(0,T_e;H_1(\Omega,\mathbb{R}^3))$. *Finally, assume that* $z_0^{(0)} \in L^2(\Omega \times Y_{\text{per}},\mathbb{R}^N)$ *and that there is* $\zeta \in L^2(\Omega \times Y_{\text{per}},\mathbb{R}^N)$ *such that*

$$\zeta(x,y) \in g\big(y, B^T T^{(0)}(x,y) - L z_0^{(0)}(x,y)\big)\,, \quad \text{a.e. in } \Omega \times Y_{\text{per}}\,, \tag{32}$$

where $(u_0^{(0)}, u_1^{(0)}, T_\infty^{(0)}, T_0^{(0)})$ *is a weak solution of the linear problem* (24) – (28) *to the data* $\hat b = b(0)$, $\hat\varepsilon_p = B z_0^{(0)}$, $\hat\gamma_D = \gamma_D(0)$.

Then to every $T_e > 0$ *there are solutions*

$$\begin{aligned}(u_0,u_1,T_\infty,T_0,z_0) &\in L^2(0,T_e;H_1(\Omega,\mathbb{R}^3)) \times L^2(\Omega_{T_e},H_1(Y_{\text{per}},\mathbb{R}^3))\\ &\times L^2(\Omega_{T_e},\mathcal{S}^3) \times L^2((\Omega\times Y_{\text{per}})_{T_e},\mathcal{S}^3) \times C(0,T_e;L^2(\Omega\times Y_{\text{per}},\mathbb{R}^N))\end{aligned}$$

of the homogenized initial-boundary value problem (17) – (23). *If a solution is given by* $(u_0,u_1,T_\infty,T_0,z_0)$, *then all solutions are obtained in the form* $(u_0,u_1+a,T_\infty,T_0,z_0)$ *with* $a \in L^2(\Omega_{T_e},\mathbb{R}^3)$.

We are able to show that the function $(u_0,\hat T_\eta,\hat z_\eta)$ is asymptotic to the solution of the microscopic problem and thus justify the homogenized problem only when the solution of this homogenized problem is of higher regularity as given in this theorem. When g is a single valued function, i.e. $g : Y_{\text{per}} \times \mathbb{R}^N \to \mathbb{R}^N$, and when g, $\mathcal{D}$, the domain Ω and the data are regular, one can show that up to a certain time the solution is regular. This result is formulated in the following

Theorem 3 (Higher regularity locally) *Let* $n \geq 1$ *be an integer, suppose that the assumptions of the preceding theorem are satisfied and that* $\Omega \in C^n$, $g \in C^n(Y_{\text{per}} \times \mathbb{R}^N,\mathbb{R}^N)$, $b \in C^n(\overline{\Omega} \times [0,\infty),\mathbb{R}^3)$, $\gamma_D \in C^n(\partial\Omega \times [0,\infty),\mathbb{R}^3)$, $z_0^{(0)} \in C^n(\overline{\Omega} \times Y_{\text{per}},\mathbb{R}^N)$. *Furthermore, suppose that* $y \mapsto \mathcal{D}[y]$ *is* n*–times continuously differentiable on* Y_{per} . *Then there exists a time* $T_r > 0$ *such that the solution of the homogenized problem satisfies* $(u_0,T_\infty) \in C^n(\overline{\Omega}_{T_r},\mathbb{R}^3\times\mathcal{S}^3)$ *and* $(u_1,T_0,z_0) \in C^n(\overline{\Omega}\times Y_{\text{per}})_{T_r},\mathbb{R}^3\times\mathcal{S}^3\times\mathbb{R}^N)$.

Remark. $y \mapsto \mathcal{D}[y]$ is n–times continuously differentiable on Y_{per} if the coefficients of the tensorial representation of $\mathcal{D}[y]$ belong to $C^n(Y_{\text{per}})$.

Theorem 4 (Justification of the homogenized problem)
(i) *Suppose that* Ω, L, $\mathcal{D}$, g, b, γ_D *and* $z_0^{(0)}$ *satisfy the assumptions of Theorem 2. Let* $T_e > 0$ *and* $\eta > 0$. *Then there is a unique solution*

$$(u_\eta,T_\eta,z_\eta) \in L^2(0,T_e;H_1(\Omega,\mathbb{R}^3)) \times L^2(\Omega_{T_e},\mathcal{S}^3) \times C(0,T_e;L^2(\Omega,\mathbb{R}^N))$$

of the microscopic problem (8) – (12) *to the inital data* $z^{(0)}(x) = z_0^{(0)}(x, \frac{x}{\eta})$.

(ii) *Suppose that additionally the assumptions of Theorem 3 are satisfied with* $n = 3$. *Let* $(u_0, u_1, T_\infty, T_0, z_0)$ *be a solution of the homogenized initial-boundary value problem* (17) – (23), *and let* T_r *be the positive time given in Theorem 3 such that this solution is* 3*–times continuously differentiable for* $t \in [0, T_r]$. *Let the functions* $\hat{T}_\eta$ *and* $\hat{z}_\eta$ *be defined by the equations* (15)*and* (16). *Then* $(u_0, \hat{T}_\eta, \hat{z}_\eta)$ *is asymptotic to the solution* (u_η, T_η, z_η) *of the microscopic problem in the time interval* $[0, T_r]$, *i.e. for all* $0 \leq t \leq T_r$

$$\lim_{\eta \to 0} \left[\|u_\eta(t) - u_0(t)\|_\Omega + \|T_\eta(t) - \hat{T}_\eta(t)\|_\Omega + \|z_\eta(t) - \hat{z}_\eta(t)\|_\Omega \right] = 0 \,. \quad (33)$$

The proof of Theorem 2 is given in Sect. 2, whereas the proof of Theorem 3 is only sketched there. Theorem 4 is proved in Sect. 3. In the proof of Theorem 4 we apply a well known homogenization result for the linear boundary value problem of elasticity theory derived by the energy method of Tartar, cf. the proof of Lemma 14 in Sect. 3. At one place in this proof we need that $\partial_t \mathrm{div}_x T_0(x,y,t)\Big|_{y=\frac{x}{\eta}}$ and $\partial_t \mathrm{rot}_x \nabla_y u_1(x,y,t)\Big|_{y=\frac{x}{\eta}}$ belong to compact subsets of H^{loc}_{-1}, where T_0 and u_1 are the functions in the solution of the homogenized problem (17) – (23). The regularity of the global solution obtained from Theorem 2 is slightly too small to prove this. There are other subtleties in the proof of Theorem 4, but this is the main point why we need higher regularity and why we can only prove the convergence result (33) locally in time. We surmise, however, that a similar inequality is valid for the global solution obtaind from Theorem 2 without the regularity assumptions of Theorem 3.

2 The Homogenized Initial Boundary Value Problem

In this section we prove Theorem 2 and sketch the proof of Theorem 3. The proof of Theorem 2 is based on the reduction of the homogenized initial-boundary value problem (17) – (23) to an evolution equation in the Hilbert space L^2 with a maximal monotone evolution operator. Existence of solutions of the initial-boundary value problem follows from the standard existence theorems for such evolution equations. This proof follows in the essential details the proof of existence for the microscopic initial-boundary value problem in [4], and we thus refer several times to that proof.

We start with a definition based on Lemma 1.

Definition 5 *Let the linear operator* $P : L^2(\Omega \times Y_{\mathrm{per}}, \mathcal{S}^3) \to L^2(\Omega \times Y_{\mathrm{per}}, \mathcal{S}^3)$ *be defined by*

$$P\hat{\varepsilon}_p = \varepsilon(\nabla_x u_0 + \nabla_y u_1)$$

for every $\hat{\varepsilon}_p \in L^2(\Omega \times Y_{\mathrm{per}}, \mathcal{S}^3)$, *where* $(u_0, u_1, T_\infty, T_0)$ *is the unique weak solution of the Dirichlet boundary value problem* (24) – (28) *to* $\hat{b} = \hat{\gamma}_D = 0$

given by Lemma 1. Furthermore, with the identity I we define the linear operator $Q = I - P$.

Lemma 6 (i) *The operators P and Q are bounded projection operators orthogonal with respect to the scalar product $[\sigma, \tau]_{\Omega\times Y}$ on $L^2(\Omega \times Y_{\text{per}}, \mathcal{S}^3)$.*
(ii) *The operator $B^T\mathcal{D}QB : L^2(\Omega\times Y_{\text{per}}, \mathbb{R}^N) \to L^2(\Omega\times Y_{\text{per}}, \mathbb{R}^N)$ is selfajoint and non-negative with respect to the scalar product $(z, \hat{z})_{\Omega\times Y}$.*

Proof: (i) The boundedness of P follows from (31). To see that P is a projection, assume that $\hat{\varepsilon}_p$ belongs to the range of P, hence $\hat{\varepsilon}_p = \varepsilon(\nabla_x w_0 + \nabla_y w_1)$ for a suitable pair (w_0, w_1). Thus, by definition of P we have $P\varepsilon(\nabla_x w_0 + \nabla_y w_1) = P\hat{\varepsilon}_p = \varepsilon(\nabla_x u_0 + \nabla_y u_1)$, where $(u_0, u_1) \in \overset{\circ}{H}_1(\Omega) \times L^2(\Omega, H_1(Y_{\text{per}}, \mathbb{R}^3))$ is the unique function with $\int_Y u_1(x,y)dy = 0$ satisfying (30) for $\hat{b} = 0$. Clearly, if we insert (w_0, w_1) for (u_0, u_1) then (30) is satisfied, hence $(u_0, u_1) = (w_0, w_1)$ and $P\hat{\varepsilon}_p = \varepsilon(\nabla_x w_0 + \nabla_y w_1) = \hat{\varepsilon}_p$, which shows that P is a projection. To prove that P is orthogonal, note that for all $\tau \in L^2(\Omega\times Y_{\text{per}}, \mathcal{S}^3)$ the function $P\tau$ is of the form $\varepsilon(\nabla_x v_0 + \nabla_y v_1)$. Therefore we can plug $P\tau$ into the second argument of the scalar product in (30) and obtain by definition of P for all $\sigma \in L^2(\Omega \times Y_{\text{per}}, \mathcal{S}^3)$

$$[P\sigma - \sigma, P\tau]_{\Omega\times Y} = 0\,.$$

Interchanging the roles of σ and τ yields

$$[P\tau - \tau, P\sigma]_{\Omega\times Y} = 0\,.$$

From these two equations we conclude for all $\sigma, \tau \in L^2(\Omega \times Y_{\text{per}}, \mathcal{S}^3)$

$$[\tau, P\sigma]_{\Omega\times Y} = [P\tau, P\sigma]_{\Omega\times Y} = [\sigma, P\tau]_{\Omega\times Y} = [P\tau, \sigma]_{\Omega\times Y}\,.$$

This yields $P^* = P$, whence P is selfajoint. Therefore P is an orthogonal projection, which clearly implies that also $Q = I - P$ is an orthogonal projection.
(ii) For $z, \hat{z} \in L^2(\Omega \times Y_{\text{per}}, \mathbb{R}^n)$ we have

$$\begin{aligned}
&(B^T\mathcal{D}QBz, \hat{z})_{\Omega\times Y} = (\mathcal{D}QBz, B\hat{z})_{\Omega\times Y} = [QBz, B\hat{z}]_{\Omega\times Y} \\
&= [QBz, QB\hat{z}]_{\Omega\times Y} = [Bz, QB\hat{z}]_{\Omega\times Y} = (\mathcal{D}Bz, QB\hat{z})_{\Omega\times Y} \\
&= (Bz, \mathcal{D}QB\hat{z})_{\Omega\times Y} = (z, B^T\mathcal{D}QB\hat{z})_{\Omega\times Y}\,,
\end{aligned}$$

which implies that $B^T\mathcal{D}QB$ is selfajoint and non-negative. This completes the proof.

Since by assumption the symmetric $N \times N$–matrix L is positive definite, it follows from this lemma that the operator $L + B^T\mathcal{D}QB : L^2(\Omega\times Y_{\text{per}}, \mathbb{R}^N) \to L^2(\Omega \times Y_{\text{per}}, \mathbb{R}^N)$ is bounded, selfadjoint and positive definite. Therefore

$$\langle z, \hat{z}\rangle_{\Omega\times Y} = \big((L + B^T\mathcal{D}QB)^{-1}z, \hat{z}\big)_{\Omega\times Y}$$

defines a scalar product on $L^2(\Omega \times Y_{\text{per}}, \mathbb{R}^N)$. The associated norm

$$|z|_{\Omega\times Y} = \langle z, z\rangle_{\Omega\times Y}^{1/2}$$

is equivalent to the norm $\|z\|_{\Omega\times Y}$.

After these preparations we can reduce the initial-boundary value problem (17) – (23) to an evolution equation. To this end we note that (20) yields

$$B^T T_0 - Lz_0 = B^T \mathcal{D}\big(\varepsilon(\nabla_x u_0 + \nabla_y u_1) - Bz_0\big) - Lz_0 \,. \tag{34}$$

Assume that $(u_0, u_1, T_\infty, T_0, z_0)$ is a solution of the initial-boundary value problem (17) – (23). We fix t. If $z_0(t)$ is known, then (17) – (20), (23) is a boundary value problem for the components $u_0(t), u_1(t), T_\infty(t), T_0(t)$ of the solution, the homogenized problem from linear elasticity theory. Consequently, these functions are obtained in the form

$$\begin{aligned}\big(u_0(t), u_1(t), T_\infty(t), T_0(t)\big) &= \big(\tilde{u}_0(t), \tilde{u}_1(t), \tilde{T}_\infty(t), \tilde{T}_0(t)\big)\\ &\quad + \big(v_0(t), v_1(t), \sigma_\infty(t), \sigma_0(t)\big)\,,\end{aligned}$$

with a solution $\big(v_0(t), v_1(t), \sigma_\infty(t), \sigma_0(t)\big)$ of the Dirichlet boundary value problem (24) – (28) to the data $\hat{b} = b(t)$, $\hat{\gamma}_D = \gamma_D(t)$, $\hat{\varepsilon}_p = 0$, and with a solution $\big(\tilde{u}_0(t), \tilde{u}_1(t), \tilde{T}_\infty(t), \tilde{T}_0(t)\big)$ of the boundary value problem (24) – (28) to the data $\hat{b} = \hat{\gamma}_D = 0$, $\hat{\varepsilon}_p = Bz_0(t)$. With the projector P from Definition 5 we thus obtain

$$\varepsilon((\nabla_x u_0 + \nabla_y u_1)(t)) - Bz_0(t) = (P - I)Bz_0(t) + \varepsilon((\nabla_x v_0 + \nabla_y v_1)(t))\,.$$

We insert this equation into (34) and obtain that (21) can be written in the form

$$\frac{\partial}{\partial t} z_0(t) \in G\Big(\big(B^T\mathcal{D}(P - I)B - L\big)z_0(t) + B^T\sigma_0(t)\Big), \tag{35}$$

with the mapping $G : L^2(\Omega \times Y_{\text{per}}, \mathbb{R}^N) \to 2^{L^2(\Omega\times Y_{\text{per}}, \mathbb{R}^N)}$ defined by

$$G(z) = \{\zeta \in L^2(\Omega \times Y_{\text{per}}, \mathbb{R}^n) \mid \zeta(x, y) \in g\big(y, z(x, y)\big) \text{ a.e.}\}\,.$$

Since σ_0 is computed from the data b and γ, and thus is known, (35) is an evolution equation for z_0. If we define the evolution operator $A(t)$ by

$$A(t)z_0 = -G\big(-(B^T\mathcal{D}QB + L)z_0(t) + B^T\sigma_0(t)\big)$$

and note that $Q = I - P$, this evolution equation can be written as

$$\frac{\partial}{\partial t} z_0(t) + A(t)\, z_0(t) \ni 0\,.$$

To transform this equation to an autonomous equation, insert

$$h = -(B^T \mathcal{D} Q B + L) z_0 + B^T \sigma_0$$

into (35). This yields the evolution equation

$$\frac{\partial}{\partial t} h(t) \in -\big(B^T \mathcal{D} Q B + L\big) G(h(t)) + B^T \frac{\partial}{\partial t} \sigma_0(t) . \tag{36}$$

Definition 7 *We define the operator* $\mathcal{C} : L^2(\Omega \times Y_{\text{per}}, \mathbb{R}^N) \to 2^{L^2(\Omega \times Y_{\text{per}}, \mathbb{R}^N)}$ *and the domain* $\Delta(\mathcal{C})$ *of* $\mathcal{C}$ *by*

$$\mathcal{C} = (L + B^T \mathcal{D} Q B) G, \qquad \Delta(\mathcal{C}) = \{ h \in L^2(\Omega \times Y_{\text{per}}, \mathbb{R}^N) \mid \mathcal{C} h \neq \emptyset \} .$$

With this operator we finally write the evolution equation (36) on $L^2(\Omega \times Y_{\text{per}}, \mathbb{R}^N)$ in the form

$$h_t(t) + \mathcal{C} h(t) \ni B^T \sigma_{0t} .$$

The existence proof is now based on the following fundamental

Theorem 8 (i) *Let the mapping* $z \mapsto g(y, z) : \mathbb{R}^N \to 2^{\mathbb{R}^N}$ *be monotone for all* $y \in Y_{\text{per}}$. *Then the operator* $\mathcal{C}$ *is monotone with respect to the scalar product* $\langle z, \hat{z} \rangle_{\Omega \times Y}$.
(ii) *If* g *and* j_λ *satisfy the conditions of Theorem 2, then* $\mathcal{C}$ *is maximal monotone with respect to this scalar product.*

We omit the proof of this theorem, since it coincides essentially with the proof of Theorem 3.3 in [4].

Corollary 9 *Suppose that* g *and* j_λ *satisfy the conditions of Theorem 2. Also, let* $\sigma_0 \in H_2^1\big(0, T_e; L^2(\Omega \times Y_{\text{per}}, \mathcal{S}^3)\big)$ *and let* $h^{(0)} \in \Delta(\mathcal{C})$.

Then the evolution equation

$$h_t + \mathcal{C} h \ni B^T \sigma_{0t} \tag{37}$$

has a unique solution $h \in H_1^\infty\big(0, T_e; L^2(\Omega \times Y_{\text{per}}, \mathbb{R}^N)\big)$ *with*

$$h(0) = h^{(0)} . \tag{38}$$

This solution satisfies

$$\big\| h_t(t) \big\|_{\Omega \times Y} \le |||\mathcal{C} h^{(0)} + B^T \sigma_{0t}(0)||| + \int_0^t \big\| B^T \sigma_{0tt}(s) \big\|_{\Omega \times Y} \, ds \qquad a.e.,$$

where

$$|||\mathcal{C} h^{(0)} + B^T \sigma_{0t}(t)||| = \inf\{ \| \zeta \|_{\Omega \times Y} \mid \zeta \in \mathcal{C} h^{(0)} + B^T \sigma_{0t}(0) \} .$$

Proof: Since $\mathcal{C}$ is maximal monotone and since for $\sigma_0 \in H_2^1(0, T_e; L^2(\Omega \times Y_{\mathrm{per}}, \mathcal{S}^3))$ the function $B^T \sigma_{0t}$ belongs to $H_1^1(0, T_e; L^2(\Omega \times Y_{\mathrm{per}}, \mathcal{S}^3))$, this theorem is an immediate consequence of [12, Theorem 2.2, p. 131].

The **proof of Theorem 2** follows from this corollary, since it can be easily shown that the reduction of the initial-boundary value problem (17) – (22) to the evolution equation (37) can be reversed and that a solution of the initial value problem (37), (38) yields a solution of the initial-boundary value problem (17) – (22). The assumptions for $z_0^{(0)}$ in Theorem 2 guarantee that the inital data $h^{(0)} = -\big(B^T \mathcal{D} Q B + L\big) z_0^{(0)} + B^T \sigma_0(0)$ belong to the domain $\Delta(\mathcal{C})$. We omit the proof, since it is essentially the same as the proof of Theorem 1.3 in [4].

The **proof of Theorem 3** is based on the standard construction of local solutions to the evolution equation (35) in the Banach space $C^n(\overline{\Omega} \times Y_{\mathrm{per}}, \mathbb{R}^N)$ using contraction estimates. Since by assumption Ω belongs to the class C^n and $y \mapsto \mathcal{D}[y]$ is n–times continuously differentiable, it can be shown by the usual regularity theory for the boundary value problem (24) – (28) that the operator P from Definition 5 maps $C^n(\overline{\Omega} \times Y_{\mathrm{per}}, \mathcal{S}^3)$ boundedly into itself. Therefore $B^T \mathcal{D}(P - I)B - L$ maps $C^n(\overline{\Omega} \times Y_{\mathrm{per}}, \mathbb{R}^N)$ boundedly into itself, which together with the assumed regularity of g allows to prove local contraction estimates for the operator

$$z_0 \mapsto G\big((B^T \mathcal{D}(P - I)B - L) z_0 + B^T \sigma_0\big).$$

We omit the details of the proof.

3 Justification of the Homogenized Problem

Here we prove Theorem 4. In the proof we need a stability estimate for the microscopic problem (8) – (12), which is obtained using the framework of the proof of existence of solutions for this problem. This existence proof is given in [4,21]; it is similar to the proof of Theorem 2, as mentioned in the preceding section. To set up this framework, we first give the definitions and state the results from [4] needed in the proof of Theorem 4, which follows afterwards:

To begin with, consider the boundary value problem formed by the equations (8), (9), (12): To given functions $\hat{b} : \Omega \to \mathbb{R}^3$, $\hat{\gamma}_D : \partial\Omega \to \mathbb{R}^3$, $\hat{\varepsilon}_p : \Omega \to \mathcal{S}^3$ and to a given number $\eta > 0$ we seek solutions $(u, T) : \Omega \to \mathbb{R}^3 \times \mathcal{S}^3$ of the equations

$$-\operatorname{div}_x T(x) = \hat{b}(x)\,, \tag{39}$$

$$T(x) = \mathcal{D}[\frac{x}{\eta}]\big(\varepsilon(\nabla_x u(x)) - \hat{\varepsilon}_p(x)\big)\,, \tag{40}$$

$$u(x) = \hat{\gamma}_D(x), \quad x \in \partial\Omega\,. \tag{41}$$

This is the linear problem of elasticity theory. To define weak solutions let $\hat{b} \in H_{-1}(\Omega, \mathbb{R}^3)$, $\hat{\varepsilon}_p \in L^2(\Omega, \mathcal{S}^3)$ and $\hat{\gamma}_D \in H_1(\Omega, \mathbb{R}^3)$. A function $(u, T) \in H_1(\Omega, \mathbb{R}^3) \times L^2(\Omega, \mathcal{S}^3)$ is called weak solution of (39) – (41), if (40) is satisfied, if the equation

$$\big(D(\frac{\cdot}{\eta})\big(\varepsilon(\nabla_x u) - \hat{\varepsilon}_p\big), \varepsilon(\nabla_x v)\big)_\Omega = (\hat{b}, v)_\Omega \tag{42}$$

holds for all $v \in \overset{\circ}{H}_1(\Omega, \mathbb{R}^3)$, and if u can be represented as $u = \hat{\gamma}_D + w$ with $w \in \overset{\circ}{H}_1(\Omega, \mathbb{R}^3)$.

Since $y \mapsto \mathcal{D}[y]$ is bounded and uniformly positive definite, it follows that

$$\big(\mathcal{D}(\frac{\cdot}{\eta})\sigma, \tau\big)_\Omega$$

is a scalar product on $L^2(\Omega, \mathcal{S}^3)$, for which constants c_1, c_2 exist such that

$$c_1 \|\sigma\|_\Omega \le \big(\mathcal{D}(\frac{\cdot}{\eta})\sigma, \sigma\big)_\Omega^{1/2} \le c_2 \|\sigma\|_\Omega$$

holds for all $\sigma \in L^2(\Omega, \mathcal{S}^3)$ and all $\eta > 0$. Using this fact, we obtain by the well known theory for the boundary value problem (39) – (41) that to $\hat{b} \in H_{-1}(\Omega, \mathbb{R}^3)$, $\hat{\varepsilon}_p \in L^2(\Omega, \mathcal{S}^3)$, $\hat{\gamma}_D \in H_1(\Omega, \mathbb{R}^3)$ and $\eta > 0$ there is a unique weak solution (u, T) satisfying

$$\|u\|_{1,\Omega} + \|T\|_\Omega \le C(\|\hat{b}\|_{\Omega;-1} + \|\hat{\varepsilon}_p\|_\Omega + \|\hat{\gamma}_D\|_{1,\Omega}), \tag{43}$$

with a constant C independent of η.

Definition 10 *To $\eta > 0$ let the linear operator $P_\eta : L^2(\Omega, \mathcal{S}^3) \to L^2(\Omega, \mathcal{S}^3)$ be defined by*

$$P_\eta \hat{\varepsilon}_p = \varepsilon(\nabla_x u)$$

for every $\hat{\varepsilon}_p \in L^2(\Omega, \mathcal{S}^3)$, where (u, T) is the unique weak solution of the Dirichlet problem (39) – (41) to $\hat{\varepsilon}_p$ and to $\hat{b} = \hat{\gamma}_D = 0$. Also, we define $Q_\eta = I - P_\eta$.

Lemma 11 *Let $\eta > 0$.*
(i) The operators P_η and Q_η are projection operators bounded uniformly with respect to η and orthogonal with respect to the scalar product $\big(\mathcal{D}(\frac{\cdot}{\eta})\sigma, \tau\big)_\Omega$ on $L^2(\Omega, \mathcal{S}^3)$.
(ii) The operator $B^T \mathcal{D}(\frac{\cdot}{\eta}) Q_\eta B : L^2(\Omega, \mathbb{R}^N) \to L^2(\Omega, \mathbb{R}^N)$ is selfadjoint and non-negative with respect to the scalar product $(z, \hat{z})_\Omega$. Moreover, there is a constant $C > 0$ such that

$$\|B^T \mathcal{D}(\frac{\cdot}{\eta}) Q_\eta B z\|_\Omega \le C \|z\|_\Omega, \tag{44}$$

for all $\eta > 0$. Hence $B^T \mathcal{D}(\frac{\cdot}{\eta}) Q_\eta B$ is bounded uniformly with respect to η.

The **proof** is essentially equal to the proof of Lemma 6 if we use the estimate (43) instead of (31).

Since L is positive definite, it follows from this lemma that the operator $L + B^T\mathcal{D}(\frac{\cdot}{\eta})Q_\eta B$ is uniformly positive definite. This implies that

$$\langle z, \hat{z}\rangle_{\Omega,\eta} = \Big(\big(L + B^T\mathcal{D}[\frac{\cdot}{\eta}]Q_\eta B\big)^{-1} z, \hat{z}\Big)_\Omega$$

defines a scalar product on $L^2(\Omega, \mathbb{R}^N)$. Furthermore, together with (44) we obtain that to the associated norm

$$|z|_{\Omega,\eta} = \langle z, z\rangle_{\Omega,\eta}^{1/2}$$

there are constants $C_1, C_2 > 0$ such that for all $\eta > 0$ and all $z \in L^2(\Omega, \mathbb{R}^N)$

$$C_1\|z\|_\Omega \le |z|_{\Omega,\eta} \le C_2\|z\|_\Omega\,. \tag{45}$$

Using the projection Q_η we define an evolution operator $\mathcal{C}_\eta : \Delta(\mathcal{C}_\eta) \subseteq L^2(\Omega, \mathbb{R}^N) \to 2^{L^2(\Omega,\mathbb{R}^N)}$ by

$$\begin{aligned}&\mathcal{C}_\eta h \\ &= \{(L + B^T\mathcal{D}(\frac{\cdot}{\eta})Q_\eta B)\zeta \mid \zeta \in L^2(\Omega, \mathbb{R}^N),\ \zeta(x) \in g(\frac{x}{\eta}, h(x)) \text{ a.e. in } \Omega\}\,.\end{aligned} \tag{46}$$

Theorem 12 *If g satisfies the conditions of Theorem 2, then the operator $\mathcal{C}_\eta$ is maximal monotone with respect to the scalar product $\langle z, \hat{z}\rangle_{\Omega,\eta}$.*

The **proof** is obtained by a slight modification of the proof of Theorem 3.3 in [4].

Corollary 13 *For all $F_i \in H_1^1(0, T_e; L^2(\Omega, \mathbb{R}^N))$ and $h_i^{(0)} \in \Delta(\mathcal{C}_\eta)$, $i = 1, 2$, the initial value problem*

$$\frac{\partial}{\partial t} h_i + \mathcal{C}_\eta h_i \ni F_i\,, \tag{47}$$

$$h_i(0) = h_i^{(0)}\,, \tag{48}$$

has unique weak solutions $h_i \in H_1^\infty(0, T_e; L^2(\Omega, \mathbb{R}^N))$. These solutions satisfy

$$|h_1(t) - h_2(t)|_{\Omega,\eta} \le |h_1^{(0)} - h_2^{(0)}|_{\Omega,\eta} + \int_0^t |F_1(s) - F_2(s)|_{\Omega,\eta}\, ds. \tag{49}$$

Proof: Cf. [14, Lemma 3.1 and Theorem 3.4, pp. 64, 65].

After these preparations we come to the

Proof of Theorem 4: We assume that the data b, γ_D and $z^{(0)}(x) = z_0^{(0)}(x, \frac{x}{\eta})$ have the properties required in Theorem 4.

Let $(v_\eta(t),\, \sigma_\eta(t))$ be a solution of the boundary value problem (39) – (41) to the data

$$\hat{b} = b(t)\,, \quad \hat{\varepsilon}_p = 0\,, \quad \hat{\gamma}_D = \gamma_D(t)\,. \tag{50}$$

By the same procedure as in in the preceding section it is shown in [4] that if (u_η, T_η, z_η) is the solution of the microscopic problem (8) – (12), then the function

$$h_\eta = -\big(B^T \mathcal{D}(\frac{\cdot}{\eta})Q_\eta B + L\big) z_\eta + B^T \sigma_\eta \tag{51}$$

satisfies the initial value problem

$$h_{\eta t}(t) + \mathcal{C}_\eta h_\eta(t) = B^T \sigma_{\eta t}(t)\,, \tag{52}$$

$$h_\eta(0) = -\big(B^T \mathcal{D}(\frac{\cdot}{\eta})Q_\eta B + L\big) z_\eta(0) + B^T \sigma_\eta(0). \tag{53}$$

The approximate solution $(u_0, \hat{T}_\eta, \hat{z}_\eta)$ constructed from $(u_0, u_1, T_\infty, T_0, z_0)$, the solution of the homogenized problem (17) – (23), can be reduced to the solution of the same initial value problem, however with different data. For, observing (17) – (23), we obtain by a simple computation that $(u_0, \hat{T}_\eta, \hat{z}_\eta)$ satisfies the equations

$$-\operatorname{div}_x \hat{T}_\eta(x,t) = -\operatorname{div}_x T_0(x,y,t)\Big|_{y=\frac{x}{\eta}}\,, \tag{54}$$

$$\hat{T}_\eta(x,t) = \mathcal{D}[\frac{x}{\eta}]\Big(\varepsilon(\nabla_x u_0(x,t)) - B\hat{z}_\eta(x,t) + \varepsilon(\nabla_y u_1(x,y,t))\Big|_{y=\frac{x}{\eta}}\Big)\,, \tag{55}$$

$$\frac{\partial}{\partial t}\hat{z}_\eta(x,t) = g\big(\frac{x}{\eta},\, B^T \hat{T}_\eta(x,t) - L\hat{z}_\eta(x,t)\big)\,, \tag{56}$$

$$\hat{z}_\eta(x,0) = z_0^{(0)}(x, \frac{x}{\eta})\,, \tag{57}$$

$$u_0(x,t) = \gamma_D(x,t)\,, \quad x \in \partial\Omega\,. \tag{58}$$

Since these equations have the same form as the equations of the microscopic problem, we can again employ the procedure from the last section and obtain that if $(\hat{v}_\eta(t),\, \hat{\sigma}_\eta(t))$ is the solution of the linear boundary value problem (39) – (41) to the data

$$\hat{b}(x) = -\operatorname{div}_x T_0(x,y,t)\Big|_{y=\frac{x}{\eta}}\,, \tag{59}$$

$$\hat{\varepsilon}_p(x) = -\varepsilon\big(\nabla_y u_1(x,y,t)\big)\Big|_{y=\frac{x}{\eta}}\,, \tag{60}$$

$$\hat{\gamma}_D(x) = \gamma_D(x,t)\,, \tag{61}$$

then the function

$$\hat{h}_\eta = -\big(B^T \mathcal{D}(\frac{\cdot}{\eta}) Q_\eta B + L\big)\hat{z}_\eta + B^T \hat{\sigma}_\eta \tag{62}$$

satisfies the initial value problem

$$\hat{h}_{\eta t}(t) + \mathcal{C}_\eta \hat{h}_\eta(t) = B^T \hat{\sigma}_{\eta t}(t)\,, \tag{63}$$

$$\hat{h}_\eta(0) = -\big(B^T \mathcal{D}(\frac{\cdot}{\eta}) Q_\eta B + L\big)\hat{z}_\eta(0) + B^T \hat{\sigma}_\eta(0). \tag{64}$$

Thus, if we note that $\hat{z}_\eta(x,0) = z_\eta(x,0) = z_0^{(0)}(x, \frac{x}{\eta})$ and apply Corollary 13, it follows from (52), (53) together with (63), (64) that for all $0 \le t \le T_r$

$$\begin{aligned} \big| h_\eta(t) - \hat{h}_\eta(t) \big|_{\Omega,\eta} \le{}& \big| B^T(\sigma_\eta(0) - \hat{\sigma}_\eta(0)) \big|_{\Omega,\eta} \\ &+ \int_0^t \big| B^T(\sigma_{\eta t}(s) - \hat{\sigma}_{\eta t}(s)) \big|_{\Omega,\eta}\, ds\,. \end{aligned} \tag{65}$$

We next use that $\big(B^T \mathcal{D}[\frac{\cdot}{\eta}] Q_\eta B + L\big)^{-1}$ is uniformly bounded with respect to η. Consequently (62) and (51) yield that there is a constant C_3 such that

$$\|z_\eta(t) - \hat{z}_\eta(t)\|_\Omega \le C_3 \big(\|h_\eta(t) - \hat{h}_\eta(t)\|_\Omega + \|\sigma_\eta(t) - \hat{\sigma}_\eta(t)\|_\Omega \big)\,, \tag{66}$$

for all $0 \le t \le T_r$ and all $\eta > 0$. This estimate, (45) and (65) imply for $0 \le t \le T_r$

$$\|z_\eta(t) - \hat{z}_\eta(t)\|_\Omega \le C_4 \Big(\|\sigma_\eta(0) - \hat{\sigma}_\eta(0)\|_\Omega + \int_0^{T_r} \|\sigma_{\eta t}(s) - \hat{\sigma}_{\eta t}(s)\|_\Omega\, ds \Big)\,, \tag{67}$$

with a constant C_4 independent of t and of η. Thus, we can estimate the difference $z_\eta - \hat{z}_\eta$ if estimates for the differences $\sigma_\eta(0) - \hat{\sigma}_\eta(0)$ and $\sigma_{\eta t} - \hat{\sigma}_{\eta t}$ can be obtained. Since the functions σ_η and $\hat{\sigma}_\eta$ both are solutions of the same elliptic boundary value problem, the problem of linear elasticity theory, we obtain such estimates from the well known homogenization theory for this boundary value problem. The estimates are stated in the following lemma, whose proof is postponed:

Lemma 14 *For all* $0 \le t \le T_r$

$$\lim_{\eta \to 0} \|\partial_t^i (v_\eta(t) - \hat{v}_\eta(t))\|_\Omega = 0\,, \quad i = 0, 1\,, \tag{68}$$

$$\lim_{\eta \to 0} \|\partial_t^i (\sigma_\eta(t) - \hat{\sigma}_\eta(t))\|_\Omega = 0\,, \quad i = 0, 1\,. \tag{69}$$

Moreover, there is a constant K *such that for all* $0 \le t \le T_r$ *and all* $\eta > 0$

$$\|\sigma_{\eta t}(t) - \hat{\sigma}_{\eta t}(t)\|_\Omega \le K. \tag{70}$$

From (69) we conclude that the term $\|\sigma_\eta(0)-\hat{\sigma}_\eta(0)\|_\Omega$ tends to zero for $\eta \to 0$, and we conclude that the integrand in (67) tends to zero for $\eta \to 0$, pointwise for every s. Since this integrand is uniformly bounded, by (70), Lebesgue's convergence theorem implies that the right hand side of (67) tends to zero for $\eta \to 0$, whence

$$\lim_{\eta \to 0} \|z_\eta(t) - \hat{z}_\eta(t)\|_\Omega = 0\,, \tag{71}$$

for all $0 \le t \le T_r$.

To obtain the estimate (33) we observe that the equations (8), (9), (12) form a boundary value problem for (u_η, T_η), and that the equations (54), (55), (58) form a boundary value problem for $(u_0, \hat{T}_\eta)$. The Definition 10 of P_η and the definitions of (v_η, σ_η) and $(\hat{v}_\eta, \hat{\sigma}_\eta)$ thus yield the decomposition

$$u_\eta = w_\eta + v_\eta\,, \quad T_\eta = \mathcal{D}[\tfrac{\cdot}{\eta}](P_\eta - I)Bz_\eta + \sigma_\eta\,,$$

$$u_0 = \hat{w}_\eta + \hat{v}_\eta\,, \quad \hat{T}_\eta = \mathcal{D}[\tfrac{\cdot}{\eta}](P_\eta - I)B\hat{z}_\eta + \hat{\sigma}_\eta\,,$$

where $w_\eta(t), \hat{w}_\eta(t) \in \overset{\circ}{H}_1(\Omega, \mathbb{R}^3)$ are the unique functions from Definition 10 which satisfy $\varepsilon(\nabla_x w_\eta(t)) = P_\eta B z_\eta(t)$ and $\varepsilon(\nabla_x \hat{w}_\eta(t)) = P_\eta B \hat{z}_\eta(t)$. We thus have

$$\varepsilon\big(\nabla_x(w_\eta - \hat{w}_\eta)\big) = P_\eta B(z - \hat{z}_\eta)\,, \tag{72}$$

$$T_\eta - \hat{T}_\eta = -\mathcal{D}[\tfrac{\cdot}{\eta}]Q_\eta B(z_\eta - \hat{z}_\eta) + (\sigma_\eta - \hat{\sigma}_\eta)\,, \tag{73}$$

$$u_\eta - u_0 = (w_\eta - \hat{w}_\eta) + (v_\eta - \hat{v}_\eta)\,. \tag{74}$$

From (69), (71), (73) and the uniform boundedness of $\mathcal{D}[\frac{\cdot}{\eta}]Q_\eta B$ we infer that

$$\lim_{\eta \to 0} \|T_\eta(t) - \hat{T}_\eta(t)\|_\Omega = 0\,.$$

Since $(w_\eta - \hat{w}_\eta)(t) \in \overset{\circ}{H}_1(\Omega, \mathbb{R}^3)$, we infer from the first Korn's inequality $\|(w_\eta - \hat{w}_\eta)(t)\|_{\Omega;1} \le c\|\varepsilon(\nabla_x(w_\eta - \hat{w}_\eta)(t))\|_\Omega$ and from (71), (72) that $\|w_\eta(t) - \hat{w}_\eta(t)\|_{\Omega;1} \to 0$ for $\eta \to 0$; from (74) and (68) we thus conclude

$$\lim_{\eta \to 0} \|u_\eta(t) - u_0(t)\|_\Omega = 0$$

for all $0 \le t \le T_r$. These two relations and (71) together yield (33).

To finish the proof of Theorem 4 it thus remains to verify Lemma 14. This lemma is a consequence of the following well known result from homogenization theory; similar results can be found in many places, cf. for example [7,30]:

Lemma 15 *Let the functions $\tau \in L^2(\Omega, \mathcal{S}^3)$, $b \in L^2(\Omega, \mathbb{R}^3)$ satisfy*

$$b - \operatorname{div} \tau = 0,$$

and let the families $\{\tau_\eta\}_{\eta>0}$ and $\{\kappa_\eta\}_{\eta>0}$ with $\tau_\eta \in L^2(\Omega, \mathcal{S}^3)$ and $\kappa_\eta \in L^2(\Omega, \mathbb{R}^{3\times 3})$ have the following properties:

(i) $\tau_\eta \rightharpoonup \tau$ *for* $\eta \to 0$, *weakly in* $L^2(\Omega, \mathcal{S}^3)$,

(ii) *The set* $\{\operatorname{div} \tau_\eta\}_{\eta>0}$ *is a subset of a compact subset of* $H_{-1}^{\mathrm{loc}}(\Omega, \mathbb{R}^3)$,

(iii) $\kappa_\eta \rightharpoonup 0$ *for* $\eta \to 0$, *weakly in* $L^2(\Omega, \mathbb{R}^{3\times 3})$,

(iv) *The set* $\{\operatorname{rot} \kappa_\eta\}_{\eta>0}$ *is a subset of a compact subset of* $H_{-1}^{\mathrm{loc}}(\Omega, \mathbb{R}^{3\times 3})$.

Let $(\overline{v}_\eta, \overline{\sigma}_\eta) \in \overset{\circ}{H}_1(\Omega, \mathbb{R}^3) \times L^2(\Omega, \mathcal{S}^3)$ be a weak solution of the boundary value problem formed by the equations

$$-\operatorname{div} \overline{\sigma}_\eta = b - \operatorname{div} \tau_\eta \,, \tag{75}$$

$$\overline{\sigma}_\eta = \mathcal{D}[\frac{\cdot}{\eta}]\big(\varepsilon(\nabla \overline{v}_\eta) + \varepsilon(\kappa_\eta)\big), \tag{76}$$

which must hold in Ω, and by the boundary condition

$$\overline{v}_\eta(x) = 0\,, \quad x \in \partial\Omega\,. \tag{77}$$

Then

$$\lim_{\eta\to 0} (\|\overline{v}_\eta\|_\Omega + \|\overline{\sigma}_\eta\|_\Omega) = 0. \tag{78}$$

For completeness we present the short **proof**, which is based on the energy method of Tartar:

We first observe that the symmetry of the matrix $\overline{\sigma}_\eta(x)$ and the equations (75) – (77) yield

$$\begin{aligned} \int_\Omega (\mathcal{D}[\frac{x}{\eta}]^{-1} \overline{\sigma}_\eta) \cdot \overline{\sigma}_\eta \, dx &= \int_\Omega \big(\varepsilon(\nabla \overline{v}_\eta) + \varepsilon(\kappa_\eta)\big) \cdot \overline{\sigma}_\eta \, dx \\ &= \int_\Omega (\nabla \overline{v}_\eta + \kappa_\eta) \cdot \overline{\sigma}_\eta \, dx = \int_\Omega \overline{v}_\eta \cdot b + \nabla \overline{v}_\eta \cdot \tau_\eta \, dx + \int_\Omega \kappa_\eta \cdot \overline{\sigma}_\eta \, dx\,. \end{aligned} \tag{79}$$

Condition (i) of the lemma implies that the set $\{\tau_\eta \mid \eta > 0\}$ is bounded in $L^2(\Omega)$, hence the set of functions $\{b - \operatorname{div} \tau_\eta \mid \eta > 0\}$ on the right hand side of (75) is bounded in $H_{-1}(\Omega)$. Moreover, condition (iii) implies that the set $\{\varepsilon(\kappa_\eta) \mid \kappa > 0\}$ is bounded in $L^2(\Omega)$. Since the problem (75) – (77) coincides with the boundary value problem (39) – (41), we thus obtain from (43) that there is C with

$$\|\overline{\sigma}_\eta\|_\Omega + \|\overline{v}_\eta\|_{\Omega;1} \le C$$

for all $\eta > 0$. Consequently, we can choose a sequence $\{\eta_k\}_{k=1}^\infty$ with $\eta_k \to 0$ such that $\{\overline{v}_k\}_{k=1}^\infty = \{\overline{v}_{\eta_k}\}_{k=1}^\infty$ converges strongly in $L^2(\Omega, \mathbb{R}^3)$ to a function $v \in \overset{\circ}{H}_1(\Omega, \mathbb{R}^3)$, and such that

$$\nabla \overline{v}_k \rightharpoonup \nabla v , \quad \overline{\sigma}_k \rightharpoonup \tilde{\sigma} , \tag{80}$$

weakly in $L^2(\Omega, \mathbb{R}^3)$, with a suitable function $\tilde{\sigma}$. Equation (75) and condition (ii) of the lemma imply that $\{\operatorname{div} \overline{\sigma}_k\}_{k=1}^\infty$ belongs to a compact subset of H_{-1}^{loc}. Furthermore, $\operatorname{rot}(\nabla \overline{v}_k) = 0$. These properties, the properties (i) - (iv) of τ_k and κ_k, and the rot-div-Lemma imply that we can pass to the limit on the right hand side of (79) and obtain with a constant $c > 0$

$$\begin{aligned} c\|\overline{\sigma}_k\|_\Omega^2 &\le \int_\Omega (\mathcal{D}[\frac{x}{\eta_k}]^{-1} \overline{\sigma}_k) \cdot \overline{\sigma}_k \, dx \\ &\to (v, b)_\Omega + (\nabla v, \tau)_\Omega + (0, \tilde{\sigma})_\Omega = (v, b - \operatorname{div} \tau)_\Omega = 0 . \end{aligned} \tag{81}$$

Observe next that (76), (81) and the property (iii) of the lemma together yield

$$\varepsilon(\nabla \overline{v}_k) \rightharpoonup 0 \qquad \text{for } k \to \infty ,$$

weakly in $L^2(\Omega, \mathcal{S}^3)$. Since $\nabla \overline{v}_k \rightharpoonup \nabla v$, by (80), it follows that $\varepsilon(\nabla v) = 0$. Using that $v \in \overset{\circ}{H}_1(\Omega, \mathbb{R}^3)$ we conclude from Korn's first inequality that $v = 0$. Relation (78) follows from this result, from the fact that $\overline{v}_k$ converges to v strongly in $L^2(\Omega, \mathbb{R}^3)$, and from (81). The proof of Lemma 15 is complete.

Proof of Lemma 14: We fix t and set

$$\tau_\eta(x) = -T_0(x, \frac{x}{\eta}, t) , \quad \kappa_\eta(x) = -\nabla_y u_1(x, y, t)\Big|_{y=\frac{x}{\eta}} , \quad \tau = -T_\infty(t) ,$$

$$\overline{v}_\eta = v_\eta(t) - \hat{v}_\eta(t) , \quad \overline{\sigma}_\eta = \sigma_\eta(t) - \hat{\sigma}_\eta(t) , \quad b = b(t) ,$$

and verify that under the assumptions of Theorem 4 these functions satisfy the hypotheses of Lemma 15.

Note first that (19) yields

$$\operatorname{div} \tau_\eta(x) = -\operatorname{div}_x T_0(x, y, t)\Big|_{y=\frac{x}{\eta}} .$$

Since by definition $\big(\hat{v}_\eta(t), \hat{\sigma}_\eta(t)\big)$ is a solution of the boundary value problem (39) – (41) to the data

$$\hat{b}(x) = -\operatorname{div}_x T_0(x, y, t)\Big|_{y=\frac{x}{\eta}} = \operatorname{div} \tau_\eta(x), \tag{82}$$

$$\hat{\varepsilon}_p(x) = -\varepsilon(\nabla_y u_1(x, y, t)\Big|_{y=\frac{x}{\eta}}) = \varepsilon(\kappa_\eta(x)), \tag{83}$$

$$\hat{\gamma}_D = \gamma_D(t), \tag{84}$$

cf. (59) – (61), and since $\big(v_\eta(t), \sigma_\eta(t)\big)$ is a solution of the same boundary value problem to the data $\hat{b} = b(t)$, $\hat{\varepsilon}_p = 0$, $\hat{\gamma}_D = \gamma_D(t)$, cf. (50), it follows that $(\overline{v}_\eta, \overline{\sigma}_\eta)$ is a solution of the boundary value problem (75) – (77). Moreover, (17) implies

$$b - \operatorname{div} \tau = b(t) + \operatorname{div}_x T_\infty(t) = 0.$$

It thus remains to verify the conditions (i) – (iv) of Lemma 15. The condition (i) is satisfied, since by assumption $T_0 \in C^3\big(\overline{(\Omega \times Y_{\mathrm{per}})_{T_r}}, \mathcal{S}^3\big)$, from which it can be shown by a modification of the proof given in [20, pp. 21] that for $w \in L^2(\Omega, \mathcal{S}^3)$

$$\lim_{\eta \to 0} (\tau_\eta, w)_\Omega = -\lim_{\eta \to 0} \int_\Omega T_0(x, \frac{x}{\eta}, t) w(x) dx$$
$$= -\frac{1}{|Y|} \int_\Omega \int_Y T_0(x, y, t) dy\, w(x) dx = -(T_\infty, w)_\Omega = (\tau, w)_\Omega\,.$$

Clearly, this relation implies (i). Also, $\operatorname{div}_x T_0 \in C^2\big(\overline{(\Omega \times Y_{\mathrm{per}})_{T_r}}, \mathbb{R}^3\big)$ yields

$$\|\operatorname{div} \tau_\eta\|_\Omega = \|\operatorname{div}_x T_0(\cdot, y, t)\Big|_{y=\frac{\cdot}{\eta}}\|_\Omega \le \|\operatorname{div}_x T_0(t)\|_{\Omega\times Y,\infty} |\Omega|^{1/2}\,,$$

from which we conclude that condition (ii) holds. To prove (iii) we note that $\nabla_y u_1 \in C^2\big(\overline{(\Omega \times Y_{\mathrm{per}})_{T_r}}, \mathbb{R}^{3\times 3}\big)$ implies for $w \in L^2(\Omega, \mathbb{R}^{3\times 3})$ that

$$\lim_{\eta \to 0} (\kappa_\eta, w)_\Omega = -\lim_{\eta \to 0} \int_\Omega \nabla_y\, u_1(x, \frac{x}{\eta}, t) w(x) dx$$
$$= -\frac{1}{|Y|} \int_\Omega \int_Y \nabla_y\, u_1(x, y, t) dy\, w(x) dx = 0,$$

hence $\kappa_\eta \rightharpoonup 0$, weakly in $L^2(\Omega, \mathbb{R}^{3\times 3})$. Again this is shown by a modification of the proof in [20, pp. 21]. Finally, to verify (iv) we note that

$$\operatorname{rot}_x \kappa_\eta(x) = [\operatorname{rot}_x \nabla_y u_1(x, y, t)]_{y=\frac{x}{\eta}}\,.$$

Thus, $\operatorname{rot}_x(\nabla_y u_1) \in C^1\big(\overline{(\Omega \times Y_{\mathrm{per}})_{T_r}}, \mathbb{R}^3\big)$ implies

$$\|\operatorname{rot} \kappa_\eta\|_\Omega = \|[\operatorname{rot}_x \nabla_y u_1(x, y, t)]_{y=\frac{x}{\eta}}\|_\Omega \le \|\operatorname{rot}_x\big(\nabla_y u_1(t)\big)\|_{\Omega\times Y,\infty} |\Omega|^{1/2}\,.$$

Condition (iv) is a consequence of this estimate.

Thus, we can apply Lemma 15 and obtain from (78)

$$\lim_{\eta \to 0} (\|v_\eta(t) - \hat{v}_\eta(t)\|_\Omega + \|\sigma_\eta(t) - \hat{\sigma}_\eta(t)\|_\Omega) = 0,$$

which yields (68) and (69) for $i = 0$. To obtain these relations for $i = 1$, we replace the functions $v_\eta, \hat{v}_\eta, \sigma_\eta, \hat{\sigma}_\eta, T_0, u_1, T_\infty, b$ by their time derivatives and argue in exactly the same way.

Finally, to prove (70) we note that $(\hat{v}_{\eta t}(t), \hat{\sigma}_{\eta t}(t))$ and $(v_{\eta t}(t), \sigma_{\eta t}(t))$, respectively, are solutions of the boundary value problem (39) – (41) to the data

$$\hat{b} = -\operatorname{div}_x T_{0t}(t)\Big|_{y=\frac{x}{\eta}}, \quad \hat{\varepsilon}_p = -\varepsilon\big(\nabla_y u_{1t}(t)\big)\Big|_{y=\frac{x}{\eta}}, \quad \hat{\gamma}_D = \gamma_{Dt}(t),$$

cf. (59) – (61), and

$$\hat{b} = b_t(t), \quad \hat{\varepsilon}_p = 0, \quad \hat{\gamma}_D = \gamma_{Dt}(t),$$

cf. (50), respectively. Since by the regularity assumptions of Theorem 4 we have $\operatorname{div}_x T_{0t} \in C^1\big((\overline{\Omega} \times Y_{\mathrm{per}})_{T_r}\big)$, $\nabla_y u_{1t} \in C^1\big((\overline{\Omega} \times Y_{\mathrm{per}})_{T_r}\big)$, $b_t \in C^2(\overline{\Omega}_{T_r})$, $\gamma_{Dt} \in C^2(\partial\Omega_{T_r})$, we conclude from (43) that there is a constant K with

$$\|\sigma_{\eta t}(t) - \hat{\sigma}_{\eta t}(t)\|_\Omega \le \|\sigma_{\eta t}(t)\|_\Omega + \|\hat{\sigma}_{\eta t}(t)\|_\Omega \le K$$

for all $0 \le t \le T_r$ and all $\eta > 0$. This completes the proof of Lemma 14.

Summary

After the formulation of the microscopic problem and of the homogenized problem we stated existence, uniqueness and regularity theorems for the homogenized problem. The homogenized problem is justified if the difference of the exact solution of the microscopic problem and of an approximate solution constructed from the solution of the homogenized problem converges asymptotically to zero for $\eta \to 0$. Here η is the scaling parameter of the microstructure. This convergence result was stated in Theorem 4. Existence and uniqueness of the solution for the homogenized problem was shown in Sect. 2, the asymptotic convergence result was proved in Sect. 3. An important ingredient of the convergence proof is the energy method of Tartar.

Acknowledgement. The author thanks Waldemar Pompe for helpful suggestions.

References

1. Alber, H.-D. (1998) Materials with memory. Initial-boundary value problems for constitutive equations with internal variables. Lecture Notes in Mathematics **1682**. Springer, Berlin
2. Alber, H.-D. (2000) Evolving microstructure and homogenization. Continuum Mech. Thermodyn. **12**, 235–286
3. Alber, H.-D. (2001) Existence of solutions to a class of quasistatic problems in viscoplasticity theory. Menaldi, J.; Rofman, E.; Sulem, A. (eds.), Optimal control and partial differential equations 95–104. IOS Press, Amsterdam
4. Alber, H.-D., Chełmiński, K. (2002) Quasistatic problems in viscoplasticity theory. Preprint Fachbereich Mathematik, TU Darmstadt **2190**, (43 pages). Submitted to Math. Z.

5. Allaire, G. (1991) Homogenization of the Navier-Stokes equations in open sets perforated with tiny holes. I: Abstract framework, a volume distribution of holes. Arch. Ration. Mech. Anal. **113**, No.3, 209–259
6. Allaire, G. (1992) Homogenization and two-scale convergence. SIAM J. Math. Anal. **23**, No.6, 1482–1518
7. Allaire, G. (2002) Shape optimization by the homogenization method. Applied Mathematical Sciences **146**. Springer, New York
8. Allaire, G., Bal, G. (1999) Homogenization of the criticality spectral equation in neutron transport. M2AN, Math. Model. Numer. Anal. **33**, No.4, 721–746
9. Allaire, G., Briane, M. (1996) Multiscale convergence and reiterated homogenisation. Proc. R. Soc. Edinb., Sect. A **126**, No.2, 297–342
10. Allaire, G., Conca, C., Vanninathan, M. (1999) Spectral asymptotics of the Helmholtz model in fluid-solid structures. Int. J. Numer. Methods Eng. **46**, No.9, 1463–1504
11. Allaire, G., Kohn, R. V. (1993) Optimal bounds on the effective behavior of a mixture of two well-ordered elastic materials. Q. Appl. Math. **51**, No.4, 643–674
12. Barbu, V. (1976) Nonlinear semigroups and differential inclusions in Banach spaces. Editura Academiei, Bucharest; Noordhoff, Leyden
13. Braides, A., Defranceschi, A. (1998) Homogenization of multiple integrals. Oxford Lecture Series in Mathematics and its Applications **12**. Clarendon Press, Oxford
14. Brézis, H. (1973) Operateurs maximaux monotones. North Holland, Amsterdam
15. Chełmiński, K. (1999) On monotone plastic constitutive equations with polynomial growth condition. Math. Meth. in App. Sci. **22**, 547–562
16. Chełmiński, K. (2001) Coercive approximation of viscoplasticity and plasticity. Asymptotic Analysis **26**, 105–133
17. Chełmiński, K., Naniewicz, Z. (2002) Coercive limits for constitutive equations of monotone-gradient type. Nonlinear Analysis TMA **48**, No.8, 1197–1214
18. Chełmiński, K., (2001) Coercive and self-controlling quasistatic models of the gradient type with composite inelastic constitutive equations. Preprint Fachbereich Mathematik, TU Darmstadt **2164**. Submitted to J. nonlinear Sci.
19. Chełmiński, K., (2002) Monotone constitutive equations in the theory of the inelastic behaviour of metals - A summary of results. This volume
20. Dacorogna, B. (1989) Direct methods in the calculus of variations. Applied Mathematical Sciences **78**. Springer, Berlin
21. Ebenfeld, S. (2001) Remarks on the quasistatic problem of viscoelasticity. Existence, uniqueness and homogenization. Preprint Fachbereich Mathematik, TU Darmstadt **2174**. To appear in Continuum Mech. Thermodyn.
22. Giaquinta, M., Modica, G., Soucek, J. (1998) Cartesian currents in the calculus of variations II. Variational integrals. Ergebnisse der Mathematik und ihrer Grenzgebiete, 3. Folge **38**. Springer, Berlin
23. Halphen, B., Nguyen Quoc Son (1975) Sur les matériaux standards généralisés. J. Méc. **14**, 39–63
24. Hornung, U. (ed.) (1997) Homogenization and porous media. Interdisciplinary Applied Mathematics **6**. Springer, New York
25. Hornung, U., Jäger, W. (1991) Diffusion, convection, adsorption, and reaction of chemicals in porous media. J. Differ. Equations **92**, No.2, 199–225

26. Hornung, U., Jäger, W., Mikelic, A. (1994) Reactive transport through an array of cells with semi-permeable membranes. RAIRO, Modélisation Math. Anal. Numér. **28**, No.1, 59–94
27. Jäger, W., Mikelic, A. (1998) On the effective equations of a viscous incompressible fluid flow through a filter of finite thickness. Commun. Pure Appl. Math. **51**, No.9–10, 1073–1121
28. Jäger, W., Mikelic, A., Neuss, N. (2001) Asymptotic analysis of the laminar viscous flow over a porous bed. SIAM J. Sci. Comput. **22**, No.6, 2006–2028
29. Jäger, W., Oleinik, O. A., Shaposhnikova, T. A (1997) On homogenization of solutions of the Poisson equation in a perforated domain with different types of boundary conditions on different cavities. Appl. Anal. **65**, No.3-4, 205–223
30. Jikov, V., Kozlov, S., Oleinik, O. (1995) Homogenization of differential operators and integral functions. Springer, Berlin
31. Kröner, E. (1977) Bounds for effective elastic moduli of disordered materials. J. Mech. Phys. Solids **25**, 137–155
32. Miehe, Ch., Schröder, J. (1999) Computational micro-macro-transitions in thermoplastic analysis of finite strains. Bruhns, O. T., Stein, E. (eds.), Proceedings of the IUTAM Symposium on Micro- and Macrostructural Aspects of Thermoplasticity held in Bochum, Germany, 25.–29. August 1997, 137–146
33. Miehe, Ch., Schröder, J., Schotte, J. (1999) Computational homogenization analysis in finite plasticity. Simulation of texture development in polycrystalline materials. Comput. Methods Appl. Mech. Eng. **171**, No.3-4, 387–418
34. Oleinik, O. A., Shamaev, A. S., Yosifian, G. A. (1992) Mathematical problems in elasticity and homogenization. Studies in Mathematics and its Applications **26**. North-Holland, Amsterdam
35. Ortiz, M. (1996) Computational micromechanics. Comput. Mech. **18**, No.5, 321–338
36. Pankov, A. (1997) G-convergence and homogenization of nonlinear partial differential operators. Kluwer, Dordrecht
37. Ponte Castaneda, P. (1997) Nonlinear composite materials: Effective constitutive behavior and microstructure evolution. Suquet, P. (ed.), Continuum micromechanics. CISM Courses Lect. **377**, 131–195. Springer, Wien
38. Ponte Castaneda, P., Zaidman, M. (1994) Constitutive models for porous materials with evolving microstructure. J. Mech. Phys. Solids **42**, No.9, 1459–1497
39. Suquet, P. (1997) Effective properties of nonlinear composites. Suquet, P. (ed.), Continuum micromechanics. CISM Courses Lect. **377**, 197–264. Springer, Wien
40. Talbot, D. R. S., Willis, J. R. (1997) Bounds of third order for the overall response of nonlinear composites. J. Mech. Phys. Solids **45**, No.1, 87–111
41. Zaoui, A. (1997) Structural morphology and constitutive behaviour of microheterogeneous materials. Suquet, P. (ed.), Continuum micromechanics. CISM Courses Lect. **377**, 291–347. Springer, Wien

Part IV

Damage and Fracture

Models and Experiments Describing Deformation, Roughness and Damage in Metal Forming

Sven Thomas[1], Stefan Jung[2], Michael Nimz[1], Clemens Müller[1], Peter Groche[2], and Eckart Exner[1]

[1] Department of Materials Science, Institute of Physical Metallurgy
Darmstadt University of Technology, D–64287 Darmstadt, Germany
heexn@phm.tu-darmstadt.de

[2] Department of Mechanical Engineering
Institute of Production Engineering and Forming Machines
Darmstadt University of Technology, D–64287 Darmstadt, Germany
info@ptu.tu-darmstadt.de

received 29 Jul 2002 — accepted 12 Nov 2002

Abstract. The simulation of forming processes aiming at a realistic description requires a feasible model, accurate input parameters to set the model alive and a comparison of the simulation results with carefully designed experiments. A brief overview of constitutive equations and approaches to model anisotropy during sheet metal forming and crack propagation is presented. Material parameters are determined in model experiments. A modified nucleation term for the GURSON model [13] is used, which is determined from the experiment. Results from specifically designed experiments illustrate the techniques discussed and demonstrate the application of deformation models. In addition, the feasibility of different experimental techniques to determine local strain is discussed.

1 Constitutive Equations for Sheet Forming Processes

Many of the constitutive equations that describe the processes leading to plastic deformation are based on the VON MISES [25] yield equation or reduce to it in the simplest case. In the following, a brief overview on approaches for describing the materials behaviour with respect to hardening and damage is presented focussing on treatments in stress space. The EINSTEIN convention for repeated indices is applied in order to simplify reading.

1.1 Hardening

The VON MISES yield equation [25] describes the yield criterion of an elastic/ideal plastic material. In order to describe real materials, hardening has to be considered, as for example in form of the HUBER-MISES (see e. g. in [38]) relation. Further extensions and, in turn, additional parameters are needed to consider different yield stresses under tension and compression, well known as BAUSCHINGER effect. This is modelled by kinematic hardening and has been worked out in the simplest form by PRAGER [28] using just one additional parameter. However, this is not sufficient for describing real materials as PRAGER only considers a rigid translation of the yield surface, where the increment is proportional to the plastic strain increment. A more detailed description is needed to assess different materials with a single model, including the distortion of the yield surface (anisotropy). The convexity of the yield surface must also be preserved following the postulate of DRUCKER [8] and IL'IUSHIN [19]. A very general description of the yield surface F has the following form (see e. g. [30]):

$$F = (\sigma_{ij} - \alpha_{ij}) A_{ijkl} (\sigma_{ij} - \alpha_{ij}) - R\,. \tag{1}$$

R and α_{ij} describe the isotropic and the kinematic hardening, respectively. A_{ijkl} allows a distortion of the yield surface. For these three types of variables (which are symmetric with respect to the indices), evolution equations depending e. g. on strain, are needed. The models proposed by CHABOCHE [5] and JIANG [20] give a complete description of the evolution of the hardening parameters. Equation (1) shows that a large number of parameters has to be determined in experiments. Therefore, simpler but less precise models, e.g. that proposed by HILL [16], are used in practical work. Neglecting stresses in the thickness direction of the sheet, HILL's model uses the following yield equation:

$$F = (G + H)\sigma_{11}^2 + (F + H)\sigma_{22}^2 - 2H\sigma_{11}\sigma_{22} + 2N\sigma_{12}^2 - 2K^2\,. \tag{2}$$

The number of parameters is further reduced by taking $H = 2 - G$ [21].

Another fairly simple model, the ICT (Isotropic-Centre-Translation) model proposed by MAZILU and MEYERS [24], is used further below to demonstrate the determination of parameters and the need of taking anisotropy into account. Invariants of this yield equation are formulated following an earlier model due to EDELMAN and DRUCKER [9]. Based on a rheological model LUO [23] introduced three tensors which shift the argument of the invariants. He obtains:

$$\overset{*}{I}_2 = \frac{1}{2}\left(\sigma_{ij}^D - \overset{*}{\sigma}{}^D{}_{ij}\right)\left(\sigma_{ij}^D - \overset{*}{\sigma}{}^D{}_{ij}\right), \tag{3}$$

$$\overset{**}{I}_2 = \frac{1}{2}\left(\sigma_{ij}^D - \overset{**}{\sigma}{}^D{}_{ij}\right)\left(\sigma_{ij}^D - \overset{**}{\sigma}{}^D{}_{ij}\right) \text{ and} \tag{4}$$

$$\overset{**}{I}_3 = \frac{1}{3}\left(\sigma_{ij}^D - \overset{**}{\sigma}{}^D{}_{ij}\right)\left(\sigma_{jk}^D - \overset{**}{\sigma}{}^D{}_{jk}\right)\left(\sigma_{ki}^D - \overset{**}{\sigma}{}^D{}_{ki}\right) \tag{5}$$

where $*$ and $**$ describe the shift by $\overset{*}{\sigma}_{ij}$ and $\overset{**}{\sigma}_{ij}$, respectively. The superscript D in σ_{ij}^D designates the deviatoric part of σ_{ij}. $\overset{*}{\sigma}{}_{ij}^D$ describes the BAUSCHINGER effect similarly to α_{ij} in (1). The other two translation tensors describe the distortion of the yield surface. In this way, a flexible representation of yield surfaces is possible. A simplification, neglecting the third anisotropy tensor, results in the following yield equation:

$$F = \overset{*}{I}_2 - K^2 \left(1 + \alpha_1 \frac{\overset{**}{I}_3}{\overset{**}{I}_2{}^{3/2}} + \alpha_2 \frac{\overset{**}{I}_3{}^2}{\overset{**}{I}_2{}^3} \right) \; . \tag{6}$$

K determines the size of the yield surface, α_1 and α_2 are numerical parameters. In spite of these simplifications, the model allows a good description of the material's anisotropy.

1.2 Damage

When a critical deformation (globally or just locally) is applied to the material, damage will occur. A description of damage has to be implemented in the model in order to describe the material behaviour correctly. In early work, damage was focused on cavity growth (see e. g. [17]). In general there are three approaches to describe this type of damage: The simplest way is to check if a critical value of a damage parameter (e. g. equivalent strain or energy) is exceeded (for a comparison of different criteria see e. g. [12]). Alternatively, a damage parameter can be implemented either by using a representative volume element (RVE) or an effective stress concept.

The effective stress approach introduces a damage tensor D_{ijkl} [29] which increases the (effective) stress $\tilde{\sigma}_{ij}$ of a damaged sample compared with that of an undamaged sample σ_{ij}. The effective stress

$$\tilde{\sigma}_{ij} = D^{-1}_{ijkl} \sigma_{kl} \tag{7}$$

replaces the stress σ_{ij} in the yield equation for the undamaged material. Using a tensor D_{ijkl} enables a description of damage. Frequently, an isotropic damage has been considered with just a scalar measure D. Evolution equations for the damage parameters D_{ijkl} or D are needed. Examples of scalar damage are shown in [4]. In [15], an example for anisotropic damage is mentioned.

The Gurson model [13] is often used to describe ductile fracture. This model considers a RVE with a spherical or other simply shaped cavity. The derived yield equation was extended by Tvergaard and Needleman [37] for considering the interaction between cavities. This allows a better description of material behaviour by introducing additional parameters q_i and substituting the void fraction f by an effective void fraction f^* in the yield equation

$$F = \frac{\sigma_{\text{eq}}^2}{\sigma_{\text{M}}^2} + 2\, q_1 f^* \cosh\left(\frac{q_2\, \sigma_{kk}}{2\sigma_{\text{M}}} \right) - \left(1 + q_3\, (f^*)^2\right) \; . \tag{8}$$

σ_M represents the yield flow stress of the undamaged material in form of a true stress-strain-curve as obtained from tension tests. In the literature, the parameters are often set equal to $q_3 = q_1^2$, $q_2 = 1$ and $q_1 = 1.5$. For details on the three-dimensional implementation in a finite element (FE) code, see [1]. GOLOGANU et al. [11] find that q_2 is determined by the shape of the cavities, whereas in [31] q_2 depends on the hardening coefficient. An extension to include anisotropy is proposed by HUANG et al. [18] by introducing the back-stress tensor in σ_{eq} and σ_{kk} and by DOEGE et al. [7] by introducing the anisotropy tensor A_{ijkl} in σ_{eq}.

The effective void fraction f^* in (8) increases in the same way as f until a critical fraction f_c is reached. f^* grows more rapidly than f for $f > f_c$. This includes the interaction between cavities. The load carrying capacity of an element is lost at a critical value f_f. The evolution equation of $\dot{f}$ is decomposed into a growth term of existing voids (depending on equivalent strain) and a nucleation term $\dot{f}_{Nucl}$. As proposed by CHU and NEEDLEMAN [6], $\dot{f}_{Nucl}$ is strain controlled by the distribution function

$$\mathcal{A} = \frac{1}{s_N\sqrt{2\pi}} \exp\left[-\frac{1}{2}\left(\frac{\varepsilon_M^{pl} - \varepsilon_N}{s_N}\right)^2\right] \quad \text{and} \quad \dot{f}_{Nucl} = \mathcal{A} f_N \dot{\sigma}_M \left(\dot{\varepsilon}_{eq}^{pl}\right) . \quad (9)$$

ε_N and s_N are free material parameters which are obtained by fitting the simulation to experimental results and f_N is the volume fraction of features provoking void nucleation (e. g. inclusions). It will be shown in section 2.1 that the function $\mathcal{A}$ with its freely adjustable parameters can be substituted by an expression which can be accessed experimentally.

2 Experimental Techniques and Methods

2.1 Parameter Identification

Anisotropy The tension test is the simplest way to obtain material parameters for FE-simulations. By testing the undeformed material the hardening coefficient, the elastic modulus and the yield strength are determined.

Pre-strained samples are tested under different directions and loadings in order to obtain the anisotropy parameters. Pre-straining can be done by tension [23] as well as by compression [21]. In the present work, the compression in the flange region during deep drawing was used as pre-straining. The amount of pre-straining is determined by the drawing depth.

Commercially available deep drawing sheet steel (DIN EN 10130, DC05) was used. Small samples were taken from the locations shown in Fig. 1a and tested in tension and compression. To prevent effects due to inhomogeneous strain, the size of the specimen is kept to the minimum handable dimensions (Fig. 1b).

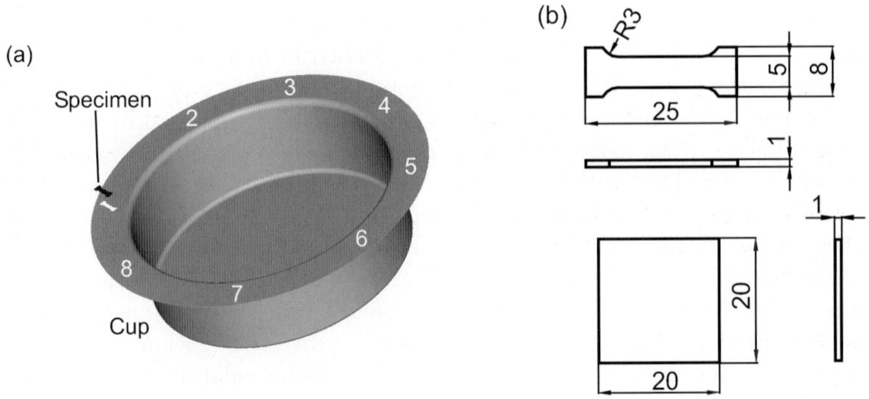

Fig. 1. **(a)** Locations in the flange (eight locations) where test specimens were taken. **(b)** Dimensions in mm of secondary tensile and compression specimens

The samples were taken at different angles (every 15 °) with respect to the rolling direction and at different locations of the deep drawn cup (Fig. 1a). During the tests the strain in loading and transverse directions were measured and yield points in 3D stress space are determined. Pre-deformation of 0, 3, 7 and 14% compressive strain was applied and controlled by the deep-drawing depth of the cup.

Typical stress strain diagrams for different pre-deformation are shown in Fig. 2. From these diagrams, the yield strength and hardening were obtained for the differently orientated samples. The yield strength depends strongly on the pre-deformation, but varies only slightly with the specimen orientation (Fig. 3). The complete analysis of all specimen locations and angles allows the

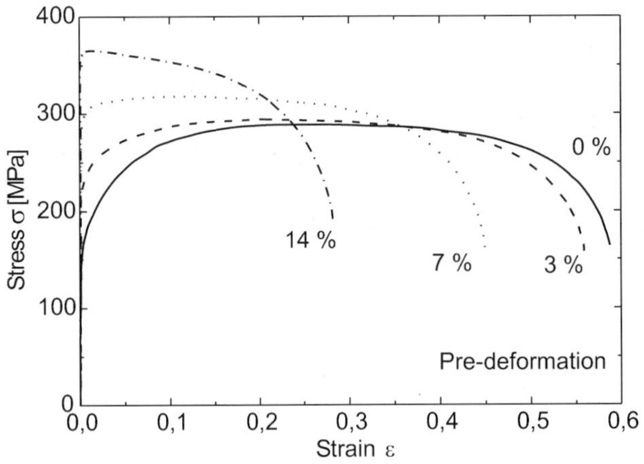

Fig. 2. Stress-strain diagrams for different amounts of pre-deformation

determination of the yield surface in the initial state and its evolution with increasing deformation. A self programmed Fortran routine [21] was used to fit the yield equation (6) to the experimental data.

In Fig. 4, the yield surface of the as-received sheet material in principal stress space σ_{11}-σ_{22} determined with the ICT theory is compared with that of VON MISES. The results reveal that there is an initial anisotropy owning to residual stresses and structural anisotropy in the sheet. The anisotropic yield surface (ICT) is shifted and slightly elongated with respect to the VON MISES yield surface. A similar yield surface was obtained for other stress directions [21]. This development of the yield surface with strain (Fig. 4b) demonstrates the need of extensive experimental investigations as carried out here to determine the input parameters for the simulation of anisotropic material behaviour. A shift and a distortion of the yield surface was obtained. This behaviour is neither adequately explained by the VON MISES nor by HILL's model.

Damage The extended damage model of GURSON (8) was used to simulate the crack propagation in Compact-Tension (CT) specimens. The aluminium alloy EN AW-7475 in an over-aged state was chosen to obtain a highly ductile behaviour. After a homogenisation process (T = 465 °C , t = 2 h) the material was tempered (T = 180 °C , t = 2 h). The true stress-strain curve is needed for the simulation and was obtained in tension tests. Further investigations on crack propagation were performed using compact-tension (CT) specimens with dimensions of 31 mm x 15 mm x 29.5 mm (geometry according to ASTM-399). The difference between the over-aged and the under-aged

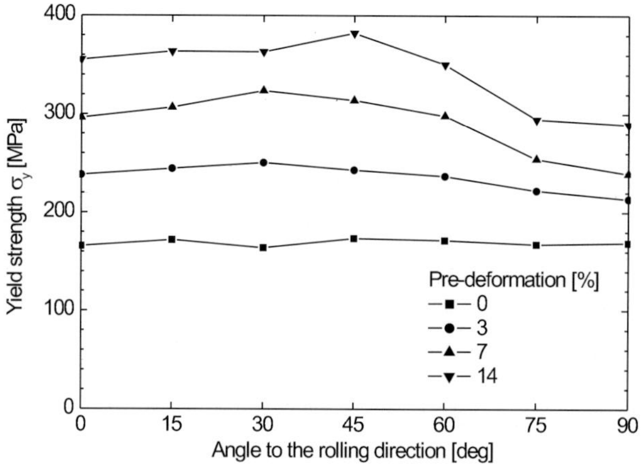

Fig. 3. Dependence of the yield stress on the pre-deformation and orientation

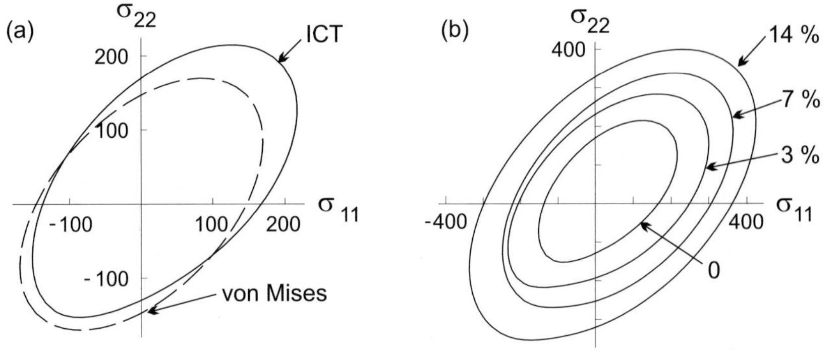

Fig. 4. (a) Comparison of the yield surface in the undeformed state, as determined for the von Mises model (dashed line) and for the ICT model (solid line). **(b)** Evolution of the yield surface at location 1 as fitted with the ICT model after pre-deformations by 0, 3, 7 and 14 %, respectively

state can be seen by comparing Fig. 5(a) and Fig. 5(c). In the over-aged state pore formation is much more pronounced than in the under-aged state.

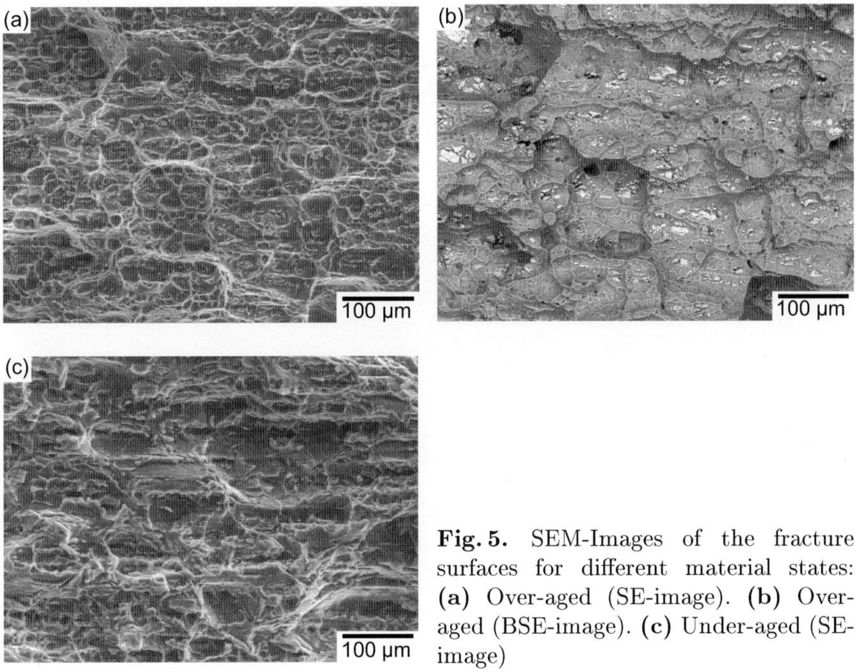

Fig. 5. SEM-Images of the fracture surfaces for different material states: **(a)** Over-aged (SE-image). **(b)** Over-aged (BSE-image). **(c)** Under-aged (SE-image)

The GURSON model describes damage using the void volume fraction f. Its evolution is calculated by using (8) which considers void growth and nucleation of new pores. Nucleation is expressed as a function of the strain distribution $\varepsilon_{\mathrm{M}}^{\mathrm{pl}}$ of the undamaged material. The nucleation term does not consider formation of pores at specific locations as e.g. grain or phase boundaries or inclusions. The aluminium alloy under investigation has iron-rich intermetallic inclusions at which pores form. This can be seen by comparison of fracture surface images in back-scattered-electron (BSE) and secondary-electron (SE) contrast (Figs. 5a and 5b). The fraction of inclusions on the fracture surface is at least 4 times higher than in cross sections. Pores formed at inclusions are dominant, but there are also smaller pores forming in the inclusion free regions.

Cross sections of the crack tip were made in order to assess the fracture mechanism. Pores form at inclusions in front of the crack tip (Fig. 6). As these pores are usually not connected to the crack tip, the smaller pores nucleate and grow. By joining these small pores, the crack propagates from the crack tip to the large pores in front of the crack.

The following assumptions are made to establish a quantitative model: The crack stops at pores which previously have formed at inclusions. New pores develop in front of the crack tip at inclusions. When a critical size is reached, the crack tip connects to these pores by formation and growth of smaller (secondary) pores. The distance of nearest-neighbour inclusions is, therefore, a direct measure of the probability of pore nucleation in front of the crack tip.

The distribution of the distance of inclusions in the direction of the crack growth was measured in cross sections. From profile measurements it was found that the crack deviates from a linear track by angles up to 45 °. Therefore, inclusions within this angle to both sides of the crack growth direction

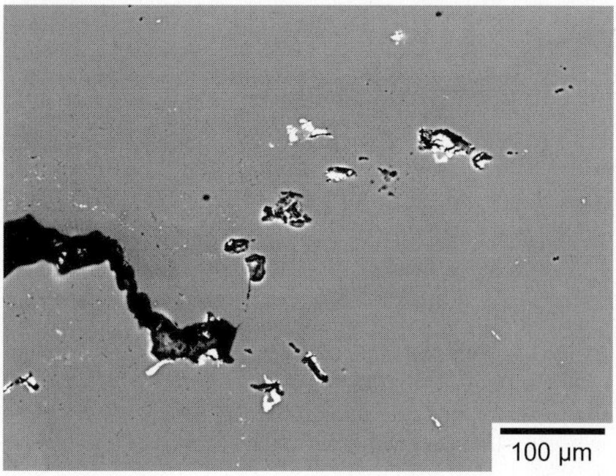

Fig. 6. Cross section of a crack tip showing pore formation at inclusions in front of the crack tip

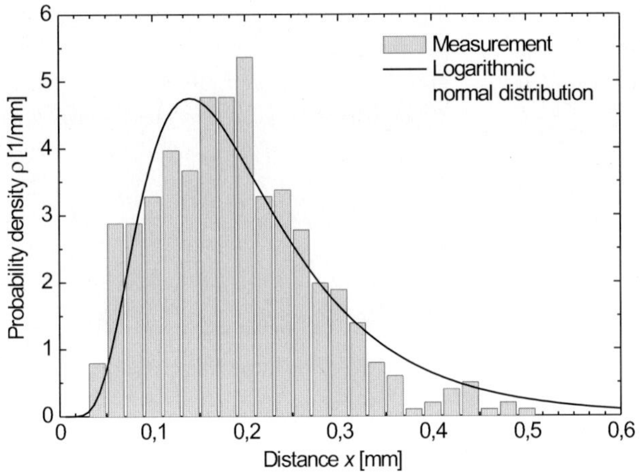

Fig. 7. Distance distribution of nearest neighbour distances of intermetallic inclusions

were considered for the distance measurement. The resulting distribution (probability density versus distance) is shown in Fig. 7.

The distribution density has a maximum at about 0.18 mm and can be reasonably fitted by the GAUSSian distribution using a logarithmic distance scale (logarithmic normal function):

$$H = \frac{1}{\sqrt{2\pi}\, w\, x} \exp\left[-\frac{1}{2} \left(\frac{\ln\{x/x_c\}}{w} \right)^2 \right] , \tag{10}$$

so H is the relative number of distances per unit length, w is the (logarithmic) standard deviation and x_c is the (geometric) mean of distances. Equation (10) replaces the expression $\mathcal{A}$ in (9). This enables experimental measurement of parameters which, in the extended Gurson model had to be fitted by adjusting the simulated curve to experimental results.

2.2 Characterization of Local Strain

For a critical analysis of deep drawn parts and for comparison of the results of experiments and simulations, characterization of local deformation is essential. A useful and well-known method to quantify the local strain is to measure the deformation of a grid imprinted on the surface [3]. Other techniques have been developed in the course of this work.

The following six different methods were compared in order to check their accuracy and feasibility:

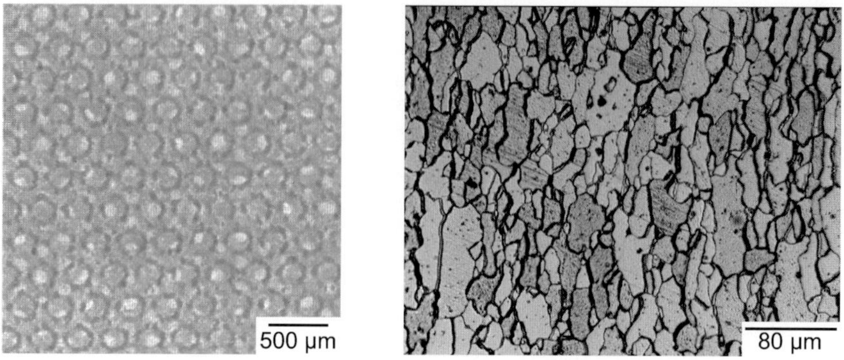

Fig. 8. Examples for microscale features used to determine local strain: **(a)** RSSE. **(b)** Microstructure (grain boundaries)

(1) Change of 3-dimensional roughness
(2) Deformation of regular surface structure elements (RSSE)
(3) Change of grain dimensions
(4) Comparison of individual RSSE
(5) Comparison of distances of microstructural features
(6) Change of hardness

Examples of RSSE (lubricant pockets or other regular pattern imposed to the sheet surface) and the microstructure in the undeformed state are shown in Fig. 8a and 8b, respectively.

A detailed description of a large variety of 3-dimensional roughness parameters is given in [33]. Measurements were carried out on surfaces by mechanical profilometry using a Hommel tester on the commercial ferritic deep-drawing sheet steel DC05. The size of the area was 1 mm x 1 mm. All roughness parameters show a similar exponential increase with plastic strain (Fig. 9). The error in strain was estimated to be $\Delta\varepsilon = \pm 0.05$ as outlined by bold lines in the figure. As an example, the estimation of the experimental error is shown in Fig. 9 for S_{a} which is the 3-dimensional arithmetic mean deviation. The dashed lines indicate the upper and lower limit of the scatter band of S_{a} values. Projection onto the strain axis (Fig. 9) yields $2\Delta\varepsilon$ with $\pm\Delta\varepsilon$ designating the experimental error of the estimated local strain ε.

Another method is to measure the change of microstructural features as described in [10]. Etched measuring grids typically have a line distance of 1 mm. A finer grid is available using the RSSE (method 2) and even finer, though irregular, is made up by the microstructure (method 3) inherent in essentially any engineering material. The area needed is determined by the size of the objects and the required accuracy. This area is almost identical for grains and lubricant pockets (exponentially decreasing curves in Fig. 10), i. e., for an area of 2 mm^2 the experimental (standard) error is 0.4 and 0.03 respectively [27]. The lubricant pockets are more regular than the grains. Therefore,

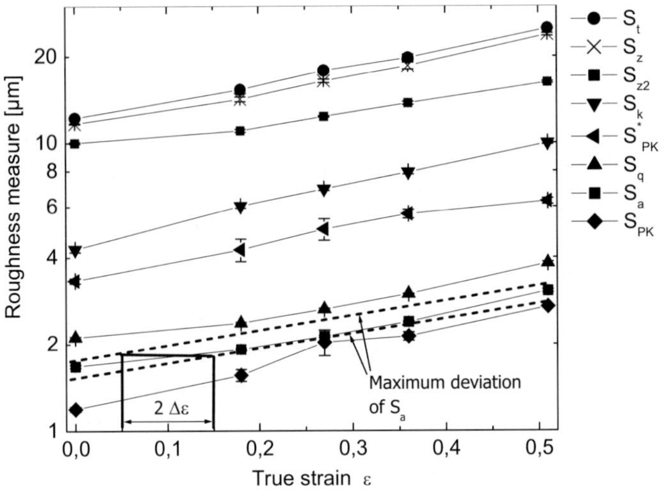

Fig. 9. Roughness parameters (as specified by [33]) evaluated by using the program Topograf [36] of data by Hommel-tester versus plastic strain for a sheet with regular (deterministic) surface structure elements (lubricant pockets) The dashed lines show the maximum deviation of S_a from a linear correlation

using the grains to determine the strain requires a large area (compared with the grain size) to obtain the same accuracy as for the lubricant pockets.

The most accurate method is the direct comparison of surface features before and after deformation, i. e. lubricant pockets (method 4) and etched grain boundaries (method 5). Both have an experimental error of $\varepsilon = 0.01$, while the needed area is 0.12 mm^2 using the grain boundaries and 2.06 mm^2 using the lubricant pocket.

The experimental error of methods (1) to (6) is shown in Fig. 10. As expected, the errors of methods (2) and (3) depend inversely on the measurement area as the number of measured features (RSSE or grains) increases quadratically with the area. The smallest error was obtained by direct comparison of surface features, independent of their size and from the considered area.

The disadvantage of methods (1) to (5) is their limitation to unidirectional deformation without contact (e. g. between sheet material and die). During deep drawing contact between the sheet and the tool occurs. This can be seen in Fig. 12 where the roughness (core roughness depth S_K) was measured on the inner and the outer side of a deep drawn cup (Fig. 1a).

The roughness S_K on the inner side is higher than on the outer due to irregular wear. During cup drawing the material is bent and re-bent at the flange. Therefore, methods (2) to (5) do not work. For the use of method (1) a detailed study for the relation between strain history and roughness is needed.

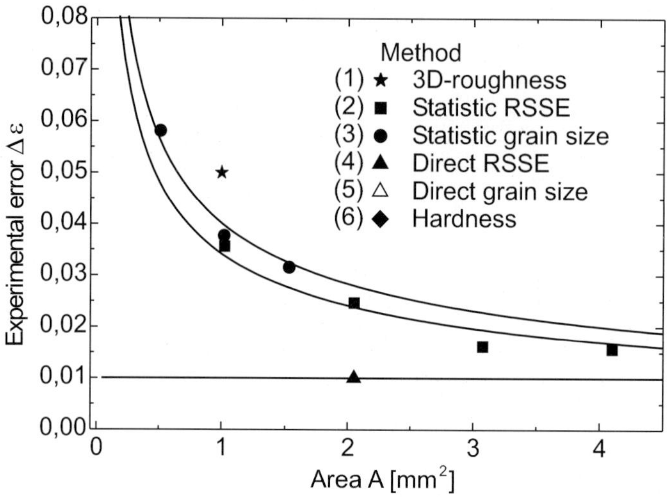

Fig. 10. Experimental error of strain using different methods

In order to obtain local strains for more complicated strain histories and problems with contact the dependence of hardness on strain was investigated. Deep drawing with bending taking place in the flange region was taken as model experiment. A schematic scetch of the experimental set-up is shown in Fig. 13a. The sheet is drawn between the blank holder and the die. A normal pressure of 2 MPa was applied to the blank holder. Two clamps were used to pull the strip. Lubricant oil was used to reduce the friction

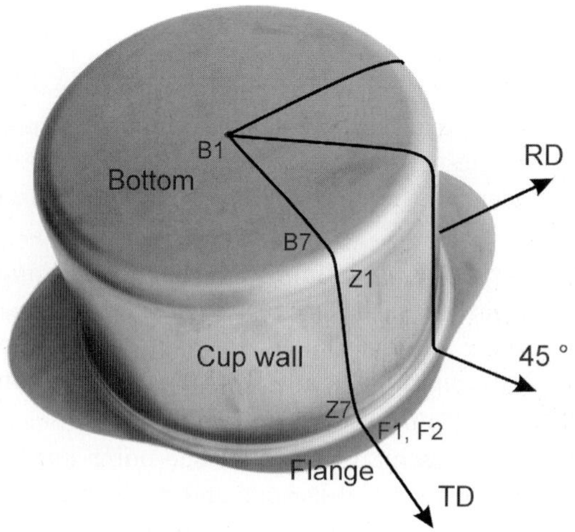

Fig. 11. Arc lengths along which the sheet thickness is measured (RD: Rolling direction, TD: Transverse direction) and locations of roughness measurements

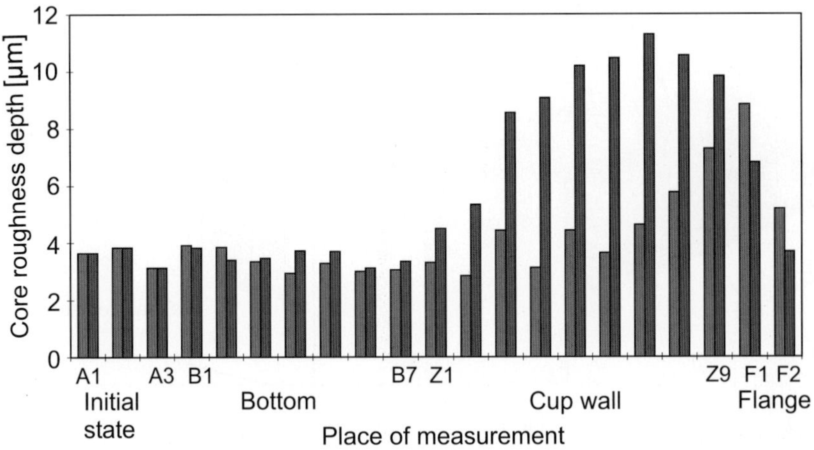

Fig. 12. Core roughness depth S_K [32] of a deep drawn cup on the inner side (left columns) and on the outer side (right columns). The locations of roughness measurements are indicated in Fig. 11.

coefficient to less than 0.05. The commercial aluminium alloy EN AW-5083 was used as sheet material. The material is bent when drawn around the edge and re-bent when leaving the edge curvature (Fig. 13b). The experiment was modeled by using the FE-program ABAQUS in order to compare the experimentally determined strains with those from the simulation. The model geometry is shown in Fig. 13b. Die and blank holder were represented by rigid shell elements and the sheet material as deformable volume elements. As a measure of strain the equivalent plastic strain

$$\varepsilon_{\mathrm{eq}}^{\mathrm{pl}} = \int \sqrt{\frac{2}{3}\dot{\varepsilon}_{ij}\dot{\varepsilon}_{ij}}\ dt \tag{11}$$

was calculated in the simulation. For uniaxial loads it reduces to the strain in the loading direction.

The strain distribution over the sheet thickness results from the following deformation stages: On the outer side the material is first elongated and then compressed and on the inner side the sequence is compression/elongation. A strain-hardening relationship is needed in order to correlate the measured hardness over the sheet thickness after bending and re-bending with the results from the FE-simulation.

The influence of strain history was studied in unidirectional experiments. Tension, compression and alternate loadings (tension/compression and compression/tension) were applied. The results are presented in Fig. 14. For the alternating loading the equivalent strain, i. e. the sum of the plastic strain in tension and compression, was chosen as strain measure. The specimens were deformed to certain strain levels and then deformed back to their original

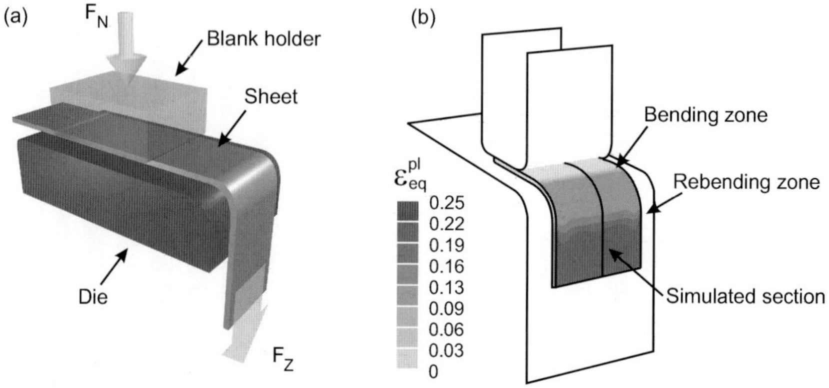

Fig. 13. Set-up for strip drawing: **(a)** Schematic view of the experiment (F_N =load applied to the blank holder, F_Z = applied force for drawing the material). **(b)** Geometry used in finite element modelling with calculated equivalent strain values indicated

length. This results in a true strain equal to zero, but an equivalent strain of about twice the unidirectional strain.

Three different curves were found which were fitted with the HOLLOMON-LUDWIK equation

$$HV = HV_0 + k\varepsilon^n \ , \tag{12}$$

which is commonly used to describe stress-strain relations [34]. HV_0 is constant and was set equal to the hardness of the annealed and undeformed mate-

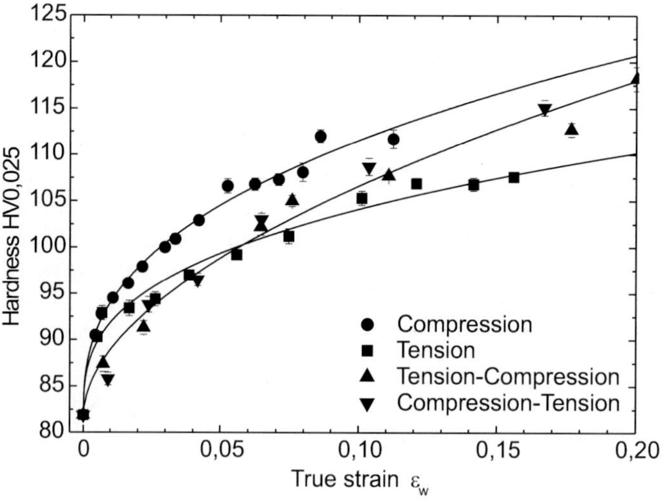

Fig. 14. Hardness of uniaxially deformed specimens as a function of true strain

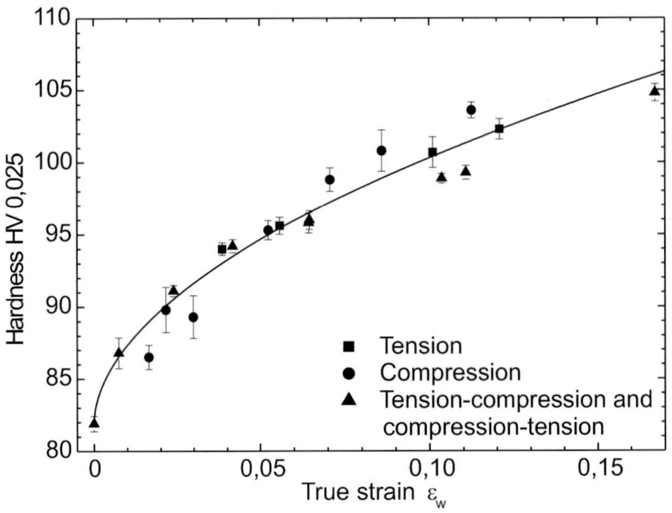

Fig. 15. Hardness vs. strain after stress relieve

rial. The highest hardness was found after compression. For both types of alternating loadings one single curve describes the data points. This curve runs below the curves after unidirectional deformation in the beginning, crosses that for tension and approaches that for compression at high strains.

After deformation the specimens were heat treated. Interestingly and conveniently for further assessment, the hardness values of a given equivalent strain become independent of the type of deformation. A common curve can be fitted to the data points (Fig. 15) with $HV_0 = 81.9$, $k = 62.4$ and $n = 0.54$. The heat treatment eliminates residual stresses in the specimen and removes the difference between hardness readings (Fig. 14) after different loadings.

$$\varepsilon = \left(\frac{HV - 81.9}{62.4}\right)^{1/0.54} \tag{13}$$

The curve shown in Fig. 15 or the corresponding relationship can now be used to transform the measured hardness values into strain values. This method has been applied to the strip bending. The hardness over the sheet thickness was measured after re-bending. The resulting strain versus distance of the surface curve is compared with the results from the FE-simulation (Fig. 16).

Both, experimental and simulated results show a similar strain distribution across the sheet thickness (Fig. 16). The material is only slightly deformed in the middle of the sheet and the deformation increases to the rim. On the inner side the true strain is higher than on the outer side, as easily shown by a simple analytical estimate of the development of strain in a

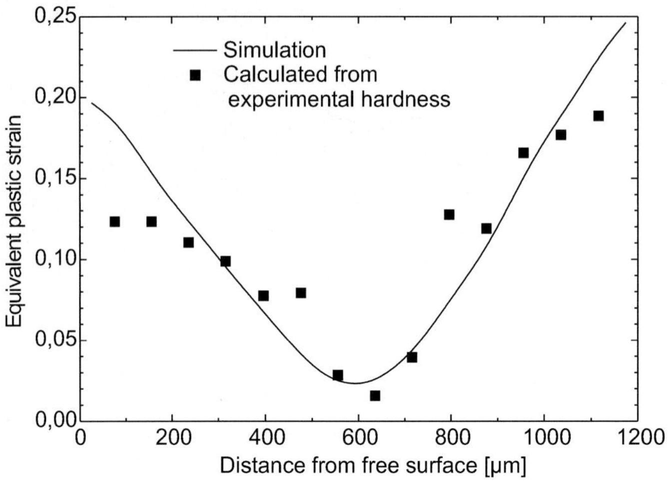

Fig. 16. Comparison of simulation results and strain calculated from hardness for the re-bent strip

plate during bending [35]. Reasonable agreement exists in the region of small strains.

The error of strain determined from hardness readings depends on the number of measurements and on the accuracy of the stress-strain curve. The error is proportional to the inverse of the square root of the number of measurements which is proportional to the area squared. The behaviour is therefore similar to that of method (2) and (3), as shown in Fig. 10. An error of $\Delta HV = 2$ corresponds to an error in strain of 0.012 (Fig. 10). However, taking into account the uncertainty of the fitted curve, a higher error must be assumed. As work hardening is less pronounced in the range of high strain, strain estimates derived from hardness increase are more prone to large scatter at large true strain. At a strain of 0.1 the error is estimated to about 0.05 [35].

2.3 Miniature Deep Drawing Facility

A miniature deep drawing facility was built in order to perform in-situ observations. The facility can be used in either a scanning electron (SEM) or a stereo light microscope. The punch is fixed and the die holder moves down. This keeps the bottom part of the sheet at a constant position and the image in focus. Figure 17 shows the device and the experimental set-up for the stereo microscope.

Circular blanks of up to 26 mm in diameter and thickness up to 1 mm can be drawn to circular cups. The drawing gap has a width of 1.2 mm. The radius of the flange and the punch are 2 mm. The loading device is driven by

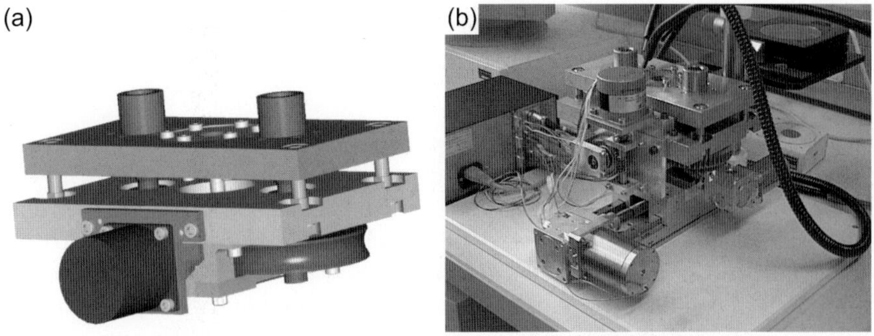

Fig. 17. Miniature deep drawing facility allowing in-situ observation of damage evolution on surfaces: (**a**) Schematic drawing. (**b**) Photograph of the set up for stereo microscopy

a step motor which allows a maximum force of $F_{\max} = 13$ kN. The drawing speed is set to $v = 0.045$ mm s^{-1}.

With a punch diameter of 10 mm the maximum drawing ratio is 2.6. By specific variation of the circular blank diameter, the maximal ratio of the commercial deep drawing steel DC04 of 2.25 [22] is exceeded. This yields a crack at the bottom of the cup. The whole deep drawing facility can be moved by table in two perpendicular directions in order to find the initiated crack and to follow its growth.

The outcome of a typical experiment is shown in Fig. 18. The crack follows the circular edge at the bottom and has the shape of a sickle. A SEM image (Fig. 18) shows that pores form in the middle of the sheet. Shear occurs towards the surface. Even though the sheet is very thin a triaxial stress occurs in the middle of the sheet. It was found that this stress state is favored by necking in thickness direction which occurs prior to the pore formation [27].

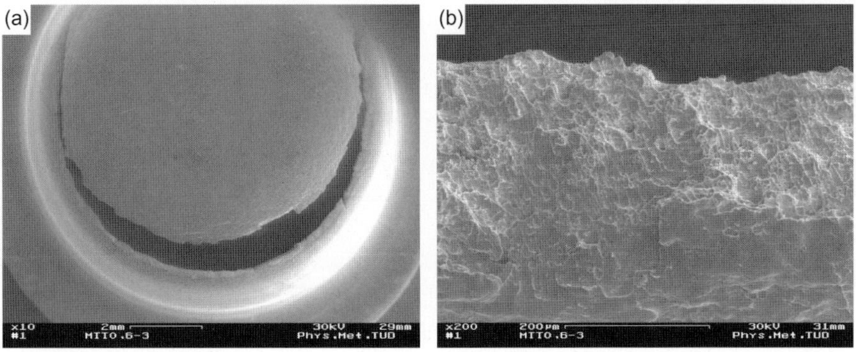

Fig. 18. Miniature deep drawn cup: (**a**) Cup with fracture near the bottom. (**b**) Fracture surface

3 Modeling of Anisotropy and Damage

3.1 Anisotropy

The simulation of the deep drawing was performed by using the commercial FE-program Pam-Stamp [14]. Three models were compared: (i) the von Mises model for isotropic deformation, (ii) the Hill model (2) for anisotropic deformation and (iii) the ICT-theory (6) which is specially designed to describe anisotropy. The set-up of the simulation was as close as possible to the experimental conditions. The sheet material and the punch is represented by shell elements and the blank holder and the die by volume elements.

The accuracy of the different models was checked by comparing the sheet thickness and the contour of the sheet in the flange region. The thickness over the distance along the arc length is measured in different directions (indicated in Fig. 11). In Fig. 19, the thickness in the rolling direction is compared with the results from the simulation. The thickness distribution is described unsatisfactorely by the models. The VON MISES model shows the largest deviations from the experimental results. In the region of the bottom, the ICT and the HILL models show approximately the thickness as observed in the experiment. The results derived with the HILL model are closer to the experimental findings in the cup wall region as well as in the flange region. Even though the ICT model has more adjustable parameters, the deviation of the predicted sheet thickness is higher than that obtained by using HILL's model. This fact can be ascribed to a deficiency in the evolution equations for the ICT theory. This shows the importance of appropriate evolution equations which incorporate the material behaviour and consideration of all pertinent

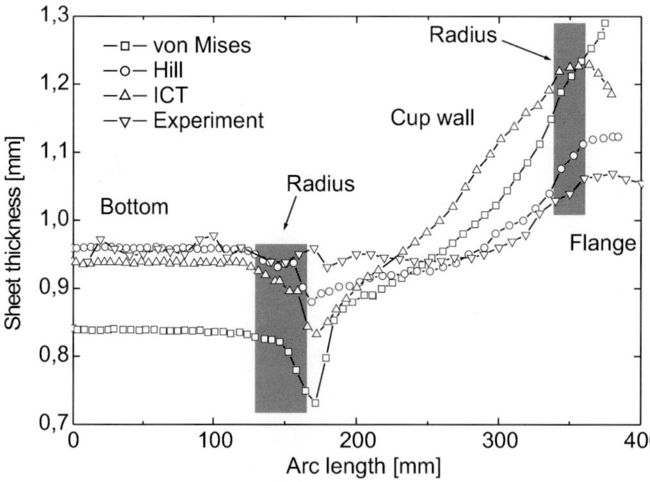

Fig. 19. Comparison of experimental and simulated sheet thickness (original thickness 1 mm) parallel to the rolling direction (Fig. 11)

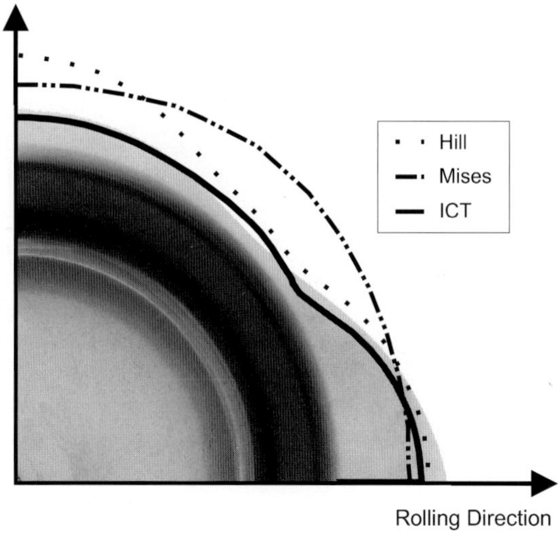

Fig. 20. Comparison of experimental and simulated contours of the flange region of the cup

physical effects. Another point of possible improvement is the calculation of the sheet thickness from the 2-dimensional deformation of shells which are used for the sheet material. The sheet thickness is calculated using the given algorithms of the program.

Comparing the contours of the cup in the flange region (Fig. 20) with the simulation results we find that the ICT model describes the shape accurately. Larger deviations are observed with the HILL model. The results obtained with the VON MISES model are not adequate as the isotropic description can obviously not predict the formation of ears.

3.2 Damage

The crack propagation in the CT-specimen was investigated in order to compare the results obtained with the standard Gurson model and its modification with the experimental findings (Fig. 21).

Apart from deviations at small crack opening displacements (COD) both simulated curves describe the experimental results very well. However, when using the nucleation term of CHU and NEEDLEMAN, a too stiff material response is predicted. In order to optimize the description, i. e. to account for the force needed for larger COD, the free parameters ε_{N}, s_{N}, f_{c} and f_{f} were modified. The stiffness could not be further reduced without numerical instabilities. Thus, the fit shown is the best result to be obtained by this procedure.

The modification presented here (10) predicts a softer material behaviour. The material parameters were adjusted to obtain an optimum approximation

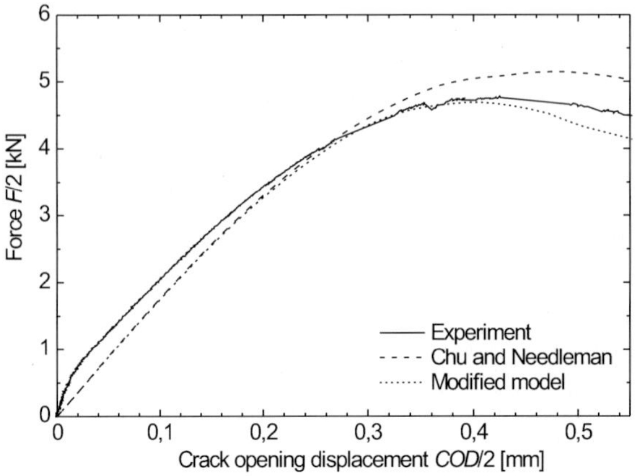

Fig. 21. Load-crack opening displacement as measured and simulated

of the experimental curve. This model yields a better fit and at the same time it is numerically more stable to variations of the input parameters. Due to the substitution of the strain dependent distribution of pore nucleation by an experimental measurable distance distribution, the number of parameters is reduced (i. e. ε_N and s_N do not need to be fitted any longer as parameters of the distribution function can be measured). The modified model can be applied to materials where the location of nucleation is experimentally accessible.

4 Conclusions

(1) Available approaches to formulate constitutive equations for hardening and damage need to be adjusted to the actual mechanisms acting in real deformation processes as, e.g., sheet forming.
(2) By introducing a modified expression for the distance distribution of voids two fitting parameters in an earlier damage model can be substituted by parameters accessible by simple experiments.
(3) Experimental techniques for parameter identification used for modelling deformation and damage for assessing the reliability of theoretical predictions must be adjusted to each material and forming technique. Adequate sensitivity and accuracy can only be obtained if available techniques are carefully modified.
(4) The comparison of simulation using various models is possible by commercially available FE codes. For the deep drawing process considered in this paper, the Hill and the ICT models describe the changes of sheet thickness and cup contours reasonably well while the von Mises model is not able to describe such anisotropy effects.

References

1. Baaser, H. (1999) Dreidimensionale Simulation duktiler Schädigungsentwicklung und Rißausbreitung. PHD Thesis, Darmstadt University of Technology, Darmstadt
2. Banabic, D., Pöhlandt, K. (2001) Yield Criteria for Anisotropic Sheet Metal. UTF Science, Bamberg
3. Brewer, R. C., Alexander, J. M. (1960) A New Technique for Engraving and Measuring Grids in Experimtal Plasticity. J. Mech. Phys. Solids **8** 76-80
4. Chaboche, J. L. (1988) Continuum Damage Mechanics: Part I and Part II. J. Appl. Mech. **55** 59-72
5. Chaboche, J. L. (1989) Constitutive Equations for Cyclic Plasticity and Cyclic Viscoplasticity. Int. J. Plast. **5** 283-295
6. Chu, C. C., Needleman, A. (1980) Void Nucleation Effects in Biaxially Streched Sheets. J. Eng. Mater. Technol. **102** 249-256
7. Doege, E., Bagaviev, A.,Dohrmann, H. (1997) Application of an Anisotropic Extension of Gurson Model to Practical Engineering Problems. In: Owen, D. R. J. and Onate, E. and Hinton, E. (Eds), Sheet Forming Processes, 5th International Conference on Computational Plasticity, Barcelona, 17-20 March 1997. CIMNE, Barcelona,1453-1458
8. Drucker, D. C. (1951) A More Fundamental Approach to Plastic Stress-Strain Relations. Proc. US Nat. Con. Appl. Mech., ASME, 487-491
9. Edelman, F., Drucker, D. C. (1951) Some Extensions of Elementary Plasticity Theory. J. Franklin Inst. **251** 581-605
10. Exner, H. E. (1993) Quantitative Description of Microstructural Geometry – A Practical Guide to Manual Processes, Part III. Pract. Metallogr. **30** 287-293
11. Gologanu, M., Leblond, J.-B. (1994) Approximate Models for Ductile Metals Containing Nonsperical Voids - Case of Axisymmetric Oblate Ellipsoidal Cavaties. J. Eng. Mater. Techol. **116** 290-297
12. Groche, P. (1991) Bruchkriterien für die Blechumformung. Fortschritts-Berichte VDI No. 229, VDI-Verlag Düsseldorf
13. Gurson, A. L. (1977) Continuum Theory of Ductile Rupture by Void Nucleation and Growth: Part I - Yield Criteria and Flow Rules for Porous Ductile Media. J. Eng. Mater. Technol.**99**2-15
14. Haug, E., Pasquale, E. di, Pickett, A. K., Ulrich, D. (1991) Industrial Sheet Metal Forming Simulation Using Explicit Finite Element Methods. In: FE-Simulation of 3-D Sheet Metal Forming Processes in Automotive Industry. VDI-Bericht 894, VDI-Verlag, Düsseldorf, 259-291
15. Hayakawa, K., Murakami, S., Liu, Y. (1998) An Irreversible Thermodynamics Theory for Elastic-Plastic-Damage Materials. Eur. J. Mech. A **17** 13-32
16. Hill, R. (1948) A Theory of the Yielding and Plastic Flow of Anisotropic Metals. Proc. R. Soc. A **193** 281-297
17. Huang, Y., Hutchinson, W., Tvergaard, V. (1991) Cavitation Instabilities in Elastic-Plastic Solids. J. Mech. Phys. Solids **39** 223-241
18. Huang, H.-M., Pan, J., Tang, S. C. (2000) Failure Prediction in Anisotropic Sheet Metals Under Forming Operations With Consideration of Rotating Principal Stretch Directions. Int. J. Plast. **16** 611-633
19. Il'Iushin, A. A. (1961) On the Postulate of Plasticity. PMM **25**, 503-507

20. Jiang, Y., Sehitoglu, H. (1996) Modeling of Cyclic Ratcheting Plasticity, Part I: Development of Constitutive Relations. J. Appl. Mech. **63** 720-725
21. S. Jung (2002) Bestimmung der beim Tiefziehen von kaltgewalztem Stahl induzierten Anisotropie. Berichte aus Produktion und Umformtechnik, Band 53. Shaker, Aachen
22. Lange, K. (1990) Umformtechnik, Bd 3: Blechbearbeitung, Springer-Verlag, Berlin
23. Luo, S. (1996) Entwicklung eines anisotropen plastischen Fließgesetzes für die Blechumformung. Berichte aus der Produktion und Umformtechnik, Band 33. Shaker, Aachen
24. Mazilu, P., Meyers, A. (1985) Yield Surface Description of Isotropic Materials After Cold Prestrain. Ingenieur Archiv **55** 213-220
25. von Mises, R. (1928) Mechanik der plastischen Formänderung von Metallen. Z. Angew. Math. Mech. **8**, 161-185
26. Nimz, M., Jung, S., Müller, C., Pompe, O. (2001) Characterisation of Local Deformations in Deep-Drawing Sheet Steel. Progress in Metallography (G. Petzow Ed.) 65-68
27. Nimz, M. (2002) Charakterisierung der lokalen Verformung und Schädigung bei der Blechumformung. PhD Thesis, TU Darmstadt
28. Prager, W. (1956) A New Method of Analyzing Stresses and Strains in Work-Hardening Plastic Solids. J. Appl. Mech. **23** 493-496
29. Rabotnov, Y. (1968) Creep Rupture. In: Hetenyi, M., Vincenti, W. (Eds.) Proceedings of the 12th International Congress of Applied Mechanics, Stanford. Springer, Berlin, 342-349
30. Rees, D. W. A. (1982) Yield Functions that Account for the Effects of Initial and Subsequent Plastic Anisotropy. Acta Mech. **43** 223-241
31. Ristinmaa, M. (1997) Void Growth in Cyclic Loaded Porous Plastic Solid. Mech. Mat. **26** 227-245
32. Staeves, J. (1998) Beurteilung der Topografie von Blechen im Hinblick auf die Reibung bei der Umformung. Berichte aus der Produktion und Umformtechnik, Band 41. Shaker, Aachen
33. Stout, K. J., Sullivan, P. J., Dong, W. P., Mainsah, E., Luo, N., Mathia, T., Zahouani, H. (1993) The Development of Methods for the Characterisation of Roughness in Three Dimensions, Publication No. EUR 15178 EN, Commission of the European Communities, Bruxelles
34. Thomas, S., Müller, C., Exner, H. E. (2001) Prediction of Local Strain and Hardness in Sheet Forming. Z. Metallkd. **92**, 830-833
35. Thomas, S. (2001) Konstitutive Gleichungen und numerische Methoden zur Beschreibung von Deformation und Versagen. PhD Thesis, TU Darmstadt
36. Topograf, version 1/2000, http://www.ptu.tu-darmstadt.de
37. Tvergaard, V., Needleman, A. (1984) Analysis of the Cup-Cone Fracture in a Round Tensile Bar. Acta Metall. **32** 157-169
38. Wriggers, P., Eberlein, R., Gruttmann, F.(1995) An Axisymmetrical Quasi-Kirchoff-Type Shell Element for Large Plastic Deformations. Arch. Appl. Mech. **65** 465-477

Remarks on the Use of Continuum Damage Models and on the Limitations of their Applicability in Ductile Fracture Mechanics

Herbert Baaser and Dietmar Gross

Institute of Mechanics, Darmstadt University of Technology
Hochschulstr. 1, D–64289 Darmstadt, Germany
{baaser, gross}@mechanik.tu-darmstadt.de

received 23 Aug 2002 — accepted 29 Oct 2002

Abstract. Ductile crack initiation and growth is investigated by applying a three–dimensional finite element analysis in conjunction with a nonlinear damage model. This numerical procedure is formally restricted by the so–called *loss of ellipticity* due to a type change of the differential equations. A time–independent finite strain formulation is applied, based on a multiplicative decomposition of the deformation gradient in an elastic and a plastic part leading to an efficient integration scheme. This formulation can be used as a general interface for the implementation of different constitutive models describing *damage* by a scalar quantity in an isotropic manner. As examples, we show some details of the thermomechanics of the Rousselier and the Gurson damage models .
We discuss the description of microscopically strongly inhomogeneous material behaviour by discretization methods using macroscopic mechanical field quantities, which are microscopic averages. Often, typical discretization length scales fall below the intrinsic material length scales, making the averaging structure of the applied numerical method doubtful. We present a detailed analysis of their limitations and a comparative study of the investigated damage models for the simulation of ductile fracture problems.

1 Introduction

During the past years application of models based on continuum damage mechanics (CDM) to ductile fracture mechanics became very popular. The computational evaluation and simulation of damage occurrence by different models implemented in the framework of the finite element method (FEM) seems to be very promising, since many contributions have followed the first, fundamental publications such as [6,12,17]. It is commonly believed that application of the FEM taking finite deformations into account in conjunction with advanced constitutive damage models with a softening regime may be capable of simulating crack initiation and growth in typical fracture specimens under test conditions. A well–known disadvantage of the numerical treatment of solid mechanics problems, where softening material behaviour occurs, is the so–called *mesh–dependence* of numerical results. In a considerable number of investigations different methods have been proposed to overcome this mesh–dependence. The common idea is to introduce a *characteristic* or *internal* length (scale) into the constitutive model or its evaluation, see [2,4,5] and references therein, where a summary of different regularization techniques is outlined.

In this paper some limitations and restrictions of the ductile damage mechanics analysis are pointed out in the scope of the finite element method and its nonlinear solution procedures at the example of a so–called COMPACT TENSION (CT) specimen . In order to resolve the highly nonlinear effects of stress and strain concentration occurring near notches, crack tips or due to shear band localization, the numerical discretization in these regions usually is refined without respecting minimal length scales limited by *inhomogeneities* on the microscale of the material. Typical physically based length scales of ductile materials such as structural steel or aluminium alloys are in the magnitude of about 50μm – 200μm. However, very often, length scales resulting from FE discretizations of detailed simulations of damage and crack initiation and growth problems fall below these "natural" barriers. As a consequence, the basic assumptions of continuum mechanics such as continuity of mechanical quantities on the macroscale are definitely violated and the numerical results are highly questionable.

In this study, 20–noded brick elements are used with quadratic shape functions along the element edges. As constitutive models, the ductile damage models of ROUSSELIER *et al.* [12] and GURSON [6] in the formulation of TVERGAARD [17] are used in the scope of isotropic finite strain plasticity. An advantage of the first model is the description of material softening due to damage by the influence of solely three material parameters, while the GURSON model incorporates essentially a larger parameter set sometimes leading to non–definite solutions, see [19]. A second advantage of the ROUSSELIER model is related to the numerical implementation of the constitutive law by means of an implicit integration scheme. The type of constitutive equations leads to symmetric tangent material moduli, which is advantageous in computing and storing the matrix expressions. To this end, we will remark for both models on internal variables and their thermomechanical relations with respect to the dissipation inequality.

Simultaneously, a localization analysis is performed during the iteration on each integration point by evaluation of the so called *localization* or *acoustic tensor* for all possible directions of localization in three dimensions. The fundamental derivation of the acoustic tensor for finite strains is described in [15]. An essential result of localization analysis for damage occurrence in a CT specimen under monotonic loading is the fact that a detailed representation of the behaviour of local strain softening is possible obviously beyond the peak load, but is restricted as long as no loss of ellipticity occurs.

2 Three–dimensional Finite Element Formulation

Starting point of any finite element discretization is the weak form of equilibrium, given here in a spatial description as

$$g(\boldsymbol{u}, \delta\boldsymbol{u}) := \int_{\mathcal{B}} \boldsymbol{\sigma}\, \text{grad}\delta\boldsymbol{u}\, \text{d}v - \int_{\partial\mathcal{B}_\sigma} \boldsymbol{t}_L\, \delta\boldsymbol{u}\, \text{d}a = 0\,, \tag{1}$$

where $\boldsymbol{u} = \boldsymbol{x} - \boldsymbol{X}$ denotes the displacement vector of a material point represented by $\boldsymbol{X}$ in the reference configuration toward a position $\boldsymbol{x}$ of the same point in the current configuration, $\delta\boldsymbol{u}$ is the first variation of the displacement field. With $\boldsymbol{\sigma}$ the CAUCHY stress tensor is characterized and $\boldsymbol{t}_L = \boldsymbol{\sigma} \cdot \boldsymbol{\mathcal{N}}$ are the prescribed tractions acting on the loaded boundary $\partial\mathcal{B}_\sigma$ of the body with (outer) normal vector $\boldsymbol{\mathcal{N}}$ in the current configuration $\mathcal{B}$.

Linearization at a known position $\boldsymbol{X} + \hat{\boldsymbol{u}}$ with respect to the current deformation state and rearrangement leads to the following representation of the element stiffness

$$\Delta g^{elmt}(\hat{\boldsymbol{u}}, \delta\boldsymbol{u}) = \int_{\mathcal{B}} (\Delta\boldsymbol{\sigma} + \text{grad}\Delta\boldsymbol{u} \cdot \boldsymbol{\sigma}) : \text{grad}\,\delta\boldsymbol{u}\ \text{d}v^{elmt}\,, \tag{2}$$

where $\Delta(\bullet)$ denotes the linearization operator and $\Delta\boldsymbol{u} = \boldsymbol{u} - \hat{\boldsymbol{u}}$ the increment of the displacement field $\boldsymbol{u}$. The right part of (2) results in the element stiffness *matrix* $\boldsymbol{K}^{elmt}$ for the discretized setting, where $\boldsymbol{K}^{elmt}$ obviously consists of two parts. The first part is obtained from the consistent linearization of the material model getting $\Delta\boldsymbol{\sigma}$ and the second part comes solely from the linearization of the used strain measure at the computed stress state $\boldsymbol{\sigma}$. For further details on the implementation of the consistent linearization of the used algorithm see [14]. The discretization chosen in this paper is based

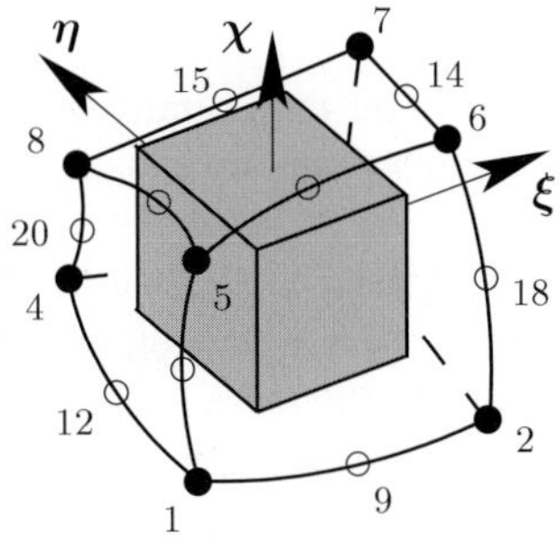

Fig. 1. 20–noded brick type element with corner nodes • and edge nodes ∘

on a 20–node–displacement element formulation with shape functions N_i,

($i = 1, 2, ..., 20$), so that quadratic functions describe the element edges. As in [8], a $2\times2\times2$ integration scheme is used, which means an *underintegration* with respect to the quadratic shape functions N_i. It shall be pointed out that no *hourglassing* modes were detected as for an 8–node–displacement element formulation and a $1 \times 1 \times 1$ integration scheme, see [2].

3 Treatment of Continuum Damage Models in the Scope of Isotropic Finite Strain Plasticity

3.1 Continuum Damage Models

Following [1] we write the KIRCHHOFF stress tensor $\boldsymbol{\tau}$ as the weighted CAUCHY stress tensor as

$$\boldsymbol{\tau} = J\,\boldsymbol{\sigma} = -p^{\tau}\boldsymbol{I} + 2/3\,q^{\tau}\hat{\boldsymbol{n}}\ , \tag{3}$$

where $J := \det \boldsymbol{F}$ is the determinant of the deformation gradient $\boldsymbol{F} = \partial\boldsymbol{x}/\partial\boldsymbol{X}$ mapping material points $\boldsymbol{X}$ onto the current configuration $\boldsymbol{x}$. The scalar $p^{\tau} = -\tau_{ij}\delta_{ij}/3$ defines the hydrostatic pressure, $q^{\tau} = \sqrt{3/2\,t_{ij}t_{ij}}$ is the equivalent KIRCHHOFF stress and $t_{ij} = \tau_{ij} + p^{\tau}\delta_{ij}$ are the components of the KIRCHHOFF stress deviator. These quantities can also be obtained for the CAUCHY stress tensor, whose deviatoric stress is $\boldsymbol{s} = \boldsymbol{\sigma} + p^{\sigma}\boldsymbol{I}$. In this notation, an additional important quantity is the normalized and dimensionless stress deviator

$$\hat{\boldsymbol{n}} = 3/(2q^{\tau})\boldsymbol{t} = 3/(2q^{\sigma})\boldsymbol{s}\ . \tag{4}$$

The second order unit tensor $\boldsymbol{I}$ is defined as the KRONECKER symbol by its components δ_{ij} in the cartesian frame. Analogous, the plastic strain rate can be written as

$$\Delta\boldsymbol{\epsilon}^{p} = \frac{1}{3}\Delta\varepsilon_{p}\boldsymbol{I} + \Delta\varepsilon_{q}\hat{\boldsymbol{n}}\ , \tag{5}$$

where $\Delta\varepsilon_p$ and $\Delta\varepsilon_q$ describe scalar rate quantities which are defined below. Note, that again the dimensionless tensor quantities $\boldsymbol{I}$ and $\hat{\boldsymbol{n}}$ are used in this notation.

The first constitutive model used in this study is the damage model proposed by ROUSSELIER *et al.* [12] . Here, taking ductile damage processes into account, the yield function is written as

$$\Phi^{Rouss} = q^{\tau} - \underbrace{\sigma_0\left[\frac{\varepsilon^{p}_{eqv}}{\sigma_0}E + 1\right]^{1/N}}_{\sigma_M} + B(\beta)D\exp(-\frac{p^{\tau}}{\sigma_1}) = 0\,, \tag{6}$$

where σ_M represents the material hardening in terms of a power law, and the last part of (6) describes the damage (softening) behaviour through the function $B(\beta)$ and an exponential expression. Furthermore, E is the YOUNG modulus, σ_0 is the initial yield stress, N is the material hardening exponent, and D and σ_1 are *damage* material parameters. The function $B(\beta)$ is the *conjugate force* to the damage quantity β, defined by

$$B(\beta) = \frac{\sigma_1 f_0 \exp(\beta)}{1 - f_0 + f_0 \exp(\beta)} \,. \tag{7}$$

Here, the initial void volume fraction f_0 is the third damage–depending material parameter used in this constitutive set of equations. The evolution equation for the damage parameter β is given by $\Delta\beta = \Delta\varepsilon_q D \exp(-p^\tau/\sigma_1)$, which is obviously dependent on the deviatoric part of the strain rate $\Delta\varepsilon_q$ and the actual hydrostatic pressure p^τ.

In the following some details of the constitutive model are pointed out:

1. The current value of the void volume fraction f can be determined by $f = B(\beta)/\sigma_1$. This quantity enables a comparison with other damage models, like [17].
2. The yield function (6) can be regarded as the classical yield condition of the VON MISES plasticity with the hardening function $\sigma^*(\varepsilon^{pl}_{eqv})$ expanded by a third term responsible for the material softening.
3. Note that for a vanishing initial void volume fraction $f_0 \equiv 0$ the yield condition (6) indicates a VON MISES yield condition without any influence of damage. However, in that case no increase of the damage quantity β will take place. This is in contrast to common formulations of GURSON's damage model, where especially a nucleation of microvoids is taken into account by an extra term in addition to the evolution equation of the void volume fraction f, see [17]. The constitutive formulation used here describes "damage" by the parameter β, which has no direct correlation to a measurable quantity. In so far no specific term for the nucleation of microvoids is assumed. With $f_0 \equiv 0$ there will be no influence of the evolution equation $\Delta\beta$ on $B(\beta)$ and (6). Furthermore, the algebraic reformulation of (12) will break down because of the vanishing derivative $\partial\Phi/\partial p$ in $(12)_1$. In view of such a numerically based argumentation it is necessary to define at least a small initial void volume fraction f_0 in order to get an evolution of the damage parameter β.
4. Considering the HELMHOLTZ free energy

$$\Psi(\boldsymbol{b}_e, \varepsilon^{pl}_{eqv}, \beta) = W(\boldsymbol{b}_e) + H_1(\varepsilon^{pl}_{eqv}) + H_2(\beta) \,, \tag{8}$$

which can additively be split into the elastic part W and the parts H_1 and H_2 connected to the internal variables ε^{pl}_{eqv} and β; the conjugate forces $q = -\partial H_1/\partial\varepsilon^{pl}_{eqv}$ and $B(\beta) = -\partial H_2/\partial\beta$ can be identified via differentiation. Alternatively, one can obtain the "damage potential"

$$H_2 = -\int_0^\beta B(\bar\beta)\mathrm{d}\bar\beta = -\left[\sigma_1 \ln(1 - f_0 + f_0 \exp(\beta))\right] \tag{9}$$

from an explicit integration of (7). The relation (9) is graphically depicted in Fig. 2, where the shaded area represents the dissipated "damage work" for a considered interval $[0, \beta]$. This is in contrast to the following GURSON model, where such a potential cannot be found, see [16].

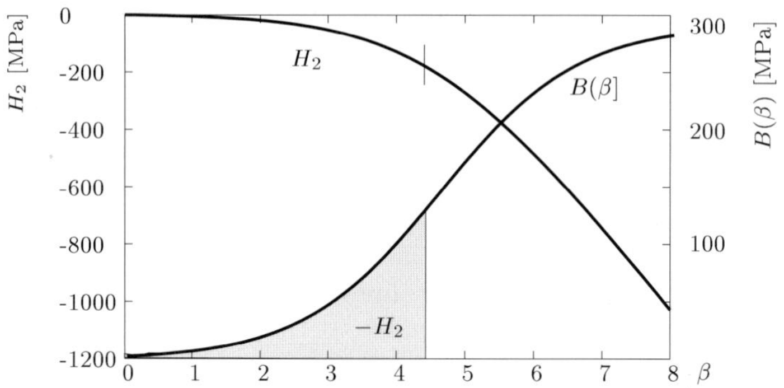

Fig. 2. Damage potential H_2 for the ROUSSELIER model and *damage force* $B(\beta)$ vs. damage parameter β (parameter $\sigma_1 = 300$ MPa, $f_0 = 0.01$ from Table 2)

The second constitutive model adopted here is the damage model of GURSON in the formulation of [17], the numerical treatment follows [3]. In contrast to (6) the yield function appears in a slightly different form as

$$\Phi^{Gurson} = \left(\frac{q^\sigma}{\sigma_M}\right)^2 + 3f^*(f)\cosh(-\frac{3}{2}\frac{p^\sigma}{\sigma_M}) - \left[1 + (\frac{3}{2}f^*(f))^2\right] = 0\ , \quad (10)$$

while the numerical treatment is identical for both models. In order to enable a comparison with the ROUSSELIER model, we assume an evolution of void volume fraction f solely by growth of existing voids and neglect the term describing void nucleation. The hardening of the matrix material is again described by σ_M and the actual void volume fraction f enters the yield function (10) as f^*, where $f^*(f)$ is defined by a bilinear function following [17].

The macroscopic plastic strain rate $\dot{\boldsymbol{\epsilon}}^p$ is determined by the classical associated flow rule

$$\dot{\boldsymbol{\epsilon}}^p = \gamma\frac{\partial\Phi}{\partial\boldsymbol{\tau}} = \gamma\left\{\frac{\partial\Phi}{\partial q^\tau}\frac{\partial q^\tau}{\partial\boldsymbol{\tau}} + \frac{\partial\Phi}{\partial p^\tau}\frac{\partial p^\tau}{\partial\boldsymbol{\tau}}\right\} \quad (11)$$

with the plastic multiplier γ. Note that $\dot{\boldsymbol{\epsilon}}^p$ coincides with the plastic increment $\Delta\boldsymbol{\epsilon}^p$ for the algorithmic setting written in principal axes and is identified as the plastic part of the spatial deformation velocity tensor $\boldsymbol{d}$ in the finite strain regime later on. The bracket on the right–hand side of (11) shows a further advantage of this formulation following [1], since it is easy to determine the derivatives of Φ with respect to the scalar quantities q^{τ} and p. It can be seen with (11) that

$$\Delta\varepsilon_p = -\gamma \frac{\partial \Phi}{\partial p^{\tau}} \quad \text{and} \quad \Delta\varepsilon_q = \gamma \frac{\partial \Phi}{\partial q^{\tau}} \,. \tag{12}$$

These two equations allow the algebraic elimination of the factor γ. Thus, the increment of the plastic strain can be expressed by the two scalar quantities $\Delta\varepsilon_p$ and $\Delta\varepsilon_q$. Furthermore, the equivalent plastic strain ε^p_{eqv} can be incremented directly by $\Delta\varepsilon_q$.

With this, the set of constitutive equations is completed. The evaluation of the material model on the local level of integration points for a given load is realized by an implicit EULER backward integration scheme for the unknowns $\Delta\varepsilon_p$, $\Delta\varepsilon_q$ and $\Delta\beta$. The exact linearization of the set of equations follows the description in [1]. The variational expression

$$\delta\boldsymbol{\tau} = \mathbb{C}_e : \left(\delta\boldsymbol{\epsilon}^{tr} - \frac{1}{3}\delta\Delta\varepsilon_p \boldsymbol{I} - \delta\Delta\varepsilon_q \hat{\boldsymbol{n}} - \Delta\varepsilon_q \frac{\partial \hat{\boldsymbol{n}}}{\partial \boldsymbol{\tau}} : \delta\boldsymbol{\tau} \right) = \mathbb{D}^{\text{Material}}_{Algo} : \delta\boldsymbol{e} \tag{13}$$

leads, after some extended algebraic manipulations as described in [1], to the expressions $\delta\Delta\varepsilon_p$ and $\delta\Delta\varepsilon_q$, where $\mathbb{C}_e$ characterizes the elastic material modulus. Finally, we obtain the algorithmic material modulus $\mathbb{D}^{\text{Material}}_{Algo}$ at the end of the considered time interval $[t, t + \Delta t]$, which is necessary to compute the complete stiffness in (2). Note, that the consistent linearization of the algorithmic modulus in (13) leads to a unconditionally symmetric matrix representation for the ROUSSELIER model, and to a symmetric representation for the GURSON model just for neglecting void nucleation as treated here.

3.2 Finite Strain Plasticity

At least in the crack tip region of elastic–plastic solids under sufficiently high load, finite deformations occur where the plastic part of the strains usually is large compared with the elastic part. The framework of multiplicative elasto-plasticity is used. Its kinematic key assumption is the multiplicative split of the deformation gradient

$$\boldsymbol{F} = \boldsymbol{F}_e \cdot \boldsymbol{F}_p \tag{14}$$

into an elastic and a plastic part, providing the basis of a geometrically exact theory and avoiding linearization of any measure of deformation. Note that

$$\mathrm{d}\hat{\boldsymbol{x}} = \boldsymbol{F}_p \cdot \mathrm{d}\boldsymbol{X} = \boldsymbol{F}_e^{-1} \cdot \mathrm{d}x \tag{15}$$

introduces a so-called *intermediate configuration*, which quantities are labeled by $\hat{(\bullet)}$. As a further advantage, fast and numerically stable iterative algorithms, proposed and described in [14], can be used. In the following, only a brief summary of the integration algorithm for a time step $[t_n; t_{n+1}]$ in the context of a FE–implementation is given. Note that in the following the index $n+1$ is suppressed for brevity if misunderstanding is unlikely to occur.

The essential aspect of the multiplicative decomposition is the resulting additive structure of the current logarithmic principal strains within the return mapping scheme as $\boldsymbol{\epsilon}^e = \boldsymbol{\epsilon}^{tr} - \Delta\boldsymbol{\epsilon}^p$. Here, $\boldsymbol{\epsilon}^e$ and $\boldsymbol{\epsilon}^{tr}$ stand for a vector representation with the components $\epsilon_i^e = \ln \mu_i^e$ and $\epsilon_i^{tr} = \ln \mu_i^{tr}$, respectively, strictly connected with the spectral decomposition of the elastic left CAUCHY–GREEN tensor.

The elastic left CAUCHY–GREEN tensor can be specified with the multiplicative decomposition as

$$\boldsymbol{b}_e = \boldsymbol{F}_e \cdot \boldsymbol{F}_e^{\mathrm{T}} = \boldsymbol{F} \cdot \boldsymbol{C}_p^{-1} \cdot \boldsymbol{F}^{\mathrm{T}} \ , \tag{16}$$

where the superscripts "-1" and "T" denote the inverse and the transpose of a tensor, respectively. That relation clearly shows the "connection" between the elastic and plastic deformation measure by the occurence of the plastic right CAUCHY–GREEN tensor $\boldsymbol{C}_p = \boldsymbol{F}_p^{\mathrm{T}} \cdot \boldsymbol{F}_p$. By means of the relative deformation gradient, see [14],

$$\boldsymbol{f} = \partial \boldsymbol{x}_{n+1} / \partial \boldsymbol{x}_n = \boldsymbol{F}_{n+1} \cdot \boldsymbol{F}_n^{-1} \ , \tag{17}$$

which relates the current configuration $\boldsymbol{x}_{n+1}$ to the configuration belonging to the previous time step at t_n, an elastic *trial*–state $\boldsymbol{b}_e^{tr} = \boldsymbol{f} \cdot \boldsymbol{b}_n \cdot \boldsymbol{f}^{\mathrm{T}}$ is calculated for the current configuration with frozen internal variables at state t_n.

In the considered case of isotropy, $\boldsymbol{b}_e$ commutes with $\boldsymbol{\tau}$, see [11,14]. We assume to fix the principle axes of $\boldsymbol{b}_e$ during the return mapping scheme described in the previous section, so that the spectral decomposition

$$\boldsymbol{b}_e^{tr} = \sum_{i=1}^{3} \mu_i^{tr\,2}\, \boldsymbol{n}_i^{tr} \otimes \boldsymbol{n}_i^{tr} \tag{18}$$

is given and the eigenvectors $\boldsymbol{n}_i^{tr}$ can also be used to compose the stress tensor $\boldsymbol{\tau} = \sum_{i=1}^{3} \tau_i\, \boldsymbol{n}_i^{tr} \otimes \boldsymbol{n}_i^{tr}$. That motivates the evaluation of the constitutive equations given in the previous section in principle axes, which means additionally a time saving compared to an evaluation of all six (symmetric) tensor components.

Furthermore, for the elastic part of the material description compressible Neo–HOOKE behaviour is used, where the plastic strain corrector $\Delta\boldsymbol{\epsilon}^p$ is obtained by the normality rule of plastic flow (11).

The general concept of LIE time derivative $\mathcal{L}_{\boldsymbol{v}}(\bullet)$ characterizing the change of a spatial field in the direction of the vector $\boldsymbol{v}$ and known to yield objective spatial fields, see [7], leads in this case to the OLDROYD rate of the elastic left CAUCHY–GREEN tensor

$$\mathcal{L}_{\boldsymbol{v}}\boldsymbol{b}_e = \overset{\triangledown}{\boldsymbol{b}}_e = \dot{\boldsymbol{b}}_e - \boldsymbol{l}\cdot\boldsymbol{b}_e - \boldsymbol{b}_e\cdot\boldsymbol{l}^{\mathrm{T}} \tag{19}$$

where $\dot{\boldsymbol{b}}_e$ denotes the material time derivative and $\boldsymbol{l} = \operatorname{grad}\dot{\boldsymbol{x}} = \dot{\boldsymbol{F}}\cdot\boldsymbol{F}^{-1}$ the spatial velocity gradient. In this case $\boldsymbol{v}$ is identified as velocity vector $\boldsymbol{v} = \dot{\boldsymbol{x}} = \partial\boldsymbol{x}/\partial t$. The decomposition of $\boldsymbol{l} = \boldsymbol{d} + \boldsymbol{w}$ in its symmetric part $\boldsymbol{d} = \operatorname{sym}(\boldsymbol{l}) = \frac{1}{2}(\boldsymbol{l} + \boldsymbol{l}^{\mathrm{T}})$ and its antimetric part $\boldsymbol{w}$, known as spin tensor, respectively, plays a crucial role in the definition of the plastic flow rule. Some basic algebraic manipulations let us also obtain the expressions in (19) as

$$\mathcal{L}_{\boldsymbol{v}}\boldsymbol{b}_e = -2\boldsymbol{F}_e\cdot\operatorname{sym}(\hat{\boldsymbol{l}}_p)\cdot\boldsymbol{F}_e^{\mathrm{T}} = -2\operatorname{sym}(\boldsymbol{l}_p\cdot\boldsymbol{b}_e)\ , \tag{20}$$

where $\hat{\boldsymbol{l}}_p$ is defined by $\hat{\boldsymbol{l}}_p = \dot{\boldsymbol{F}}_p\cdot\boldsymbol{F}_p^{-1}$ acting on the intermediate configuration. Please note, that we do *not* make any assumption concerning the antimetric part $\boldsymbol{w}$ of $\boldsymbol{l}$. Because of the restriction to isotropic material behaviour, the focus is just directed to the symmetric part $\boldsymbol{d}$ of $\boldsymbol{l}$. So, the additive decomposition $\boldsymbol{d} = \boldsymbol{d}_e + \boldsymbol{d}_p$ results from the multiplicative decomposition $\boldsymbol{F} = \boldsymbol{F}_e\cdot\boldsymbol{F}_p$.

The definition of the associated flow rule (see also (11)) in that finite strain regime as

$$\boldsymbol{d}_p := \gamma\,\frac{\partial\Phi}{\partial\boldsymbol{\tau}} \tag{21}$$

enables with (19) and (20) the formulation

$$\mathcal{L}_{\boldsymbol{v}}\boldsymbol{b}_e = -2\operatorname{sym}(\gamma\,\frac{\partial\Phi}{\partial\boldsymbol{\tau}}\cdot\boldsymbol{b}_e)\ . \tag{22}$$

If the condition $\Phi \leq 0$ (see (6) and (10)) is fulfilled by the current stress state $\boldsymbol{\tau}$, this state is possible and it is the solution. If, on the other hand, $\Phi \leq 0$ is violated by the trial–state, the trial stresses must be projected back on the yield surface $\Phi = 0$ in an additional step, often called "exponential return mapping".

In that case, $\boldsymbol{x} = \boldsymbol{x}_{n+1}$ is fixed and (22) results in

$$\overset{\triangledown}{\boldsymbol{b}}_e = \dot{\boldsymbol{b}}_e = -2\gamma\operatorname{sym}(\frac{\partial\Phi}{\partial\boldsymbol{\tau}}\cdot\boldsymbol{b}_e)\ , \tag{23}$$

with $\boldsymbol{l} \equiv \boldsymbol{0}$. The solution of the first order differential equation (23) is given by

$$\boldsymbol{b}_{e\,n+1} = \sum_{i=1}^{3}\underbrace{\exp\left[2\epsilon_i^e\right]}_{\mu_i^{e\,2}}\,\boldsymbol{n}_{i\,n+1}^{tr}\otimes\boldsymbol{n}_{i\,n+1}^{tr}\ , \tag{24}$$

where the elastic logarithmic strains $\boldsymbol{\epsilon}^e$ are obtained in principle axes, see (11–12), so that $\boldsymbol{b}_{e\,n+1}$ is known and $\boldsymbol{C}_p^{-1} = \boldsymbol{F}^{-1}\cdot\boldsymbol{b}_{e\,n+1}\cdot\boldsymbol{F}^{-\mathrm{T}}$ can be stored as *history variable* for the next time step.

4 Localization Analysis

4.1 Acoustic Tensor

A steady evaluation of the "spatial localization tensor" $\boldsymbol{Q}$ is performed on each integration point during the iteration to check the material stability. The spatial localization tensor $\boldsymbol{Q} = \boldsymbol{n} \cdot \mathbb{D} \cdot \boldsymbol{n}$ is the contraction of the current fourth order material tensor $\mathbb{D}$ by the spatial surface unit normal vector $\boldsymbol{n}$ with respect to its second and fourth index. This derivation, introduced in [15], is motivated by the assumption of a spatially continuous incremental equilibrium across an arbitrary band of discontinuity, which implies that the nominal traction rate inside and outside the band is the same:

$$\overset{\circ}{\boldsymbol{t}}(\boldsymbol{x}^{out}) = \overset{\circ}{\boldsymbol{t}}(\boldsymbol{x}^{band}) \; . \tag{25}$$

With the definition of the nominal traction rate $\overset{\circ}{\boldsymbol{t}} = J^{-1} \overset{\circ}{\boldsymbol{\tau}} \cdot \boldsymbol{n}$, the nominal rate of the KIRCHHOFF stress tensor $\overset{\circ}{\boldsymbol{\tau}}$ can be related to the spatial velocity gradient $\boldsymbol{l} := \dot{\boldsymbol{F}} \cdot \boldsymbol{F}^{-1}$ via

$$\overset{\circ}{\boldsymbol{\tau}} = \mathbb{D} : \boldsymbol{l} \; . \tag{26}$$

Note that the material tensor splits off into $\mathbb{D} = \mathbb{D}^{Material} + \mathbb{D}^{Geometry}$, where the first part results from the linearization of the constitutive equations and the second part is obtained by the linearization of the geometrical setting. This is in strong equivalence to the formulation of the element stiffness matrix for the discretized representation of (2) in Sec. 2. A detailed discussion is given in [10]. These relations become more evident by reformulation of the material rate of the first PIOLA–KIRCHHOFF stress tensor $\dot{\boldsymbol{P}}$ in terms of the KIRCHHOFF stress tensor $\boldsymbol{\tau}$ into

$$\dot{\boldsymbol{P}} = \boldsymbol{F}^{-1} \cdot \dot{\boldsymbol{\tau}} - \boldsymbol{F}^{-1} \cdot \boldsymbol{l} \cdot \boldsymbol{\tau} \; . \tag{27}$$

Relating $\dot{\boldsymbol{P}}$ to the rate of the deformation gradient, $\dot{\boldsymbol{F}}$, via the tangent map $\mathbb{D}^{\mathrm{P}}$ yields

$$\overset{\circ}{\boldsymbol{\tau}} = \dot{\boldsymbol{\tau}} - \boldsymbol{l} \cdot \boldsymbol{\tau} = \boldsymbol{F} \cdot \mathbb{D}^{\mathrm{P}} : \dot{\boldsymbol{F}} \, , \tag{28}$$

which is related to the known *frame–invariant* (*objective*) OLDROYD rate $\overset{\nabla}{\boldsymbol{\tau}}$ by $\overset{\circ}{\boldsymbol{\tau}} = \overset{\nabla}{\boldsymbol{\tau}} + \boldsymbol{\tau} \cdot \boldsymbol{l}^{\mathrm{T}}$. Note that the derivations and argumentation in [13] about the loss of material stability by real or imaginary wave speeds very illustratively denote the term "acoustic tensor". The double contraction $\boldsymbol{Q} = \boldsymbol{n} \cdot \mathbb{D} \cdot \boldsymbol{n}$ of the material tensor $\mathbb{D}$ by the normal vector $\boldsymbol{n}$ indicates a possible wave propagation direction. By this the normal of a possible failure plane is characterized. The condition for obtaining well set numerical results is the positive definiteness of the second order tensor $\boldsymbol{Q}$, which is checked by a positive value of the determinant $q = \det [\boldsymbol{Q}]$.

4.2 Numerical treatment

In the numerical analysis $q = \det[\boldsymbol{Q}] = \det[\boldsymbol{n} \cdot \mathbb{D} \cdot \boldsymbol{n}]$ has to be evaluated for all possible directions $\boldsymbol{n}$ at every location $\boldsymbol{x}$. For that reason the vector $\boldsymbol{n} = [\cos\lambda\ \cos\varphi, \sin\lambda\ \cos\varphi, \sin\varphi]^{\mathrm{T}}$ is parameterized by spherical coordinates with the angles λ and φ characterizing the longitude and the latitude, respectively. Note, that — as a simplification — we use the algorithmic material tensor $\mathbb{D}^{\mathrm{Material}}_{Algo}$ as part of $\mathbb{D}$, knowing that both, $\mathbb{D}^{\mathrm{Material}}$ and $\mathbb{D}^{\mathrm{Material}}_{Algo}$, differ from one another for sufficiently large time steps. This difference should be subjected to further investigations.

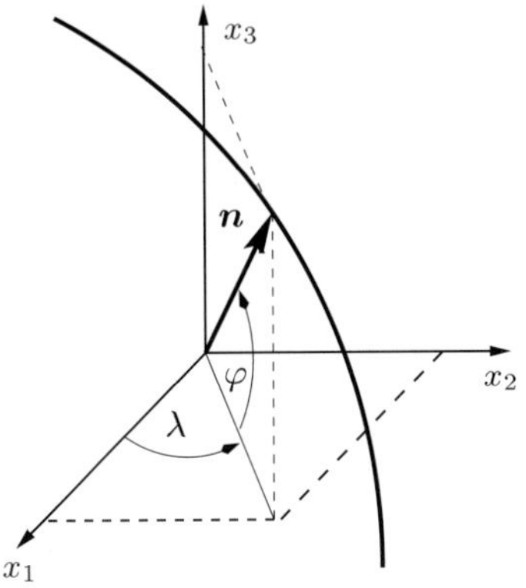

Fig. 3. Parametrization of $\boldsymbol{n}$ with longitude angle λ and latitude angle φ

To detect a possible critical direction, where q may vanish, one has to compute $q = \det[\boldsymbol{Q}] \to min$ for a set $\{\lambda, \varphi\}$. As a remark, it should be mentioned here, that for the general 3D case q is a function of terms in the power of six, e.g. $\sin^6\lambda$ or $\cos^6\varphi$. This minimization procedure is equivalent to the evaluation of $\nabla q(\lambda, \varphi) = \mathbf{0}$, for which we propose a classical Newton iteration scheme through

$$\begin{bmatrix}\lambda\\ \varphi\end{bmatrix}_{k+1} = \begin{bmatrix}\lambda\\ \varphi\end{bmatrix}_{k} - \underbrace{\begin{bmatrix}\dfrac{\partial^2 q}{\partial\lambda^2} & \dfrac{\partial^2 q}{\partial\lambda\,\partial\varphi}\\[2ex] \dfrac{\partial^2 q}{\partial\lambda\,\partial\varphi} & \dfrac{\partial^2 q}{\partial\varphi^2}\end{bmatrix}^{-1}}_{=:\ \boldsymbol{H}^{-1}} \cdot \begin{bmatrix}\dfrac{\partial q}{\partial\lambda}\\[2ex] \dfrac{\partial q}{\partial\varphi}\end{bmatrix}, \tag{29}$$

and suitable initial conditions, e.g. $[\lambda, \varphi]_0^{\mathrm{T}} = [10°, 80°]_0^{\mathrm{T}}$, where k indicates the iteration loop number. Note, that the choice of the initial conditions is of specific interest for solving this problem and is still in discussion, see [9] or [18].

Because of the large number of operations needed especially for determing the HESSE matrix $\boldsymbol{H}$ in (29), the expressions of the related FORTRAN code are obtained by the algebraic manipulation program MATHEMATICA exploiting some advanced methods for code generation. Still, in this case (3D, $\mathbb{D}^{\text{Material}}$ symmetric) evaluation of more than 6100 multiplications for determining *one* of the inverse of the HESSE matrix $\boldsymbol{H}$ during the iteration loop in the integration points would be necessary, which is comparable to the inversion of a 35×35–matrix by GAUSS elimination. By sophisticated substitutions of different terms in q it is possible to reduce the maximum exponent from 6 to 2, which is much more accessible for compiler optimization procedures. Furthermore, the total number of multiplications is reduced to about 670, so that the amount of compiling $\boldsymbol{H}$ is minimized.

5 Example and Results

5.1 Model of a CT specimen

As an example, a three–dimensional model of a CT specimen discretized by 20–node solid elements as shown in Fig. 4(a) is examined. Due to symmetry,

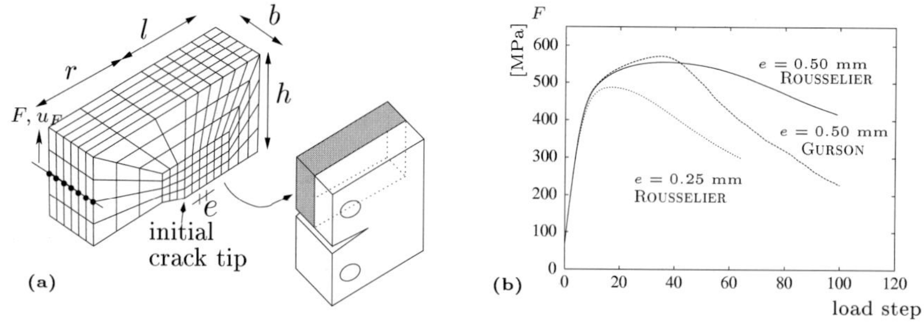

Fig. 4. (a) Model of CT specimen and **(b)** typical load–deflection curve for different discretizations

just a quarter of the structure is modeled, for length dimensions see Table 1. The loading is applied by a prescribed displacement u_F (by 0.01 mm/step) of the nodes lying on the marked line, see Fig. 4(a). The chosen discretization is characterized by the typical element edge length e in front of the crack tip. In this investigation the element edge lengths $e = 0.50$ mm and $e = 0.25$ mm are used. In addition, the typical mesh sensitive results for a classical, local FE simulation using different discretizations are plotted as load–deflection curve in Fig. 4(b). The dependence of the reaction force on the finite element mesh can clearly be seen. The set of geometry and material parameters used is shown in Table 1, where the four parameters describing the elastic and hardening behaviour can be obtained by simple tensile tests. The parameters of Table 2 are responsible for the damage representation of the constitutive models. For this contribution we fitted the parameter f_c of the GURSON model to the fixed set of parameters used for the ROUSSELIER damage description by trying to obtain sufficient agreement between the global load–deflection curves in Fig. 4(b) for both models using e.g. $e = 0.5$ mm. The parameter f_F representing the final void volume fraction of the GURSON model does not effect the respective load–deflection curves in the first part, where no crack advance occurs.

Table 1. Geometry and (hardening) material parameters

r [mm]	l [mm]	b [mm]	h [mm]	E [MPa]	ν	σ_0 [MPa]	N
6	5	3	5	210000	0.2	460	7

Table 2. Damage material parameters for both models

both	ROUSSELIER		GURSON	
f_0	D	σ_1 [MPa]	f_c	f_F
0.01	3	300	0.008	0.19

5.2 Results

A result for the load–displacement curves for different discretizations is plotted in Fig. 4(b) and can also be found in [4]. The mesh sensitivity is obvious, if no additional regularization technique is applied. In the following, we concentrate on computations resulting from the evaluation of the localization tensor $\boldsymbol{Q}$ and its determinant. The representation of the results is focused on the FE–integration point being located directly in front of the crack tip in the center of the specimen, which is the point with the largest load and the highest damage parameter. Figure 5 shows the normalized determinant of $\boldsymbol{Q}$ vs. the two spatial angles λ and φ, parameterizing the normal vector $\boldsymbol{n}$ in each case by 20 steps (ROUSSELIER parameter set of Table 1 and $e = 0.50$ mm). Displayed is the situation for load steps 2 and 8 ($u_F = 0.02, 0.08$ mm), which

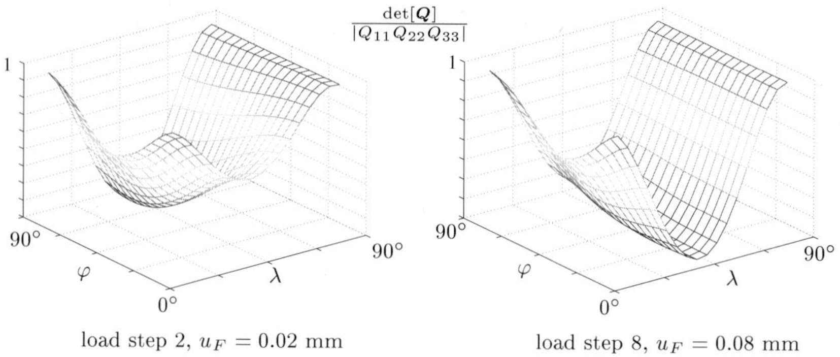

Fig. 5. Decrease of $q = \det[\boldsymbol{Q}]$ for $u_F = 0.02$ mm and $u_F = 0.08$ mm

represents directly the situation before the onset of localization ($q \to 0$). Obviously, the decrease of $q/|Q_{11}Q_{22}Q_{33}|$ during load steps 2 to 8 can be seen.

Because of the numerical costs determining these quantities during the iteration, we apply a NEWTON iteration scheme following (29) for finding the minimum of these surfaces. For the mentioned integration point, Fig. 6(a) shows $\det[\boldsymbol{Q}]$ vs. 30 load steps for two different discretizations with $e = 0.50$ mm and $e = 0.25$ mm and ROUSSELIER material, respectively. In addition, the situation for a standard (non damaging) VON MISES material with the same power–hardening law σ^* and $e = 0.50$ mm is plotted. As expected, for the VON MISES material, the values of q decrease rapidly during the load incrementation over 30 steps, but never reach $q = 0$ indicating a possible localization. In contrast, the curves for the ROUSSELIER damage material show a zero–crossing and thus a localization occurrence. Again, the mesh sensitivity is obvious through the results for $e = 0.25$ mm at load step 4, while the discretization with $e = 0.50$ mm reaches zero at load level 9.

In Fig. 6(b) the equivalent stress of the integration point in front of the crack tip is plotted vs. the applied load steps for the ROUSSELIER material and $e = 0.50$ mm. Note that the equivalent stress at the critical load level 9 ($u_F = 0.09$ mm) appears in the decreasing part of the load curve obviously after reaching the peak load.

In Fig. 7 the contours of integration points are depicted, where the determinant of the acoustic tensor reaches zero for the load steps 10, 20 and 30 in front of the crack tip. The contour lines in crack propagation direction are plotted over the discretized width b of the specimen using $e = 0.50$ mm and the ROUSSELIER material set of Table 1 demonstrating local loss of ellipticity in front of the crack tip during the computation. It should be emphasized, that the contour lines $q = 0$ do not coincide with the contour lines of lost load carrying capacity (crack growth) defined by a critical damage parameter.

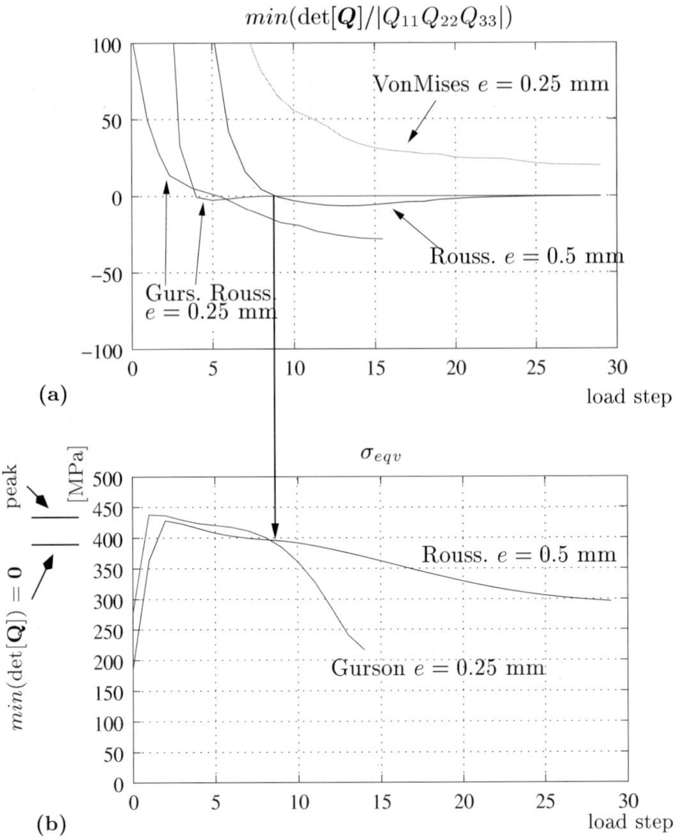

Fig. 6. **(a)** Determinant of acoustic tensor $\boldsymbol{Q}$ and **(b)** equivalent stress in front of the crack tip. Difference between *peak* load and level of zero crossing is marked at the vertical axis of **(b)**

These crack growth contours follow the $q = 0$ contours far behind (at higher load steps), indicating that they are determined in inadmissible situations.

In Fig. 8 the angles λ and φ characterizing the normal $\boldsymbol{n}$ of the failure plane of the highest loaded integration point in front of the crack tip are plotted. Obviously, the computations show nearly constant results $\lambda = 0°$ and $\varphi = 90°$, which describes the classical *mode I* failure regime for the considered CT specimen.

The results show impressingly the close limits of the continuum damage mechanics using the FE method without additional regularization avoiding a type change of the leading differential equation. Mesh refinements resulting in the typical mesh sizes in the magnitude of the intrinsic material length scales can not represent the real, potentially inhomogeneous, material structure on the microlevel.

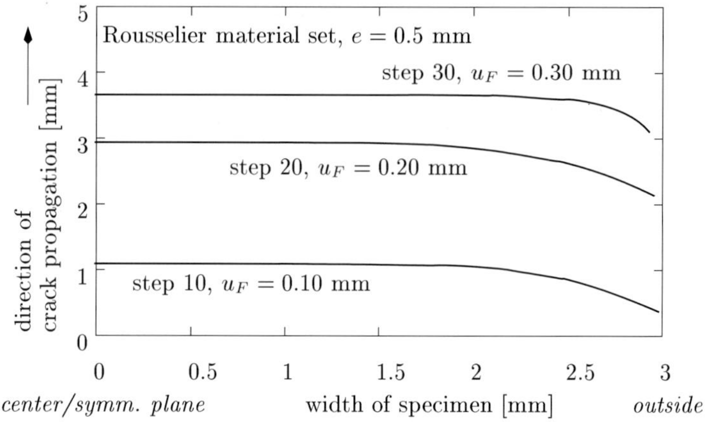

Fig. 7. Contours of integration points on the ligament reaching $q = 0$ for load steps 10, 20, 30 using the ROUSSELIER material set and $e = 0.50$ mm

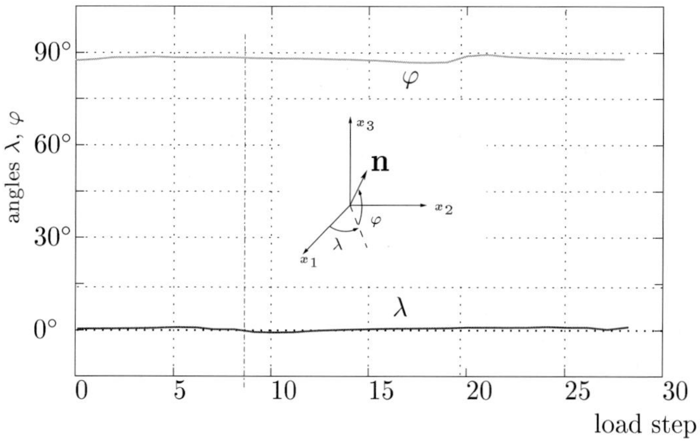

Fig. 8. Orientation of $\boldsymbol{n}$ in front of the crack tip. The vertical line marks the load step of loosing ellipticity for the considered problem, cf. Fig. 6

6 Summary

In this contribution we present a study on ductile damage analysis by a 3D simulation of CT specimen using the ROUSSELIER damage model within a finite element formulation based on 20–node–solid elements. The main attention was focused on the limitations of the finite element method discretizing mechanical field equations by piecewise continuous functions, which are used to represent inhomogeneous constituents of material on the microscale.

Typical FE analyses, resolving the situations in front of crack tips or in shear band regions as detailed as possible, are known to produce mesh sensitive results because of the changing type of the basic differential equations. This "loss of ellipticity" is checked by a steady evaluation of the acoustic tensor and a stop of the overall computation reaching such a point of stability. It is worth mentioning that this critical situation is reached early during the nonlinear iteration process, so that the subsequently determined numerical results become questionable, if no method of regularization is applied.

References

1. Aravas N. (1987) On the numerical integration of a class of pressure–dependent plasticity models. International Journal for Numerical Methods in Engineering, 24:1395–1416
2. Baaser H., Gross D. (1998) Damage and strain localisation during crack propagation in thin–walled shells. In A. Bertram and F. Sidoroff, editors, Mechanics of Materials with Intrinsic Length Scale, number ISBN 2-86883-388-8, pages 13–17, Magdeburg, Germany. EDP Sciences. Journal de Physique IV, **8**
3. Baaser H., Gross D. (2000) Crack analysis in ductile cylindrical shells using Gursons model. International Journal of Solids and Structures, 37:7093–7104
4. Baaser H., Gross D. (2001) 3D Nonlocal Simulation of Ductile Crack Growth — A Numerical Realization. European Journal of Finite Elements, 10(2–3–4):353–367. ISBN 2-7462-0260-3
5. Baaser H., Tvergaard V. (2003) A new algorithmic approach treating nonlocal effects at finite rate-independent deformation using the Rousselier damage model. Computer Methods in Applied Mechanics and Engineering, 192 (1–2), 107–124
6. Gurson A.L. (1977) Continuum theory of ductile rupture by void nucleation and growth: Part I - yield criteria and flow rules for porous ductile media. Journal for Engineering Materials in Technology, 99:2–15
7. Holzapfel G.A. (2000) Nonlinear Solid Mechanics. Number ISBN 0-471-82304. Wiley
8. Mathur K.K., Needleman A., Tvergaard V. (1994) Ductile failure analyses on massively parallel computers. Computer Methods in Applied Mechanics and Engineering, 119:283–309
9. Ortiz M., Leroy Y., Needleman A. (1987) A finite element method for localized failure analysis. Computer Methods in Applied Mechanics and Engineering, 61:189–214
10. Petryk H. (1997) Plastic instability: criteria and computational approaches. Archives of Computational Methods in Engineering, 4(2):111–151
11. Reese S., Wriggers P. (1997) A material model for rubber–like polymers exhibiting plastic deformation: computational aspects and a comparison with experimental results. Computer Methods in Applied Mechanics and Engineering, 148:279–298
12. Rousselier G., Devaux J.-C., Mottet G., Devesa G. (1989) A methodology for ductile fracture analysis based on damage mechanics: An illustration of a local approach of fracture. Nonlinear Fracture Mechanics, 2:332–354

13. Schreyer H.L., Neilsen M.K. (1996) Analytical and numerical tests for loss of material stability. International Journal for Numerical Methods in Engineering, 39:1721–1736
14. Simo J.C. (1992) Algorithms for static and dynamic multiplicative plasticity that preserve the classical return mapping schemes of the infinitesimal theory. Computer Methods in Applied Mechanics and Engineering, 99:61–112
15. Steinmann P., Larsson R., Runesson K. (1997) On the localization properties of multiplicative hyperelasto–plastic continua with strong discontinuities. International Journal of Solids and Structures, 34(8):969–990
16. Steinmann P., Miehe C., Stein E. (1994) Comparison of different finite deformation inelastic damage models within multiplicative elastoplasticity for ductile materials. Computational Mechanics, 13:458–474
17. Tvergaard V. (1989) Material failure by void growth to coalescence. Advances in Applied Mechanics, 27:83–151
18. Wells G.N., Sluys L.J. (2001) Analysis of slip planes in three–dimensional solids. Computer Methods in Applied Mechanics and Engineering, 190:3591–3606
19. Zhang Z.L., Hauge M. (1999) On the Gurson micromechanical parameters. Fatigue and Fracture Mechanics, 29:364–382

Ductile Crack Growth in Metallic Foams

Ingo Schmidt[1], Christoph Richter[2], and Dietmar Gross[1]

[1] Institute of Mechanics, University of Technology
Hochschulstr. 1, D–64289 Darmstadt, Germany
{i.schmidt, gross}@mechanik.tu-darmstadt.de
[2] Siemens AG, Power Generation
D–45466 Mülheim/Ruhr, Germany
christoph-richter@siemens.com

received 18 Jul 2002 — accepted 24 Oct 2002

Abstract. This article deals with the modelling and simulation of mode I crack growth in metal foams. Crack growth initiation and subsequent resistance to propagation are explored numerically using two- and three-dimensional beam networks as models for the foam microstructure around the crack tip. The cell wall material is characterised by an uniaxial elasto-plastic stress-strain relation with linear hardening, a critical fracture stress and a fracture energy per unit area. A special finite element is developed for the simulation of the fracture of individual cell walls, allowing for the simulation of diffuse cracking of cell walls in the course of (macro-) crack propagation. Parameter studies of the influence of cell wall material parameters on crack resistance curves under small scale yielding conditions are presented, including a comparison of the two- and three-dimensional models.

1 Introduction

Metal foams are highly porous materials with pore volume fractions of up to and above 90 % and a sponge-like microstructure. The cell wall or -edge material is typically an aluminium alloy but many solid materials can nowadays be 'foamed' to produce such microstructures. Advances in manufacturing of the past years have made metallic foams available for industrial applications that exploit the unique mechanical, thermal and acoustic properties of these materials: potential applications include energy absorbing components, load carrying sandwich structures and heat exchangers. As with other new materials, the successful application of metallic foams requires a thorough understanding of the material properties. A detailed review of these can be found in [1] (see also [2] for current research topics).

Regarding the fracture properties of foams, previous theoretical investigations have been confined mainly to brittle cell wall materials as in, e.g., [3] [4] [5] [6] [7]. In the present study we are concerned with the simulation of mode I crack growth under small scale yielding conditions. The term 'small scale yielding' (SSY) signifies situations in which the zone in which inelastic processes like plasticity and material separation take place is small compared to all other dimensions of the cracked body's geometry. This implies the existence of a region where the linear elastic K-field solution, specified by a single parameter – the stress intensity factor K – dominates. Under these circumstances, the external loading of the specimen is transmitted to the crack tip through only this parameter and, consequently, the initiation of crack growth is governed by a critical stress intensity factor criterion $K = K_c$. Equivalently, this can be expressed through Griffith's energy release rate criterion, $G = G_c$, which states that crack growth initiates when the elastic strain energy that would be released upon increasing the crack length at fixed boundary displacements equals the energy that is necessary to separate the newly created crack faces. For ductile materials it is found that the critical stress intensity factor K_c, i.e. the resistance to crack growth, initially increases with the crack extension to reach a steady state value after a certain amount of crack propagation; this is called R-curve behaviour. The explanation for this phenomenon is that only part of the energy stored in the cracked body is available to advance the crack because the other part is dissipated through plastic deformation in the crack tip region. For practical purposes, the knowledge of the so called R(esistance)-curve, i.e. the critical stress intensity factor K_c (or energy release rate G_c) as a material dependent function of the crack advance Δa, is of great importance.

In order to elucidate the role of plasticity in mode I crack growth in fully dense, ductile materials under SSY conditions, a cohesive zone embedded in a homogeneous continuum was used in [8] to compute R-curves and their dependence upon certain material parameters. There, crack propagation is modelled by specifying a relation between the traction on the crack faces and the separation of these. This relation is also called 'bridging law' because it can be thought of as representing the effect of material elements bridging the gap between what is about to become the crack faces. This relation is characterised by a peak value of the traction, marking the initiation of decohesion, and a critical separation for which the traction drops to zero; it is implemented along the ligament, i.e. the line ahead of the crack tip in the direction of the straight crack, and the solid is described by a v. Mises type flow theory. These studies showed that a crucial parameter for the steady state value of the toughness, i.e. the increase of resistance to crack growth, is the ratio of the peak stress in the traction-separation relation and the yield strength of the solid; at a critical value of this ratio, crack propagation becomes impossible, and the toughness rises sharply just before that value.

Aiming at metal foams, that same approach was followed in [9], where the surrounding solid was characterised by a constitutive law that incorporates the pressure dependent yielding exhibited by metallic foams. Here, too, a pronounced dependence of the steady state toughness on the ratio of peak stress to the yield strength of the foam has been found.

Experimental measurements of the toughness of various metal foams together with a characterisation of the local deformations near the crack tip have been published in [10] and [11]. These investigations revealed in particular that the process zone for a variety of investigated foams always extends over approximately 4 to 7 cells. In view of its aforementioned importance, the bridging law has been estimated in [10] from tests on deep edge notch specimens. Here, the bridging stress is taken as the average traction on the net section of the specimen, where the net section extends over $\approx 12 - 24$ cells. In this range, the average traction exhibits a 30 % variation and the question arises which value would be appropriate to use in the cohesive zone model. Moreover, the cohesive zone concept becomes somewhat questionable when it is applied to situations in which the process zone consists of only a few, say five, cell walls – some of which are already fractured and, hence, do not carry any load. Bearing in mind that, in the cohesive zone model, the traction varies from zero to its peak value over a distance equivalent to only so few cells, the concept of an 'average traction' at each point on the cohesive surface is difficult to justify. From what has been said, the cohesive zone model then becomes a merely phenomenological model; its ingredients are difficult to identify separately, i.e. other than through the simulation of the R-curve itself – which is what should actually be predicted. Nevertheless, cohesive zone models have been applied successfully in a variety of circumstances where model parameters for a specific material (-combination) have been determined from one test and test results for different geometries could be reproduced with these parameters (cf. the examples in [12]). However, this approach leaves open the question of how microstructural details affect the macroscopic crack growth resistance – a question so important for the design of materials.

An alternative way of modelling crack propagation is to resolve, to some extent, the microstructure of the material around the crack tip and to rely on a (possibly) more profound knowledge about the behaviour of the constituent materials. Such an approach has been followed in, among many others, [13] in the context of creeping polycrystals, or in [14] to study the interaction of one or more voids with a macroscopic crack tip. In materials with an open cell microstructure it is clear that macroscopic crack growth takes place by successive fracturing of individual cell walls or -edges. Recent experimental investigations confirmed the viewpoint advocated in [15] that these discrete fracture events are bending dominated processes, as illustrated in Fig. 1; it shows the deformed microstructure of a tensile Duocel foam specimen with a bent cell edge failing close to a vertex. With regard to metal foams, a two dimensional study in the above mentioned spirit has been presented in [16]; there, the cellular microstructure was represented by a network of slender beams composed of an elasto plastic material and crack propagation was simulated by an 'element removal technique': To simulate the fracture of individual cell walls, small elements at the vertices have been taken out of the finite element model in a manner consistent with energy considerations. The outcome of this study was a quantitative prediction of how the cell

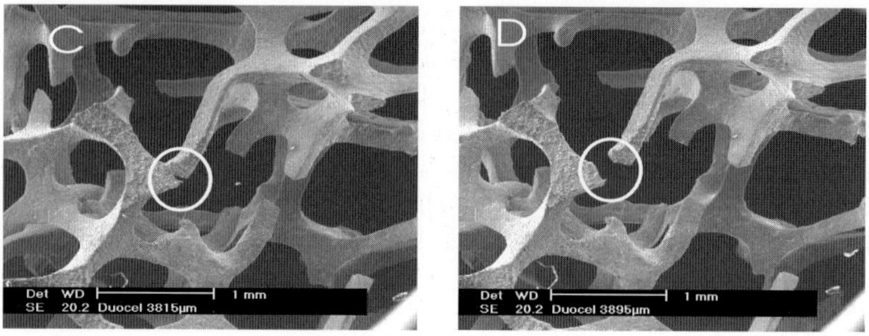

Fig. 1. Cell edge failure in a tensile Duocel foam. (from [20])

wall material parameters influence the macroscopic fracture response of the foam. For technical reasons, only one element at a time could be removed from the model; this restricted the applicability of the procedure to a certain range of material parameters. Moreover, the procedure is computationally involved, because it requires repeated interruptions and postprocessings of the computation.

The purpose of the present study is, firstly, to overcome these restrictions by introducing a 'separation element' in the numerical treatment of the model, which allows for the simultaneous cracking of many cell walls without interrupting the calculation and, secondly, the application of the model to three dimensional systems.

2 Specification of the Model

The mechanical model used here has been described in [16] for the 2D case. For the sake of completeness, the main ingredients are repeated in the following. We consider a macroscopic crack in two- and three dimensional cellular structures as depicted in Fig. 2 for the 2D case. The structure is composed of beams with average length l which are assumed to be sufficiently slender so that they can be modelled as Euler-Bernoulli beams. The cross section of the beams is taken as rectangular in the 2D – and circular in the 3D case with their thickness and diameter denoted by t respectively. The cell wall material is described by the uniaxial, bilinear stress-strain relation

$$\begin{array}{ll} \epsilon = \sigma/E & \sigma < \text{for}\sigma_{\mathrm{y}} \quad , \\ \epsilon = \sigma_{\mathrm{y}}/E + (\sigma - \sigma_{\mathrm{y}})/H & \text{for}\sigma > \sigma_{\mathrm{y}} \end{array} \tag{1}$$

in terms of the true stress σ, true strain ϵ, Young's modulus E, yield strength σ_{y} and constant hardening modulus H. This relation governs the behaviour of the cell wall material up to the point where a critical fracture strength σ_{f} is reached, which marks the beginning of localised deformation and subsequent fracture at a point in the structure. The details of this process are not modelled here; rather, it is assumed that the fracture of an individual cell wall has a certain energy cost, given by the cross sectional area of the beam times the specific fracture energy of the cell wall material Γ_0. The latter quantity is assumed to completely describe the influence of the cell wall material's fracture properties upon the macroscopic fracture response; the precise manner in which this assumption is incorporated will be described later. Under

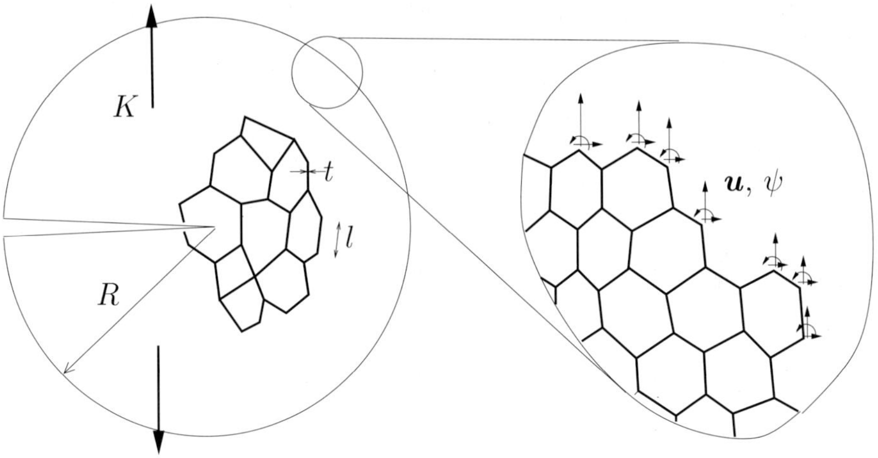

Fig. 2. Cellular structure with a macroscopic crack under small scale yielding conditions

conditions of small-scale yielding, the displacements on a boundary remote from the crack tip are specified by the mode-I K-field of the linear elasticity theory. These displacements serve as boundary data with the applied K acting as a loading parameter. Macroscopic crack propagation is simulated by first incrementally increasing the displacements of the remote boundary (which is taken to be of circular shape) until the stress in one of the beams near the crack tip attains the fracture strength. Since there is no transverse distributed loading on the beams, the bending moment is linear over each beam and fracture occurs at one of the end points. Subsequently, the 'separation element', to be described shortly, is activated. This element is located at the beam end points and connects two nodes with translational and rotational degrees of freedom and identical initial positions; the corresponding forces and moments exhibit a 'softening' dependence on the difference in the displacements and rotations of the two nodes in such a way that *i)* all section forces have dropped to zero at the same time and *ii)* the work done by these forces equals the prescribed value $\Gamma_0 A$, where A is the cross section of the beam.

2.1 Foam Topology

To represent the microstructure of the foam, irregular cellular structures have been generated by randomly disturbing the vertex positions of a regular structure which are composed of identical unit cells that pack to fill space. Specifically, we used hexagonal cells for the 2D case and Tetrakaidecahedra for the 3D case as depicted in Fig. 3. The irregularity is an essential feature because it induces bending of the cell walls also under macroscopically hydrostatic stress states (cf. [17])– which are dominant in front of a crack tip.

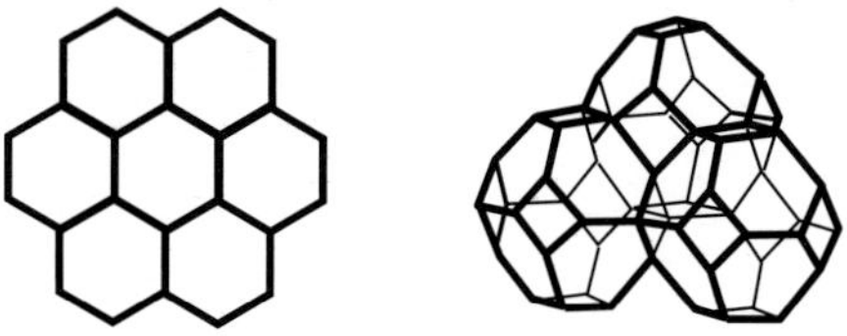

Fig. 3. Two and three dimensional idealised microstructures: Regular arrays of hexagons and tetrakaidecahedra

3 Normalisation

It follows from a dimensional analysis that the suitably normalised solution of the above described problem depends only upon the following dimensionless combinations of material parameters:

$$\rho = t/l\ ,\ \ \sigma_{\mathrm{y}}/E\ ,\ \ \sigma_{\mathrm{f}}/\sigma_{\mathrm{y}}\ ,\ \ H/E\ ,\ \ \tilde{\Gamma}_0 = \Gamma_0 E/(\sigma_{\mathrm{y}}^2 l)\quad , \tag{2}$$

where $\rho = t/l$ is a measure for the relative density, i.e. the solid volume fraction, the latter being proportional to the first and second power of the former in the two and three dimensional case respectively; $\tilde{\Gamma}_0$ is, up to a constant factor, the work required to break a beam divided by the elastic strain energy contained in a beam of length l under pure bending when the yield strain is attained at the outer fibres.

Based on the linear elastic K-field displacements, Euler-Bernoulli beam theory and the scaling law for the effective stiffness with the relative density, a simple order-of-magnitude estimate for the toughness of a 2D cellular material can be derived as (c.f. [16])

$$K_{\mathrm{Y}}^{\mathrm{2D}} = \sqrt{\pi l}\,\sigma_{\mathrm{y}}(t/l)^2\quad . \tag{3}$$

In like fashion, the only difference being the scaling of the stiffness with the relative density [15], the derivation for the 3D case gives

$$K_{\mathrm{Y}}^{\mathrm{3D}} = \sqrt{\pi l}\,\sigma_{\mathrm{y}}(t/l)^3\quad . \tag{4}$$

The values (3) and (4) will be used to normalise the stress intensity factor in the two- and three dimensional case respectively.

$$\tilde{K} = K/K_{\mathrm{Y}} \tag{5}$$

4 Numerical Technique

4.1 The Finite Element Model

The commercial finite element code ABAQUS is used to calculate stresses and strains in the cell walls of the cellular structure subjected to prescribed displacements and rotations

$$\begin{aligned} u_1(r,\varphi) &= \frac{K\sqrt{r}}{2\mu^*\sqrt{2\pi}}\cos\varphi/2(\alpha-\cos\varphi) \quad , \\ u_2(r,\varphi) &= \frac{K\sqrt{r}}{2\mu^*\sqrt{2\pi}}\sin\varphi/2(\alpha-\cos\varphi) \quad , \\ \psi(r,\varphi) &= \frac{1}{2}(u_{2,1}-u_{1,2}) \end{aligned} \tag{6}$$

on a circular boundary remote from the crack tip where $\alpha = (3-\nu^*)/(1+\nu^*)$ for plane stress in 2D and $\alpha = 3 - 4\nu^*$ for plane strain in 3D. The parameters μ^*, ν^* are the effective shear modulus and Poisson ratio of the cellular structure. Due to the irregularity, they are not available in closed form and have to be determined from finite element calculations. This is done here using a 'representative volume element' and applying periodic boundary conditions to compute the stress response to macroscopically homogeneous deformations (cf. [17] [18]).

The cell walls are modelled as Timoshenko beams[1] which are discretised with up to seven linear elements whose material behaviour is characterised by (1).

To reduce computational cost in the 3D case, the idealised microstructure of the foam is resolved and fully discretised only in a circular region around the crack tip which is embedded in a ring of solid linear 8 node elements (see Fig. 4) and only the displacements are prescribed on the outer boundary of that ring. The solid elements are supposed to 'fill up' a region of K-dominance where the effective elastic properties sufficiently describe the material behaviour. Inner and outer radii of the ring are chosen as $R_i \approx 40l$ and $R_o = 2R_i$ so that for both the two- and three dimensional case, the remote boundary where the K-field displacements are prescribed is located at a distance $\approx 80l$ away from the crack tip.

The transition from beam- to solid elements is realised through a kinematic constraint requiring the displacements and rotations of a beam node

[1] The finite shear flexibility is, due to the slenderness of the beams, of minor importance

hitting the surface of a solid element to be equal to those given by the interpolation of the displacement field of the solid element at the respective position. Since the rotations are small at the position in question, these are calculated from the linear kinematic relations as in $(6)_3$.

To mimic an infinite thickness – corresponding to plane strain conditions along the crack front – periodic boundary conditions are applied at the side faces of the model: Two nodes initially opposite to each other are forced to remain opposite to each other and to also keep their distance. In addition, for two such nodes with rotational degrees of freedom, the latter are forced to be equal. The thickness of the structure is set to zl corresponding to n tetrakaidecahedral cells. As already mentioned earlier, an irregularity in the cellular structure is introduced by adding a random perturbation to the vertex positions which is uniformly distributed between $\pm 30\%$ of the average beam length l.

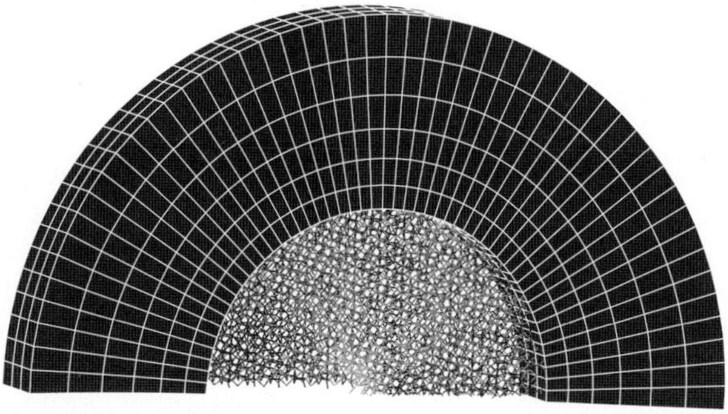

Fig. 4. Three dimensional finite element model

4.2 The Separation Element

As mentioned earlier, it is assumed that localised deformation and subsequent failure of a cell wall initiates when the stress attains a critical value. When modelling a cell wall as a one dimensional structure with translational and rotational degrees of freedom, such a local failure results in a discontinuous distribution of the displacements and rotations along the beam axis with jumps of these quantities at the point of fracture. In the spirit of a cohesive surface model, the task is then to define a constitutive relation for the forces and moments transmitted between the separating cross sections of the beam, i.e. a relation between the section forces $\boldsymbol{f}$ and -moments $\boldsymbol{m}$ and the jumps in the displacements $[\![\boldsymbol{u}]\!]$ and rotations $[\![\boldsymbol{\varphi}]\!]$ (see Fig. 5). Based on the assumption that the separation of the cell wall material requires a fixed amount of work per unit area, given by Γ_0, and with the requirement that the forces and moments in the cross section vanish simultaneously, a 'force-separation' law is proposed in the form

$$\begin{aligned} \boldsymbol{f}([\![\boldsymbol{u}]\!],[\![\boldsymbol{\varphi}]\!]) &= f_0(1-D)[\![\boldsymbol{u}]\!]/u \\ \boldsymbol{m}([\![\boldsymbol{u}]\!],[\![\boldsymbol{\varphi}]\!]) &= m_0(1-D)[\![\boldsymbol{\varphi}]\!]/\varphi \end{aligned} \quad , \quad D = \frac{\mu - \mu_0}{2\Gamma_0 A - \mu_0} \tag{7}$$

$$\text{where} \quad \mu = f_0 u + m_0 \varphi \quad , \quad \mu_0 = f_0 u_0 + m_0 \varphi_0$$

Here, the subscript $()_0$ denotes the value of the respective quantity at the

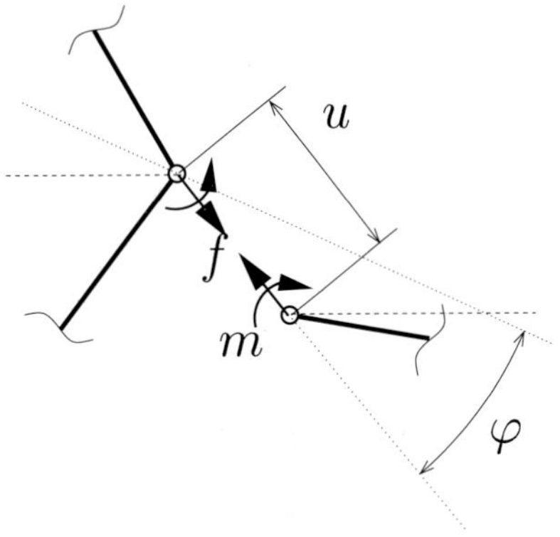

Fig. 5. Separation element connecting beam node and vertex node for a plane deformation. Dashed and dotted lines represent initial and current orientation of the nodes respectively

instant when the fracture stress is attained, A is the cross sectional area of the beam and u, φ, f, m denote the absolute values of the vectorial quantities $[\![\boldsymbol{u}]\!]$, $[\![\boldsymbol{\varphi}]\!]$, $\boldsymbol{f}$, $\boldsymbol{m}$ respectively. The specific structure of (7) assures that the above mentioned requirements are met: The damage parameter D is initially zero and increases with the separation; when $D = 1$, both force and moment are zero, i.e. the separation process is completed and the work done by the section forces and moments during this process, $W_{\text{sep}} = \int_0^T (\boldsymbol{f}\cdot[\![\dot{\boldsymbol{u}}]\!]+\boldsymbol{m}\cdot[\![\dot{\boldsymbol{\varphi}}]\!])dt$, equals $\Gamma_0 A$ regardless of the particular history of the separation $[\![\boldsymbol{u}(t)]\!]$ and $[\![\boldsymbol{\varphi}(t)]\!]$ (see Appendix). In truth, this work of separation is likely to depend – to some degree – upon the deformation history; the idea is here that in the foam structure, the failing cell walls will experience comparable deformation histories during a macroscopic crack growth, so that we may postulate an equal W_{sep} for all the local fracture events. Of course, the jumps in displacement and rotation before and at the beginning of the process, $[\![\boldsymbol{u}/\boldsymbol{\varphi}]\!]_0$, should be zero. However, one would then have to switch from a kinematic constraint to a constitutive relation at a certain point which is difficult to implement. Therefore, large but finite initial stiffnesses c_{f0} and c_{m0} are introduced for the section forces and moments respectively so that, until the fracture strength is reached, the constitutive relation simply reads

$$\boldsymbol{f} = c_{f0}[\![\boldsymbol{u}]\!] \quad , \quad \boldsymbol{m} = c_{m0}[\![\boldsymbol{\varphi}]\!] \tag{8}$$

and a small but non-zero separation is present whenever $\boldsymbol{f}$ or $\boldsymbol{m} \neq \boldsymbol{0}$.

When the fracture strength is reached the values of $f/m = f/m_0$ and $u/\varphi = u/\varphi_0$ are recorded and are used to rewrite relation (7) in a form akin to elastic-plastic constitutive relations which is summarized in the table below.

	section force	section moment
elasticity relation	$\boldsymbol{f} = c_f \llbracket \boldsymbol{u} \rrbracket$	$\boldsymbol{m} = c_m \llbracket \boldsymbol{\varphi} \rrbracket$
elastic domain	$\Phi_f = f - f_0(1-D) \leq 0$	$\Phi_m = m - m_0(1-D) \leq 0$
evolution equations	if $\Phi_f \cdot \Phi_m = 0$: $\dot{D} = \langle \dfrac{f_0 \dot{u} + m_0 \dot{\varphi}}{2\Gamma_0 A - \mu_0} \rangle$	
	if $\Phi_f = 0$: $\dot{c}_f = -\langle f_0 \dot{D} + c_f \dot{u} \rangle / u$	if $\Phi_m = 0$: $\dot{c}_m = -\langle m_0 \dot{D} + c_m \dot{\varphi} \rangle / \varphi$

(9)

where use is made of the notation $\langle x \rangle = 1/2(x + |x|)$. The separation law in the form (9) is equivalent to (7) when both u and φ increase monotonically but it allows also for elastic 'unloading', i.e. a temporary decrease of the absolute value of the displacement and/or rotation jump independent of each other and without an accompanying unphysical decrease of the damage parameter D. Strictly speaking, the above form results in a path dependence of the work of separation W_{sep} mentioned earlier but this found to be of minor significance in the actual calculations.

Let $\boldsymbol{F} = [\boldsymbol{f}_1 \, , \, \boldsymbol{f}_2 \, , \, \boldsymbol{m}_1 \, , \, \boldsymbol{m}_2]$ and $\boldsymbol{U} = [\boldsymbol{u}_1 \, , \, \boldsymbol{u}_2 \, , \, \boldsymbol{\varphi}_1 \, , \, \boldsymbol{\varphi}_2]$ denote the vector of nodal forces and displacements respectively. Then, with $\llbracket \boldsymbol{u} \rrbracket = \boldsymbol{u}_2 - \boldsymbol{u}_1$ and $\llbracket \boldsymbol{\varphi} \rrbracket = \boldsymbol{\varphi}_2 - \boldsymbol{\varphi}_1$, the tangent matrix

$$\bar{\boldsymbol{D}} = \frac{\partial \boldsymbol{F}_{n+1}}{\partial \boldsymbol{U}_{n+1}} \tag{10}$$

is readily obtained from (9).

The above described element is used to connect beam elements to the corresponding adjacent vertices in the discretisation of the cellular structure and the calculation is performed in a small strain, finite rotation context. In the finite element code, the direction of the rotation vector $\boldsymbol{\varphi}$ is defined as that of the axis about which the rotation from the initial to the current orientation of the node is performed; the magnitude $|\boldsymbol{\varphi}|$ is the rotation angle in radians. It can readily be shown that the vector $\llbracket \boldsymbol{\varphi} \rrbracket$ then has as its direction the axis about which the rotation taking the orientation of node 1 to that of node 2 is performed, and that its magnitude is given by the rotation angle of that rotation. The elasticity relations $(9)_1$ can therefore be understood as the effect of a translational and rotational spring attached to the nodes where the first is directed along the connecting line of the two nodes and the second is oriented along the above described rotation axis.

Numerical Stabilisation If, during the calculation, several separation elements are active, instabilities can occur in the sense that the global stiffness matrix becomes singular and the Newton algorithm breaks down. This is so in particular when the value of Γ_0 is small, in which case the elastic energy released from the structure can exceed the work needed to complete the separation of an individual beam. This corresponds to a 'snap back' instability and similar problems have been reported in [19] in the context of modelling the delamination in layered composite structures by means of a cohesive surface. In order to avoid this effect, a regularisation is performed following the procedure described in [19]. In essence, this procedure introduces an artificial viscosity η and results in a modification of the update for the nodal forces and the tangent matrix according to

$$\begin{aligned} \boldsymbol{F}_{n+1} &= \beta \boldsymbol{F}_n + (1-\beta)\bar{\boldsymbol{F}}_{n+1} + \frac{1-\beta}{\delta}\boldsymbol{C}(\boldsymbol{U}_{n+1} - \boldsymbol{U}_n) \\ \boldsymbol{D} &= \frac{1-\beta}{\delta}\boldsymbol{C} + (1-\beta)\bar{\boldsymbol{D}} \end{aligned} \tag{11}$$

where $\delta = \Delta t/\eta$, $\beta = e^{-\delta}$, $\boldsymbol{C}$ is the elastic tangent matrix (containing only the stiffnesses c_f and c_m on the diagonal) and $\bar{()}$ denotes quantities of the rate independent solution. The first term in $(11)_2$ adds positive entries to the diagonal of the tangent matrix, which increase with η. On the other hand, large values for η result in a significant alteration of the rate independent solution. Therefore, the parameter η must be chosen so as to ensure the regularising effect but not to cause significant changes of, say, the resulting work of separation. The value of η is determined from trial calculations with only a few separation elements and depends primarily on the parameter Γ_0. Moreover, the resulting work of separation W_{sep} is checked for each separation element individually during the calculation. It is found that the regularising effect can be achieved with only a few percent 'error' in W_{sep}.

5 Results

In this section, predicted resistance curves and indicators for the damage in front of the crack tip are presented where attention is directed to the influence of the cell wall material properties.

5.1 Material Parameters

The cell wall material properties for typical aluminium alloys from which metal foams are made result in the following values for the dimensionless parameters (2):

$$\sigma_y/E \;= 0.1\%\;, \quad \sigma_f/\sigma_y \;= 2.0\;, \quad H/E = 0.11 \quad . \tag{12}$$

Moreover, a typical slenderness ratio of the cell walls has been chosen as $t/l = 0.05$. If not stated otherwise, the parameters assume the above mentioned values. A rough estimate for the work of separation according to $\Gamma_0 \approx K_c^2/E$, where K_c is the critical stress intensity factor, leads to $\tilde{\Gamma}_0 \approx (K_c/\sigma_y)^2/l$ with values ranging from 1 to 100 for cell sizes of the order of 1 mm. More insight can be gained from experimental measurements of force-displacement curves on a single cell wall extracted from the foam structure. Such tests have been performed in [20] and the result of their measurement for pure tension is depicted in Fig. 6. According to the approach of modelling the fracture process of an individual cell wall described in the last section, the work of separation should contain the whole work dissipated in the course of localised plastic deformation and subsequent rupture. Since the peak value in the curve Fig. 6 marks the beginning of necking, the area under the force-displacement curve from that point on should equal $\Gamma_0 A$. This leads to a value for $\tilde{\Gamma}_0 \approx 30$ which is considerably larger than the value used in the previous, two dimensional study [16] and hence gives additional motivation to explore the influence of the parameter $\tilde{\Gamma}_0$. Note that a numerical calculation of this single strut experiment with one separation element reproduces more or less precisely the response of Fig. 6 because a linear decrease of the force with the end displacement is built in into the separation law (9).

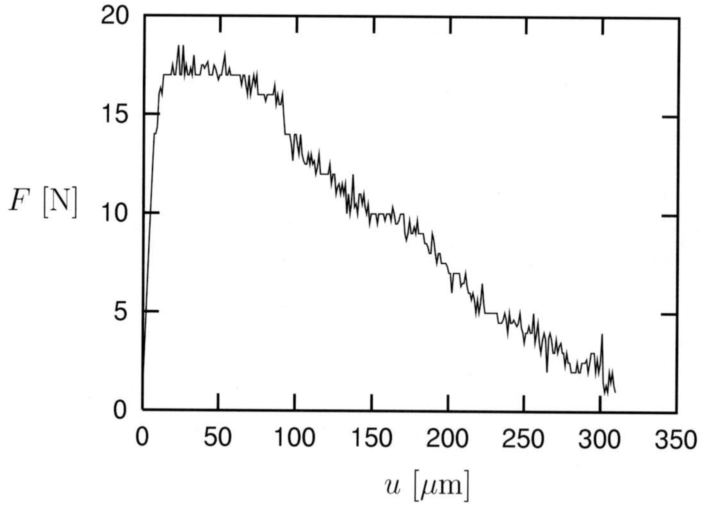

Fig. 6. Experimentally observed force displacement curve for pure tension of a single cell edge of a Duocel Alu foam (from [20])

5.2 The Two Dimensional Case

The influence of the cell wall toughness measure is illustrated in Fig. 7 which shows crack resistance curves for three different values[2] of $\tilde{\Gamma}_0$. These have been obtained by computing the R-curves for four different realisations of the microstructure and taking their average for each discrete value of the crack propagation Δa. The error bars indicate the corresponding standard deviation. It is seen that increasing Γ_0 increases primarily the slope of the R-curve while the initiation toughness is much less affected. Qualitatively, this trend is expected: Increasing the toughness of the cell wall material will increase the macroscopic toughness of the foam, but Fig. 7 shows that the latter does not just scale linearly with the former.

In Fig. 8, the crack resistance curves for different values of the relative density measure $\rho = t/l$ are compared for two values of $\tilde{\Gamma}_0$. The solid curves are the results of Fig.7 ($\rho = 0.05$), drawn here without error bars, and the dashed curves show the corresponding results for one realisation with $\rho = 0.1$. Despite the large difference in the corresponding values of ρ, the dimensionless crack growth resistance is virtually unaffected for both $\tilde{\Gamma}_0$ values. This means that the influence of the relative density is essentially captured in the normalisation (3) and confirms the conclusion drawn in the previous study [16] for $\tilde{\Gamma}_0 = 3$. This is all the more remarkable, because the arguments leading to (3) were based on the assumption of linear elastic material behaviour, but in the cases considered here there is much more plasticity involved than there is for smaller $\tilde{\Gamma}_0$.

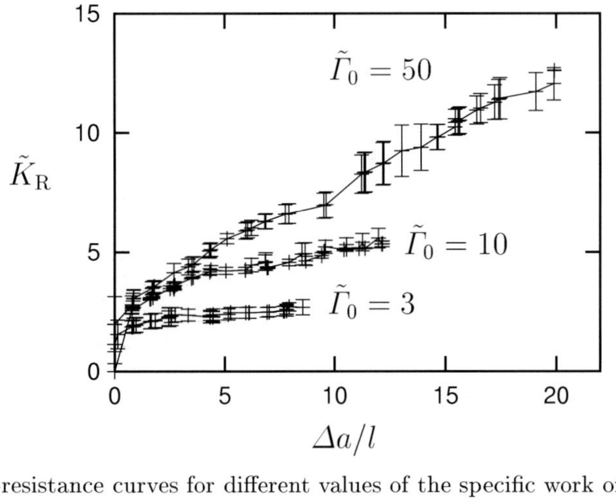

Fig. 7. K-resistance curves for different values of the specific work of separation

[2] The result for $\tilde{\Gamma}_0 = 3$ is taken from [16].

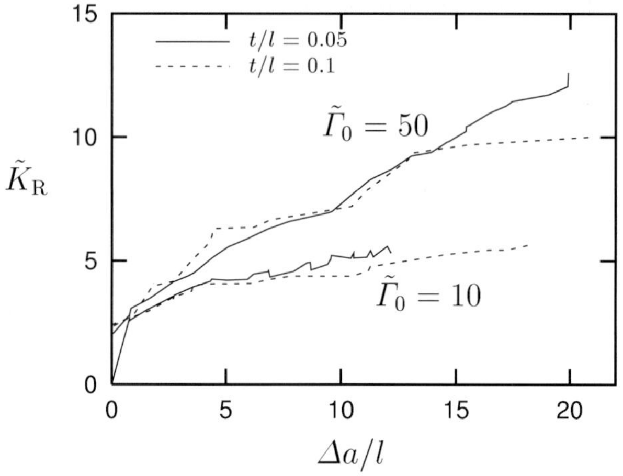

Fig. 8. K-resistance curves for altered relative density

Next, the influence of the cell wall fracture strength is explored by comparing resistance curves for two values of $\sigma_{\rm f}/\sigma_{\rm y}$. For small $\tilde{\Gamma}_0$ it has been found in [16] that increasing $\sigma_{\rm f}/\sigma_{\rm y}$ leads to a resistance curve which is essentially shifted to higher values by a constant amount. In contrast, for $\tilde{\Gamma}_0 = 50$ the resistance curve shows also a greater slope (Fig.9) and increasing $\sigma_{\rm f}/\sigma_{\rm y}$ thus appears to have a qualitatively different effect. The picture becomes clearer when plotting the ratio of the corresponding $K_{\rm R}$ values against the crack advance (Fig. 10), which shows that there is a constant factor ≈ 1.5 between the two curves. This suggests that the normalised crack growth resistance scales linearly with the ratio of fracture- to yield strength of the cell wall material. This is again perfectly consistent with the findings in [16] because there the R-curve effect was much less pronounced (cf. Fig.7) – i.e. a steady state was reached after only little crack advance – and the ratio of the steady state $K_{\rm R}$ values is found to be ≈ 1.5 also for $\tilde{\Gamma}_0 = 3$

5.3 The Three Dimensional Case

Here we present the first results for the three dimensional case in terms of crack growth resistance curves for the case $\tilde{\Gamma}_0 = 50$ and two different relative density measures ρ (Fig. 11a). These curves represent the results for one particular realisation of the microstructure, but it is fully expected that the deviation from the mean is comparable with that of the two dimensional case. For comparison with the latter, the corresponding result from Fig. 7 is also included. The negligible difference between the two 3D resistance curves for $t/l = 0.05$ and $t/l = 0.1$ shows that the scaling proposed in (4) captures the influence of the relative density equally well as in the 2D case. Moreover, the normalised values of the crack growth resistance coincide, to a good

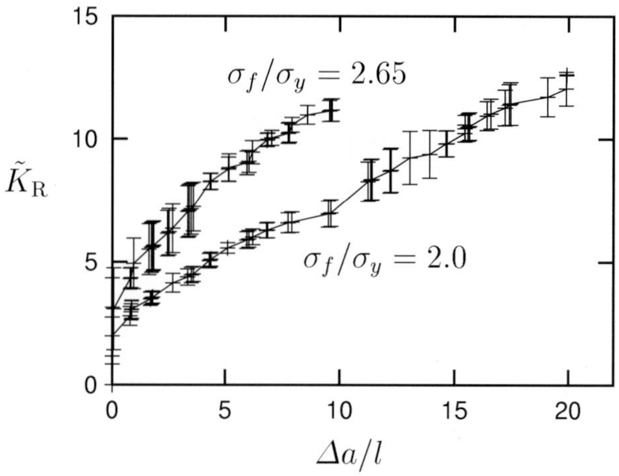

Fig. 9. K-resistance curves for altered fracture strength ($\tilde{\Gamma}_0 = 50$)

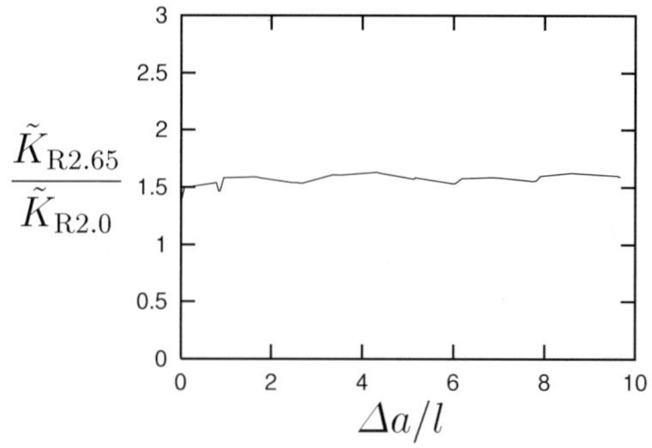

Fig. 10. Ratio of K_{R} values for $\sigma_{\mathrm{f}}/\sigma_{\mathrm{y}} = 2.65$ and $\sigma_{\mathrm{f}}/\sigma_{\mathrm{y}} = 2.0$

approximation, for the two and three dimensional model. The same is true for the results in Fig. 11b for $\tilde{\Gamma}_0 = 10$ and $\sigma_{\mathrm{f}}/\sigma_{\mathrm{y}} = 2.0$. This figure shows also the influence of increasing the fracture strength $\sigma_{\mathrm{f}}/\sigma_{\mathrm{y}}$ which leads to an elevated resistance. We conclude that a two dimensional model appears to not only give qualitative insight but provides also quantitative information. In view of the large computation times needed for the 3D calculations, this is a valuable finding. To illustrate the distribution of 'damage' around the crack tip, Fig.12 shows the location, projected onto the x_1-x_2 plane, of separation elements with $D > 0$. To understand this rather diffuse distribution, one has to observe that the chosen threshold 0.1 is rather small and that not all of the marked elements are active: some of them have undergone unloading and will not continue to separate.

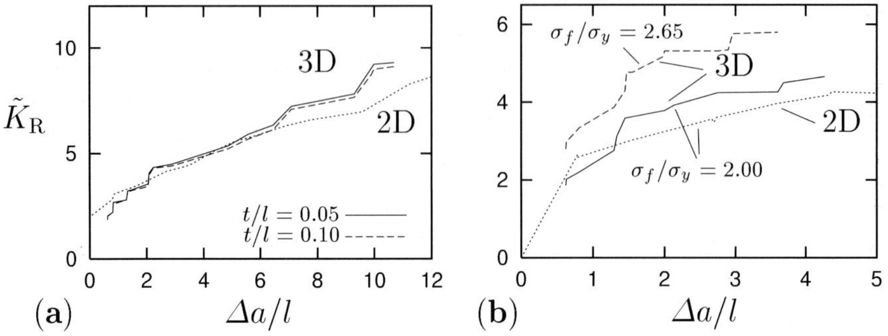

Fig. 11. Comparison of K-resistance curves for the two- and three dimensional model: (**a**) influence of the relative density measure $\rho = t/l$ ($\Gamma_0 = 50$) (**b**) effect of fracture strength σ_f/σ_y ($\Gamma_0 = 10$)

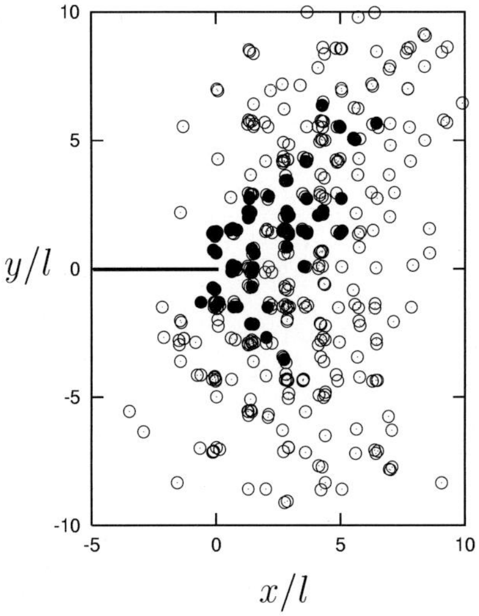

Fig. 12. Damage distribution in front of the crack tip: The location of separation elements with $0 < D < 0.1$ and $D > 0.1$ are marked with hollow and filled circles respectively.The crack tip is located at the origin

5.4 Concluding Remarks

Crack growth under mode I loading and small scale yielding conditions in two- and three dimensional cellular structures has been simulated using the finite element method. A separation element has been developed that simulates the fracture of a cell wall in an approximate manner. This element allows for the simultaneous fracture of cell walls and in particular for the investigation of the effect of higher values of the cell wall toughness. The scaling of the crack growth resistance with the fracture strength of the cell wall material found in a previous study has been confirmed to hold also for more realistic values of the cell wall toughness. Moreover, the influence of the relative density is found to be adequately described with a simple power law which can be derived on the basis of elementary considerations.

The results for the two dimensional, hexagonal and the three dimensional tetrakaidecahedral microstructure with the same level of imperfection coincide to a good approximation. This gives confidence in results obtained from analyses of two dimensional models but more 3D calculations are needed to support this conclusion.

Appendix

We firstly note that, because the directions of $\boldsymbol{f}$ / $\boldsymbol{m}$ coincide with those of $[\![\boldsymbol{u}]\!]$ / $[\![\boldsymbol{\varphi}]\!]$ due to (7), the power of the section forces can be written as

$$\boldsymbol{f} \cdot [\![\dot{\boldsymbol{u}}]\!] = f\dot{u} \quad , \quad \boldsymbol{m} \cdot [\![\dot{\boldsymbol{\varphi}}]\!] = m\dot{\varphi} \tag{13}$$

The work done by the section forces during the separation process can thus be expressed as

$$\begin{aligned} W_{\text{sep}} &= \int_0^T (\boldsymbol{f} \cdot [\![\dot{\boldsymbol{u}}]\!] + \boldsymbol{m} \cdot [\![\dot{\boldsymbol{\varphi}}]\!]) dt \\ &= \int_0^{t_0} (f\dot{u} + m\dot{\varphi}) dt + \int_{t_0}^T (f\dot{u} + m\dot{\varphi}) dt \\ &= \frac{1}{2}(f_0 u_0 + m_0 \varphi_0) + \int_{t_0}^T (1-D)(f_0 \dot{u} + m_0 \dot{\varphi}) dt \\ &= \frac{1}{2}\mu_0 + \int_{t_0}^T (1-D)\dot{D}(2\Gamma_0 A - \mu_0) dt \\ &= \frac{1}{2}\mu_0 + (2\Gamma_0 A - \mu_0) \int_0^1 (1-D) dD \\ &= \frac{1}{2}\mu_0 + (2\Gamma_0 A - \mu_0)\frac{1}{2} = \Gamma_0 A \end{aligned} \tag{14}$$

and, hence, is independent of the histories $[\![\boldsymbol{u}(t)]\!]$ / $[\![\boldsymbol{\varphi}(t)]\!]$.

References

1. Ashby, M., A. Evans, N. Fleck, L. Gibson, J. Hutchinson, and N. Wadley: 2000, Metal foams – a design guide. Boston, USA: Butterworth, Heinemann.
2. Banhart, J., Ashby, M.F. and Fleck, N.A. (Eds) (2001) Cellular Metals and Metal Foaming Technology. MIT Publishing, Bremen.
3. Huang, J. S., Gibson, L. J. (1991) Fracture toughness of brittle honeycombs. Acta Metall. Mater. **39**(7), 1617–1626.
4. Huang, J. S., Gibson, L. J. (1991) Fracture toughness of brittle foams. Acta Metall. Mater. **39**(7), 1627–1636.

5. Choi, J. B., Lakes, R. S. (1996) Fracture toughness of re-entrant foam materials with a negative Poisson's ratio: Experiment and analysis. Int. J. Fracture **80**, 73–83.
6. Hallström, S., Grenestedt, J. (1997) Mixed mode fracture of cracks and wedge shaped notches in expanded PVC foam. Int. J. Fracture **88**, 343–358.
7. Chen, J. Y., Huang, J. S. (1998) Fracture analysis of cellular materials: A strain gradient model. J. Mech. Phys. Solids **46**(5), 789–828.
8. Tvergaard, V. and Hutchinson, J. W. (1992) The relation between crack growth resistance and fracture process parameters in elastic-plastic solids. J. Mech. Phys. Solids **40**(6), 1377–1397
9. Chen, C., Fleck N. A. and Lu, T. J. (2001) The mode I crack growth resistance of metallic foams. J. Mech. Phys. Solids **49**(2), 231–259
10. McCullough, K.Y.G., Fleck, N.A., Ashby, M.F. (1999) Toughness of aluminum alloy foams. Acta Mater. **47**(8), 2331–2343
11. Motz, C., Pippan, R. (2002) Fracture behaviour and fracture toughness of closed cell metallic foams. Acta Mater. **50**(8), 2013–2033
12. Hutchinson, J. W., Evans, A. G. (2000) Mechanics of materials: Top-down approaches to fracture. Acta Mater. **48**(1), 125–135
13. Onck, P.R., van der Giessen, E. (1999) Growth of an initially sharp crack by grain boundary cavitation. J. Mech. Phys. Solids **47**, 99–139
14. Tvergaard, V., Hutchinson, J.W. (2002) Two mechanisms of ductile fracture: void by void growth versus multiple void interaction. Int. J. Solids Structures **39**, 3581–3597
15. Gibson, L. J., Ashby, M. F. (1998) Cellular Solids. Pergamon Press.
16. Schmidt, I., Fleck, N. A. (2001) Ductile fracture of two-dimensional cellular structures. Int. J. Fracture **111**, 327–342
17. Chen, C., Lu, T. J., Fleck, N. A. (1999) Effect of imperfections on the yielding of two-dimensional foams. J. Mech. Phys. Solids **47**(11), 2235–2272.
18. Schmidt, I. (2001) On plasticity and deformation induced anisotropy in metallic foams. in: Cellular Metals and Metal Foaming Technology. Banhart, J., Ashby, M.F. and Fleck, N.A. (Eds), MIT Publishing, Bremen.
19. Wagner, W., Gruttmann, F., Sprenger, W. (2001) A finite element formulation for the simulation of propagating delaminations in layered composite structures. Int. J. Num. Meth. Eng. **51**(11), 1337–1359
20. van Merkerk, R., Onck, P.R. (2002) Internal Report, University of Groningen, Dept. of Applied Physics

The Principle of Generalized Energy Equivalence in Continuum Damage Mechanics

Dirk Reckwerth and Charalampos Tsakmakis

Institute of Mechanics, Darmstadt University of Technology
Hochschulstrasse 1, D–64289, Darmstadt, Germany
{reckwerth, tsakmakis}@mechanik.tu-darmstadt.de

received 12 Sep 2002 — accepted 12 Nov 2002

Abstract. Using the methods of continuum damage mechanics, constitutive models for elasto-plastic response and isotropic damage are derived. The approach is based on the concept of effective stress combined with a principle of generalized energy equivalence as explained in the paper. A family of yield functions parameterized by n, is obtained. If $n = 1$, then the yield function is identical to that commonly used in the damage model based on the strain equivalence hypothesis. Results of the analysis are so interpreted, that $n = 1$ should be favored, when modeling the damage response of metallic materials. This case is further investigated, by comparing predicted responses with those due to the model established in the framework of the strain equivalence principle. To this end the numerical method of finite elements is employed.

Keywords: Energy equivalence, continuum damage mechanics, strain equivalence

1 Introduction

Continuum damage mechanics models rely upon the works of Kachanov [1] and Rabotonov [2], who considered creep rupture of metals under uniaxial loading. These works have been later extended in the framework of irreversible thermodynamics in order to describe general three-dimensional loading processes (see e.g. the literature cited in Skrzypek [3] and Chaboche [4]). Generally, there are two concepts in common use when modeling damage effects within the continuum damage mechanics approach. The first one is the concept of effective stress combined with the strain equivalence hypothesis, which can be attributed to Lemaitre and Chaboche (see e.g. [4]-[9]). The second concept, first introduced by Cordebois and Sidoroff (see e.g. [10]), makes use of the notions of effective strain and effective stress and requires the hypothesis of energy equivalence. The present work is concerned with elasto-plastic material response exhibiting isotropic and kinematic hardening effects. The aim is to introduce a new version of the energy equivalence principle, which in the paper is referred to as generalized energy equivalence principle. For simplicity, the theory is presented for the case of isotropic damage. As a consequence, we obtain a family of yield functions parameterized by n. If $n = 1$, then the yield function coincides with that one commonly used by Chaboche (see e.g. [4]) in the framework of the strain-equivalence principle. This case is further investigated by comparing results according to both, the model derived from the proposed generalized energy equivalence principle and the model established in the framework of the strain equivalence principle. The results are established numerically by employing the finite element method. Generally, noticeable quantitative differences between the predicted responses may become visible. This is true, even if the material parameters are so chosen that the predicted uniaxial tensile strain-stress responses are nearly identical.

2 Preliminaries — Elasto–Plastic Materials Models

Throughout the paper, the underlying deformations are assumed to be small. We denote by $\mathbf{E}$ the linearized strain tensor, while the Cauchy stress tensor is denoted by $\mathbf{T}$. Only isothermal deformation processes with homogeneous temperature distribution will be considered, and the second law of thermodynamics is assumed in the form of the Clausius-Duhem inequality. Since the formulation is not affected by a space dependence, an explicit reference to space will be dropped. Commonly the same symbol is used to designate a function and the value of that function at a point. However, if we deal with different representations of the same function, then use will often be made of different symbols. We write $\dot{\varphi}(t)$ for the material time derivative of a function $\varphi(t)$, where t is the time. Moreover, second-order tensors are denoted by bold-face letters, whereas fourth-order tensors are represented by bold-face calligraphic letters. For two second-order tensors $\mathbf{A}$ and $\mathbf{B}$, we write $tr\ \mathbf{A}$ for the trace of $\mathbf{A}$, $\mathbf{A}^T$ for the transpose of $\mathbf{A}$, $\mathbf{A} \cdot \mathbf{B} = tr(\mathbf{A}\mathbf{B}^T)$ for the inner product between $\mathbf{A}$ and $\mathbf{B}$, $||\mathbf{A}|| = \sqrt{\mathbf{A} \cdot \mathbf{A}}$ for the Euclidian norm of $\mathbf{A}$, as well as $\mathbf{A} \otimes \mathbf{B}$ for the tensor product between $\mathbf{A}$ and $\mathbf{B}$. We will use the symbol $\mathbf{1}$ for the second-order identity tensor, so that $\mathbf{A}^D = \mathbf{A} - \frac{1}{3}(\mathrm{tr}\mathbf{A})\,\mathbf{1}$ is the deviator of $\mathbf{A}$, while $\boldsymbol{\mathcal{E}}$ is the fourth-order identity tensor operating in the space of all symmetric second-order tensors. The (undamaged) elasto-plastic materials with isotropic and kinematic hardening we deal with in the present paper are characterized by the following equations:

$$\mathbf{E} = \mathbf{E}_e + \mathbf{E}_p \ , \tag{1}$$

$$\Psi = \Psi_e + \Psi_p \ , \tag{2}$$

$$\Psi_e = \bar{\Psi}_e(\mathbf{E}_e) = \frac{1}{2\rho}\mathbf{E}_e \cdot \boldsymbol{\mathcal{C}}[\mathbf{E}_e] \ , \tag{3}$$

$$\boldsymbol{\mathcal{C}} = 2\mu\boldsymbol{\mathcal{E}} + \lambda \mathbf{1} \otimes \mathbf{1} \ , \tag{4}$$

$$\Psi_p = \Psi_p^{(kin)} + \Psi_p^{(is)} \tag{5}$$

$$\Psi_p^{(kin)} = \bar{\Psi}_p^{(kin)}(\mathbf{Y}) = \frac{c}{2\rho}\mathbf{Y}\cdot\mathbf{Y} \ , \tag{6}$$

$$\Psi_p^{(is)} = \bar{\Psi}_p^{(is)}(r) = \frac{\gamma}{2\rho}\left(r^2 + 2r_0 r\right) \ , \tag{7}$$

$$\boldsymbol{\xi} = \boldsymbol{\xi}(\mathbf{Y}) := \rho\frac{\partial\bar{\Psi}_p^{(kin)}}{\partial\mathbf{Y}} = c\mathbf{Y} \ , \tag{8}$$

$$R = R(r) := \rho\frac{\partial\bar{\Psi}_p^{(is)}}{\partial r} = \gamma(r + r_0) \ , \tag{9}$$

$$F = \bar{F}(\mathbf{T},\boldsymbol{\xi},R) := \bar{f}(\mathbf{T},\boldsymbol{\xi}) - k \quad \text{: yield function} \tag{10}$$

$$f = \bar{f}(\mathbf{T},\boldsymbol{\xi}) := \sqrt{\tfrac{3}{2}\left(\mathbf{T}-\boldsymbol{\xi}\right)^D\cdot\left(\mathbf{T}-\boldsymbol{\xi}\right)^D} \ , \tag{11}$$

$$k := R + k_0 = \gamma(r + r_0) + k_0 \ , \tag{12}$$

$$F = 0 \Leftrightarrow f = k \quad \text{: yield condition} \ , \tag{13}$$

$$\text{plastic loading} \ \Leftrightarrow \ F = 0 \ \& \ \left(\dot{F}\right)_{\mathbf{E}_p=\text{const}} > 0 \ , \tag{14}$$

Flow rule:

$$\dot{\mathbf{E}}_p = \begin{cases} \Lambda\dfrac{\partial\bar{f}}{\partial\mathbf{T}} = \Lambda\dfrac{\partial\bar{F}}{\partial\mathbf{T}} = \dfrac{3\Lambda}{2k}\left(\mathbf{T}-\boldsymbol{\xi}\right)^D & \text{for plastic loading} \\ 0 & \text{otherwise} \end{cases} \ , \tag{15}$$

$$\dot{s} := \sqrt{\frac{2}{3}\dot{\mathbf{E}}_p\cdot\dot{\mathbf{E}}_p} = \Lambda \ , \tag{16}$$

Clausius-Duhem inequality:

$$\mathcal{D}_{C-D} := \mathbf{T}\cdot\dot{\mathbf{E}} - \rho\dot{\Psi} = \mathbf{T}\cdot\dot{\mathbf{E}}_e + \mathbf{T}\cdot\dot{\mathbf{E}}_p - \rho\dot{\Psi}_e - \rho\dot{\Psi}_p \geq 0 \ , \tag{17}$$

$$\mathbf{T} = \rho\frac{\partial\bar{\Psi}_e}{\partial\mathbf{E}_e} = \boldsymbol{\mathcal{C}}\left[\mathbf{E}_e\right] \quad \text{: elasticity law,} \tag{18}$$

Dissipation inequality:

$$\begin{aligned} \mathcal{D}_d &:= \mathbf{T}\cdot\dot{\mathbf{E}}_p - \rho\dot{\Psi}_p = \mathbf{T}\cdot\dot{\mathbf{E}}_p - \boldsymbol{\xi}\cdot\dot{\mathbf{Y}} - R\dot{r} \\ &= \left(\mathbf{T}-\boldsymbol{\xi}\right)\cdot\dot{\mathbf{E}}_p - R\dot{r} + \boldsymbol{\xi}\cdot\left(\dot{\mathbf{E}}_p - \dot{\mathbf{Y}}\right) \geq 0 \ , \end{aligned} \tag{19}$$

$$\dot{\mathbf{Y}} = \dot{\mathbf{E}}_p - b\Lambda\boldsymbol{\xi} = \dot{\mathbf{E}}_p - b\dot{s}\boldsymbol{\xi} \quad \text{: kinematic hardening rule} \ , \tag{20}$$

$$\dot{r} = \left(1 - \beta r\right)\Lambda = \left(1 - \beta\frac{R - \gamma r_0}{\gamma}\right)\dot{s} \quad \text{: isotropic hardening rule} \ . \tag{21}$$

Here, (1) is the decomposition of the strain into elastic and plastic parts, Ψ is the specific free energy, ρ is the mass density and μ, λ are the elasticity parameters. Isotropic and kinematic hardening are described by the internal variables r and $\mathbf{Y}$, which represent a scalar strain ($r \geq 0$) and a second order strain tensor, respectively. From (8), (9), we recognize that $\boldsymbol{\xi}$ and R are thermodynamically conjugated stresses to $\mathbf{Y}$ and r, respectively. Often, $\boldsymbol{\xi}$ is called the back stress tensor. Equations (10)-(13) define the yield condition, with k denoting the yield stress. In (7)-(12), γ, R_0, c, k_0 are material parameters, while r_0 represents an initial strain, which causes the initial scalar stress $R_0 = \gamma r_0$. Such initial stresses are important when modeling the energy stored in some metallic materials due to rearrangements of dislocations during plastic flow (cf. also Chaboche [11], [12]). The response of the plastic strain $\mathbf{E}_p$ is governed by the flow rule, which has the form of an associated normality rule. For the case of rate-independent plasticity, we deal with in this paper, the scalar multiplier Λ has to be determined from the so-called consistency condition $\dot{F} = 0$. Plastic flow is involved if the loading criteria (14) are satisfied. (As shown in [13], these loading criteria apply for both strain and stress controlled loading processes.) Using standard arguments, one may prove that the elasticity law (18) and the dissipation inequality (19) are sufficient conditions for the validity of the Clausius-Duhem inequality (17). Finally, (20) and (21) are evolution equations governing the response of the kinematic and isotropic hardening, respectively, with β, b being nonnegative material parameters.

To see that (20) and (21) does not contradict the dissipation inequality, we insert in (19) the material time derivative of (6) and (7) to obtain

$$\mathcal{D}_d = \dot{s} k_0 + R\,(\dot{s} - \dot{r}) + \boldsymbol{\xi} \cdot \left(\dot{\mathbf{E}}_p - \dot{\mathbf{Y}}\right) \geq 0\ . \tag{22}$$

Since $\dot{s} \geq 0$, we have $\dot{s}k_0 \geq 0$. We recall that $r \geq 0$ has to hold, which implies $R \geq 0$, in view of (9). Therefore, the relations

$$\dot{s} - \dot{r} \geq 0\ , \tag{23}$$

$$\boldsymbol{\xi} \cdot \left(\dot{\mathbf{E}}_p - \dot{\mathbf{Y}}\right) \geq 0\ , \tag{24}$$

are sufficient conditions for (22) to hold. On the other hand, these inequalities are satisfied if the relations

$$\dot{\mathbf{E}}_p - \dot{\mathbf{Y}} = \Lambda b \boldsymbol{\xi}\ , \tag{25}$$

$$\dot{s} - \dot{r} = \Lambda \beta r \tag{26}$$

apply, the latter being identical to (20), (21). Note that the evolution equation (20), together with homogeneous initial conditions renders $\mathbf{Y}$, and therefore $\boldsymbol{\xi}$ too, to be deviatoric.
Alternatively, and equivalently, one may introduce a so-called dissipation potential

$$\varphi = \bar{\varphi}(\mathbf{T}, \boldsymbol{\xi}, R) = \bar{F}(\mathbf{T}, \boldsymbol{\xi}, R) + \frac{1}{2} b\, \boldsymbol{\xi} \cdot \boldsymbol{\xi} + \frac{1}{2} \frac{\beta}{\gamma} (R - \gamma r_0)^2\ , \tag{27}$$

where $\bar{\varphi}(\mathbf{T}, \boldsymbol{\xi}, R)$ is a continuously differentiable, convex, scalar valued function, enclosing the origin of the space of its variables. Then, the dissipation inequality (19) is automatically satisfied if the normality conditions

$$\dot{\mathbf{E}}_p = \Lambda \frac{\partial \bar{\varphi}}{\partial \mathbf{T}} = \Lambda \frac{\partial \bar{F}}{\partial \mathbf{T}} , \tag{28}$$

$$\dot{\mathbf{Y}} = -\Lambda \frac{\partial \bar{\varphi}}{\partial \boldsymbol{\xi}} = -\Lambda \left(-\frac{\partial \bar{f}}{\partial \mathbf{T}} + b\boldsymbol{\xi} \right) = \dot{\mathbf{E}}_p - b\Lambda\boldsymbol{\xi} , \tag{29}$$

$$\dot{r} = -\Lambda \frac{\partial \bar{\varphi}}{\partial R} = -\Lambda \left(-1 + \frac{\beta}{\gamma}(R - R_0) \right) = (1 - \beta r)\Lambda , \tag{30}$$

hold. Actually, this is the approach advocated by Lemaitre and Chaboche [8].

Before going any further, it should be mentioned, that the plasticity model defined by (1)-(21) has been discussed intensively by Chaboche (see e.g. [11], [14]) and is often attributed to him. Also, the evolution equation (20) is equivalent to the so-called Armstrong-Frederick hardening rule (see [15]).

3 Concepts of Continuum Damage Mechanic

3.1 Basic Relations

The continuum damage approach relies upon the assumption that the set of internal state variables given in Sect. 2, is amplified by a damage variable. In the case of isotropic damage, we deal with in the present paper, the damage variable is scalar valued and is denoted by D. It is assumed that $D \in [0, 1]$. The values $D = 0$ and $D = 1$ correspond to the undamaged state and the complete local rupture, respectively, while $D \in (0, 1)$ reflects a partially damaged state. In what follows we shall use the same nomenclature as in Sect. 2.

Common features in the theories of continuum damage mechanics, as regarded here, are the decomposition of strain

$$\mathbf{E} = \mathbf{E}_e + \mathbf{E}_p , \tag{31}$$

as well as the existence of a specific free energy, which admits the representations

$$\Psi = \hat{\Psi}(\mathbf{E}_e, \mathbf{Y}, r, D) = \Psi_e + \Psi_p \ , \ \Psi_p = \Psi_p^{(kin)} + \Psi_p^{(is)} , \tag{32}$$

$$\Psi_e = \hat{\Psi}_e(\mathbf{E}_e, D) \ , \ \Psi_p^{(kin)} = \hat{\Psi}_p^{(kin)}(\mathbf{Y}, D) \ , \ \Psi_p^{(is)} = \hat{\Psi}_p^{(is)}(r, D) . \tag{33}$$

Sufficient conditions for the Clausius-Duhem inequality

$$\mathcal{D}_{C-D} = \mathbf{T} \cdot \dot{\mathbf{E}} - \rho\dot{\Psi}_e - \rho\dot{\Psi}_p^{(kin)} - \rho\dot{\Psi}_p^{(is)} \geq 0 \tag{34}$$

to hold are the elasticity law

$$\mathbf{T} = \rho \frac{\partial \hat{\Psi}_e}{\partial \mathbf{E}_e} , \tag{35}$$

together with the dissipation inequality

$$\mathcal{D}_d := \mathbf{T} \cdot \dot{\mathbf{E}}_p - \boldsymbol{\xi} \cdot \dot{\mathbf{Y}} - R\dot{r} - \rho \frac{\partial \hat{\Psi}}{\partial D} \dot{D} \geq 0 , \tag{36}$$

where

$$\boldsymbol{\xi} := \rho \frac{\partial \hat{\Psi}_p^{(kin)}}{\partial \mathbf{Y}} , \quad R := \rho \frac{\partial \hat{\Psi}_p^{(is)}}{\partial r} . \tag{37}$$

Clearly,

$$\begin{aligned} \mathcal{D}_{dp} &:= \mathbf{T} \cdot \dot{\mathbf{E}}_p - \boldsymbol{\xi} \cdot \dot{\mathbf{Y}} - R\dot{r} \\ &= (\mathbf{T} - \boldsymbol{\xi}) \cdot \dot{\mathbf{E}}_p - R\dot{r} + \boldsymbol{\xi} \cdot (\dot{\mathbf{E}}_p - \dot{\mathbf{Y}}) \geq 0 , \end{aligned} \tag{38}$$

$$\mathcal{D}_{dd} := -\rho \frac{\partial \hat{\Psi}}{\partial D} \dot{D} \geq 0 , \tag{39}$$

are sufficient conditions for the validity of (36). Inequalities (38), (39) describe the dissipation parts due to plastic flow and damage evolution, respectively. As in the theory without damage (see Sect. 2), (38) may be satisfied by requiring the normality conditions

$$\dot{\mathbf{E}}_p = \Lambda \frac{\partial \hat{\varphi}}{\partial \mathbf{T}} , \quad \dot{\mathbf{Y}} = -\Lambda \frac{\partial \hat{\varphi}}{\partial \boldsymbol{\xi}} , \quad \dot{R} = -\Lambda \frac{\partial \hat{\varphi}}{\partial r} , \tag{40}$$

where

$$\varphi = \hat{\varphi}(\mathbf{T}, \boldsymbol{\xi}, R, D) \tag{41}$$

represents a convex dissipation potential. Plastic flow is defined to occur if the condition for plastic loading

$$\text{plastic loading} \Leftrightarrow F = 0 \ \& \ (\dot{F})_{\mathbf{E}_p = \text{const}} \geq 0 , \tag{42}$$

$$F = \hat{F}(\mathbf{T}, \boldsymbol{\xi}, R, D) = \hat{f}(\mathbf{T}, \boldsymbol{\xi}, D) - \hat{k}(R, D) , \tag{43}$$

applies, where $\hat{F}(\mathbf{T}, \boldsymbol{\xi}, R, D)$ denotes the yield function. Of course, in the case

$$\frac{\partial \hat{\varphi}}{\partial \mathbf{T}} = \frac{\partial \hat{F}}{\partial \mathbf{T}} \tag{44}$$

the flow rule (40) represents an associated normality rule.

As mentioned in the introduction, the concepts of effective stress combined with the strain equivalence principle and the energy equivalence principle are often utilized in continuum damage mechanics. When isotropic damage is regarded, the theory based on to the strain equivalence principle has been proved to be a very efficient tool for modeling the constitutive behaviour of ductile materials. Therefore, it is of interest here to compare responses predicted by this theory with those predicted by the constitutive theory according to the generalized energy equivalence, to be established below. To this end, we shall sketch briefly in the next section a well established damage model, which results from the strain equivalence principle and is attributed to Chaboche (see. e.g. [4]). In order to render the present work self-contained we shall derive this model from the constitutive relations given in Sect. 2.

3.2 Principle of Strain Equivalence

Continuum damage models which rest upon the strain equivalence principle, and are coupled with plastic material behaviour, have been initiated and intensively investigated by Lemaitre and Chaboche (see e.g. [4], [7]). As mentioned, we shall be concerned here with the theory proposed by Chaboche.

With reference to the plasticity model in Sect. 2, the essential features of this theory can be summarized as follows.

Let the free energy parts $\hat{\Psi}_e$, $\hat{\Psi}_p^{(kin)}$, $\hat{\Psi}_p^{(is)}$ in (32), (33) be represented by

$$\hat{\Psi}_e\left(\mathbf{E}_e, D\right) = \frac{(1-D)}{2\rho}\mathbf{E}_e \cdot \boldsymbol{C}\left[\mathbf{E}_e\right] \ , \tag{45}$$

$$\hat{\Psi}_p^{(kin)}(\mathbf{Y}, D) = \frac{c(1-D)}{2\rho}\mathbf{Y} \cdot \mathbf{Y} \ , \tag{46}$$

$$\hat{\Psi}_p^{(is)}(r, D) = \frac{\gamma(1-D)}{2\rho}\left(r^2 + 2r_0 r\right) \ . \tag{47}$$

Hence, in view of (35), (37),

$$\mathbf{T} = (1-D)\boldsymbol{C}[\mathbf{E}_e] \ , \ \boldsymbol{\xi} = (1-D)c\mathbf{Y} \ , \ R = (1-D)\gamma(r+r_0) \ , \tag{48}$$

which yield

$$\tilde{\mathbf{T}} := \frac{\mathbf{T}}{1-D} = \boldsymbol{C}[\mathbf{E}_e] \ , \ \tilde{\boldsymbol{\xi}} := \frac{\boldsymbol{\xi}}{1-D} = c\mathbf{Y} \ , \ \tilde{R} := \frac{R}{1-D} = \gamma(r+r_0) \ . \tag{49}$$

The variables $\tilde{\mathbf{T}}$, $\tilde{\boldsymbol{\xi}}$, $\tilde{R}$ are called effective stress variables. By comparing $(49)_1$, $(49)_2$, $(49)_3$ with (18), (8), (9), respectively, we recognize that the strain-stress relations (49), for the damaged material, result from those in Sect. 2 for the undamaged material, if there the stress variables $\mathbf{T}$, $\boldsymbol{\xi}$, R are replaced by the corresponding effective stress variables $\tilde{\mathbf{T}}$, $\tilde{\boldsymbol{\xi}}$, $\tilde{R}$, whereas the strain variables

$\mathbf{E}_e$, $\mathbf{Y}$, r and the parameter r_0 remain the same (strain equivalence). This motivates to assume the yield function $\hat{F}$ in (43) and the dissipation potential $\hat{\varphi}$ in (41) to be equal to $\bar{F}$ in (10) and $\bar{\varphi}$ in (27), respectively, but with $\mathbf{T}$, $\boldsymbol{\xi}$, R replaced by $\tilde{\mathbf{T}}$, $\tilde{\boldsymbol{\xi}}$, $\tilde{R}$:

$$\begin{aligned} F = \hat{F}(\mathbf{T}, \boldsymbol{\xi}, R, D) &= \bar{F}(\tilde{\mathbf{T}}, \tilde{\boldsymbol{\xi}}, \tilde{R}) = \bar{f}(\tilde{\mathbf{T}}, \tilde{\boldsymbol{\xi}}) - \tilde{R} - k_0 \\ &= \hat{f}(\mathbf{T}, \boldsymbol{\xi}, D) - \frac{R}{1-D} - k_0 \,, \end{aligned} \tag{50}$$

$$\begin{aligned} \hat{f}(\mathbf{T}, \boldsymbol{\xi}, D) &= \bar{f}(\tilde{\mathbf{T}}, \tilde{\boldsymbol{\xi}}) = \sqrt{\frac{3}{2} \left(\tilde{\mathbf{T}} - \tilde{\boldsymbol{\xi}}\right)^D \cdot \left(\tilde{\mathbf{T}} - \tilde{\boldsymbol{\xi}}\right)^D} \\ &= \sqrt{\frac{3}{2} \frac{(\mathbf{T} - \boldsymbol{\xi})^D}{1-D} \cdot \frac{(\mathbf{T} - \boldsymbol{\xi})^D}{1-D}} \,, \end{aligned} \tag{51}$$

$$\begin{aligned} \hat{\varphi}(\mathbf{T}, \boldsymbol{\xi}, R, D) &= \bar{\varphi}(\tilde{\mathbf{T}}, \tilde{\boldsymbol{\xi}}, \tilde{R}) \\ &= \bar{F}(\tilde{\mathbf{T}}, \tilde{\boldsymbol{\xi}}, \tilde{R}) + \frac{1}{2} b\, \tilde{\boldsymbol{\xi}} \cdot \tilde{\boldsymbol{\xi}} + \frac{1}{2} \frac{\beta}{\gamma} \left(\tilde{R} - \gamma r_0\right)^2 \\ &= \hat{F}(\mathbf{T}, \boldsymbol{\xi}, R, D) + \frac{b}{2(1-D)^2} \boldsymbol{\xi} \cdot \boldsymbol{\xi} \\ &\quad + \frac{1}{2} \frac{\beta}{\gamma} \left(\frac{R}{1-D} - \gamma r_0\right)^2 \,. \end{aligned} \tag{52}$$

Actually, it is not difficult to prove that $\hat{F}$ in (50) and $\hat{\varphi}$ in (52) are convex functions of their arguments. Thus the evolution laws (40) become

$$\dot{\mathbf{E}}_p = \frac{3\Lambda}{2(1-D)\hat{f}} \frac{(\mathbf{T} - \boldsymbol{\xi})^D}{1-D} = \frac{3\dot{s}}{2\hat{f}} \frac{(\mathbf{T} - \boldsymbol{\xi})^D}{1-D} \,, \tag{53}$$

$$\dot{s} := \sqrt{\frac{3}{2} \dot{\mathbf{E}}_p \cdot \dot{\mathbf{E}}_p} = \frac{\Lambda}{1-D} \,, \tag{54}$$

$$\dot{\mathbf{Y}} = \dot{\mathbf{E}}_p - b\dot{s} \frac{\boldsymbol{\xi}}{1-D} = \dot{\mathbf{E}}_p - bc\dot{s}\mathbf{Y} = \dot{\mathbf{E}}_p - bc \frac{\Lambda}{1-D} \mathbf{Y} \,, \tag{55}$$

$$\dot{r} = (1 - \beta r)\dot{s} = (1 - \beta r) \frac{\Lambda}{1-D} \,. \tag{56}$$

It still remains to check inequality (39). Keeping in mind (45)-(47),

$$\begin{aligned} -\rho \frac{\partial \hat{\Psi}}{\partial D} &= \frac{1}{2} \left(\mathbf{E}_e \cdot \boldsymbol{\mathcal{C}} [\mathbf{E}_e] + c\, \mathbf{Y} \cdot \mathbf{Y} + \gamma (r^2 + 2 r_0 r)\right) \\ &= \rho \left(\bar{\Psi}_e (\mathbf{E}_e) + \bar{\Psi}_p^{(kin)} (\mathbf{Y}) + \bar{\Psi}_p^{(is)} (r)\right) \end{aligned} \tag{57}$$

which is always nonnegative. This means that (39) will be satisfied in every admissible process if

$$\dot{D} \geq 0 \,. \tag{58}$$

In other words, any evolution equation rendering D to be a monotonically increasing function of time, will be compatible with (39). Now, it is a straight forward matter to show that for uniaxial, tensile loading conditions, the model established in this section implies the following equations :

$$\varepsilon = \varepsilon_e + \varepsilon_p \ , \tag{59}$$

$$\sigma = (1 - D)E\varepsilon_e \ , \tag{60}$$

$$\sigma - \frac{3}{2}\xi = R + (1 - D)k_0 \ , \tag{61}$$

$$\xi = (1 - D)cy \ , \tag{62}$$

$$R = (1 - D)\gamma(r + r_0) \ , \tag{63}$$

$$\dot{y} = (1 - bcy)\dot{\varepsilon}_p \ , \tag{64}$$

$$\dot{r} = (1 - \beta r)\dot{\varepsilon_p} \ . \tag{65}$$

Here, $E = \frac{\mu(3\lambda+2\mu)}{\lambda+\mu}$ is the Young's modulus, while σ, ε, ε_e, ε_p, ξ, y are the uniaxial components of $\mathbf{T}$, $\mathbf{E}$, $\mathbf{E}_e$, $\mathbf{E}_p$, $\boldsymbol{\xi}$, $\mathbf{Y}$, respectively. In order to obtain an insight into essential features of the model, it is sufficient to accomplish (59)-(65) by the simple damage law

$$D = \left(\frac{\varepsilon_p}{\alpha}\right)^2 \ , \tag{66}$$

as indicated by Chaboche [4]. Fig. 1. shows the strain-stress response according to these equations, for the material parameters given in Table 1. Moreover, the case of ideal plasticity ($R = y = 0$) is represented, for which (59)-(65) imply

$$\varepsilon_e = \frac{k_0}{E} = \text{const} \tag{67}$$

during plastic flow. It is readily seen, that for plasticity with isotropic and

Table 1. Material parameters used in (59)-(65) in order to produce the strain-stress response with isotropic and kinematic hardening illustrated in Fig 1 .

E[MPa]	k_0[MPa]	r_0	c[MPa]	γ[MPa]	b[MPa]	β[MPa]	α
200000	200	0	30000	30000	10	10	0,02

kinematic hardening, the elastic strain during plastic flow becomes

$$\varepsilon_e = \frac{3cy}{2E} + \frac{\gamma r}{E} + \frac{k_0}{E} \ . \tag{68}$$

From (64), (65), we recognize that r, y are monotonically increasing functions of ε_p. Consequently, ε_e is a monotonically increasing function of ε_p as well,

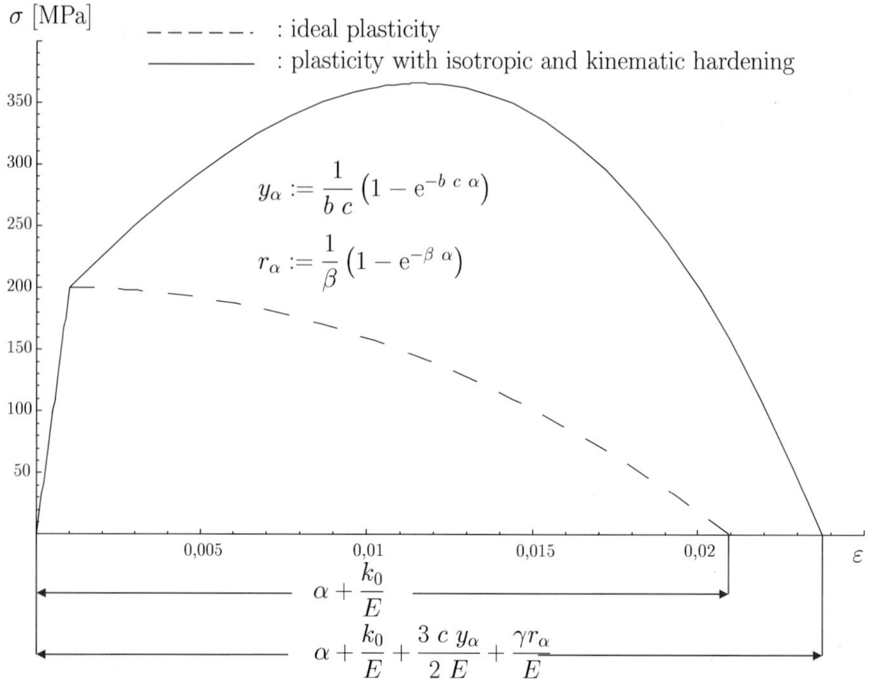

Fig. 1. Strain-stress graphs for uniaxial tensile loading. The constitutive theory is based on the strain equivalence principle according to Chaboche [4].

independent of the material parameters involved in the hardening rules.

Before going any further, we remark that the model in this section has been discussed by Lämmer and Tsakmakis [16] with reference to further models derived from the strain equivalence principle. Moreover, these authors have shown how the theory may be extended to finite deformations in a thermodynamically consistent way.

3.3 Proposed Principle of Generalized Energy Equivalence

3.3.1 Thermomechanical Framework

Damage theories based on energy equivalence have been initiated by Cordebois and Sidoroff (see [10]). These authors proposed a continuum damage mechanics theory for an elasto-plastic material which is embedded in the framework of irreversible thermodynamics with internal state variables. Energy equivalence with an undamaged model material is taken for the elastic part of the free energy only. Saanouni et al. [17] and Chow and Lu [18] extended the theory of Cordebois and Sidoroff to obtain also energy equivalence with respect to the whole energy stored in the material and the plastic part of the stress power, respectively. Moreover, Saanouni et al. [17] derived the evolution equations for the internal state variables by using a dissipation potential expressed in terms of appropriately defined effective stress variables.

In this section we will present a new version of continuum damage theory based on an energy equivalence principle, which we refer to as the generalized energy equivalence principle. According to this approach the unknown constitutive law for the real, damaged material is assumed to depend on the damage variable D, as explained in Sect. 3.1. At every material point, we assign to the real material a fictitious, undamaged material with a known constitutive law. The latter does not include directly damage variables, but is expressed in terms of so-called effective state variables, which are suitably defined. The unknown constitutive law governing the strain-stress response of the real material is then determinable on the basis of this attribution.

In the following, the constitutive equations for the associated fictitious material are assumed to be the same as those for the elastic plastic material given in Sect. 2. If X is a state variable for the real material, then we will write $\tilde{X}$ for the corresponding effective state variable related to the fictitious material. Especially, we shall be concerned with the effective variables $\tilde{\mathbf{E}}$, $\tilde{\mathbf{E}}_e$, $\tilde{\mathbf{E}}_p$, $\tilde{\mathbf{Y}}$, $\tilde{r}$, $\tilde{\mathbf{T}}$, $\tilde{\boldsymbol{\xi}}$, $\tilde{R}$, and the effective initial strain $\tilde{r}_0$. Also, we will designate (constitutive) functions by the symbol (f) to denote that these are referred to the fictitious material. (No use of particular symbols is made for the functions attributed to the real material). This way, the fictitious elastic plastic model material, we deal with, is defined by the following equations (cf. Sect.2):

$$\tilde{\mathbf{E}} = \tilde{\mathbf{E}}_e + \tilde{\mathbf{E}}_p \; , \tag{69}$$

$$\Psi^{(f)} = \Psi_e^{(f)} + \Psi_p^{(f)} \; , \quad \Psi_p^{(f)} = \Psi_p^{(f)kin} + \Psi_p^{(f)is} \; , \tag{70}$$

$$\Psi_e^{(f)} = \bar{\Psi}_e^{(f)}(\tilde{\mathbf{E}}_e) = \frac{1}{2\rho}\tilde{\mathbf{E}}_e \cdot \boldsymbol{C}[\tilde{\mathbf{E}}_e] \; , \tag{71}$$

$$\Psi_p^{(f)kin} = \bar{\Psi}_p^{(f)kin}(\tilde{\mathbf{Y}}) = \frac{c}{2\rho}\tilde{\mathbf{Y}} \cdot \tilde{\mathbf{Y}} \; , \tag{72}$$

$$\Psi_p^{(f)is} = \bar{\Psi}_p^{(f)is}(\tilde{r}) = \frac{\gamma}{2\rho}(\tilde{r}^2 + 2\tilde{r}_0\tilde{r}) \; , \tag{73}$$

$$\tilde{\mathbf{T}} = \rho\frac{\partial \bar{\Psi}_e^{(f)}}{\partial \tilde{\mathbf{E}}_e} = \boldsymbol{C}[\tilde{\mathbf{E}}_e] \; , \tag{74}$$

$$\tilde{\boldsymbol{\xi}} = \rho\frac{\partial \bar{\Psi}_p^{(f)kin}}{\partial \tilde{\mathbf{Y}}} = c\tilde{\mathbf{Y}} \; , \tag{75}$$

$$\tilde{R} = \rho\frac{\partial \bar{\Psi}_p^{(f)is}}{\partial \tilde{r}} = \gamma(\tilde{r} + \tilde{r}_0) \; , \tag{76}$$

$$\begin{aligned}\varphi^{(f)} &= \bar{\varphi}^{(f)}(\tilde{\mathbf{T}}, \tilde{\boldsymbol{\xi}}, \tilde{R}) \\ &= \bar{F}^{(f)}(\tilde{\mathbf{T}}, \tilde{\boldsymbol{\xi}}, \tilde{R}) + \frac{1}{2} b \tilde{\boldsymbol{\xi}} \cdot \tilde{\boldsymbol{\xi}} + \frac{1}{2} \frac{\beta}{\gamma} (\tilde{R} - \gamma \tilde{r}_0)^2 \ , \end{aligned} \tag{77}$$

$$\bar{F}^{(f)} = \bar{f}^{(f)}(\tilde{\mathbf{T}}, \tilde{\boldsymbol{\xi}}) - \tilde{R} - k_0 \ , \tag{78}$$

$$\bar{f}^{(f)}(\tilde{\mathbf{T}}, \tilde{\boldsymbol{\xi}}) = \sqrt{\frac{3}{2} (\tilde{\mathbf{T}} - \tilde{\boldsymbol{\xi}})^D \cdot (\tilde{\mathbf{T}} - \tilde{\boldsymbol{\xi}})^D} \ , \tag{79}$$

$$\dot{\tilde{\mathbf{E}}}_p = \Lambda^{(f)} \frac{\partial \bar{\varphi}^{(f)}}{\partial \tilde{\mathbf{T}}} = \Lambda^{(f)} \frac{\partial \bar{F}^{(f)}}{\partial \tilde{\mathbf{T}}} = \frac{3 \Lambda^{(f)}}{2 \bar{f}^{(f)}} (\tilde{\mathbf{T}} - \tilde{\boldsymbol{\xi}})^D \ , \tag{80}$$

$$\dot{\tilde{s}} = \sqrt{\frac{2}{3} \dot{\tilde{\mathbf{E}}} \cdot \dot{\tilde{\mathbf{E}}}} = \Lambda^{(f)} \ , \tag{81}$$

$$\dot{\tilde{\mathbf{Y}}} = -\Lambda^{(f)} \frac{\partial \bar{\varphi}^{(f)}}{\partial \tilde{\boldsymbol{\xi}}} = \dot{\tilde{\mathbf{E}}}_p - b \dot{\tilde{s}} \tilde{\boldsymbol{\xi}} \ , \tag{82}$$

$$\dot{\tilde{r}} = (1 - \beta \tilde{r}) \dot{\tilde{s}} \ , \tag{83}$$

$$\mathcal{D}_d^{(f)} = \tilde{\mathbf{T}} \cdot \dot{\tilde{\mathbf{E}}}_p - \tilde{\boldsymbol{\xi}} \cdot \dot{\tilde{\mathbf{Y}}} - \tilde{R} \dot{\tilde{r}} \ , \tag{84}$$

where $\tilde{\rho} \equiv \rho$ has been used. Following Cordebois and Sidoroff [10], as well as Saanouni et al. [17], we define the effective stress variables by

$$\tilde{\mathbf{T}} := \frac{\mathbf{T}}{\sqrt{1 - D}} \ , \ \tilde{\boldsymbol{\xi}} := \frac{\boldsymbol{\xi}}{\sqrt{1 - D}} \ , \ \tilde{R} := \frac{R}{\sqrt{1 - D}} \ . \tag{85}$$

The generalized energy equivalence principle assumed in the present paper postulates two energy like quantities for the real and the fictitious material to be related to each other. On the one hand, similar to Saanouni et al. [17], the free energy parts for the real material $\hat{\Psi}_e$, $\hat{\Psi}_p^{(kin)}$, $\hat{\Psi}_p^{(is)}$ (cf. (32),(33)) are supposed to be equal to the free energy parts $\bar{\Psi}_e^{(f)}$, $\bar{\Psi}_p^{(f)kin}$, $\bar{\Psi}_p^{(f)is}$, respectively, the latter being given by (71)-(73):

$$\hat{\Psi}_e(\mathbf{E}_e, D) = \bar{\Psi}_e^{(f)}(\tilde{\mathbf{E}}_e) = \frac{1}{2\rho} \tilde{\mathbf{E}}_e \cdot \boldsymbol{\mathcal{C}}[\tilde{\mathbf{E}}_e] \ , \tag{86}$$

$$\hat{\Psi}_p^{(kin)}(\mathbf{Y}, D) = \bar{\Psi}_p^{(f)kin}(\tilde{\mathbf{Y}}) = \frac{c}{2\rho} \tilde{\mathbf{Y}} \cdot \tilde{\mathbf{Y}} \ , \tag{87}$$

$$\hat{\Psi}_p^{(is)}(r, D) = \bar{\Psi}_p^{(f)is}(\tilde{r}) = \frac{\gamma}{2\rho} (\tilde{r}^2 + 2 \tilde{r}_0 \tilde{r}) \ . \tag{88}$$

These equations will enable to determine effective strains $\tilde{\mathbf{E}}_e = \tilde{\mathbf{E}}_e(\mathbf{E}, D)$, $\tilde{\mathbf{Y}}_e = \tilde{\mathbf{Y}}_e(\mathbf{Y}, D)$, $\tilde{r} = \tilde{r}(r, D)$. On the other hand, the dissipation power for the real material $\mathcal{D}_{dp}$ is assumed to be connected, by means of an appropriate constitutive function, with the dissipation power for the fictitious material $\mathcal{D}^{(f)}$. From this connection it will be possible to obtain the yield function for

the real material, as well as evolution equations for the internal strains $\mathbf{E}_p$, $\mathbf{Y}$, r.
To elaborate, let $\tilde{\mathbf{E}}_e$ satisfy a relation of the form $\tilde{\mathbf{E}}_e = \lambda \mathbf{E}_e$, where λ is a scalar valued function of state. Then, by virtue of (35), (74), (86),

$$\tilde{\mathbf{T}} = \frac{\mathbf{T}}{\sqrt{1-D}} = \rho \frac{\partial \bar{\Psi}_e^{(f)}}{\partial \tilde{\mathbf{E}}_e} = \frac{\rho}{\lambda} \frac{\partial \hat{\Psi}_e}{\partial \mathbf{E}_e} = \frac{1}{\lambda} \mathbf{T} , \tag{89}$$

which furnishes $\lambda = \sqrt{1-D}$, and hence

$$\left. \begin{aligned} \tilde{\mathbf{E}}_e &= \sqrt{1-D}\mathbf{E}_e , \\ \hat{\Psi}_e(\mathbf{E}_e, D) &= \tfrac{(1-D)}{2\rho} \mathbf{E}_e \cdot \mathcal{C}[\mathbf{E}_e] , \\ \mathbf{T} &= (1-D)\mathcal{C}[\mathbf{E}_e] . \end{aligned} \right\} \tag{90}$$

This means, the elasticity laws for the two stress tensors $\mathbf{T}$ and $\tilde{\mathbf{T}}$, together with energy equality (86), enables us to determine the effective strain tensor $\tilde{\mathbf{E}}_e$. Using similar steps, one may find out the relations

$$\left. \begin{aligned} \tilde{\mathbf{Y}} &= \sqrt{1-D}\mathbf{Y} , \\ \hat{\Psi}_p^{(kin)}(\mathbf{Y}, D) &= \tfrac{c(1-D)}{2\rho} \mathbf{Y} \cdot \mathbf{Y} , \\ \boldsymbol{\xi} &= (1-D)c\mathbf{Y} , \end{aligned} \right\} \tag{91}$$

$$\left. \begin{aligned} \tilde{r} = \sqrt{1-D}r \ , \ \tilde{r}_0 &= \sqrt{1-D}r_0 , \\ \hat{\Psi}_p^{(is)}(r, D) &= \tfrac{\gamma(1-D)}{2\rho}(r^2 + 2r_0 r) , \\ R &= (1-D)\gamma(r + r_0) . \end{aligned} \right\} \tag{92}$$

Next we have to stipulate, with reference to the fictitious material, how the real material will dissipate energy. We recall from (75), $(37)_1$ and the equality(87), that a relation of the form $\tilde{\boldsymbol{\xi}} \cdot \dot{\tilde{\mathbf{Y}}} = \boldsymbol{\xi} \cdot \dot{\mathbf{Y}}$ cannot be required generally in the framework of an energy equivalence principle. Indeed, such a relation implies $\dot{D} \equiv 0$, i.e. the damage state will remain constant always. Similar arguments hold for isotropic hardening. Because of this, we generally assume the existence of a nonnegative scalar function χ, which may depend on state variables and their rates, so that

$$\tilde{\mathbf{T}} \cdot \dot{\tilde{\mathbf{E}}}_p = \chi \mathbf{T} \cdot \dot{\mathbf{E}}_p \ , \ \tilde{\boldsymbol{\xi}} \cdot \dot{\tilde{\mathbf{Y}}} = \chi \boldsymbol{\xi} \cdot \dot{\mathbf{Y}} \ , \ \tilde{R}\dot{\tilde{r}} = \chi R \dot{r} , \tag{93}$$

and therefore

$$\mathcal{D}_{dp} = \frac{1}{\chi}\left(\tilde{\mathbf{T}} \cdot \dot{\tilde{\mathbf{E}}}_p - \tilde{\boldsymbol{\xi}} \cdot \dot{\tilde{\mathbf{Y}}} - \tilde{R}\dot{\tilde{r}}\right) = \frac{1}{\chi}\mathcal{D}_d^{(f)} . \tag{94}$$

In other words, (93) controls the way, with respect to the fictitious material, in which the real material will dissipate plastic power. There are just these equations which characterize our generalized energy equivalence principle.

3.3.2 Yield Function-Flow Rule

We recall that the yield function for the real material is expected to have the form (43), but for the present it is not specified further. Now, we suppose $\bar{f}^{(f)}(\tilde{\mathbf{T}}, \tilde{\boldsymbol{\xi}}, \tilde{R})$ in (79) to be only a plastic potential for the strain $\tilde{\mathbf{E}}_p$, which is in agreement with (80). Also we suppose $\mathbf{E}_p$ to obey an associated normality rule, i.e.

$$\dot{\mathbf{E}}_p = \Lambda \frac{\partial \hat{F}(\mathbf{T}, \boldsymbol{\xi}, R, D)}{\partial \mathbf{T}} \, . \tag{95}$$

Then, by virtue of $(85)_1$, $(93)_1$, (80), (78), (95)

$$\begin{aligned} \tilde{\mathbf{T}} \cdot \dot{\tilde{\mathbf{E}}}_p &= \frac{\mathbf{T}}{\sqrt{1-D}} \cdot \left(\Lambda^{(f)} \frac{\partial \bar{f}^{(f)}}{\partial \tilde{\mathbf{T}}} \right) = \Lambda^{(f)} \mathbf{T} \cdot \frac{\partial \hat{f}}{\partial \mathbf{T}} \\ &= \chi \mathbf{T} \cdot \dot{\mathbf{E}}_p = \chi \Lambda \mathbf{T} \cdot \frac{\partial \hat{F}}{\partial \mathbf{T}} \, , \end{aligned} \tag{96}$$

where use has been made of the relations

$$\begin{aligned} \hat{f}(\mathbf{T}, \boldsymbol{\xi}, D) := \bar{f}^{(f)}(\tilde{\mathbf{T}}, \tilde{\boldsymbol{\xi}}) &= \sqrt{\frac{3}{2} \left(\tilde{\mathbf{T}} - \tilde{\boldsymbol{\xi}} \right)^D \cdot \left(\tilde{\mathbf{T}} - \tilde{\boldsymbol{\xi}} \right)^D} \\ &= \sqrt{\frac{3}{2} \frac{(\mathbf{T} - \boldsymbol{\xi})^D}{\sqrt{1-D}} \cdot \frac{(\mathbf{T} - \boldsymbol{\xi})^D}{\sqrt{1-D}}} \, , \end{aligned} \tag{97}$$

$$\frac{\partial \bar{f}^{(f)}(\tilde{\mathbf{T}}, \tilde{\boldsymbol{\xi}})}{\partial \tilde{\mathbf{T}}} = \sqrt{1-D} \, \frac{\partial \hat{f}(\mathbf{T}, \boldsymbol{\xi}, D)}{\partial \mathbf{T}} \, . \tag{98}$$

Equation (96) is satisfied if

$$\frac{\partial \hat{F}}{\partial \mathbf{T}} = \frac{\Lambda^{(f)}}{\chi \Lambda} \frac{\partial \hat{f}}{\partial \mathbf{T}} \, . \tag{99}$$

It is emphasized, that plastic flow (for both, the real and the fictitious material) occurs only if plastic loading for the real material holds, which is defined by

$$\text{plastic loading} \Leftrightarrow \hat{F} = 0 \quad \& \left(\frac{d}{dt} \hat{F} \right)_{\mathbf{E}_p = \text{const}} > 0 \, . \tag{100}$$

In particular, an equation of the form $\bar{f}^{(f)}(\tilde{\mathbf{T}}, \tilde{\boldsymbol{\xi}}) - (\tilde{R} + k_0) = 0$ does not represent necessarily a yield condition for the real material. Unfortunately, $\hat{F}(\mathbf{T}, \boldsymbol{\xi}, R, D)$ cannot be determined uniquely from (99), since e.g. $\Lambda^{(f)}$ is unknown too. To overcome this ambiguity, we assume the scalar multiplier $\Lambda^{(f)}/(\chi\Lambda)$ to be a given constitutive function. Because of the fact that $\hat{F}(\mathbf{E},\boldsymbol{\xi},R,D)$ and $\hat{f}(\mathbf{T},\boldsymbol{\xi},D)$ are functions of the state variables $\mathbf{T}$, $\boldsymbol{\xi}$, R, D, (99) indicates that $\Lambda^{(f)}/(\chi\Lambda)$ has to be a function of these variables as well (but not of their rates). In what follows, we suppose $\Lambda^{(f)}/(\chi\Lambda)$ to be a nonnegative function of the damage variable D:

$$\frac{\Lambda^{(f)}}{\chi\Lambda} =: g = g(D) \ . \tag{101}$$

Hence, the unknown yield function $\hat{F}$ can be calculated from the differential equation

$$\frac{\partial \hat{F}}{\partial \mathbf{T}} = g\frac{\partial \hat{f}}{\partial \mathbf{T}} \ , \tag{102}$$

which follows from (99). A solution of this differential reads

$$\begin{aligned}\hat{F}(\mathbf{T},\boldsymbol{\xi},R,D) &= g\ \hat{f}(\mathbf{T},\boldsymbol{\xi},D) - \hat{k}(R,D)\\ &= g\ \bar{f}^{(f)}(\tilde{\mathbf{T}},\tilde{\boldsymbol{\xi}}) - \hat{k}(R,D) \ ,\end{aligned} \tag{103}$$

with $\hat{k}(R,D)$ being a constant of integration. We remark that the term $g\ \bar{f}^{(f)}$ appears on the right side of (103). This motivates to suppose for $\hat{k}$ the form

$$\hat{k}(R,D) = g\tilde{R} + k_0 \ , \tag{104}$$

Then, from (103),(104), (97),$(85)_3$,

$$\begin{aligned}\hat{F}(\mathbf{T},\boldsymbol{\xi},R,D) &= g\ \bar{f}^{(f)}(\tilde{\mathbf{T}},\tilde{\boldsymbol{\xi}}) - g\ \tilde{R} - k_0\\ &= g\sqrt{\frac{3}{2}\frac{(\mathbf{T}-\boldsymbol{\xi})^D}{\sqrt{1-D}}\cdot\frac{(\mathbf{T}-\boldsymbol{\xi})^D}{\sqrt{1-D}}} - g\ \frac{R}{\sqrt{1-D}} - k_0 \ .\end{aligned} \tag{105}$$

Having established the yield function, the flow rule (95) becomes

$$\dot{\mathbf{E}}_p = \frac{3\Lambda g}{2\hat{f}\sqrt{1-D}}\ \frac{(\mathbf{T}-\boldsymbol{\xi})^D}{\sqrt{1-D}} \ , \tag{106}$$

from which

$$\dot{s} := \sqrt{\frac{2}{3}\dot{\mathbf{E}}_p\cdot\dot{\mathbf{E}}_p} = \frac{\Lambda g}{\sqrt{1-D}} \ . \tag{107}$$

We recall from (81) that $\dot{\tilde{s}} = \Lambda^{(f)}$, so that

$$\frac{\dot{\tilde{s}}}{\dot{s}} = \frac{\Lambda^{(f)}}{\Lambda\ g}\sqrt{1-D} = \chi\sqrt{1-D} \tag{108}$$

Similarly, it is readily seen that

$$\dot{\tilde{\mathbf{E}}}_p = \chi\sqrt{1-D}\dot{\mathbf{E}}_p \ . \tag{109}$$

We recognize from (105) and (106) that for determining the yield function, and hence the flow rule, it is sufficient to know the function $g(D)$ only. Especially, it is not necessary to know $\Lambda^{(f)}$ and χ explicitly, while Λ, as usually, has to be calculated from the consistency condition $\frac{d}{dt}\hat{F}(\mathbf{T}, \boldsymbol{\xi}, R, D) = 0$.

3.3.3 Hardening Rules We insert in $(93)_2$ the relations $(85)_2$, (82), keeping in mind $(91)_1$:

$$\tilde{\boldsymbol{\xi}} \cdot \dot{\tilde{\mathbf{Y}}} = \frac{\boldsymbol{\xi}}{\sqrt{1-D}} \cdot \left(\dot{\tilde{\mathbf{E}}}_p - b\,\dot{\tilde{s}}\,\tilde{\boldsymbol{\xi}}\right) = \chi\,\boldsymbol{\xi} \cdot \dot{\mathbf{Y}} \ . \tag{110}$$

A solution of this equation reads

$$\dot{\mathbf{Y}} = \frac{1}{\chi\,\sqrt{1-D}} \left(\dot{\tilde{\mathbf{E}}}_p - b\,\dot{\tilde{s}}\,\tilde{\boldsymbol{\xi}}\right) \ , \tag{111}$$

or, with the aid of (108), (109), $(85)_2$, (107), $(91)_3$

$$\dot{\mathbf{Y}} = \dot{\mathbf{E}}_p - b\,\dot{s}\frac{\boldsymbol{\xi}}{\sqrt{1-D}} = \dot{\mathbf{E}}_p - b\,c\,\dot{s}\sqrt{1-D}\;\mathbf{Y} = \dot{\mathbf{E}}_p - b\,c\Lambda\,g\mathbf{Y} \ . \tag{112}$$

This represents the kinematic hardening rule for the real material. The corresponding equation for isotropic hardening may be gained from $(93)_3$ in a similar way:

$$\tilde{R}\dot{\tilde{r}} = \frac{R}{\sqrt{1-D}} \left(1-\beta\sqrt{1-D}\;r\right)\left(\dot{s}\chi\sqrt{1-D}\right) = \chi R\dot{r} \ . \tag{113}$$

The latter will be satisfied if

$$\dot{r} = \left(1-\beta\sqrt{1-D}\;r\right)\dot{s} = \left(1-\beta\sqrt{1-D}\;r\right)\frac{\Lambda\;g}{\sqrt{1-D}} \ , \tag{114}$$

which is the evolution equation governing the response of isotropic hardening for the real material.

3.3.4 Dissipation Inequality - Dissipation Potential

First, we shall verify compatibility of the evolution equations, derived in Sects. 3.3.2 and 3.3.3, with the dissipation inequality (38). For doing this, we need the expressions $(\mathbf{T}-\boldsymbol{\xi}) \cdot \dot{\mathbf{E}}_p$, $R\dot{r}$, $\boldsymbol{\xi} \cdot (\dot{\mathbf{E}}_p - \dot{\mathbf{Y}})$, for which, after some algebraic manipulations, we obtain

$$(\mathbf{T}-\boldsymbol{\xi}) \cdot \dot{\mathbf{E}}_p = R\dot{s} + \frac{\sqrt{1-D}}{g}k_0\dot{s} \ , \tag{115}$$

$$R\dot{r} = R\dot{s} - \beta\sqrt{1-D}Rr\dot{s} \ , \tag{116}$$

$$\boldsymbol{\xi}\cdot(\dot{\mathbf{E}}_p - \dot{\mathbf{Y}}) = \frac{b}{\sqrt{1-D}}(\boldsymbol{\xi}\cdot\boldsymbol{\xi})\dot{s} \ . \tag{117}$$

On substituting these into $(38)_2$,

$$\mathcal{D}_{dp} = \left(\frac{\sqrt{1-D}}{g}k_0 + \beta\sqrt{1-D}Rr + \frac{b}{\sqrt{1-D}}(\boldsymbol{\xi}\cdot\boldsymbol{\xi})\right)\dot{s} \ , \tag{118}$$

which is always nonnegative. Note in passing that, in view of (94), the latter implies $\mathcal{D}_d^{(f)} \geq 0$ for the fictitious material, provided $\chi > 0$ holds. To prove also inequality (39), we calculate, from (90)-(92),

$$-\rho\frac{\partial\hat{\Psi}}{\partial D} = \frac{1}{2}\left(\mathbf{E}_e\cdot\boldsymbol{C}[\mathbf{E}_e] + c\,\mathbf{Y}\cdot\mathbf{Y} + \gamma(r^2 + 2r_0 r)\right) \geq 0 \ . \tag{119}$$

Thus, it is sufficient to require

$$\dot{D} \geq 0 \tag{120}$$

in order to ensure the validity of (39), and therefore the validity of the whole dissipation inequality (36).

It is probably of interest to remark that the flow rule (106), the kinematic hardening rule (112) and the isotropic hardening rule (114) may alternatively be derived from the convex plastic dissipation potential

$$\begin{aligned}
\hat{\varphi}(\mathbf{T},\boldsymbol{\xi},R,D) &:= g\bar{f}^{(f)}(\tilde{\mathbf{T}},\tilde{\boldsymbol{\xi}}) - g\tilde{R} - k_0 + \frac{g\beta}{2\gamma}(\tilde{R} - \gamma\tilde{r}_0)^2 + \frac{1}{2}gb\tilde{\boldsymbol{\xi}}\cdot\tilde{\boldsymbol{\xi}} \\
&= g\hat{f}(\mathbf{T},\boldsymbol{\xi},D) - g\frac{R}{\sqrt{1-D}} - k_0 \\
&\quad + \frac{g\beta}{2\gamma}\left(\frac{R}{\sqrt{1-D}} - \gamma\sqrt{1-D}\;r_0\right)^2 \\
&\quad + \frac{1}{2}gb\frac{\boldsymbol{\xi}}{\sqrt{1-D}}\cdot\frac{\boldsymbol{\xi}}{\sqrt{1-D}} \ .
\end{aligned} \tag{121}$$

Actually, it is not difficult to show that

$$\dot{\mathbf{E}}_p = \Lambda\frac{\partial\hat{\varphi}}{\partial\mathbf{T}} = \frac{3\Lambda g}{2\hat{f}\sqrt{1-D}}\;\frac{(\mathbf{T}-\boldsymbol{\xi})^D}{\sqrt{1-D}} \ , \tag{122}$$

$$\dot{\mathbf{Y}} = -\Lambda\frac{\partial\hat{\varphi}}{\partial\boldsymbol{\xi}} = \dot{\mathbf{E}}_p - b\dot{s}\frac{\boldsymbol{\xi}}{\sqrt{1-D}} \ , \tag{123}$$

$$\dot{r} = -\Lambda\frac{\partial\hat{\varphi}}{\partial R} = (1 - \beta\sqrt{1-D}\;r)\dot{s} \ , \tag{124}$$

which are in agreement with (106), (112) and (114), respectively.

3.3.5 Discussion of the Model - Uniaxial Tensile Loading

In Sects. 3.3.1-3.3.4, constitutive relations for elasto-plastic materials coupled with damage have been derived from the generalized energy equivalence principle proposed. The resulting constitutive model is now summarized:

$$\mathbf{E} = \mathbf{E}_e + \mathbf{E}_p \ , \tag{125}$$

$$\mathbf{T} = (1-D)\mathcal{C}[\mathbf{E}_e] \ , \ \boldsymbol{\xi} = (1-D)c\mathbf{Y} \ , \ R = (1-D)\gamma(r+r_0) \ , \tag{126}$$

$$F = \hat{F}(\mathbf{T}, \boldsymbol{\xi}, R, D) = g(D)\hat{f}(\mathbf{T}, \boldsymbol{\xi}, D) - \hat{k}(R, D) \ , \tag{127}$$

$$\hat{f}(\mathbf{T}, \boldsymbol{\xi}, D) = \sqrt{\frac{3}{2}\frac{(\mathbf{T}-\boldsymbol{\xi})^D}{\sqrt{1-D}} \cdot \frac{(\mathbf{T}-\boldsymbol{\xi})^D}{\sqrt{1-D}}} \ , \tag{128}$$

$$\hat{k}(R, D) = g(D)\frac{R}{\sqrt{1-D}} + k_0 \ , \tag{129}$$

$$\text{plastic loading} \Leftrightarrow \hat{F} = 0 \quad \& \left(\frac{d}{dt}\hat{F}\right)_{\mathbf{E}_p = \text{const}} > 0 \ , \tag{130}$$

$$\dot{\mathbf{E}}_p = \Lambda\frac{\partial \hat{F}}{\partial \mathbf{T}} = \frac{3\Lambda g}{2\hat{f}\sqrt{1-D}}\ \frac{(\mathbf{T}-\boldsymbol{\xi})^D}{\sqrt{1-D}} \ , \ \dot{s} = \frac{\Lambda g}{\sqrt{1-D}} \ , \tag{131}$$

$$\dot{\mathbf{Y}} = \dot{\mathbf{E}}_p - bc\dot{s}\sqrt{1-D}\ \mathbf{Y} = \dot{\mathbf{E}}_p - bc\Lambda g\mathbf{Y} \ , \tag{132}$$

$$\dot{r} = (1-\beta\sqrt{1-D}\ r)\dot{s} = (1-\beta\sqrt{1-D}\ r)\frac{\Lambda g}{\sqrt{1-D}} \ . \tag{133}$$

For definiteness, in the remainder of the paper we set

$$g(D) = \frac{1}{(1-D)^{n-\frac{1}{2}}} \ , \tag{134}$$

where n is a material parameter. Characteristic properties of the model may be highlighted by discussing uniaxial tensile loading. For this case, one finds from (125)-(134)

$$\varepsilon = \varepsilon_e + \varepsilon_p \ , \tag{135}$$

$$\sigma = (1-D)E\varepsilon_e \ , \tag{136}$$

$$\sigma - \frac{3}{2}\xi = R + (1-D)^n k_0 \ , \tag{137}$$

$$\xi = (1-D)cy \ , \tag{138}$$

$$R = (1-D)\gamma(r+r_0) \ , \tag{139}$$

$$\dot{y} = (1-bc\sqrt{1-D}\ y)\dot{\varepsilon}_p \ , \tag{140}$$

$$\dot{r} = (1-\beta\sqrt{1-D}\ r)\dot{\varepsilon}_p \ , \tag{141}$$

where we have used the same nomenclature as in Sect. 3.2. To make the analysis as simple as possible, we accomplish (135)-(141) with the damage law (66),

$$D = \left(\frac{\varepsilon_p}{\alpha}\right)^2 . \tag{142}$$

Plots of strain-stress responses according to these equations, for varying values of n and $c = \gamma$, are shown in Figs. 2-6, The other material parameters are taken to be the same as in Table 1. These figures indicate that for $n < 1$ the strain ε is unbounded as $D \to 1$. To see this also analytically, we use (136)-(138) to solve for ε_e :

$$\varepsilon_e = \frac{\sqrt{3c}}{2E} y + \frac{\gamma}{E}(r + r_0) + \frac{k_0}{E}(1 - D)^{(n-1)} . \tag{143}$$

Since $\varepsilon_p \to \alpha < \infty$ as $D \to 1$, the response of the strain ε will be controlled by the corresponding response of ε_e. On the other hand, the response of ε_e is described by (143). Because of (140),(141), the strains y, r are bounded as $D \to 1$. Thus, we conclude from (143) that ε_e, and hence ε too, will be bounded as $D \to 1$ only if $n \geq 1$.
On the other hand, we believe that $n > 1$ is not a realistic assumption for metallic materials. To support this assessment, we consider once more the

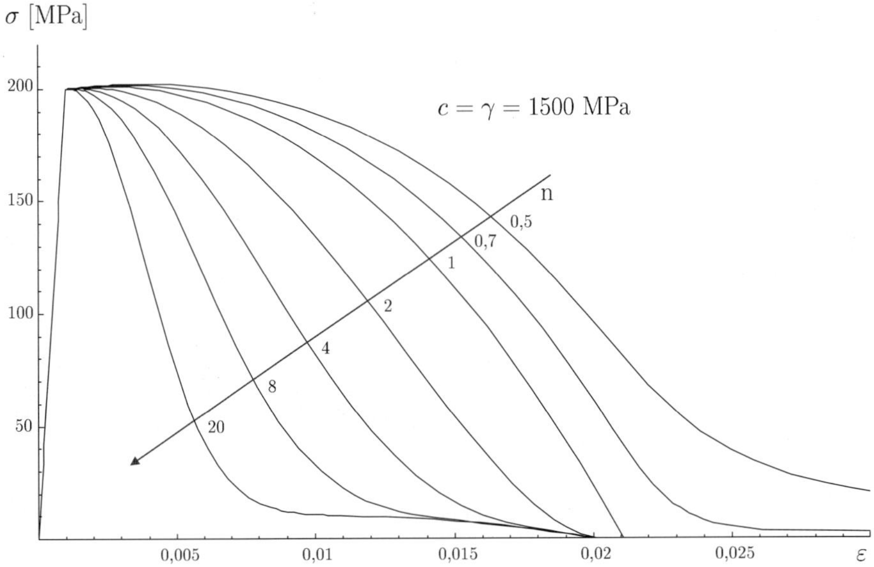

Fig. 2. Effect of the parameter n on the strain stress response for $c = \gamma = 1500$MPa. The remaining material parameters are taken from Table 1.

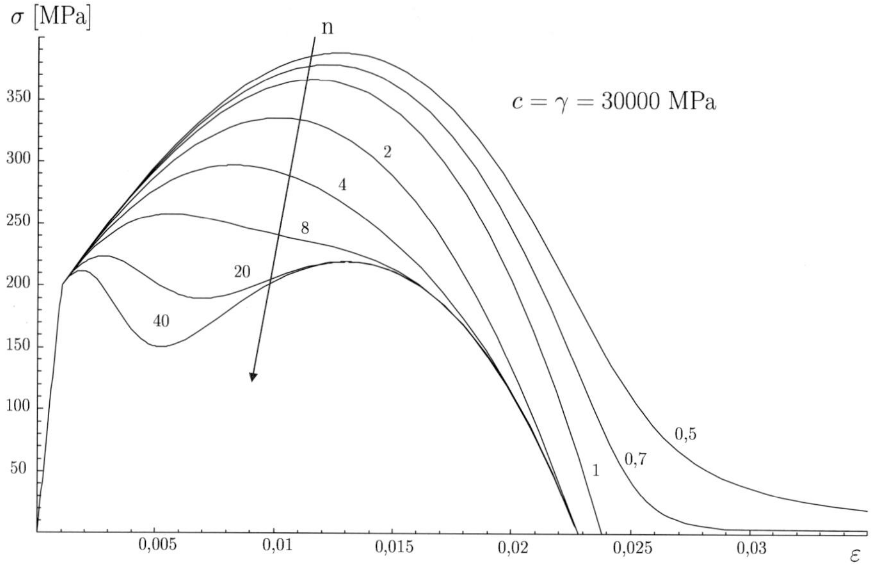

Fig. 3. Effect of the parameter n on the strain stress response for $c = \gamma =$ 30000MPa. The remaining material parameters are taken from Table 1.

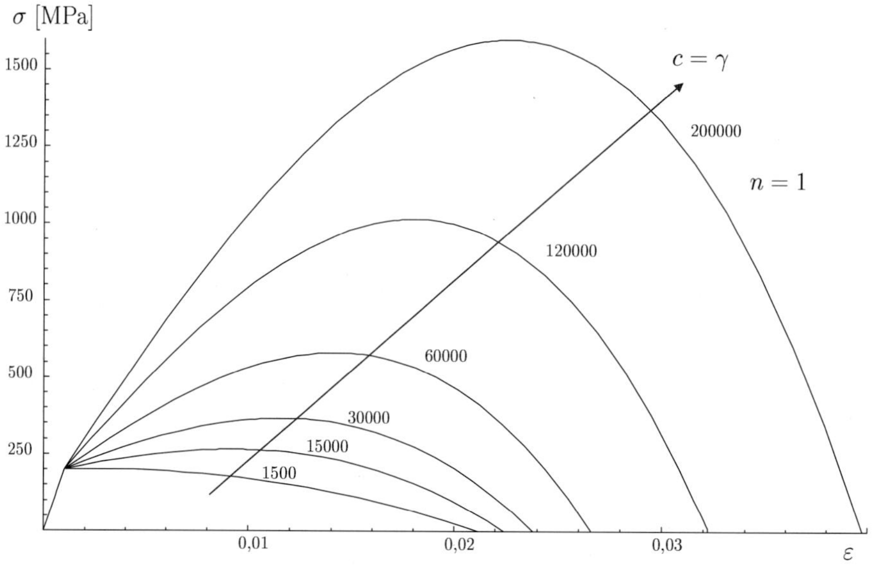

Fig. 4. Effect of the hardening parameters $c = \gamma$ on the strain-stress response for $n = 1$. Unit of c, γ is MPa and the remaining material parameters are taken from Table 1.

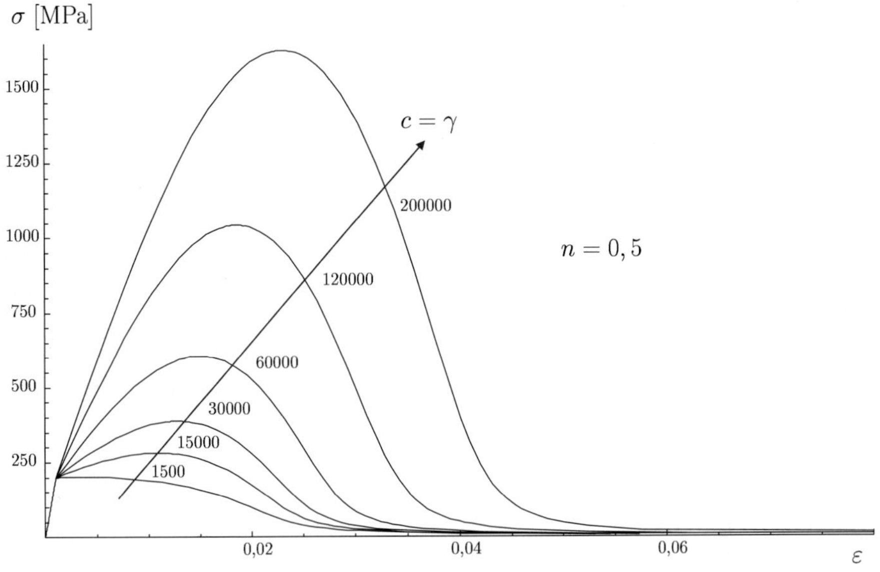

Fig. 5. Effect of the hardening parameters $c = \gamma$ on the strain-stress response for $n = 0,5$. Unit of c, γ is MPa and the remaining material parameters are taken from Table 1.

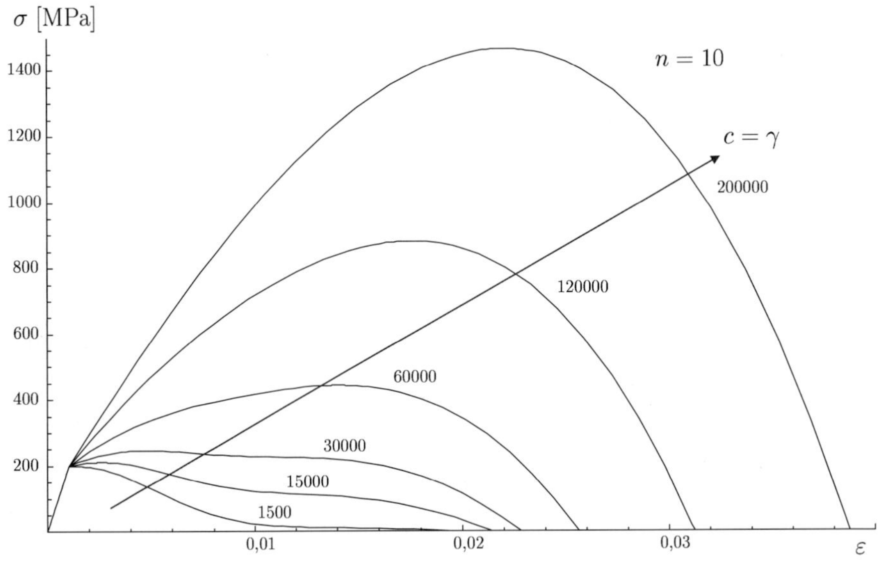

Fig. 6. Effect of the hardening parameters $c = \gamma$ on the strain-stress response for $n = 10$. Unit of c, γ is MPa and the remaining material parameters are taken from Table 1.

case of ideal plasticity. From (143),

$$\varepsilon_e = \frac{k_0}{E}(1-D)^{n-1} \; . \tag{144}$$

If $n > 1$, then $\lim \varepsilon_e \to 0$ as $D \to 1$. This contradicts our expectation that ε_e should be constant until local rupture. In other words, we believe that the lattice distortion of the elastic ideal plastic material will remain constant during plastic flow, independent of the damage evolution. Hence, we conclude that for metallic materials $n = 1$ should be chosen.
It must be remarked, that for $n = 1$ the yield function is identical to that used in the model derived from the strain equivalence principle. The same is also true for the parts of the free energy, the elasticity law and the relations $\boldsymbol{\xi} = \boldsymbol{\xi}(\mathbf{Y}, D)$, $R = R(r, D)$. The only differences are in the evolution equations for $\mathbf{Y}$ and r. Also, for $n = \frac{1}{2}$ we have $g = 1$ and the model reduces to that proposed by Saanouni et. al. [17].

4 Concluding Remarks - Notched Specimen under Tension

We have proposed a generalized energy equivalence principle to establish the constitutive model governing the response of a real, damaged material, from a known model describing the response of an undamaged elasto-plastic material. Within the assumptions made in the paper, a family of yield functions, parameterized by n, is obtained. Results of the analysis are so interpreted that $n = 1$ should be chosen when modeling the damage response of metallic materials. In this case the resulting yield function is identical to that used in the damage theory according to the strain equivalence principle as assumed by Chaboche. Generally, for $n = 1$ the uniaxial responses predicted by the two theories are nearly identical. The question may then be arised whether this is true for arbitrary loading processes indicating inhomogeneous deformations. To clarify this question, we assume equal material parameters for both models, as given in Table 2. The evolution law for damage is adopted

Table 2. Material parameters used for the calculated responses in Fig. 7, 8

E[MPa]	k_0[MPa]	r_0	c[MPa]	γ[MPa]	b[MPa]	β[MPa]	α_1	p	q
200000	400	0	10000	10000	10	10	1	1	0

from Lemaitre [7]:

$$\dot{D} = \alpha_1 \frac{\left(-\rho \frac{\partial \Psi}{\partial D}\right)}{(1-D)^q} \dot{s} \; . \tag{145}$$

To avoid the well known problem of mesh dependencies in the softening regime, we focus attention to viscoplasticity, which regularizes the equations of plasticity. A simple case of viscoplasticity arises by defining $\dot{s}$ to be given by

$$\dot{s} = \frac{\langle F \rangle^m}{\eta} \tag{146}$$

rather to be determined from the consistency condition $\dot{F} = 0$. In (146), $\langle x \rangle$ denotes the function

$$\langle x \rangle := \begin{cases} x \text{ if } x \geq 0 \\ \\ 0 \text{ if } x < 0 \end{cases}, \tag{147}$$

for real x. In the ensuing analysis, the material parameters m and η are chosen to be $m = 2,5$, $\eta = 2 \cdot 10^8 (MPa)^m s$. As can be seen in Fig. 7, the uniaxial responses for the two constitutive models are nearly identical. Next we consider a notched circular cylinder tensile specimen, subjected to prescribed displacement along the upper boundaries. The predicted responses have been calculated by employing the ABAQUS finite element code. This provides a user subroutine, in which the two models have been implemented. The finite element mesh and the imposed loading used, are shown in Fig. 9. Because of various symmetry conditions, only a quarter of the specimen has been meshed with 117 eight-node axial symmetric solid elements. Figure 8 illustrates for the two models the radial distribution of the damage variable D, with r being the radius in the plane through the notch root. The results are referred to the overall resulting strain (global strain) $\mathrm{e}^* = (\mathrm{L} - \mathrm{L}_0)/\mathrm{L}_0$, where L, L_0 are the current and initial length of the inhomogeneously deformed specimen. It can be seen that for $\mathrm{e}_0^* = 0,2\%$ both models predict identical radial distributions. However, noticeable quantitative differences between the responses predicted by the two models may be recognized for $\mathrm{e}_2^* = 0,356\%$ and $\mathrm{r}/\mathrm{r}_0 = 0,6$.

We close this section with the remark that energy equivalence principles have an advantage over strain energy principles. Namely, they allow to use the same effective stress definition for both, the Cauchy stress tensor and the back-stress tensor even if anisotropic damage is regarded. Actually, this is the reason for developing the generalized energy equivalence principle.

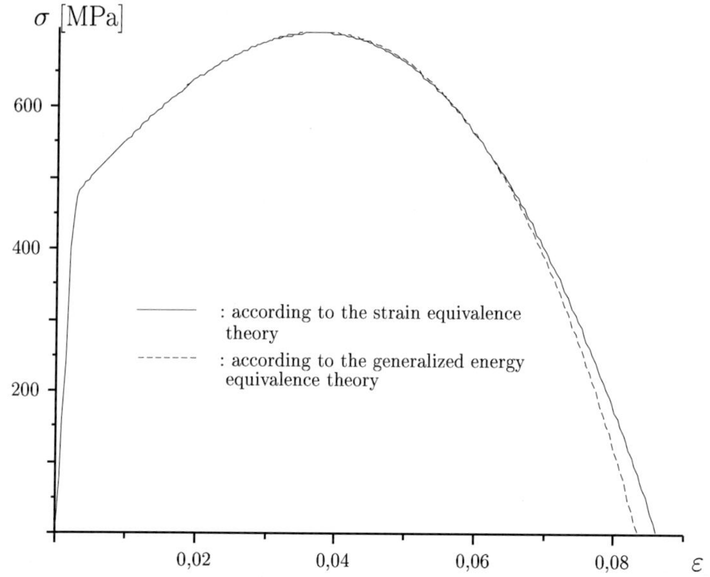

Fig. 7. Uniaxial tensile responses predicted by the two models.

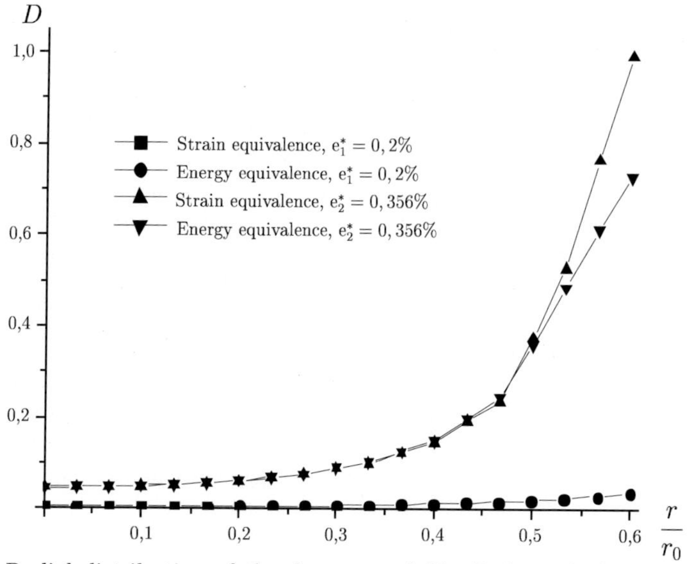

Fig. 8. Radial distribution of the damage variable D through the notch root for $e_1^* = 0,2\%$ and $e_2^* = 0,356\%$.

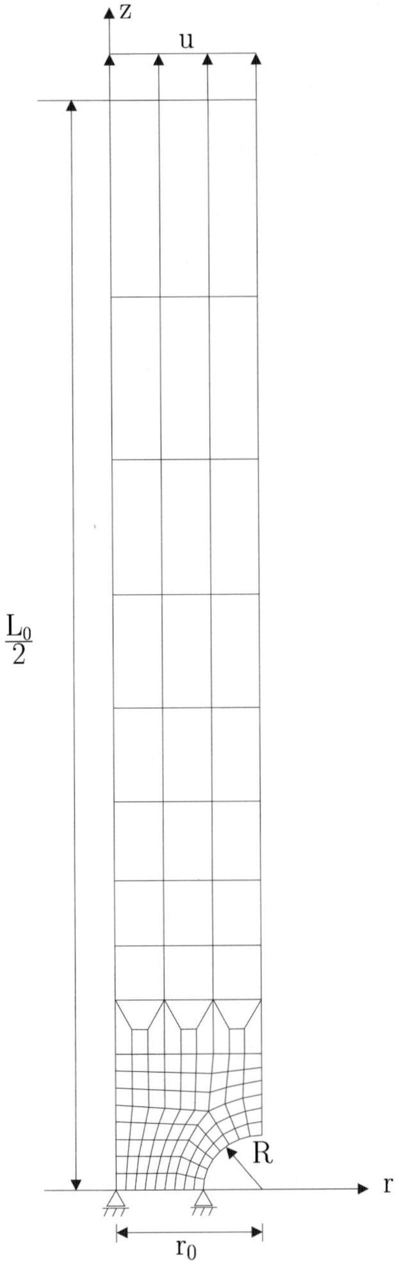

Fig. 9. Circular notched specimen. The assigned material parameters are given in Table 2.

References

1. Kachanov L.M. (1958) Time of the rupture process under creep conditions, Isv. Akad. Nauk. SSR. Otd Tekh. Nauk., 26–31
2. Rabotonov Y.N. (1969) Creep problems in structural members, North-Holland
3. Skrzypek J.J. (1999) Material damage models for creep failure analysis and designs of structures, in : Creep and damage in materials and structures, CISM No. 399, edited by Altenbach h., Skrzypek J.J., ISBN: 3-211-83321-8, Springer Verlag New York, 97–160
4. Chaboche J.-L. (1999) Thermodynamically founded CDM models for creep and other conditions, in: Creep and damage in materials and structures, CISM No. 399, edited by Altenbach h., Skrzypek J.J., ISBN: 3-211-83321-8, Springer Verlag New York, 209–278
5. Lemaitre J. (1984) How to use damage mechanics, Nuclear Engineering and Design 80, 233-245
6. Lemaitre J. (1985) A continuous damage mechanics model for ductile fracture, Journal of Engineering Materials and Technology, Vol. 107, 83–89
7. Lemaitre J. (1996) A course on damage mechanics (Second Ed.), Berlin, , Springer Verlag New York, ISBN: 3-540-60980-6
8. Lemaitre J., Chaboche J.-L. (1990) Mechanics of solid materials, ISBN: 0-521-32853-5
9. Chaboche J.-L. (1988) Continuum damage mechanics: Part 1 - General concepts, Journal of Applied Mechanics, Vol. 55, 59–72
10. Cordebois J.P. , Sidoroff F. (1983) Damage induced elastic anisotropy, in: Mechanical Behavior of Anisotropic Solids (Ed. Boehler J.P.), Martinus Nijhoff, Boston (1983), 761–774
11. Chaboche J.-L. (1993) Cyclic viscoplastic constitutive equations, Part 2: Stored energy - comparison between models and experiments, Journal of Applied Mechanics, Vol. 60, 822–828
12. Chaboche J.-L. (1993) Cyclic viscoplastic constitutive equations, Part 1: A thermodynamically consistent formulation, Journal of Applied Mechanics, Vol. 60, 813–821
13. Tsakmakis Ch. (1991) On the loading conditions and the decomposition of deformation, in: Anisotropy and Localization of Plastic Deformation, Proceedings of Plasticity '91 (Ed. Boehler J.P., Khan A.S.), ISBN: 1-85166-688-5, 353–356
14. Chaboche J.-L. (1986) Time-independent constitutive theories for cyclic plasticity, International Journal of Plasticity, Vol. 2, 149–188
15. Armstrong P.J., Frederick C.O. (1966) A mathematical representation of the Multiaxial Bauschinger Effect, G.E.G.B. Report RD/B/N 731
16. Lämmer H., Tsakmakis Ch. (2000) Discussion of coupled elastoplasticity and damage constitutive equations for small and finite deformations, International Journal of Plasticity, 495–523
17. Forster Ch., Hatira F.B., Saanouni K. (1994) On anelastic flow with damage, International Journal of Damage Mechanics, Vol. 3, 140–169
18. Chow C.L., Lu T.J. (1992) An analytical and experimental study of mixed-mode ductile fracture under nonproportional loading, International Journal of Damage Mechanics, Vol. 1, 191–236

Index

3D element, 346, 370

acoustic tensor, 346, 354
action functional, 281
additive decomposition, 255
anisotropy, 55, 108, 169, 324
Armstrong-Frederick, 79, 84, 246, 385
artificial data, 153

balance
– of energy, 283
– of linear momentum, 283
– of mass, 283
Banach's fixed point theorem, 289
Bauschinger effect, 324
BDF-2 scheme, 42
Bodner-Partom model, 134, 140, 245

cavity
– bounds of growth, 209
– critical stress-state, 205, 207, 216
– ductile material, 204
– elastic-plastic material, 204
– formation of, 204, 205
– growth due to thermal load, 206
– hyper-elastic material, 204
– in layered composites, 215
– micro-, 220
– radius of, 206, 208
– region of nucleation, 216
– simulation of growth, 208
– theory, 204
cell problem, 299
ceramic-metal composites, 203
characteristic length, 271, 346
Clausius-Duhem inequality, 383
coercive
– approximation, 232
– limit property of initial data, 233
– models, 231
cohesive zone, 364
Compact Tension (CT), 213, 328, 346, 356
compressible fluid flow, 283
consistency condition, 384
constitutive relation of monotone type, 296
constitutive theory, 169
continuum damage, 345, 348, 381
contractivity, 20
convergence order, 17
crack
– branching, 204
– growth, 328, 341, 345, 365
– initiation, 345
crack growth, 363
cup, 327, 333, 339, 341

damage, 325, 385
damage evolution law, 402
damage model, 220
deep drawing, 73, 326
deformation induced anisotropy, 169
deformation theory, 205, 208
design rules, 219
differential-algebraic equation (DAE), 7, 39
differentiation index, 7
diffusion, 275
directional hardening, 34
dislocation density, 253
dissipation inequality, 54, 109, 257, 346, 383
distortion of yield surface, 169
distortional hardening, 175
dual problem, 13
ductile damage, 346, 364, 381

earing, 73
effective state variable, 391
effective stress, 325, 387, 391
eigenstress, 205, 206
elastic predictor, 52, 79, 89, 352
elasticity, 172, 281
elasto-plastic materials, 51, 107, 351, 382
elasto-viscoplasticity, 31
elliptic, 259, 346
energy

– equivalence, 390, 391
– method, 314
entropy inequality, 277
Eshelby, 206
– tensor, 207, 254
Euler–Lagrange equations, 281, 282
experimental data of SS316, 140
experimental determination of yield surfaces, 170
exponential map, 60, 119, 353
extended Korn's inequality, 269
external load, 206

failure modes, 219, 354
FEM, 36, 65, 87, 126, 345, 370
finite strain, 51, 79, 81, 108, 345, 351
flow rule, 173, 350, 383
fourth-order tensor, 170
free energy, 281

generalized standard materials, 297
Gurson model, 220, 325, 350

Hamilton's principle, 281
hardening rule, 175, 396
hardness, 334
heat
– equation, 276
– flow, 283
– flow of Müller type, 285
– production, 283
higher gradient materials, 281
Hill model, 324
homogenized problem, 298
– of linear elasticity theory, 301
hourglassing, 348
hydrostatic tension, 205
hyperbolic
– –hyperbolic problem, 290
– –parabolic problem, 279, 287, 291
– problem, 275
hyperelasticity, 87

incompatible configuration, 252
inhomogeneities, 346
initial boundary value problem, 275
inter-penetrating network, 203
interface
– strong, 204
– weak, 204
– with precipitates, 204
interface strength, 219
intermediate configuration, 58, 110, 351
internal lenght, 271
internal length, 346
invariance requirement, 58
isotropic damage, 385
isotropic hardening, 34, 79, 81
isotropic tensor functions, 55

Kato's direct method, 289
kinematic hardening, 79, 81, 173, 324

layered composite
– crack-branching, 214
– failure of, 210
– processing, 211
– simulation of, 215
– with precipitates, 212
– with cavities, 215
Legendre–Hadamard, 286
linear self-controlling, 234
Lipschitz continuity condition, 285, 286
liquid-phase-bonding, 211
local strain, 331
localization, 346, 354
locking, 65, 348

Mandel stress tensor, 80, 83
material parameters, 144, 209, 216, 346
Melan-Prager model, 229
mesh–dependence, 346
metal layer, 203
metallic foams, 363
micro-cavities, 220
microscopic problem, 296
microstructure, 363
models
– of convex composite type, 246
– of monotone type, 228
– of pre-monotone type, 227
multiindex notation, 276
multiplicative decomposition, 53, 252, 352
multipolar compressible fluid flow, 285

Newton's law of motion, 275
Norton-Hoff model, 228
numerical integration, 79, 87

Oldroyd rate, 354
operator split, 87, 352
optimal composite, 219
overall stress, 299

parabolic
– –parabolic problem, 292
– problem, 276
parameter identification, 150, 326
periodicity cell, 300
phase angle, 213
phase transitions
– of Cahn–Allen type, 285
– of Cahn–Hilliard type, 285
plastic corrector, 79, 92, 352
plastic incompressibility, 60, 79, 94
plasticity, 79, 108, 169, 177, 352
plate theories
– of Kirchhoff–Love type, 285
– of Reissner–Mindlin type, 285
Poincaré's lemma, 287
polyconvex, 259
Prandtl-Reuss model, 229
precipitate
– real, 216
– stiff, 216
predictor polynomial, 45
pressure infiltration, 203, 206
principal component analysis, 161
principle of equipresence, 277

quasiperiodic, 298

R–curves, 212, 214, 364
radial stress component, 209
rate-independent
– models, 244
– plasticity, 256
representative volume element, 299
residual stress, 337
return mapping, 24, 60, 119, 352
Ritz-Galerkin, 38
robustness of stochastic simulations, 157
roughness, 332
Rousselier model, 26, 348
Runge-Kutta method
– algebraically stable, 21
– coercive, 19
– definition, 16

safe-load condition, 242
second law of thermodynamics, 277
self-controlling, 234
sheet forming, 323
shell, 67, 109
simultaneous parameter fits, 150
singular perturbation, 8
small elastic strains, 267
small scale yielding, 363
softening, 346, 381
space discretization, 36
spin tensor, 116, 353
stochastic optimization methods, 150
stochastic simulation, 153
strain equivalence principle, 387
stress intensity factor, 213, 364
stress-strain relation, 206
strong $\mathbb{L}^2$ solutions, 230
structural tensors, 55

thermal
– expansion, 205
– induced strain, 207
– induced stress, 206
thermoelasticity, 275
thermomechanics, 345
time integration, 39
transformation of internal variables, 244
translation of yield surface, 169
two-level Newton method, 24

underintegration, 348

validation of stochastic simulations, 156
variable stepsize, 42
variational formulation, 268
viscoelasticity, 281
viscoplasticity, 83, 169, 173, 235
void, 204, 325, 326, 330, 349

weak safe-load condition, 235
weak solution, 309
weak-type solutions, 241

yield function, 170
yield locus, 171
yield surface, 169, 353
Yosida approximation, 258

Printing (Computer to Plate): Saladruck Berlin
Binding: Stürtz AG, Würzburg